ASTRONOMY AND
ASTROPHYSICS LIBRARY

Series Editors: I. Appenzeller, Heidelberg, Germany
G. Börner, Garching, Germany
A. Burkert, München, Germany
M. A. Dopita, Canberra, Australia
T. Encrenaz, Meudon, France
M. Harwit, Washington, DC, USA
R. Kippenhahn, Göttingen, Germany
J. Lequeux, Paris, France
A. Maeder, Sauverny, Switzerland
V. Trimble, College Park, MD, and Irvine, CA, USA

James Lequeux

The Interstellar Medium

With the Collaboration
of Edith Falgarone and Charles Ryter

With 151 Figures
Including 32 Color Plates

Professor James Lequeux
Observatoire de Paris
LERMA
Avenue de l'Observatoire 61
75014 Paris, France

Cover picture: N 44 in the Large Magellanic Cloud. A bright association of about 40 young, hot stars ionizing the surrounding interstellar gas and digging a hole within it. European Southern Observatory, 2.2 m MPG/ESO telescope with wide field imager.

Translation from the French language edition of: *Le Milieu Interstellaire*. By James Lequeux
© 2003 Editions EDP Sciences/CNRS Éditions, Paris, France

Library of Congress Control Number: 2004103366

ISSN 0941-7834
ISBN 3-540-21326-0 Springer Berlin Heidelberg New York

This work is subject to copyright. All rights are reserved, whether the whole or part of the material is concerned, specifically the rights of translation, reprinting, reuse of illustrations, recitation, broadcasting, reproduction on microfilm or in any other way, and storage in data banks. Duplication of this publication or parts thereof is permitted only under the provisions of the German Copyright Law of September 9, 1965, in its current version, and permission for use must always be obtained from Springer. Violations are liable for prosecution under the German Copyright Law.

Springer is a part of Springer Science+Business Media

springeronline.com

© Springer-Verlag Berlin Heidelberg 2005
Printed in Germany

The use of general descriptive names, registered names, trademarks, etc. in this publication does not imply, even in the absence of a specific statement, that such names are exempt from the relevant protective laws and regulations and therefore free for general use.

Typesetting: LE-TeX, Leipzig
Cover design: *design & production* GmbH, Heidelberg

Printed on acid-free paper   55/3141/ba - 5 4 3 2 1 0

To Geneviève

Foreword

In spite of its generally low density, the medium between the stars plays a very important role in astrophysics. Stars are born from this interstellar matter when it undergoes gravitational collapse. During their lifetime and in particular in the final stages of their evolution, the stars return matter to the interstellar medium. This material is enriched in heavy elements arising from the thermonuclear processes that occurred inside the stars. New stars form from the interstellar matter that has been enriched by previous stellar generations. The evolution of the Universe is thus characterized by a continual mass exchange between stars and the surrounding medium. Mass exchange also takes place between the interstellar medium of galaxies and the intergalactic medium.

The interstellar matter consists of atomic, molecular and ionized gas at various temperatures, and also of dust grains which contain a large fraction of the elements heavier than hydrogen and helium. These grains are formed in the circumstellar shells around stars at the end of their lifetime. They are also continuously destroyed, altered and reformed in interstellar space. Interstellar grains play two important roles. Firstly, they absorb a large fraction of the photons emitted by stars and reemit the corresponding energy thermally as mid- and far-infrared photons. Secondly, molecules form on the surfaces of interstellar grains, in particular the most abundant hydrogen molecule H_2. Interstellar physics and chemistry differ markedly from those of terrestrial laboratories due to the very different conditions. The elementary processes are often more visible in interstellar space because of the low densities. The study of the interstellar medium is thus of fundamental interest for understanding and elucidating these basic processes.

The purpose of this book is to describe interstellar matter in our Galaxy in all of its various forms, and to consider the physical and chemical processes that are occuring within this matter. Like the authors, the reader will come up with difficulties related to the extreme complexity of the interstellar medium, which makes impossible a linear description. Moreover, our very conception of this medium experiences a deep evolution at the time of writing, as we progressively realise that it is most often turbulent and out of equilibrium. However, we do not yet know enough on this to find it necessary to abandon the relatively simple concepts that have governed its study up to now. These concepts stay at the basis of this book, and their pedagogical value will stand for a long time.

The first seven chapters of this book present the various components making up the interstellar matter and detail the ways that we are able to study them. The following seven chapters are devoted to the physical, chemical and dynamical processes that control the behaviour of interstellar matter. This includes the instabilities and cloud collapse processes that lead to the formation of stars. The last chapter summarizes the transformations that can occur between the different phases of the interstellar medium.

This book is written for graduate students, for young astronomers and also for researchers who develop an interest in the interstellar medium. It may seem somewhat ambitious to write a new handbook after the classic text by Spitzer, *Physical processes in the interstellar medium* [490]. However, since this book appeared a quarter of a century ago there have been quite a number of new developments. It has been our experience that Spitzer's book, although very pedagogical, is somewhat concise and sometimes difficult for the non-specialist to follow. In this book we have reproduced some of Spitzer's demonstrations, but often with more detail in order to help the reader to better follow the subject under consideration[1].

We choose to focus on methods rather than results, because methods evolve relatively slowly with time while results may become obsolete rather fast. We cannot pretend to completeness in such a complex domain, but we hope to have succeeded in providing the reader with the main tools that are required for a successful approach to interstellar matter. To this end, we give tables containing useful data and a large number of references to research papers. These references have been selected mainly for their pedagogical interest. They are not complete in any way: this book is not a review and cannot give credit to all scientists who developed the field and paved the way for future generations. The illustrations in the text are generally reproduced from research articles. We have also gathered a large number of colour plates which illustrate the wide variety of aspects of the interstellar medium.

In general we use the c.g.s. system of units, rather than the International System (S.I.) m.k.s. This may seem old-fashioned, but the vast majority of the research papers that the reader will consult make use of the c.g.s. system. A conversion table between the two systems is printed at the end of the book, as well as tables giving the most useful constants and physical quantities.

We heartily thank the colleagues who read and criticized the manuscript: Patrick Boissé, François Boulanger and Guillaume Pineau des Forêts, and especially Laurent Verstraete for his pertinent and constructive criticisms. We also thank several colleagues who supplied us with illustrations, in particular Jean-Charles Cuillandre from the Canada–France–Hawaii Telescope Corporation. We wish to express our gratitude to the NASA Astrophysics Data System (ADS), which has been of great

---

[1] After publication of the original french edition of our book, another textbook on the same subject has appeared: Dopita M.A., Sutherland R.S. 2003, *Astrophysics of the Diffuse Universe*, Springer Verlag, Berlin. The two books are rather different, thus complementary: the latter puts more emphasis than ours on the atomic physics and on the results, but does not treat of high-energy and turbulent phenomena and does not cover in detail the photodissociation regions.

help in our bibliographical searches. Finally, we have enormously appreciated the help of Anthony Jones who revised the manuscript.

Charles Ryter thanks the Service d'Astrophysique of the CEA and his director, Laurent Vigroux, for hospitality during the redaction of this book.

Paris, March 2004 *James Lequeux*

Contents

1 Our Galaxy, Host of the Interstellar Medium 1
   1.1 Our Galaxy: Orders of Magnitude 2
   1.2 Stellar Populations 5
      1.2.1 Generalities 5
      1.2.2 The Disk Populations 6
      1.2.3 The Bulge and the Halo 8
   1.3 Distribution of Interstellar Matter 9

2 Radiations and Magnetic Fields 13
   2.1 Radiation Fields 13
      2.1.1 Extragalactic Radiation 13
      2.1.2 Galactic Radiation 15
   2.2 The Interstellar Magnetic Fields 20
      2.2.1 Magnetic Field Measurements Using the Zeeman Effect 21
      2.2.2 Measurement of the Magnetic Fields
           Using Faraday Rotation 22
      2.2.3 Estimate of the Magnetic Fields
           from the Galactic Synchrotron Radiation 23
      2.2.4 Estimate of the Direction of the Magnetic Fields
           from the Linear Polarization of Light 24
      2.2.5 Results 25

3 Radiative Transfer and Excitation 27
   3.1 The Transfer Equation 27
      3.1.1 Demonstration 27
      3.1.2 The Rayleigh–Jeans Approximation:
           Radioastronomy Notations 30
      3.1.3 Excitation Temperature 32
   3.2 Two-Level System out of LTE 33
      3.2.1 General Relations 33
      3.2.2 Pure Radiative Equilibrium 35
      3.2.3 The Coupling of Excitation and Transfer;
           the LVG Approximation 35
   3.3 The General Case; Masers 39

4 The Neutral Interstellar Gas ... 45
4.1 The Atomic Neutral Gas ... 45
- 4.1.1 The 21-cm Line of Atomic Hydrogen ... 45
- 4.1.2 Fine-Structure Lines in the Far-Infrared ... 51
- 4.1.3 Interstellar Absorption Lines ... 55

4.2 The Molecular Component ... 63
- 4.2.1 Introduction ... 63
- 4.2.2 Electronic Transitions ... 66
- 4.2.3 Vibrational Transitions ... 71
- 4.2.4 Rotational Transitions ... 73
- 4.2.5 The Diffuse Interstellar Bands ... 83

5 The Ionized Interstellar Gas ... 87
5.1 H II Regions ... 87
- 5.1.1 Theory of Photoionization: the Strömgren Sphere ... 87
- 5.1.2 Continuous Emission ... 91
- 5.1.3 The Recombination Lines ... 96
- 5.1.4 The Radio Recombination Lines ... 101
- 5.1.5 The Forbidden Lines ... 105
- 5.1.6 Abundance Determinations in H II Regions ... 108

5.2 The Diffuse Ionized Gas ... 110
5.3 The Hot Gas ... 112
- 5.3.1 Collisional Ionization by Electrons at High Temperatures ... 113
- 5.3.2 The Emission of X-Ray Lines ... 115
- 5.3.3 The Thermal X-Ray Continuum ... 116
- 5.3.4 Results ... 116

5.4 The X-Ray Absorption ... 117

6 The Interstellar Medium at High Energies ... 119
6.1 Cosmic Rays ... 119
- 6.1.1 The Origin of Cosmic Rays ... 120
- 6.1.2 Solar Cosmic Rays and Solar Modulation ... 120
- 6.1.3 Galactic Cosmic Rays ... 122
- 6.1.4 Very High-Energy Cosmic Rays ... 128
- 6.1.5 Cosmic Electrons ... 129
- 6.1.6 Confinement of Cosmic Rays in the Galaxy ... 130

6.2 The Gamma-Ray Continuum ... 133
- 6.2.1 Gamma-Ray Production by Nuclear Interactions ... 133
- 6.2.2 Gamma-Ray Production by Bremsstrahlung ... 136
- 6.2.3 Gamma-Ray Production by the Inverse Compton Effect ... 137

6.3 The Mass of the Interstellar Medium ... 137
- 6.3.1 The Use of Gamma-Ray Observations to Determine the Mass of the Interstellar Medium in the Galaxy ... 138
- 6.3.2 Use of the Virial Mass of Molecular Clouds ... 141

		6.3.3	A Comparison Between W_{CO} and Extinction 142
		6.3.4	A Comparison Between W_{CO} and Millimetre/Submillimetre Dust Emission 143
		6.3.5	X-Ray and Mid-Infrared Absorptions 144
	6.4	The Gamma-Ray Lines 145

7 Interstellar Dust 149
    7.1 Interstellar Reddening and Extinction 150
        7.1.1 General Ideas 150
        7.1.2 Extinction and Dust Models 155
        7.1.3 X-Ray Scattering by Dust 158
    7.2 Interstellar Dust Emission 159
        7.2.1 Grains in Thermal Equilibrium 159
        7.2.2 Small Grains out of Thermal Equilibrium 162
        7.2.3 The Aromatic Emission Bands in the Mid-Infrared 167
        7.2.4 The Very Small Grains 172
        7.2.5 The Big Grains 173
    7.3 Global Dust Models 173
    7.4 Infrared Absorptions and Ice Mantles 174
    7.5 The Infrared Fluorescence 177

8 Heating and Cooling of the Interstellar Gas 179
    8.1 Heating Processes 180
        8.1.1 Generalities, Thermalization Time 180
        8.1.2 Heating by Low-Energy Cosmic Rays 183
        8.1.3 Photoelectric Heating from Grains 187
        8.1.4 Photoelectric Heating by the Photoionization of Atoms and Molecules 191
        8.1.5 X-Ray Heating 192
        8.1.6 Chemical Heating 194
        8.1.7 Heating by Grain-Gas Thermal Exchange 194
        8.1.8 Hydrodynamic and Magnetohydrodynamic Heating 196
    8.2 Cooling Processes 197
        8.2.1 Fine-Structure Line Cooling 197
        8.2.2 Cooling by the Collisional Excitation of Permitted Lines 199
        8.2.3 Cooling by Electron–Ion Recombination 202
        8.2.4 Cooling by Dust 203
    8.3 Thermal Equilibrium and Stability 203
        8.3.1 The Atomic Medium 203
        8.3.2 The Hot Ionized Gas 207
        8.3.3 H II Regions 207
        8.3.4 Molecular Clouds 208

9 Interstellar Chemistry ... 209
9.1 Gas-Phase Chemistry ... 209
9.1.1 Ion–Molecule Reactions ... 210
9.1.2 Radiative Association ... 211
9.1.3 Dissociative Recombination ... 212
9.1.4 Neutral–Neutral Reactions ... 213
9.1.5 Photodissociation and Photoionization ... 214
9.2 Chemistry on Dust Grains ... 215
9.2.1 H_2 Formation on Grains ... 216
9.2.2 Formation of Other Molecules on Grains ... 218
9.3 Equilibrium Chemistry and Chemical Kinetics ... 219
9.4 Some Results ... 221
9.4.1 Chemistry in the Diffuse Interstellar Medium ... 221
9.4.2 Chemistry in Dense Molecular Clouds ... 223

10 Photodissociation Regions ... 227
10.1 General Presentation ... 227
10.2 Physico-Chemistry ... 229
10.2.1 The Penetration of Far-UV Radiation and Photodissociation ... 229
10.2.2 Chemistry ... 230
10.2.3 Heating Processes ... 230
10.2.4 Cooling Processes ... 232
10.3 Stationary Models ... 233
10.4 Out of Equilibrium Models ... 240

11 Shocks ... 241
11.1 The Equations of Gas Dynamics ... 241
11.1.1 A Single-Fluid Medium ... 241
11.1.2 A Multi-Fluid Medium ... 243
11.2 Different Types of Shocks ... 244
11.2.1 Shocks with no Magnetic Field ... 245
11.2.2 Shocks in a Magnetized Medium ... 249
11.2.3 Multi-Fluid Shocks in a Weakly Ionized Gas ... 250
11.3 Non-Stationary Shocks ... 252
11.4 Physico-Chemistry in Shocks ... 254
11.5 Radiation and the Diagnosis of Shocks ... 257
11.6 Instabilities in Shocks ... 259

12 Shock Applications ... 263
12.1 Supernova Remnants ... 263
12.1.1 The Free Expansion Phase ... 264
12.1.2 The Adiabatic Phase ... 265
12.1.3 The Isothermal, or Radiative, Expansion Phase ... 267
12.1.4 The Evolution of Plerions ... 268

12.1.5 The Expansion of Supernova Remnants
in an Inhomogeneous Medium 269
12.1.6 Non-Thermal Radiation of Supernova Remnants 270
12.2 Bubbles.. 272
12.3 The Dynamics of H II Regions 274
12.3.1 The Ionization Front.................................. 275
12.3.2 The Shock 278
12.3.3 Neutral Globules in a H II Region 279
12.3.4 The Evolution of H II Regions......................... 283
12.4 The Acceleration of Cosmic Rays 284
12.4.1 Propagation of Charged Particles in a Magnetic Field 285
12.4.2 Diffusion of Charged Particles in a Disordered Medium 288
12.4.3 Energy Losses 291
12.4.4 The Acceleration of Charged Particles 292

13 Interstellar Turbulence.. 301
13.1 Velocity Structure and Fragmentation 301
13.2 Incompressible Turbulence................................... 304
13.2.1 The Birth of Turbulence.............................. 304
13.2.2 The Developed Kolmogorov Turbulence 305
13.2.3 Turbulent Viscosity and Pressure 307
13.2.4 Intermittency.. 308
13.3 Turbulence in the Interstellar Medium 309
13.4 Some Effects of Interstellar Turbulence 312
13.4.1 Turbulent Transport and Interstellar Chemistry 312
13.4.2 Intermittency of Turbulence Dissipation
as a Gas Heating Source 313

14 Equilibrium, Collapse and Star Formation 321
14.1 Stability and Instability: the Virial Theorem 321
14.1.1 A Simple Form of the Virial Theorem
with No Magnetic Field nor External Pressure 321
14.1.2 The Jeans Length and Jeans Mass 323
14.1.3 The General Form of the Virial Theorem............... 326
14.1.4 The Stability of the Virial Equilibrium.................. 327
14.1.5 The Density Distribution in a Spherical Cloud
at Equilibrium....................................... 331
14.1.6 Stability and Instabilities in the Presence
of a Magnetic Field 332
14.1.7 The Coupling of the Gas and Magnetic Fields:
Ambipolar Diffusion 335
14.2 Collapse and Fragmentation.................................. 340
14.2.1 The Free-Fall Time................................... 340
14.2.2 Collapse Configurations............................... 341
14.2.3 The Role of Rotation 343
14.2.4 The Role of Magnetic Fields 345

14.3 The End of Collapse: Star Formation 347
14.4 The Initial Mass Function and Its Origin 348
    14.4.1 Determinations of the Initial Mass Function,
            and Related Problems 348
    14.4.2 The Origin of the Initial Mass Function 350

15 Changes of State and Transformations 353
15.1 Atomic, Molecular and Warm Ionized Gas 354
    15.1.1 Ionized Gas and Exchanges with the Neutral Gas 354
    15.1.2 Atomic Gas–Molecular Gas Exchanges 355
15.2 Hot Gas and the Galactic Fountain 360
15.3 Gas–Dust Exchange 363
15.4 Evolution of Interstellar Dust 365
    15.4.1 Dust in Circumstellar Envelopes
            and Planetary Nebulae 366
    15.4.2 Dust in the Interstellar Medium 367
    15.4.3 Dust Around Protostars and in the Solar System 368

A Designation of the Most Used Symbols 371

B Principal Physical Constants 375

C Journal Titles Abbreviations 379

References 381

Index 395

Color Plates 403

1 Our Galaxy, Host of the Interstellar Medium

Just as any galaxy, our own Galaxy[1], the Milky way, is a complex system of stars, gas and dust particles (see Plate 1 for a spiral galaxy seen face-on and Plate 2 for a collection of images of the Milky way at various wavelengths). These elements are bathed in a magnetic field, subject to radiation covering the entire electromagnetic spectrum, and exposed to neutral and charged "cosmic-ray" particles of all energies.

Galaxies are bound by their own gravity and their various components interact strongly, exchanging mass, momentum and energy. For such a complex system it is not possible to give a linear description in which the various elements can be analysed one by one. The description can only be given by a series of approximations. Here, there is not only a difficulty with the presentation, but also a genuine intellectual difficulty. For many problems a star can be considered as an isolated system. A whole galaxy like ours can also be considered as an isolated system, as a first approximation, but this system is considerably more complex than a star. This leads us to consider simpler sub-systems like star clusters, interstellar clouds, etc., that we may attempt to describe by models in which the interactions with the rest of the Galaxy are considered as limiting conditions. These conditions are themselves described by models that isolate, to various degrees of approximation, other sub-systems. As a result, our knowledge of the physics of the Galaxy and in particular the central problem of star formation, progresses only through iterations in a description which, crude as it is, must however be global. At each step, new observations and a better understanding of any sub-system can modify this global picture and the description of the other sub-systems.

The very nature of astronomy, an observational science in which direct experiment is almost invariably impossible, prevents testing the validity of any mechanism that aims at explaining a set of observations. This is only possible in experimental physics. Most often, we cannot even have recourse (as is the case for stellar physics) to a statistical analysis of relatively homogeneous families of objects, for which we hope that a single parameter dominates the observed variations. In fact, the property we want to describe depends most often on the environment.

We will thus be forced to give, with little justification, numerical values for some parameters. The confidence, or doubts, that we may have in these values, and more generally in the basic mechanisms, depends on the extent of our understanding of the

---

[1] In this book, we always write Galaxy with a capital G for our Milky way, in order to avoid confusion with other galaxies.

1.1 Our Galaxy: Orders of Magnitude

The Galaxy is a self-gravitating system. Its most conspicuous structure is a rotating stellar *disk*. The radius of this system is about 20 kpc (kiloparsec; 1 parsec = 3.08×10^{18} cm). Its thickness is a few hundreds of parsecs. The Sun is 7–8 kpc from the centre of the Galaxy and revolves around this centre with a linear velocity of 180–200 km s^{-1}. These numbers are still somewhat uncertain: for a recent discussion see Olling & Merrifield [388]. The revolution period for the regions of the disk located at the solar radius is of the order of 240 million years.

To this disk is added a spheroidal system, not particularly flattened, that extends farther than the disk. Its central part, the *bulge*, is very luminous within a radius of 2 kpc, but its outer parts, the *halo*, are of low brightness.

All these stellar systems can be considered as collisionless. The time between successive collisions is in effect

$$\tau = (n\sigma_v 2\pi R^2)^{-1} = 10^{28} \left(\frac{n}{\text{pc}^{-3}}\right)^{-1} \left(\frac{R}{R_\odot}\right)^{-2} \left(\frac{\sigma_v}{\text{km s}^{-1}}\right)^{-1} \text{ s}, \quad (1.1)$$

where n is the number density of stars, R their average radius, σ_v their velocity dispersion and $R_\odot = 6.955 \times 10^{10}$ cm is the radius of the Sun. In the solar neighbourhood, $n < 1$ pc^{-3} and σ_v is of the order of 10–30 km s^{-1}. The typical collision time is then 10^{10} times the age of the Galaxy, which is of the order of 3×10^{17} s. Even in the central parsec of the Galaxy, where the density is 10^7 times higher, the collision time is still very large.

To a first approximation, the Galaxy can be considered as a closed system as far as matter exchanges with the surroundings are concerned. Such exchanges indeed seem to exist, but they do not lead to important mass variations on time scales much shorter than the age of the Galaxy. On the other hand, the Galaxy exchanges energy with the external world as electromagnetic radiation and neutrinos. The total radiated energy is of the order of 4×10^{10} L$_\odot$ (L$_\odot$ = 1 solar luminosity = 3.85×10^{26} W = 3.85×10^{33} erg s^{-1}).

The stars of the galactic disk have small velocities with respect to the bulk rotation velocity around the centre. We can derive the distribution of mass, and the total mass of the Galaxy, from a study of the variation of the rotation velocity with galactocentric radius. Neglecting the random motion of the stars with respect to rotation, a justified approximation except for the bulge, we can write that the centrifugal force is just balanced by gravity. Assuming for the moment spherical symmetry, this yields

$$\boxed{M(< R) = Rv(R)^2/G,} \quad (1.2)$$

or numerically

$$M(< R) = 2.32 \times 10^5 \left[\frac{v(R)}{\text{km s}^{-1}}\right]^2 \left(\frac{R}{\text{kpc}}\right) \text{M}_\odot, \quad (1.3)$$

R being the galactocentric radius, $M(< R)$ the mass inside R, $v(R)$ the rotation velocity at radius R and G the constant of gravitation.

Actually, the mass is not distributed according to spherical symmetry. If we considers a more realistic model with an infinitely thin disk plus a spherical distribution of mass, we can show that the mass within a radius R is 0.6 to 1.0 times the mass given by the above expression (Lequeux [315]). At large R, the distribution of mass becomes close to spherical and this expression is sufficient.

The total mass of the Galaxy is estimated to be $M_G \simeq 1.7 \times 10^{11}$ M$_\odot$ within a radius of 20 kpc. Outside this there are essentially no stars. However mass increases with radius at larger radii. We cannot measure the rotation velocity of the disk beyond 30 kpc, but the dynamics of globular clusters and of satellite galaxies implies that mass extends even further. The total mass of the Galaxy is poorly known but is probably greater than 10^{12} M$_\odot$ (see e.g. Kulessa & Lynden-Bell [291]). The mass/luminosity ratio inside a radius of 20 kpc is

$$M_G/L_G \simeq 5\text{M}_\odot/\text{L}_\odot, \quad (1.4)$$

but this ratio increases to at least 30 if the total mass of the Galaxy is considered. These values should be compared to the mass/luminosity ratios of stars that are given roughly by

$$M/L \simeq (M/\text{M}_\odot)^{-2.5}, \quad (1.5)$$

for stellar masses lower than about 15 M$_\odot$. The average mass/luminosity ratio of the Galaxy is thus considerably larger than that of its component stars, which have an average mass of the order of 1 M$_\odot$. We conclude that most of the mass of the Galaxy is contained in non-luminous matter, *dark matter*, which is generally considered to be distributed in a system with little flattening. The contribution of dark matter to the mass of the disk near the Sun is rather small, but it dominates the total mass at larger radii. Despite many efforts, the nature of this dark matter is still essentially unknown.

The medium between the stars – the *interstellar medium* – the subject of this book, is made of gas and dust that are generally considered to be well mixed. To zero order we may consider, in a very schematic way, that the gas exists as several distinct components. These components are designated according to the form taken by hydrogen which makes up 70% of the mass. The properties of these components are summarized in Table 1.1. In reality, the components are partly mixed and are strongly perturbed by winds from massive stars, supernova explosions and other phenomena. As a consequence, random macroscopic motions are important.

Most of this matter is confined in the disk, but some exists in the halo which contains, in particular, an important fraction of the hot gas. Some of these phases are more or less in pressure equilibrium. For the hot and warm atomic phases, the

Table 1.1. Components of the interstellar medium in the Galaxy. This is a very schematic classification. The total masses of these phases are uncertain. What we give are essentially masses proportional to the fractions of these phases within a few kpc around the Sun. The observational separation between the warm atomic medium and the diffuse ionized medium is somewhat problematic.

Phase		Density cm^{-3}	Temperature K	Total mass M$_\odot$
Atomic (H I)	Cold	$\simeq 25$	$\simeq 100$	1.5×10^9
	Warm	$\simeq 0.25$	$\simeq 8\,000$	1.5×10^9
Molecular (H$_2$)		$\geq 1\,000$	≤ 100	10^9?
Ionized	H II regions	$\simeq 1 - 10^4$	$\simeq 10\,000$	5×10^7
	Diffuse	$\simeq 0.03$	$\simeq 8\,000$	10^9
	Hot	$\simeq 6 \times 10^{-3}$	$\simeq 5 \times 10^5$	10^8?

pressure P is such that $P/k = nT \simeq 5\text{–}20 \times 10^3$ K cm^{-3}. Conversely, the pressure is considerably higher inside ionized nebulae (H II regions) and molecular clouds. H II regions are not confined by their own gravity or by that of the stars they contain, and are therefore undergoing expansion. On the other hand, molecular clouds are maintained by their own gravity and by external pressure. The interstellar medium takes part in the galactic rotation and is also agitated by random motions with a 1-dimensional velocity dispersion of 6 to more to 10 km s^{-1}, depending upon the component.

The total mass of the interstellar medium is of the order of 5% of the total stellar mass and of the order of 0.5% of the total mass of the Galaxy, which is dominated by dark matter. The composition by mass of the medium is 70% hydrogen, 28% helium and about 2% heavy elements. C, N, O, Mg, Si, S and Fe are the most abundant heavy elements (see Table 4.2). The heavy elements are distributed differently between the gas and the dust grains depending upon the physical conditions. In molecular clouds, the atoms assemble into molecules, with the exception of the noble gases. The "molecules" form a continuum of sizes from diatomic molecules to grains with diameters larger than 0.1 μm, with intermediate clusters containing hundreds of atoms.

The external physical factors which influence the interstellar medium will be detailed in Chap. 2. Here is a short summary.

The *electromagnetic radiation* contains photons of all energies. In the solar neighbourhood its total energy density is about 1 eV cm^{-3} = 1.6×10^{-12} erg cm^{-3}. To this, we must add an energy density of 0.26 eV cm^{-3} corresponding to the blackbody radiation of the Universe. However, this radiation is at millimetre and submillimetre wavelengths and interacts relatively weakly with the interstellar medium. The "ordinary" electromagnetic radiation is dominated by the ultraviolet (UV) and visible light and also by strong radiation in the mid- and far-infrared (IR). The intensity and spectral energy distribution of this radiation varies with the location.

The *magnetic field* is on average of the order of 5 microgauss (μG). It is partly organized at the galactic scale, but in general the irregular component dominates. The mean magnetic energy density is of the order of 1 eV cm^{-3}. The field is larger in denser regions of the interstellar medium.

Cosmic rays are high-energy, mostly relativistic particles, which propagate through the gas with little interaction except at low energies (a few tens of MeV at most). At a given energy there are about 100 times more protons than electrons, and heavier particles are rare. The total energy density of cosmic rays is about 1 eV cm^{-3}.

It is of interest to note that the energy densities of photons, magnetic field, cosmic-ray particles and random motions of the interstellar medium are similar, and are of the order of 1 eV cm^{-3} in the vicinity of the Sun. In the cold neutral phase, the motions are generally supersonic and correspond to a non-thermal kinetic energy far greater than the thermal energy of the gas. The coincidence between these different forms of energy is not fortuitous, but results from efficient energy transfer between them. In any case, this equality does not allow us to neglect any of these forms of energy when studying the physics of the interstellar medium. There are exceptions, fortunately, for example the physics of H II regions is dominated by the radiation of the exciting stars.

1.2 Stellar Populations

1.2.1 Generalities

A stellar population can be defined as an ensemble of stars sharing the same age, the same chemical composition and the same average kinematic properties. It can contain stars with very different masses. Stellar populations are important for studies of the structure of our Galaxy, which is directly related to its kinematics thus to its dynamics. They can also be used to study the temporal evolution of the Galaxy which is, in part, linked to the age and chemical composition of stars. The notion of stellar populations was introduced in 1944 by Baade in his comparative study of the Galaxy and the closest major spiral galaxy, the Andromeda galaxy M 31.

Baade's Population I is the major population in the disk. It contains massive, young stars. Thanks to their ultraviolet radiation, strong winds and final explosion as supernovae at the end of their lives, massive stars are those that most affect the interstellar medium. The lifetime of a star is related to its mass by the approximate relation $t \simeq 10^{10}(M/M_\odot)^{-2.5}$ year. The lifetime depends less on mass for stars more massive than 10 $M_\odot$ and is about 3×10^6 years for the most massive stars known (mass somewhat larger than that 100 $M_\odot$). The kinematics of Population I stars is dominated by the rotation of the galactic disk. Their velocity dispersion is less than 30 km s^{-1}, and is only 6–7 km s^{-1} for the youngest stars. The metallicity of these stars (the abundance of the heavy elements, or "metals", with respect to hydrogen) is large, and they present a wide dispersion in age.

Baade's Population II is characterized by a large velocity dispersion, ≥ 100 km s^{-1}, and is an old population. Its metallicity is variable, but generally small. It is distributed with an approximate spherical symmetry and constitutes the halo of the Galaxy. It is now generally considered that the stellar halo was built up through the successive captures of dwarf spheroidal galaxies by the Galaxy.

This classification is somewhat schematic and must be looked at more carefully, considering the disk and the spheroidal components of the Galaxy separately.

1.2.2 The Disk Populations

Studies of disk stars show that their formation has taken place throughout the lifetime of the Galaxy. There is a statistical relation between age and metallicity of disk stars. It obviously reflects the chemical evolution of the Galaxy, which results from the cycle, interstellar medium → stars → interstellar medium → ..., with a progressive enrichment in the heavy elements synthetized and ejected by stars. This enrichment was much faster in early stages of galactic evolution than it is now. Also, the velocity dispersion of stars is statistically higher for older stars. The origin of this variation is still somewhat controversial, but the majority of astronomers tend to believe that the thickness of the disk has decreased since its formation. The stars formed long ago are in a thicker disk and must have acquired a larger velocity dispersion in order to balance the gravitational attraction of the disk.

We will now calculate the vertical density distribution $\rho_i(z)$ of stars of species i, perpendicular to the galactic plane. We assume that these stars make an homogeneous isothermal ensemble, with a velocity dispersion perpendicular to the plane $\sigma_i = \langle v_z^2 \rangle^{1/2}$. The Poisson equation is

$$\nabla^2 \Phi = \nabla \cdot \mathbf{K} = -4\pi G \rho, \tag{1.6}$$

where Φ is the gravitation potential, $\mathbf{K}$ the gravitation force, ρ the total density and G the constant of gravitation[2]. Since the Galaxy is approximately axisymmetric, we can use a cylindrical coordinate system (R, ϕ, z), so that

$$\frac{1}{R}\frac{\partial}{\partial R}(R K_R) + \frac{\partial K_z}{\partial z} = -4\pi G \rho. \tag{1.7}$$

The rotational velocity is given by $v^2/R = -K_R$, and (1.7) can be written as

$$\rho = -\frac{1}{4\pi G}\left(\frac{\partial K_z}{\partial z} - \frac{1}{R}\frac{\partial v^2}{\partial R}\right). \tag{1.8}$$

Observation shows that near the Sun the rotational velocity is almost independent of R, so that the second term is negligible with respect to the first, and

$$\rho \simeq -\frac{1}{4\pi G}\frac{\partial K_z}{\partial z}. \tag{1.9}$$

---

[2] We neglect here for the moment the dark matter, that we assume to make only a small fraction of the mass of the disk near the Sun. But see later in this section.

1.2 Stellar Populations

K_z is a negative quantity (a return force). In the gravitation field of the disk a star executes periodic oscillations about the galactic plane. The perpendicular velocity $v_z(z)$ at altitude z is dependent upon the velocity at $z = 0$, $v_z(0)$, through the conservation of the total energy

$$v_z^2(z) = v_z^2(0) + 2\int_0^z K_z(z')\,dz'. \tag{1.10}$$

Assuming a maxwellian distribution of stellar velocities in the galactic plane, we obtain for population i

$$\rho_i[0, v_z(0)] = \rho_i(0)\frac{1}{(2\pi)^{1/2}\sigma_i}\exp\left[-\frac{v_z^2(0)}{2\sigma_i^2}\right], \tag{1.11}$$

where $\rho_i(0)$ and $\rho_i[0, v_z(0)]$ are, respectively, the total density of stars of type i and their density at velocity $v_z(0)$ in the plane. σ_i is their velocity dispersion. If the distribution of stars is "well-mixed", i.e. if there are as many stars moving toward the galactic plane as stars moving away, the Liouville theorem implies

$$\rho_i[z, v_z(z)] = \rho_i[0, v_z(0)]. \tag{1.12}$$

Combining the three preceding equations, we get

$$\rho_i[z, v_z(z)] = \rho_i(0)\frac{1}{(2\pi)^{1/2}\sigma_i}\exp\left[-\frac{v_z(0)^2}{2\sigma_i^2} + \frac{1}{\sigma_i^2}\int_0^z K_z(z')\,dz'\right], \tag{1.13}$$

then, integrating over velocity,

$$\boxed{\rho_i(z) = \rho_i(0)\exp\left[-\frac{1}{2\sigma_i^2}\int_0^z K_z(z')\,dz'\right],} \tag{1.14}$$

the *hydrostatic equilibrium equation*, which is the solution to our problem.

If we can measure the velocity dispersion and the vertical distribution of a stellar population, we can estimate $K_z(z)$ or $\Phi_z(z)$. The form of this potential is not simple, so that the actual distribution of $\rho_i(z)$ is neither exponential nor gaussian, although it is often assumed to simplify to an exponential. A more correct approximate expression is

$$\rho_i(z) = \frac{\rho_i(0)}{2z_0}\text{sech}^2\left(\frac{z}{2z_0}\right), \tag{1.15}$$

where z_0 is a scale height.

The Poisson equation allows us to calculate the total density in the plane

$$\rho(0) = \frac{1}{4\pi G}\frac{\partial^2 \Phi}{\partial z^2}, \tag{1.16}$$

as well as the projected density on the disk. We may compare the local density determined in this way to the sum of the densities of the different stellar populations and of the interstellar medium. The difference, if real, gives the contribution of dark matter. A recent work in this field is due to Crézé et al. [107], who used data obtained with the HIPPARCOS satellite to derive the density law $\rho_i(z)$ of a certain type of stars (the A stars) and also their velocity dispersion σ_i, then

$$\rho(0) = (0.076 \pm 0.015) M_\odot \, \text{pc}^{-3}, \tag{1.17}$$

to be compared with a density of about 0.043 $M_\odot$ pc^{-3} for all types of stars and stellar remnants. The difference can be due to dark matter, to interstellar matter, or to both. We know that the density of interstellar medium is very small within 100 pc of the Sun (Sfeir et al. [466]), due to the existence of the *Local bubble* that will be discussed later. This is the region studied by Crézé et al. [107]. Their result is therefore not in contradiction with the presence of a non-negligible quantity of dark matter near the Sun. However the above equations cannot yield its distribution.

The vertical velocity dispersion is about 6 km s^{-1} for the youngest stars (O and B stars, cepheids, supergiants), and increases to 8 km s^{-1} for A stars, 11 km s^{-1} for F stars, 21 km s^{-1} for G–M dwarfs and 16 km s^{-1} for cold giants (see e.g. Binney & Merrifield [41]). The corresponding scale heights are uncertain given the uncertainties in the vertical force $K_z(z)$. Here we will define the height scale h as for a gaussian distribution, a good approximation for the thinnest systems

$$\rho(z) = \rho(0) \exp(-z^2/2h^2). \tag{1.18}$$

h is of the order of 80 pc for the youngest stars and increases with increasing age of the stars, reaching some 300 pc for the oldest disk stars.

1.2.3 The Bulge and the Halo

These spheroidal components are less interesting for us since we are concerned with interstellar matter. Their rotation is slow and they are essentially supported against gravity by their large stellar velocity dispersion. The density of the stellar halo varies approximately as $R^{-3.5}$ and its total luminosity is about 4×10^7 L$_\odot$ in the visible, a very small number when compared to the luminosity of the disk, about 7×10^9 L$_\odot$.

Dark matter may also be spherically distributed in a system, also called the dark halo although this may be confused with the stellar halo. If the dark matter distribution is indeed spherical, then the constancy of the linear rotation velocity of the disk at large galactocentric radii implies that the mass $M(< R)$ inside radius R is proportional to R. At large radii the mass is dominated by dark matter and the density of this matter varies as R^{-2}, i.e. more slowly than the density of halo stars. The mass of such a distribution diverges at infinity, so that it must be truncated at some distance from the galactic centre. In any case, we saw earlier that the total mass of the Galaxy is at least 10^{12} M$_\odot$, and is largely dominated by dark matter.

1.3 Distribution of Interstellar Matter

We can directly measure the distance d to stars through their geometrical parallax. Presently an accuracy of $(d/10 \text{ pc})$ % is possible with the HIPPARCOS satellite. Other methods allow an indirect determination of stellar distances. For example, the most widely used method consists of comparing the apparent magnitude m with the absolute magnitude M obtained from the spectral type and photometry. The magnitude is a logarithmic measure of the energy E received from the star at the Earth, and is defined as $m = -2,5 \log E + const$. By definition, the absolute magnitude M is equal to the apparent magnitude of the star placed at a distance of 10 pc and observed in the absence of extinction. This is an intrinsic quantity of the star. If there is extinction, we can write

$$m - M = 5 \log \left(\frac{d}{10 \text{pc}} \right) - A, \tag{1.19}$$

where A is the *extinction* at the considered wavelength, expressed in magnitudes.

It is unfortunately impossible to determine the distance of interstellar matter in this way. If interstellar matter can be associated with particular stars, for which the distance is measurable, we can obtain its distance. This is the case for H II regions for which the ionizing stars are known. For neutral, absorbing clouds (the *dark clouds*), we can determine the distance to unabsorbed stars seen in front of such clouds, i.e. stars unextinguished by the cloud. The distances to the main dark clouds around the Sun have been determined using this method, e.g. the distance to the Taurus, ρ Ophiuchi or Chamaeleon clouds. If this means is not possible, the only available method is to use the radial velocity of the gas, obtained through observations of an emission or absorption line. This is the *kinematic distance* method that we will now briefly describe.

The rotation of the galactic disk, which contains most of the interstellar medium, is not that of a solid body. Instead, *differential rotation* occurs and each point along a line of sight in the plane of the disk generally has a different radial velocity. It is convenient to take as the origin for the radial velocities a point coinciding with the Sun but moving with the local rotation of the disk. This point is called the *Local Standard of Rest* (LSR). Letting Θ and Θ_0 be the respective linear rotation velocities of point P and of the LSR, l be the galactic longitude of P and R its distance from the galactic centre (Fig. 1.1), the radial velocity of P with respect to the LSR is

$$v_{rad} = \Theta \sin(l + \beta) - \Theta_0 \sin l = R_0 \left(\frac{\Theta}{R} - \frac{\Theta_0}{R_0} \right) \sin l = R_0(\omega - \omega_0) \sin l, \tag{1.20}$$

where R_0 is the distance of the Sun from the galactic centre and ω the angular rotation velocity. If point P is not far from the Sun, we can write, to first order

$$v_{rad} \simeq -2A(R)(R - R_0) \sin l, \tag{1.21}$$

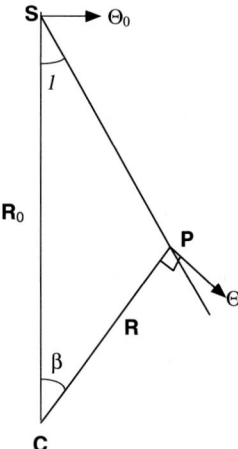

Fig. 1.1. The differential rotation of the Galaxy. S is the position of the Sun, C that of the galactic centre and P is the point under consideration. The other quantities are defined in the text.

defining

$$A(R) \equiv -\frac{1}{2} R \frac{d\omega}{dR} = \frac{1}{2} \left(\frac{\Theta}{R} - \frac{d\Theta}{dR} \right), \qquad (1.22)$$

one of the two Oort parameters. The other Oort parameter is

$$B(R) \equiv -\frac{1}{2} \left(\frac{\Theta}{R} + \frac{d\Theta}{dR} \right). \qquad (1.23)$$

We can also write

$$v_{rad} \simeq -rA(R_0) \sin 2l, \qquad (1.24)$$

r being the distance of point P, which shows that the radial velocity is zero at longitudes 0, 90, 180 and 270 degrees.

We can easily find that the *proper motion* (lateral apparent displacement of point P) is numerically given by

$$\mu = \frac{A(R_0) \cos 2l + B(R_0)}{4.74} \text{ arc second per year.} \qquad (1.25)$$

If the galactic latitude b of point P is not zero, we have to multiply the values of v_{rad} obtained above by $\cos b$.

The local value of the Oort parameters is still poorly known given the uncertainty in $d\Theta/dR$. We have $A(R_0) = 11\text{--}15$ km s^{-1} kpc^{-1} and $B(R_0) = -(12\text{--}14)$ km s^{-1} kpc^{-1}. For a recent discussion see Olling & Merrifield [388].

The kinematic distance method is thus uncertain, not only by virtue of the uncertainty in the Oort parameters, but also because the radial velocity falls to zero at some galactic longitudes and because there are local systematic departures from

1.3 Distribution of Interstellar Matter

pure rotation. At best, there is an uncertainty of about 0.5 kpc. Moreover, there is a distance ambiguity for half of the galactic disk: a given radial velocity can correspond to two possible distances. Unfortunately this method is the only one that can be used to map the distribution of neutral hydrogen H I from 21-cm line velocities. For molecular clouds, and even more so for H II regions the situation is more favorable because they are often associated with young stars for which distances can be measured more accurately through photometry (Russeil et al. [443]).

Figure 1.2 is a map of the concentrations of ionized gas (H II regions) in the Galaxy. We do not show similar maps for the neutral components because the kinematic distances are too uncertain. and molecular gas in the galactic disk. The radial distribution of the ionized gas is not very different from that of the molecular gas (see Fig. 2 of Smith et al. [477]). These different components do not have the same scale height perpendicular to the galactic plane. H II regions follow the distribution of the very young and hot ionizing stars and have the same scale height of 80 pc at half-density (Guibert et al. [214]). There is also diffuse ionized gas partly associated with atomic gas, for which the scale height reaches 1 kpc (Reynolds et al. [430]). Molecular gas, from which the stars form, has a similar distribution to

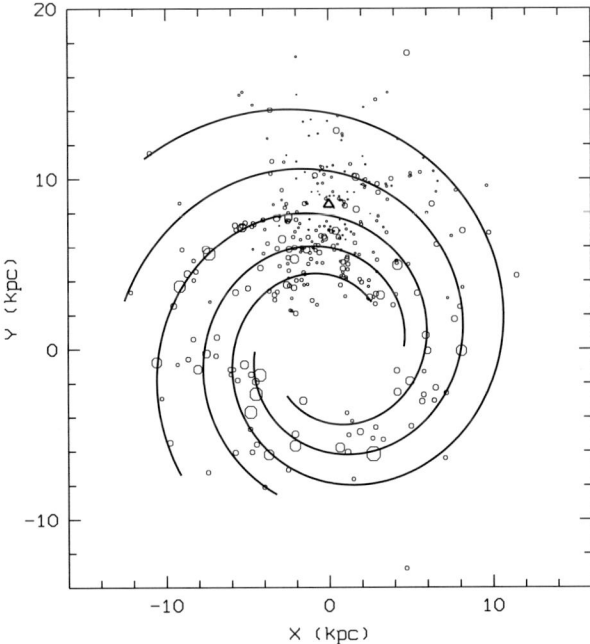

Fig. 1.2. Outline of the spiral structure of the Galaxy. The symbols correspond to H II regions with known distances, the size of the symbol indicating the relative stellar ultraviolet fluxes that ionizes the gas. The position of the Sun, assumed here to be at 8.5 kpc from the galactic centre (position 0,0) is represented by a triangle. The best fit to the positions of the H II regions is a 4-arm logarithmic spiral. By kind permission of Delphine Russeil (see [443]).

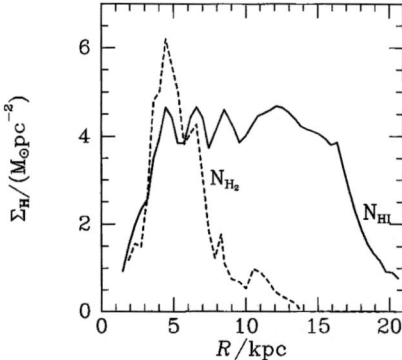

Fig. 1.3. Radial distribution of the atomic and molecular components of interstellar matter in the Galaxy, averaged over concentric rings. The distance of the Sun from the galactic centre (R_0) is taken to be 8.5 kpc. The emission from the central regions is not displayed. The surface density of the molecular component was obtained from observations of CO molecules assuming a conversion factor X of 2.3×10^{20} mol. cm^{-2} (K km s^{-1})$^{-1}$ (see (6.26)). This factor, hence the amount of molecular gas, is probably overerestimated (cf. Sect. 6.3). From Binney & Merrifield [41], reprinted by permission of the Princeton University Press.

young stars (velocity dispersion $\simeq 6$ kms) and thus the same thickness, but there also exists a faint molecular disk about 3 times thicker (Dame & Thaddeus [112]). For HI, we must distinguish the relatively cold gas with a velocity dispersion of about 9 km s^{-1} and a scale height of about 210 pc (Malhotra [346]), similar to that of the thick molecular disk just mentioned, from a warmer, more diffuse, component which makes an even thicker disk (Falgarone & Lequeux [169]; Dickey & Lockman [124]). In Sect. 2.2, we will present an example of the determination of the scale height of the gas.

2 Radiations and Magnetic Fields

The interstellar medium is a semi-open medium, i.e., in weak interaction with the external world. Its energy sources are multiple: stellar radiation, high-energy particles, and mechanical energy from supernovae, stellar winds and differential rotation of the Galaxy.

In this chapter we examine the radiation fields that affect interstellar matter. The high-energy cosmic particles have some influence on the physics of interstellar matter, but will not be examined before Chap. 6. Incidentally, these cosmic particles have improperly, and for historical reasons, been called *cosmic rays* or *cosmic radiation*. The sources of mechanical energy will be discussed in Chapters 10 to 14 together with the dynamics of the interstellar medium. For practical reasons, the last section of the present chapter will deal with the interstellar magnetic fields and their measurements. Although the magnetic fields are produced by electric currents which circulate in the interstellar matter, it is often convenient to consider them as external parameters, like radiation fields. This is why we discuss them in the present chapter.

Plate 2 shows images of the Galaxy at different wavelengths. The reader might be interested in viewing these pictures in order to illuminate the present and the following chapters.

2.1 Radiation Fields

The interstellar medium is bathed in electromagnetic radiation which include: the galactic radiation field coming directly from stars, photons produced by secondary processes in the interstellar medium itself, and extragalactic radiations emitted by other galaxies and by the Universe as a whole (the blackbody radiation at 2.726 K). Let us first discuss briefly the extragalactic radiation, which does not greatly affect the interstellar medium.

2.1.1 Extragalactic Radiation

The energy density of the blackbody radiation of the Universe is 0.26 eV cm^{-3}, a considerable number. However, given its temperature (2.726 K) this radiation is at submillimetre and millimetre wavelengths where the interstellar medium is very transparent and its effects are very limited. However, it may play a role in the population of the rotational levels of interstellar molecules because many of them have transitions in this wavelength range. We will see an example of this later in Sect. 3.2.

The radiation originating in other galaxies and in the intergalactic medium is considerably fainter and can be neglected in the galactic plane. Its energy density is 2.7×10^{-14} erg cm^{-3} or 0.017 eV cm^{-3}, of which 2/3 are in the mid- and far-infrared at $\lambda > 6\,\mu$m (Puget & Guiderdoni [418]). However, the ultraviolet (UV) and X-ray components of this radiation field probably have a large influence on the galactic

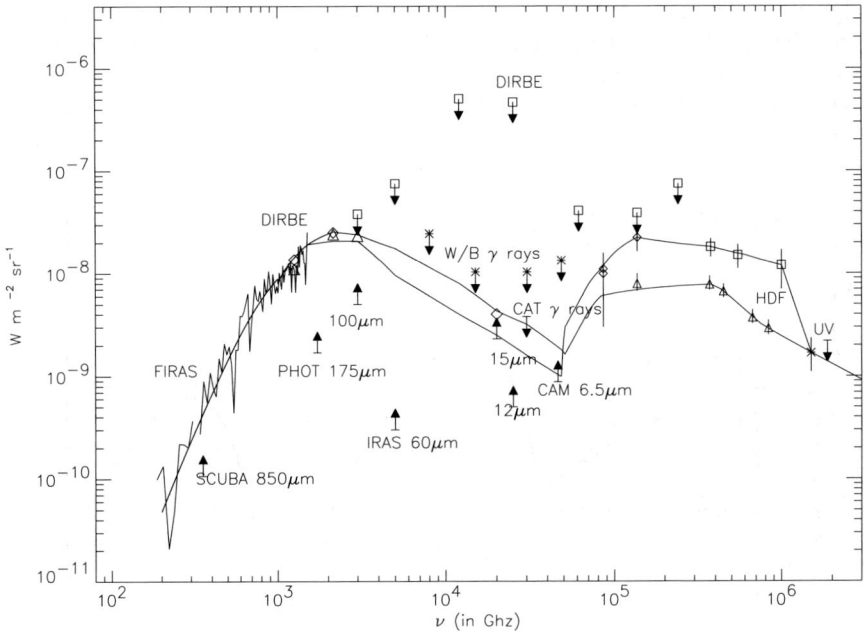

Fig. 2.1. The spectral energy distribution of the extragalactic backgound radiation. The quantity νI_ν is plotted as a function of frequency, where I_ν is the brightness of the sky outside the terrestrial atmosphere at frequency ν. The interest in using $\nu I_\nu = \lambda I_\lambda$, rather than I_ν, is that this quantity, when considered in any given logarithmic interval of frequency (or wavelength), represents the total energy in this range. The cosmologic radiation at 2.726 K dominates at millimetre wavelengths (frequencies lower than 500 GHz), but is not plotted here. The radiation shown on this figure is due to the integrated radiation of galaxies at all redshifts. The different measurements are as follows. SCUBA= bolometric measurement with the JCMT radiotelescope; FIRAS, DIRBE: measurements with the COBE satellite; IRAS: measurements with the IRAS satellite; PHOT, CAM 15 µm, 12 µm, 6.5 µm: measurements with the ISO satellite; HDF: integrated emission of galaxies in the Hubble Deep Field observed with the Hubble Space Telescope; UV: various measurements. Additionally, upper limits are derived indirectly from observations of the very high energy extragalactic gamma-rays observed with the W/B and CAT devices. References for all these determinations are given by Gispert et al. [199], from which this figure is reproduced with the permission of ESO. The lines indicate the range of possible values of the extragalactic background.

halo gas and on the external parts of the disk. Important progress has been recently made towards an understanding of the integrated radiation from galaxies. Its spectral energy distribution is shown in Fig. 2.1, and details on many aspects of this radiation can be found in Leinert et al. [313].

2.1.2 Galactic Radiation

Observations of Galactic Radiation

Figure 2.2 summarizes the spectral energy distribution of the radiation incident upon the Earth's atmosphere from high galactic and ecliptic latitudes (the night sky brightness). At the altitudes of terrestrial observatories, X-rays, UV and most IR radiation are totally absorbed, so that we must go above the lower atmosphere to observe them. There are already substantial gains in going to high mountain sites

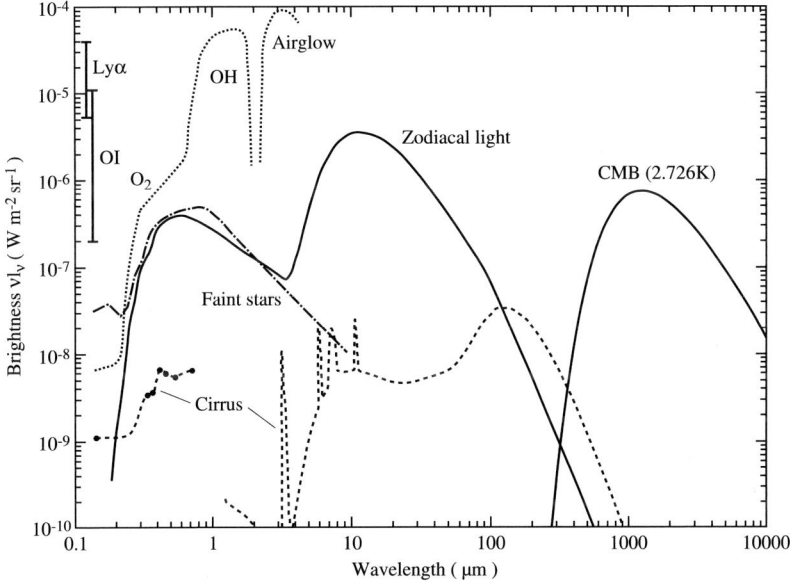

Fig. 2.2. Spectral energy distribution of the sky brightness outside the lower terrestrial atmosphere, at high galactic and ecliptic latitudes. νI_ν is plotted as a function of wavelength as for Fig. 2.1. The emission indicated as Airglow (dotted line) is due to the OH et O_2 molecules in the upper terrestrial atmosphere. We can get rid of them by observing from satellites. The zodiacal light brightness shown is that of the ecliptic poles, and is larger at lower ecliptic latitudes. The emissions indicated as Lyα and OI come from the extended atmosphere and are troublesome even for satellite observations. The indication "cirrus" relates to the light scattered by interstellar dust at high galactic latitudes in the visible, and to emission by this dust in the IR. The integrated emission of stars at high galactic latitudes is also shown in the visible and the near-IR, as well as that of the blackbody radiation of the Universe at millimetre wavelengths. From Leinert et al. [313], with the permission of ESO.

(altitudes of a few km), still higher gains using airborne observatories at about 14 km altitude and even higher gains from balloons at some 40 km altitude. At the same time we can reduce most of the atmospheric absorption and turbulence which can affect observations in the electromagnetic spectrum. However, transmission in the UV and the far IR is still poor, and the sky brightness in the visible and the near-IR is still dominated by the emission from O_2 and OH molecules in the upper atmosphere. We have to go to the usual altitude of artificial satellites, 300 km or more above the Earth, to get rid of this emission. But even here there is monochromatic emission from hydrogen and atomic oxygen in the far-UV (Lyman α line at 1 216 Å and the O I resonance line at 1 302 Å). The zodiacal light which results from scattering of the sunlight by interplanetary dust grains, and from the thermal emission by these grains in the mid- and far-IR, cannot be avoided. The only possibility is to build a model of this light and to subtract it from the observed brightness if we wish to obtain the radiation field beyond the Solar system. We can see in Fig. 2.2 that this radiation field is dominated by stellar emission in the UV (Fig. 2.3), the visible and the near-IR, and that emission by interstellar dust dominates in the mid- and far-IR. See also Fig. 2.4 from Désert et al. [120].

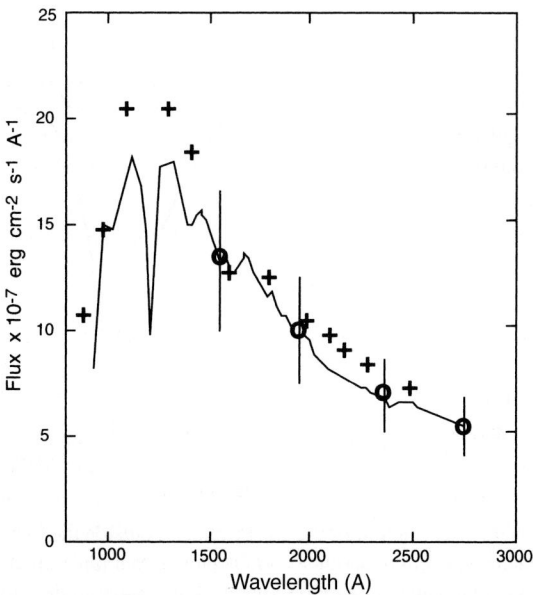

Fig. 2.3. Spectral energy distribution of the ultraviolet radiation outside the terrestrial atmosphere, integrated over the whole sky ($4\pi I_\lambda$). Circles with error bars result from direct integration with the TD1 satellite. The curve is a model by Gondhalekar based on star counts, where interstellar extinction has been corrected for in these observations. Crosses correspond to a similar model by Mathis et al. [352]. The Habing's model [217] gives very close results. Adapted from Gondhalekar et al. [206].

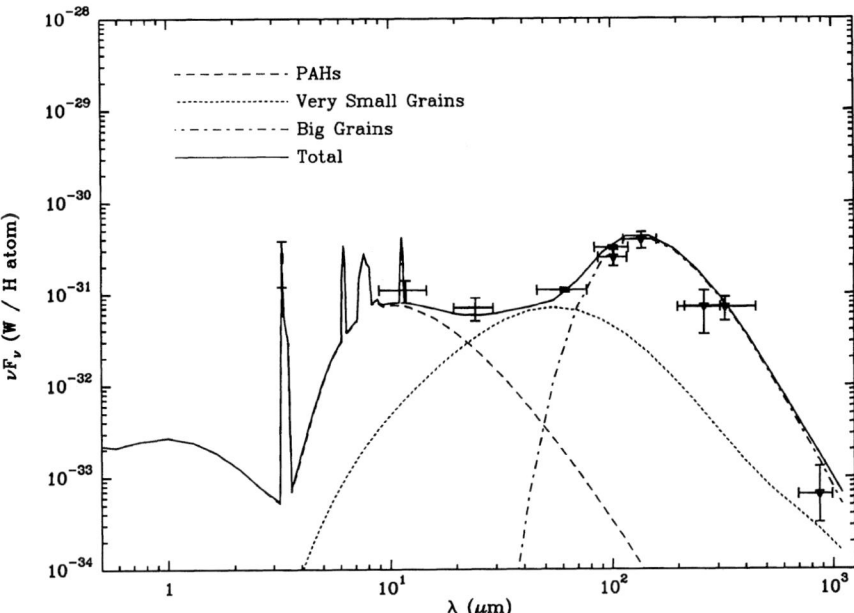

Fig. 2.4. Spectral energy distribution of the interstellar dust emission in our Galaxy. The energy $\nu F_\nu = \lambda F_\lambda$ emitted by a quantity of dust normalized to the hydrogen abundance is plotted as a function of wavelength. The observations (crosses) correspond to the "cirrus" component at high galactic latitudes. The curves present the results from a model, and shows three components and the sum of these. This model will be described in Chap. 7. Taken from Désert et al. [120].

UV, Visible and Near-IR Radiation

Before the recent direct observations described by Leinert et al. [313], it was necessary to calculate this radiation in the solar neighbourhood from the (more or less) known distribution of the stars and of the dust which affects their radiation. This has been done by several authors, in particular Mathis et al. [352]. Figure 2.3 shows their determination in the UV, compared to that of Gondhalekar et al. [206]. The model results are normalized to the photometry performed with the TD1 satellite. The agreement is good given the errors. The "typical" interstellar radiation field of Habing [217], which is often used, is also in good agreement with the other models. The radiation field of Draine [132] is about 2 times higher. The energy density of the interstellar radiation near the Sun, integrated over all wavelengths, is close to 1 eV cm^{-3}. Integration of the Habing's field in the UV from 912 to 2 066 Å (6 to 13.6 eV) corresponds to an intensity I of 1.3×10^{-4} erg s^{-1} cm^{-2} sterad^{-1}, or to an energy density $u = 4\pi I/c$ of 5.4×10^{-14} erg cm^{-3} or 0.034 eV cm^{-3} (Pak [391]). The Draine's radiation field in the same wavelength range is 1.64 times higher. It is perhaps more useful to give the integrated energy between 912 et 1 130 Å, because photons in this region are responsible for the ionization of species other than H, He, N, O and noble gases, and for photodissociation of most molecules in the neutral interstellar medium. Draine's field corresponds to an intensity over this wavelength range of 3.71×10^{-5} erg s^{-1} cm^{-2} sterad^{-1}, or to an energy density of 1.55×10^{-14} erg cm^{-3} or 0.0097 eV cm^{-3}. We must be careful about the choice of the reference field in the articles which make use of such a field. The symbol χ is usually used to designate the ratio between the UV radiation field at 1 000 Å in the studied region and the local reference field of Habing. G_0 usually represents the ratio of the radiation density from 6 to 13.6 eV relative to that of the Habing field. We will not differentiate between these two quantities in this book, and will designate both by χ. The radiation field in other regions of the

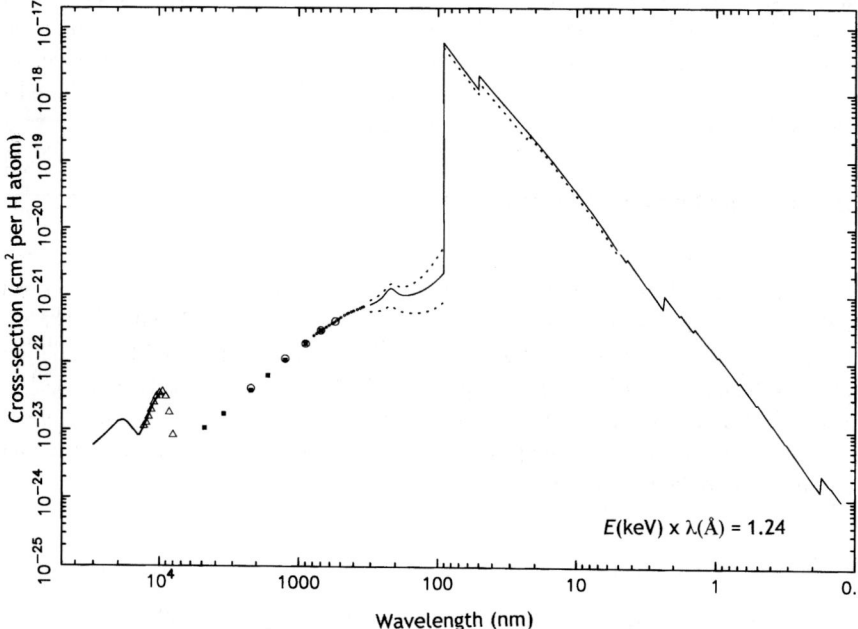

Fig. 2.5. Interstellar extinction as a function of wavelength. The ordinates give the extinction cross-section of photons normalized to one hydrogen atom. The absorption at wavelengths shorter than 91.2 nm is given for solar abundances of the elements (these abundances are almost identical to those given in Table 4.2, column 2). The relation between the wavelength and the energy of the photon is given at the bottom right (useful for X rays). The huge extinction peak near 90 nm (900 Å) is due to the Lyman continuum absorption by hydrogen atoms, which produces their ionization. The sources are by order of decreasing wavelength: *full line*: Draine & Lee [134], fig. 6a, for flattened spheroids of astronomical silicate; *triangles*: Rieke & Lebovsky [435]; *full squares*: Martin & Whittet [347]; *large circles*: Cardelli et al. [77]; *small circles*: Bastiaansen [23]; *full line*: average UV extinction curve; above and underneath, *dashed line*: extreme UV extinction curves, respectively for HD 204 827 and for HD 37 023 (Fitzpatrick & Massa [183]); *full line* from 91.2 to 5 nm: Rumph et al. [442]; *full line* from 5 to 0.12 nm: Morrison & McCammon [374]. Adapted from Ryter [444].

Galaxy can only be derived from models, the best probably being that of Mathis et al. [352].

We note that almost everywhere in the Galaxy the radiation field is truncated below the Lyman discontinuity of hydrogen at 911.7 Å. This is due to the presence of much atomic hydrogen in the Galaxy which completely absorbs radiation at wavelengths shorter than 911.7 Å. The interstellar medium becomes progressively transparent toward the much shorter wavelengths, equivalent to soft X-ray energies (Fig. 2.5). The far-UV radiation at wavelengths shorter than 911.7 Å can only propagate if hydrogen is entirely ionized, which occurs in H II regions. The situation is however more complex in the very diffuse medium which is partly ionized and where the density of H I is small enough to allow the propagation of radiation over large distances. This is the situation near the Sun, where the density of the medium is very low (this is the *Local bubble*). This partial transparency has allowed the observation of a few white dwarf stars over the UV to X-ray range (Barstow et al. [22]). In most of what follows we will assume that UV photons with wavelengths shorter than 911.7 Å can be ignored, except of course in H II regions.

Mid- and Far-Infrared Radiation

This radiation near the Sun is now rather well known at all wavelengths thanks to the IRAS (InfraRed Astronomy Satellite), the COBE (Cosmic Background Explorer) and the ISO (Infrared Space Observatory) satellites: see Fig. 2.2 and 2.3. The interstellar medium is relatively transparent at these wavelengths, and this radiation field can be calculated in different regions in the Galaxy more easily than the UV or visible radiation fields. On the other hand, IR photons have relatively little effect on the interstellar medium, except for molecular clouds which can have substantial opacity even in the far infrared.

X-Rays

The hot interstellar gas thermally emits low-energy X-rays (< 1 keV), to which we must add the contribution of individual supernova remnants and an extragalactic X-ray background which is obvious at energies > 1 keV. We have not mentioned this extragalactic background before because its influence on the interstellar medium is negligible. This is not the case for sub-keV X-rays that are easily absorbed by the interstellar medium. This absorption complicates much of the interpretation of the X-ray observations (Snowden et al. [480]). The Local bubble around the Sun is an irregular cavity of about 150 pc radius (Sfeir et al. [466]). It contains X-ray emitting hot gas, however, an important fraction of the soft X-rays comes also from the halo, far from the bubble. Soft X-rays play a role in the heating of the diffuse interstellar medium (see Chap. 8).

2.2 The Interstellar Magnetic Fields

The interstellar magnetic fields are very important in the physics of interstellar matter. At large scales, magnetic pressure and the pressure of high-energy charged particles, confined by the magnetic fields, add to the kinetic non-thermal pressure[1] of the gas to balance the gravitational attraction of the disk. This determines the vertical distribution of the gas.

These three pressures are, at $z = 0$

- The macroscopic kinetic pressure related to large-scale motions of interstellar clouds: $p_G(0) = \rho \langle v_z^2 \rangle$, where v_z is the r.m.s. 1-dimensional velocity, such that $\langle v_z^2 \rangle = \frac{1}{3} \langle v_{total}^2 \rangle$.
- The pressure of cosmic particles: $p_C(0) = \frac{1}{3} U_C$, where U_C is the energy density of these particles, assumed to be mostly relativistic
- The magnetic pressure: $p_B(0) = B^2/8\pi$.

They are all of the order of 10^{-12} dynes cm^{-2}.

We can easily obtain the equation for hydrostatic equilibrium perpendicular to the galactic plane if we assume that these three pressures are proportional to each other at all z. Then we can introduce $\alpha \equiv p_B/p_G$ and $\beta \equiv p_C/p_G$. If we assume also that $d\phi/dz = -z(dK_z/dz)$ (see Sect. 1.2) with $dK_z/dz = const$, an assumption which is approximately justified for the stellar gravitation field for $z < 250$ pc (Kuijken & Gilmore [290], Fig. 14), the equation for hydrostatic equilibrium is

$$\frac{dp}{dz} = -\rho \frac{d\Phi(z)}{dz} = \rho K_z(z), \tag{2.1}$$

with $p = \rho(1 + \alpha + \beta)\langle v_z^2 \rangle$. In this equation, ρ is the density of the gas while $K_z(z)$ is determined by all the matter in the disk.

The solution is

$$\boxed{\rho = \rho(0) \exp\left[-\frac{\Phi(z)}{(1 + \alpha + \beta)\langle v_z^2 \rangle}\right].} \tag{2.2}$$

Up to heights of a few hundred parsecs we have approximately $K_z(z) \simeq -az$ so that

$$\Phi(z) = -\int_0^z K(z)\, dz \simeq az^2/2, \tag{2.3}$$

and

$$\rho \simeq \rho(0) \exp(-z^2/2h^2), \tag{2.4}$$

with

---

[1] We will see that the thermal pressure of the gas is negligible at the scale of the galactic disk, with respect to the pressure due to random motions, except for the hot gas which makes up only a tiny portion of the total.

$$h = \left[\frac{(1+\alpha+\beta)\langle v_z^2 \rangle}{a}\right]^{1/2}. \tag{2.5}$$

Assuming $\langle v_z^2 \rangle^{1/2} = 9\,\text{km}\,\text{s}^{-1}$, independent of z, and $a = 1.8 \times 10^{-11}\,\text{cm}\,\text{s}^{-2}\,\text{pc}^{-1}$, corresponding to a total density of matter of $0.1\,\text{M}_\odot\,\text{pc}^{-3}$ in the galactic plane, we find $h = 210$ pc, in reasonable agreement with H I observations (Malhotra [346]).

At smaller scales, the magnetic fields play a major role in the physics of gravitational instability and molecular cloud collapse which may lead to the formation of stars (see later Chap. 14).

There are several possible means of measuring the interstellar magnetic fields. We will examine them now. The general conclusions which can be drawn from these measurements will be discussed afterwards.

2.2.1 Magnetic Field Measurements Using the Zeeman Effect

The Zeeman effect is the removal, by a magnetic field, of the degeneracy of a level of total non-zero angular momentum $\mathbf{J} = \mathbf{L} + \mathbf{S}$ of an atom or molecule. $\mathbf{L}$ is the sum of the orbital momenta of electrons, and $\mathbf{S}$ that of their spin angular momenta. For this effect to exist it is sufficient that the atom or molecule possesses a non-zero magnetic moment $\mathbf{N}$. In such a case there is an interaction with the magnetic field $\mathbf{B}$ which results in a precession of the total angular momentum $\mathbf{J}$ around $\mathbf{B}$. Since the orientation of $\mathbf{J}$ with respect to $\mathbf{B}$ is quantified, with corresponding values of the quantum number $M_J = J, J-1, ..., -J$, each of the M_J sublevels is displaced by a different quantity under this interaction which depends on the angle between $\mathbf{N}$ (or $\mathbf{J}$) and $\mathbf{B}$, hence on M_J. The magnetic moment can be due either to the total orbital momentum of electrons if non zero ($\mathbf{L} \neq 0$ for atoms, or $\mathbf{\Lambda} \neq 0$ for diatomic molecules), or to the spin of an unpaired electron. In this case the magnetic moment is of the order of the Bohr magneton

$$\mu_B = \frac{\hbar e}{2m_e c} = 9.2741 \times 10^{-21}\,\text{erg}\,\text{gauss}^{-1}, \tag{2.6}$$

where e and m_e are respectively the charge and the mass of the electron.

For molecules, the magnetic moment can also come from localized charges or from the nuclear spins. In this case its value is generally close to

$$\mu_p = \frac{\hbar e}{2m_p c} = \frac{\mu_B}{1\,850}, \tag{2.7}$$

m_p being the proton mass.

In the case of atoms, a magnetic field splits a level J into $2J + 1$ equidistant sublevels of energy

$$E_{J,M_J} = E_{J,0} + \mu_B B g_J M_J, \tag{2.8}$$

where $E_{J,0}$ is the energy of the level in the absence of a magnetic field. g_J is the *Landé factor*

$$g_J = 1 + \frac{J(J+1) + S(S+1) - L(L-1)}{2J(J+1)}, \tag{2.9}$$

which represents the spin-orbit coupling of electrons and differs from unity only for atoms with total spin $S \neq 0$. The selection rules are $\Delta M = 0, M \neq 0$ and $\Delta M = \pm 1$. The $J+1 \rightarrow J$ transition is split
 – either into 3 components: one π component ($\Delta M = 0$) and two σ components ($\Delta M = \pm 1$) if $S = 0$;
 – or into several components $\pi(\Delta M = 0)$ and $\sigma(\Delta M = \pm 1)$ if $S \neq 0$.

The π components are linearly polarized parallel to the magnetic field and are maximum when the line of sight is perpendicular to **B**. The σ components are circularly polarized in the plane perpendicular to **B**. Observing along the magnetic field, the π component vanishes and only the σ components, which are polarized in opposite senses, are observed. These are the components that are used in practice to determine the magnetic fields.

The frequency displacement of a Zeeman component with respect to the rest central frequency of a line $J, J-1$ is

$$\Delta \nu_{\pm} = \mu_B B[g_J M - g_{J-1}(M \mp 1)]. \tag{2.10}$$

For the hydrogen atom, $\Delta \nu = 2.8 \mu_B B_{\parallel}/(1\,\mu G)$, and the displacement is 1.4 Hz μG^{-1}. The order of magnitude of the displacement is similar for lines of OH and CH (cf. Bel & Leroy [27]). Displacements this small compared to the kHz Doppler width of the lines are obviously difficult to measure. The only possibility is to compare the frequencies of a pair of σ components taking advantage of their opposite circular polarizations. In this way only $B_{\parallel}$ is measured. Positive results have been obtained on relatively dense H I clouds, for which magnetic fields from 10 to 3 000 μG have been measured. Conversely, the general magnetic field of the Galaxy, a few μG, is too faint to be measured in this way.

2.2.2 Measurement of the Magnetic Fields Using Faraday Rotation

In the presence of a magnetic field, an ionized medium is dielectric with different refractive indices for the two opposite circular polarizations. A linearly polarized plane wave such as that emitted by galactic or extragalactic synchrotron radiosources or by pulsars is equivalent to the sum of two circularly polarized waves with opposite senses. The propagation of such a wave in a magnetized plasma leads to a phase shift between these circular components and thus to a rotation of the plane of polarization of the equivalent plane wave. This is the *Faraday rotation* (Rohlfs & Wilson [439] p. 43)

$$\begin{aligned} \Omega &= \frac{e^3}{2\pi \nu^2 m_e^2 c^2} \int_0^L n_e B_{\parallel} dl \\ &= 8.1 \times 10^5 \left(\frac{\lambda}{m}\right)^2 \int_0^L \left(\frac{n_e}{cm^{-3}}\right) \left(\frac{B_{\parallel}}{G}\right) d\left(\frac{l}{pc}\right) \text{ rad}, \end{aligned} \tag{2.11}$$

where e and m_e are respectively the charge and mass of the electron, ν the frequency of the wave, n_e the electron density, $B_\parallel$ the longitudinal component of the magnetic field and L the length of the line of sight. We often define $\Omega = \lambda^2 RM$, where RM is the *Rotation Measure*. We then have

$$RM = 8.1 \times 10^5 \int_0^L \left(\frac{n_e}{\text{cm}^{-3}}\right) \left(\frac{B_\parallel}{\text{G}}\right) d\left(\frac{l}{\text{pc}}\right) \text{ rad m}^{-2}, \qquad (2.12)$$

the wavelength λ being expressed in metres.

Only the average value of the longitudinal component of the magnetic field, $B_\parallel$, can be measured in this way. For this, we have to measure the position angle of the linearly polarized wave at 3 different frequencies in order to resolve possible ambiguities by multiples of π in the rotation. The electron density must also be known. For this, it is interesting to use pulsars as sources. Pulsars emit linearly polarized radio waves as short, periodic pulses. The interstellar plasma is dispersive, and the difference between the arrival times of the pulses at frequencies ν_1 and ν_2 ($\nu_1 > \nu_2$) is

$$\Delta t = \frac{e^2}{2\pi m_e c}\left(\frac{1}{\nu_2^2} - \frac{1}{\nu_1^2}\right) \int_0^L n_e \, dl \text{ seconds.} \qquad (2.13)$$

Measurement of this dispersion gives $\langle n_e \rangle$. We can also independently obtain the distance of the observed pulsars, thus allowing us to map the magnetic fields. Measurements of the Faraday rotation of extragalactic radiosources are less powerful in determining the value of the magnetic fields but give results in good agreement with pulsar measurements (cf. e.g. Clegg et al. [101]). The random component of the magnetic fields produces a frequency-dependent depolarisation of radio waves. Its observation should in principle allow us to measure this random component but the modelling procedure is very complex (Sokoloff et al. [483]).

2.2.3 Estimate of the Magnetic Fields from the Galactic Synchrotron Radiation

Relativistic electrons moving in the magnetic fields of the galactic interstellar medium (see Section 6.1) emit synchrotron radiation in the radio range. A very complete book on the subject of synchrotron radiation has been written by Sokolov & Ternov [482]. This radiation is strongly linearly polarized. The fact that regions with degrees of linear polarization equal to or larger than 70% are often seen in the Galaxy and in external galaxies shows that the magnetic fields can be well ordered. The direction of polarization is perpendicular to the magnetic field, and we can obtain the direction of the field provided that the measurements are corrected for the Faraday rotation. If we know, from some other means, the flux of relativistic electrons, we can get also the modulus of the magnetic field. This method is useful because it is sensitive not only to $B_\parallel$ but also to the entire magnetic fields. For an isotropic distribution of electrons of energy E, the characteristic frequency of synchrotron radiation in a magnetic field B is (Lang [299], p. 29)

$$\nu_c = \left(\frac{E}{m_e c^2}\right)^2 \frac{3eB}{4\pi m_e c}, \tag{2.14}$$

or numerically

$$\boxed{\frac{\nu_c}{\text{MHz}} \simeq 16.1 \left(\frac{E}{\text{GeV}}\right)^2 \left(\frac{B}{\mu\text{G}}\right).} \tag{2.15}$$

Thus an electron of 5 GeV (observable from satellites) emits near 2 000 MHz (observable from the ground) in the typical galactic magnetic field of 5 µG. A power-law electron energy spectrum

$$\boxed{n(E) = KE^{-\gamma}dE} \tag{2.16}$$

yields a power-law radio spectrum

$$\boxed{\begin{aligned} I(\nu) &= 0.933 \times 10^{-23} a(\gamma) LKB^{\frac{\gamma+1}{2}} \\ &\quad \times \left(\frac{\nu}{6.26 \times 10^{18}\,\text{Hz}}\right)^{-\frac{(\gamma-1)}{2}} \text{erg cm}^{-2}\,\text{s}^{-1}\,\text{sterad}^{-1}\,\text{Hz}^{-1}, \end{aligned}} \tag{2.17}$$

where $a(\gamma)$ is a factor close to 2 tabulated by Rohlfs & Wilson [439], Table 9.2. L is the length of the line of sight in cm. Near the Sun

$$K \simeq 5 \times 10^{-16} \text{ erg}^{\gamma-1}\,\text{s}^{-1}\,\text{cm}^{-3}. \tag{2.18}$$

The radio spectrum of the synchrotron radiation of the Galaxy indeed follows a power law with a slope of -0.75, in agreement with the exponent $-\gamma \simeq -2.5$ for the energy spectrum of cosmic electrons. But it is not certain that the flux of cosmic electrons observed near the Earth is representative of a more extended region of the Galaxy, so that it is dangerous to directly combine the relativistic electron flux with a (rather difficult) determination of the energy density of synchrotron radiation if we wish to derive the magnetic fields. It is better to combine the gamma-ray radiation and the synchrotron radiation of the Galaxy. The gamma-ray flux is proportional to the product of the cosmic proton flux and the density of interstellar matter (cf. Sect. 6.2), while the synchrotron flux is proportional to the product of the cosmic electron flux and $B^{(\gamma+1)/2}$. If we know the distribution of interstellar matter and the ratio of the flux of cosmic electrons and protons ($\simeq 0.01$ for the same energy, see Sect. 6.1), we can in principle obtain the magnetic fields. This method has particular problems, however, and we rather use the gamma-ray observations to determine the density of interstellar matter assuming that the cosmic-ray flux is known!

2.2.4 Estimate of the Direction of the Magnetic Fields from the Linear Polarization of Light

We observe a faint linear polarization of visible starlight, correlated with the degree of interstellar extinction due to dust grains. This suggests that polarization is due to an anisotropy and a partial alignment, along the magnetic field, of the grains that are responsible for extinction, provided they are not spherical (for a recent paper see Roberge & Lazarian [438]). The material of these grains is certainly at least diamagnetic, and perhaps partly paramagnetic or ferromagnetic if they contain elements with unpaired electrons. Also, the grains continuously experience exchanges with the interstellar gas that produce their rotation. Collisions with atoms or molecules of the gas, and also the ejection of H_2 molecules that can form on their surface (see Chap. 9), exert a torque on the grains. The grains being generally charged, their rotation contributes to their total magnetic moment **M**. An order of magnitude for this contribution is $q\omega a^2/2c$, q and a being respectively the charge and the radius of the grain and ω its angular velocity.

A magnetic field **B** exerts a torque **M** × **B** on the grain. If the angular momentum **J** of the grain and **B** are not aligned, the torque forces the grain to precess around **B**, yielding globally some alignment of the grains and an anisotropy of the medium. The degree of alignment is limited by collisions, which can change the direction and amplitude of **J**, but we can show that the precession period is a lot shorter than the interval between successive collisions, allowing some statistical anisotropy. Given the uncertainties in the properties of the grains and on the origin of their magnetic moment, a quantitative estimate is not really possible, and it is not even possible to predict if the direction of the linear polarization of the light transmitted by the medium is parallel or perpendicular to the magnetic field. Observations suggest that it is parallel, however.

More recently, linear polarization of the far-infrared radiation emitted by interstellar dust has been detected, yielding information on the direction of the magnetic fields.

2.2.5 Results

At large scales, observations of the Faraday rotation, and to a lesser extent optical and radio polarization measurements, have demonstrated the existence of ordered magnetic fields. Rand & Lyne [424] found by comparing models to the observations of many pulsars a regular magnetic field of 1.4 μG. This field is roughly directed towards galactic longitude 90°, and is hence perpendicular to the direction of the galactic centre. This suggests an orientation along spiral arms. This field reverses at 0.4 kpc towards the galactic centre, then again at 5.5 kpc. An ordered magnetic field more or less parallel to the spiral arms is seen in nearby spiral galaxies. This is clearly observed from the direction of polarization of the synchrotron radiation. On this ordered field a random field of about 5 μG is superimposed, clearly larger (Rand & Kulkarni [423]). This disordered field dominates the optical polarization of nearby stars (Leroy [320]). The energy density corresponding to the total magnetic field is close to 1 eV cm^{-3}.

The existence of radio synchrotron radiation in the halo of many galaxies including ours is direct proof of the presence of magnetic fields at a large distance from the plane of the disk. These magnetic fields have probably been driven by the ionized

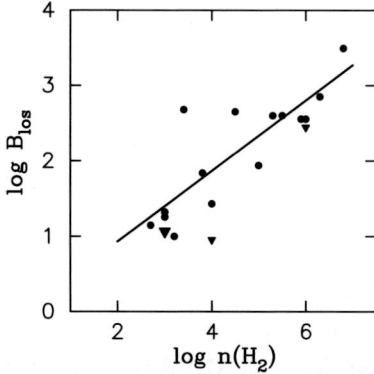

Fig. 2.6. Intensity of the magnetic fields in dark (molecular) clouds, as a function of density. The points correspond to measured values, the triangles to upper limits. The line is the relation $B \propto n(H_2)^{0.45}$. Reproduced from Crutcher [108], with the permission of the AAS.

gas ejected by supernova explosions in the disk, as can be seen by looking at the "chimneys" which emerge from the disks of galaxies (Chap. 15 and Plate 32).

At the scale of interstellar clouds in the Galaxy, the intensity of the magnetic fields increases with density, to the extent that density can be considered as reasonably well-known. However, this increase is slower than predicted from conservation of the magnetic flux during the spherical contraction of clouds (Fig. 2.6). If this flux was conserved, we would expect to see $B \propto n^{2/3}$, contrary to observations. This shows that either the magnetic flux is not conserved during the contraction, or that the contraction is not spherical. This will be discussed later in Chap. 14.

There is some confusion about the direction of magnetic fields at the cloud scale. This comes from the fact that the optical polarization measurements, which are most often the only useful ones, cannot give an unambiguous answer. However, galactic centre observations of the polarization of the far-infrared dust emission (e.g. Novak et al. [387]) and of synchrotron radio emission (e.g. Lang et al. [298]) give very important information on the magnetic fields and allow, in particular, an unambiguous determination of their direction.

3 Radiative Transfer and Excitation

In the four chapters that will follow this one, we will describe the different components of the interstellar medium and the methods that allow their study. However, in order to understand these following chapters we need to describe some elements of radiative transfer and of excitation of the energy levels of interstellar atoms and molecules. To proceed from the simple to the complex, we will first establish the equation of transfer. This involves the simple case of a system with only two energy levels, in which collisions with atoms and molecules completely determine the populations of these levels. This is the *Local Thermal Equilibrium* (LTE) case. Then we will examine the two-level system in the non-LTE case, including a discussion of the photon escape formalism and of the Large Velocity Gradient approximation. We will finally discuss multi-level systems, ending with a simplified discussion of interstellar masers. Our experience is that the notation used by radioastronomers is difficult for other astronomers to understand, so we will explain this notation. In the next chapters, the reader will find various applications of the notions developed in this one.

3.1 The Transfer Equation

3.1.1 Demonstration

Let us consider a medium consisting of atoms or molecules for which is is sufficient to consider only two energy levels: the fundamental level l *(lower)* and the excited level u *(upper)*, separated by an energy E. We will deal with the transfer of radiation in this medium at frequencies ν close to $\nu_0 = E/h$. Radiation with the *specific intensity* I_ν at frequency ν (unit: erg s^{-1} cm^{-2} sterad^{-1}) is incident on a slice through the medium. On an infinitely small length ds along the direction of propagation, the radiation at frequency ν is diminished by *absorption* in the spectral line by those atoms in the l level that can interact with the radiation. Let $n_l(\nu)$ be their density. At the same time, the radiation is increased by the *spontaneous emission* of those atoms in the u level that radiate at frequency ν, as well as by their *stimulated emission* at this frequency. The probability R_{lu} for an upward transition per unit time (*radiative excitation*) in a radiation field with a monochromatic energy density u_ν at frequency ν (unit: erg cm^{-3} Hz^{-1}) can be expressed as

$$R_{lu} = B_{lu}\frac{cu_\nu}{4\pi}. \qquad (3.1)$$

This defines the Einstein's *absorption coefficient* B_{lu}. If the radiation is assumed to be isotropic (case of the local interstellar radiation field), $u_\nu = 4\pi I_\nu/c$ and $R_{lu} = B_{lu}I_\nu$.[1]

Conversely the probability for a downward transition per unit time (*radiative de-excitation*) is

$$R_{ul} = A_{ul} + B_{ul}\frac{cu_\nu}{4\pi}, \qquad (3.2)$$

where A_{ul} is the Einstein's *spontaneous emission probability* and B_{ul} the Einstein's *stimulated emission coefficient*.

Balancing the radiative transitions gives the transfer equation, which writes

$$\frac{dI_\nu}{ds} = \frac{h\nu}{4\pi}\{n_u(\nu)A_{ul} - [n_l(\nu)B_{lu} - n_u(\nu)B_{ul}]I_\nu\}. \qquad (3.3)$$

Let us emphasize that this equation applies to a single frequency inside the spectral line. We will sometimes be lead to integrate it over the line profile. Let us define the *absorption coefficient*

$$\kappa_\nu \equiv \frac{h\nu}{4\pi}[n_l(\nu)B_{lu} - n_u(\nu)B_{ul}], \qquad (3.4)$$

the *optical depth* (or *optical thickness*)

$$\tau_\nu \equiv \int \kappa_\nu\, ds, \qquad (3.5)$$

and the *source function*

$$S_\nu \equiv \frac{n_u(\nu)A_{ul}}{n_l(\nu)B_{lu} - n_u(\nu)B_{ul}}. \qquad (3.6)$$

The transfer equation then takes the simple form

$$\frac{dI_\nu}{d\tau_\nu} = S_\nu - I_\nu. \qquad (3.7)$$

If the source function is uniform along the line of sight, the transfer equation can be easily integrated to get

$$I_\nu(\tau_\nu) = I_\nu(0)e^{-\tau_\nu} + S_\nu(1 - e^{-\tau_\nu}). \qquad (3.8)$$

---

[1] Another important case is that of a directive radiation coming from a distant source, for example a star. Then the intensity writes, for a star with radius R_* and surface brightness $B_\nu(T_{\text{eff}})$ at distance D, $I_\nu = (\pi R_*^2/D^2)B_\nu(T_{\text{eff}})$, and the radiation energy density is simply $u_\nu = I_\nu/c$.

Note that these equations are general and imply nothing about the mechanism of population of the energy levels.

Let us now assume, for the moment, that the medium is in *local thermal equilibrium* (LTE) at a temperature T. In this case, the ratio between the populations of the two levels is entirely determined by collisions and obeys Boltzmann's law[2]

$$\frac{n_u(\nu)}{n_l(\nu)} = \frac{g_u}{g_l} \exp\left(-\frac{h\nu_0}{kT}\right), \tag{3.9}$$

where g_u et g_l are the *statistical weights* (numbers of sub-levels of levels u and l respectively)[3].

If the optical depth is very large, I_ν must tend to the blackbody intensity (Planck function) $B_\nu(T)$ at the temperature T of the medium. This gives $S_\nu = B_\nu(T)$, so that

$$\frac{dI_\nu}{d\tau_\nu} = B_\nu(T) - I_\nu. \tag{3.10}$$

Let us examine what happens at the line centre ($\nu = \nu_0$). S_{ν_0} is now the Planck function $B_{\nu_0}(T)$

$$\boxed{B_{\nu_0}(T) = \frac{2h\nu_0^3}{c^2} \frac{1}{e^{\frac{h\nu_0}{kT}} - 1}.} \tag{3.11}$$

Identifying the terms in (3.6) and (3.11) and using Boltzmann's law, we can then obtain the Einstein's relations between the probabilities or coefficients of emission and absorption

$$A_{ul} = \frac{2h\nu_0^3}{c^2} B_{ul}, \tag{3.12}$$

and

$$g_l B_{lu} = g_u B_{ul}. \tag{3.13}$$

Noting that ν is very close to ν_0 inside the line, the absorption coefficient is then a function of A_{ul} only:

$$\boxed{\kappa_\nu \equiv \frac{c^2 n_l(\nu) g_u}{8\pi \nu_0^2 g_l} A_{ul} \left[1 - \frac{g_l n_u(\nu)}{g_u n_l(\nu)}\right].} \tag{3.14}$$

We see that κ_ν is proportional to the density n.

Let us define the *column density* of atoms $N = \int n\, ds$ (unit: atoms cm^{-2}). This is the number of atoms in a cylinder of unit cross-section along the line of sight. Then the optical depth τ_ν (cf. (3.5)) is proportional to N.

---

[2] To be rigorous, Boltzmann's law is $n_u/n_l = (g_u/g_l)\exp(-h\nu_0/kT)$, involving the *total* populations n_l and n_u of the two levels. In order to write (3.9), we have to assume that the shape of the spectral line is the same in absorption and in emission, or in other words to assume that the distributions of $n_l(\nu)$ and of $n_u(\nu)$ with frequency are identical.

[3] The statistical weight of a level n whose angular momentum is represented by the quantum number J is $g_n = 2J + 1$.

Note again that (3.12), (3.13) and (3.14) are general despite the assumption of LTE. However, the source function S_ν is only equal to the Planck function B_ν for a thermal radiative process in a medium at LTE, whatever its optical thickness.

3.1.2 The Rayleigh–Jeans Approximation: Radioastronomy Notations

At radio frequencies, in particular at $\nu = 1\,420$ MHz ($\lambda = 21.1$ cm), the frequency of the hydrogen line which will be discussed in the next chapter, $h\nu \ll kT$ so that the Planck function (3.11) becomes

$$B_\nu(T) \simeq \frac{2kT\nu^2}{c^2}, \tag{3.15}$$

(the *Rayleigh–Jeans approximation*). The solution of the transfer equation (3.8) becomes, at LTE

$$I_\nu(\tau_\nu) = I_\nu(0)e^{-\tau_\nu} + \frac{2kT\nu^2}{c^2}(1 - e^{-\tau_\nu}). \tag{3.16}$$

Given this solution, radioastronomers express surface brightnesses or specific intensities in terms of a *brightness temperature*, T_B, defined by

$$T_B \equiv \frac{c^2}{2k\nu^2} I_\nu. \tag{3.17}$$

The brightness temperature only has a physical significance at LTE, but it allows us to conveniently write the LTE solution to the transfer equation in the simple form

$$T_B(\tau_\nu) = T_B(0)e^{-\tau_\nu} + (1 - e^{-\tau_\nu})T. \tag{3.18}$$

This equation states that, if the intervening medium is hotter than the background brightness temperature, the line is seen in emission, and in absorption for the opposite case. This is in fact a very general property, valid even if the medium is not in LTE. In the optically thin case for which $\tau_\nu \ll 1$, and if the background $T_B(0)$ is weak, we have simply

$$T_B(\tau_\nu) = \tau_\nu T. \tag{3.19}$$

Since τ_ν is proportional to the column density of matter $N = \int n\,ds$ in our hypothesis of a uniform S_ν, the line intensity and T_B are proportional to the column density. For an optically thick medium, T_B at LTE is equal to the kinetic temperature T_K of the medium.

Radioastronomers are also in the habit of measuring the energy received by an antenna in terms of an *antenna temperature*, T_A. This is the temperature of an imaginary blackbody which would enclose the antenna completely and yield the same signal as observed. The energy received by the antenna is then $W_\nu = kT_A$ per unit interval of frequency (the *Nyquist theorem*).

If the antenna were perfect, an extended region of uniform brightness represented by a brightness temperature T_B would yield an antenna temperature $T_A = T_B$.

However there is no perfect antenna, and some antenna efficiency is involved. For interstellar sources, that are often rather extended, the most useful efficiency is the *beam efficiency*, η_b. This is the ratio between the antenna temperature and the brightness temperature of a uniform source which just covers the primary beam of the antenna, assuming no attenuation by the Earth's atmosphere. Antenna temperatures corrected by this efficiency and also for absorption by the terrestrial atmosphere are called *main-beam temperatures*, $T_{mb} = T_A/[\eta_b \exp(-\tau_{atm})]$, τ_{atm} being the optical depth of the atmosphere in the direction of the source. Antenna temperatures are generally given as main-beam temperatures in articles dealing with the interstellar medium, but the reader must be careful and should consider the actual definition used by the author. $T_{mb} = T_B$ by definition for a source covering exactly the primary beam, but there is an over-correction ($T_{mb} > T_B$) for a more extended source which gives some energy in the side lobes of the antenna. If the source observed by the antenna is smaller than the primary beam, as is frequently the case for extragalactic observations, T_A or $T_{mb} < T_B$, and we say that there is *beam dilution*. For a complete discussion of the problems of efficiency and calibration of antennae, see Rohlfs & Wilson [439] and Kutner & Ulich [294].

It is of interest to give some expressions useful for understanding articles dealing with radioastronomy and with infrared astronomy.

The energy received from a source (point-like or extended) is expressed as a *flux density* S_ν, a monochromatic quantity whose unit in the International System of units is the W m^{-2} Hz^{-1}. We most often use a sub-multiple of this unit, the jansky:

$$1 \text{ Jy} = 10^{-26} \text{ W m}^{-2} \text{ Hz}^{-1} = 10^{-23} \text{ erg cm}^{-2} \text{ Hz}^{-1} \text{ s}^{-1}. \quad (3.20)$$

For a uniform extended source with brightness I_ν, subtending a solid angle Ω, the flux density is $S_\nu = I_\nu \Omega$.

An antenna of equivalent area A_{eq} pointed on-axis at a point source with a flux density S_ν collects a power

$$P_\nu = A_{eq} S_\nu = kT_A. \quad (3.21)$$

If the source is at the angular distance θ, ϕ from the axis, the collected energy $P_\nu(\theta, \phi)$ is obviously smaller and is reduced by a factor $G_\nu(\theta, \phi)/G_\nu(0, 0)$, where G_ν is the antenna *gain* defined as

$$G_\nu(\theta, \phi) = \frac{P_\nu(\theta, \phi)}{\int \int P_\nu(\theta, \phi) d\Omega}. \quad (3.22)$$

The gain expresses the directional properties of the antenna. It is normalized such that

$$\int \int G_\nu(\theta, \phi) \, d\Omega = 1. \quad (3.23)$$

Let us now assume that we place the antenna in an enclosure radiating uniformly in all directions with a brightness temperature T. The total energy received by the antenna is then kT per unit frequency interval. The enclosure radiates a specific intensity

$$I_\nu = \frac{2kT}{\lambda^2}, \qquad (3.24)$$

but since an antenna is only sensitive to one polarization it can only collect half this energy. In the direction (θ, ϕ) it collects in the solid angle $d\Omega$

$$dP_\nu(\theta, \phi) = A_{eq} \frac{kT}{\lambda^2} \frac{G_\nu(\theta, \phi)}{G_\nu(0,0)} d\Omega. \qquad (3.25)$$

Integrating over all the solid angles and writing the total collected energy as kT, we obtain the on-axis gain

$$G_\nu(0,0) = A_{eq}/\lambda^2. \qquad (3.26)$$

Since the angular resolving power of a circular antenna of diameter D is of the order of λ/D and its surface is of the order of D^2, the on-axis gain is close to the inverse of the solid angle subtended by the primary beam of the antenna.

At millimetre and submillimetre wavelengths, radioastronomers still use the temperature notation even though the Rayleigh–Jeans approximation is no longer valid. This can be a cause of confusion. The brightness temperature is still defined by the Rayleigh–Jeans expression, but we append a star symbol to it in order to avoid confusion:

$$T_B^* \equiv \frac{c^2}{2k\nu^2} I_\nu. \qquad (3.27)$$

T_B^* is obviously different from the temperature of a blackbody emitting the same specific intensity, which is

$$T = \frac{h\nu}{k} \frac{1}{\ln[1 + 2h\nu^3/c^2 I_\nu]}. \qquad (3.28)$$

3.1.3 Excitation Temperature

If we wish to obtain the total (or integrated) intensity in a spectral line, we have to integrate the transfer equation or its solution over frequency. For this, we must know the normalized spectral distribution $\phi_{ul}(\nu)$ of the intensity of the emission line, which is such that

$$\int_{\text{line}} \phi_{ul}(\nu)\, d\nu = 1. \qquad (3.29)$$

The absorption coefficient can now be written

$$\kappa_\nu = \frac{c^2 n_l g_u}{8\pi\nu^2 g_l} A_{ul} \left[1 - \frac{g_l n_u}{g_u n_l}\right] \phi_{ul}(\nu), \qquad (3.30)$$

where n_u and n_l are the densities of atoms in levels u and l, respectively. By analogy with the Boltzmann expression for the level populations at LTE the *excitation temperature* T_{ex} can be defined by

$$\boxed{\frac{g_l n_u}{g_u n_l} = \exp\left(\frac{-h\nu_0}{kT_{ex}}\right),} \qquad (3.31)$$

where ν_0 is the central frequency of the line. The absorption coefficient is obtained by substitution of (3.31) in (3.30),

$$\boxed{\kappa_\nu = \frac{c^2 n_l g_u}{8\pi\nu^2 g_l} A_{ul} \left[1 - \exp\left(\frac{-h\nu_0}{kT_{ex}}\right)\right]\phi_{ul}(\nu),} \qquad (3.32)$$

or in the Rayleigh–Jeans approximation

$$\kappa_\nu = \frac{c^2 n_l g_u}{8\pi\nu^2 g_l} A_{ul} \frac{h\nu_0}{kT_{ex}} \phi_{ul}(\nu). \qquad (3.33)$$

In practice we often express the line shape as a function of radial velocity v rather than as a function of frequency. If we call $\Delta\nu$ (or Δv if expressed in radial velocities) the line width at half-intensity, we have $\phi_{ul}(\nu_0) \approx 1/\Delta\nu = c/(\nu_0 \Delta v)$. The optical depth τ_0 at the line centre is, in the Rayleigh–Jeans approximation, a function of the column density N_l of atoms in level l:

$$\tau_0 \approx \frac{c^3 g_u A_{ul}}{8\pi\nu^3 g_l} \frac{N_l}{\Delta v} \frac{h\nu_0}{kT_{ex}}. \qquad (3.34)$$

The excitation temperature has no physical meaning except at LTE. It can take any value, even negative. Negative excitation temperatures correspond to a population inversion of the levels, i.e. to a maser effect: see later Sect. 3.3.

3.2 Two-Level System out of LTE

3.2.1 General Relations

In the general case, radiative transitions cannot be neglected with respect to collisions which can populate the energy levels. If we assume that the level populations are statistically at equilibrium with the radiation and collisions, we can write that the number of upward transitions per unit time is equal to the number of downward transitions (statistical equilibrium):

$$n_l(R_{lu} + C_{lu}) = n_u(R_{ul} + C_{ul}). \qquad (3.35)$$

R_{lu} and R_{ul} are, respectively, the probabilities for radiative excitation and de-excitation (cf. (3.1) and (3.2)). C_{lu} and C_{ul} are, respectively, the probabilities of collisional excitation and de-excitation, which are proportional to the density n of the particles responsible for the collisions. The relation between these quantities, assuming LTE, is then

$$n_l C_{lu} = n_u C_{ul}, \qquad (3.36)$$

and since we have at LTE

$$\frac{n_u}{n_l} = \frac{g_u}{g_l}\exp\left(-\frac{h\nu}{kT_K}\right), \qquad (3.37)$$

$$C_{lu} = C_{ul}\frac{g_u}{g_l}\exp\left(-\frac{h\nu}{kT_K}\right). \qquad (3.38)$$

The latter expression contains only atomic parameters and remains valid in the general case provided that a kinetic temperature T_K can be defined.

Coming back to the general non-LTE case for a two-level system, we can then write, expressing R_{lu} and R_{ul} from the spontaneous emission probability A_{ul} and using the Einstein relations (3.12) and (3.13),

$$\boxed{\frac{n_u}{n_l} = \frac{g_u}{g_l}\frac{A_{ul}I_\nu c^2/2h\nu^3 + C_{ul}\exp(-h\nu/kT_K)}{(1 + I_\nu c^2/2h\nu^3)A_{ul} + C_{ul}}.} \qquad (3.39)$$

If the Rayleigh–Jeans approximation is valid, then this expression simplifies to

$$T_{ex} = T_K\left(\frac{T_0 + T_B}{T_0 + T_K}\right), \qquad (3.40)$$

putting

$$T_0 = \frac{h\nu}{k}\frac{C_{ul}}{A_{ul}}, \qquad (3.41)$$

a quantity with no physical meaning. We can verify that if collisions dominate, T_0 is large and $T_{ex} \simeq T_K$ (collisional equilibrium or LTE). Conversely, if collisions are negligible, T_0 is small and $T_{ex} \simeq T_B$ (radiative equilibrium).

The collisional de-excitation probability is

$$C_{ul} = n\langle\sigma_{ul}v\rangle, \qquad (3.42)$$

where n is the density of particles responsible for collisions, v their velocity and σ_{ul} the cross-section for collisional de-excitation (which generally depends on v). The symbols $\langle\ \rangle$ mean that the average is taken over all relative velocities v. There exists a critical density n_{crit} for which collisional and radiative transitions have the same importance, i.e.,

$$n_{crit} = \frac{1}{\langle\sigma_{ul}v\rangle}\left(1 + \frac{I_\nu c^2}{2h\nu^3}\right)A_{ul}, \qquad (3.43)$$

and in the Rayleigh–Jeans approximation

$$n_{crit} = \frac{1}{\langle\sigma_{ul}v\rangle}\left(1 - \frac{kT_B}{h\nu}\right)A_{ul}. \qquad (3.44)$$

3.2.2 Pure Radiative Equilibrium

An interesting case where transitions are dominated by radiation is that of the CN and CH molecules in the neutral atomic interstellar medium that we will consider in the next chapter. CN, for example, has a transition $J = 1 \to 0$ at a wavelength of 2.64 mm, with $A_{10} \simeq 10^{-5}$ s^{-1}, a relatively large value. We can show that the level population is entirely dominated by radiation as long as the density of H I is lower than 10^3 atom cm^{-3}. This is indeed generally the case in the diffuse medium. Then the excitation temperature T_{ex} is equal to the radiation temperature at 2.64 mm. It is possible to probe the populations of the 0 and 1 levels thanks to electronic transitions starting from these levels. The corresponding lines are in the near-UV near 3 874 Å (cf. the references in Meyer et al. [366]). Measuring the intensity ratio between these lines which are observed in absorption in front of bright stars, we find $T_{ex}(1\text{-}0) = 2.8 \pm 0.1$ K. The cosmological blackbody radiation of the Universe was discovered after the first observations of this phenomenon. It corresponds to a temperature of 2.726 ± 0.010 K and completely dominates the radiation field at millimetre wavelengths, except near strong continuum or line sources. Thus the observation of the absorption lines of CN (as well as those of CH and CH$^+$) allowed a pre-discovery of the blackbody radiation of the Universe. An interesting check of the prevalence of radiation in the populations of the levels of CN is the fact that the rotational line of CN at 2.64 mm is not seen superimposed on the cosmic background radiation of the Universe, neither in emission nor in absorption.

3.2.3 The Coupling of Excitation and Transfer; the LVG Approximation

We have seen that the level equilibrium at a given point is in general determined both by the local density and by the radiation field in the line. In turn this radiation field is itself determined by the properties of the emitting atoms or molecules in the other regions of the medium. In order to calculate the populations of the levels it is thus necessary, in principle, to simultaneously solve the equation of statistical equilibrium and the equation of transfer at all points of the medium. This is now feasible with work stations and even with microcomputers. However, analytical simplifications are possible in two cases.

The first of these cases is that of regions deep within an optically thick medium. Then the emitting particles are isolated from the external world, and the second law of thermodynamics tells us that the medium is in equilibrium with both collisions and radiation. This implies that $T_{ex} = T_B = T_K$ whatever the density. However, this case is only of academic interest since such regions are not observable in the line by an external observer.

The other case is much more interesting. Let us consider a medium emitting a locally narrow line, whose width δv is due to the Doppler effect from the thermal motion of the atoms and local random macroscopic motions (*microturbulence*). Let us also assume that there are important velocity gradients at large scale, due for example to a general contraction or expansion of the medium. Then at a given frequency inside the line, the emission comes only from a restricted region of the line of sight, i.e., that whose mean radial velocity corresponds to this frequency. We can then use the *large velocity gradient* (LVG) approximation. This approximation was proposed by Sobolev and developed for millimetre molecular lines by Scoville & Solomon [462] and by Goldreich & Kwan [201]. The LVG approximation also assumes that the velocity gradient, the density and the kinetic and excitation temperatures are uniform in the medium. The simple analytical approximations that allow to calculate the line intensity in this approximation are somewhat tricky to demonstrate and difficult to find in the literature, so we will now give a complete derivation, after Surdej [501].

The equation of local statistical equilibrium can be written as (see (3.35), (3.1) and (3.2))

$$n_l(\frac{c}{4\pi}\langle u(\mathbf{r})\rangle B_{lu} + C_{lu}) = n_u(A_{ul} + \frac{c}{4\pi}\langle u(\mathbf{r})\rangle B_{ul} + C_{ul}), \quad (3.45)$$

where $\langle u(\mathbf{r})\rangle$ is the mean local radiation density in the line at the current position $\mathbf{r}$. $\langle u(\mathbf{r})\rangle$ is linked to the source function $S(\mathbf{r}')$ in the medium by the relation:

$$\langle u(\mathbf{r})\rangle = \int K(\mathbf{r}-\mathbf{r}')S(\mathbf{r}')\,d\mathbf{r}', \quad (3.46)$$

where $K(\mathbf{r}-\mathbf{r}')$ is a function (kernel) which depends on the radiation field throughout the medium and on the propagation of this radiation. However, if there is a large velocity gradient, those photons that are involved in the line at the point $\mathbf{r}$ can only come from a nearby region defined by a distance l such that

$$l \simeq \delta v\, R/\Delta V, \quad (3.47)$$

where δv is the local width of the line, ΔV is the total velocity range and R is the total depth of the medium along the line of sight. Let us furthermore assume the complete redistribution of frequencies, i.e. that an absorbed photon is re-emitted (scattered) without memory of the direction, frequency and polarisation of the incoming photon, the line having the same profile in emission and absorption. This is a reasonable hypothesis. We can then consider the source function as approximately uniform in a volume of radius l around the current point, and the local radiation density can be written simply as

$$\langle u(\mathbf{r})\rangle = [1 - \beta(\mathbf{r})]S(\mathbf{r}), \quad (3.48)$$

where $\beta(\mathbf{r})$ is the *escape probability* of a line photon at point $\mathbf{r}$.

The *absorption* probability $a = 1 - \beta$ of a photon with an emission frequency ν in direction $\mathbf{s}$ is

$$a(\nu, \mathbf{s}) = \int_0^{s(\nu)} \exp\left[-\int_0^s \kappa(\nu)\phi\left(\nu - \nu_0 - \frac{\nu_0}{c}\frac{dv_s}{ds}s'\right) ds'\right]$$
$$\times \kappa(\nu)\phi\left(\nu - \nu_0 - \frac{\nu_0}{c}\frac{dv_s}{ds}s\right) ds. \quad (3.49)$$

The first term (exponential) of this equation is the probability that a photon with frequency ν reaches the point s in the current direction without being absorbed. The second term is the absorption probability for the same photon between points s and $s + ds$. ϕ is the normalized profile of the line, and the term $(v_0/c)(dv_s/ds)$ in its argument represents the Doppler effect due to the velocity gradient along the line of sight. The integral is limited in fact to the point of abscissa $s(v)$ where the photon falls outside the line profile. By hypothesis, this point is relatively close from the point where the photon has been emitted. We can simplify this equation by performing the following changes of variables:

$$x = v - v_0 - \frac{v_0}{c}\frac{dv_s}{ds}s', \tag{3.50}$$

and

$$y = v - v_0 - \frac{v_0}{c}\frac{dv_s}{ds}s. \tag{3.51}$$

By hypothesis, $\kappa(v)$ and dv_s/ds can be considered as constant in the integrals, and we can write

$$\tau_0(s) = \frac{\kappa(v)}{v_0}\frac{c}{dv_s/ds}. \tag{3.52}$$

The absorption probability is then

$$a(v, s) = \int_{y[s(v)]}^{v_0-v} \frac{d}{dy}\left\{\exp\left[\tau_0(s)\int_{v_0-v}^{y}\phi(x)\,dx\right]\right\}dy, \tag{3.53}$$

i.e.,

$$a(v, s) = 1 - \exp\left[\tau_0(s)\int_{v_0-v}^{y[s(v)]}\phi(x)\,dx\right]. \tag{3.54}$$

Integrating over the line profile, assumed to be limited to the frequencies within the interval $-\Delta v$ to $+\Delta v$ (roughly, Δv is the local width of the line at half-intensity), we get

$$a(s) = \int_{v_0-\Delta v}^{v_0+\Delta v} a(v, s)\phi(v - v_0)\,dv. \tag{3.55}$$

Using the change of variable $f = v - v_0$, from which $df = dv$, we get

$$a(s) = \int_{-\Delta v}^{+\Delta v}\left(1 - \exp\left[\tau_0(s)\int_{f}^{y[s(v)]}\phi(x)\,dx\right]\right)df. \tag{3.56}$$

Given the normalization of the line profile $\phi(v - v_0)$, we obtain

$$a(s) = 1 + \int_{-\Delta v}^{+\Delta v}\frac{d}{df}\left(\exp\left[\tau_0(s)\int_{f}^{y[s(v)]}\phi(x)\,dx\right]\right)df. \tag{3.57}$$

Note that

$$y[s(v)] = v - v_0 \mp \frac{v_0}{c}\frac{dv_s}{ds}s(v), \tag{3.58}$$

according to the sign of dv_s/ds. The Doppler effect at the point $s(v)$ after which absorption is negligible is such that

$$\frac{v_0}{c}\frac{dv_s}{ds}s(v) = v - (v_0 - \Delta v), \tag{3.59}$$

so that

$$y[s(v)] = \mp \Delta v. \tag{3.60}$$

This allows us to obtain the final expression for the absorption probability in the direction **s**

$$a(\mathbf{s}) = 1 - \frac{1 - \exp(-|\tau_0(\mathbf{s})|)}{|\tau_0(\mathbf{s})|}. \tag{3.61}$$

The escape probability is then given by

$$\beta(\mathbf{s}) = \frac{1 - \exp(-|\tau_0(\mathbf{s})|)}{|\tau_0(\mathbf{s})|}. \tag{3.62}$$

We now have to integrate the escape probability over all directions in order to get the global escape probability. Let us consider the simple plane-parallel case. Let μ be the cosine of the angle $(\mathbf{s}, \mathbf{n})$ between the line of sight **s** and the normal **n** to the layers of the medium. Writing $\tau_0(\mu) \equiv \tau_0(\mathbf{s})$ we have

$$\beta = \int \frac{1 - \exp(-|\tau_0(\mu)|)}{|\tau_0(\mu)|}\frac{d\Omega}{4\pi} = \frac{1}{2}\int \frac{1 - \exp(-|\tau_0(\mu)|)}{|\tau_0(\mu)|}d\mu. \tag{3.63}$$

Let us define

$$\tau_0 = \frac{\kappa}{v_0}\frac{c}{dv/dz}, \tag{3.64}$$

an expression integrated on the line profile, z being the coordinate normal to the layers of the medium. τ_0 is the mean optical depth in the line normal to the medium. On the other hand

$$dv_s/ds = \mu^2(dv/dz). \tag{3.65}$$

Applying the *Eddington approximation* which replaces μ^2 by the average value of $\cos^2(\mathbf{s},\mathbf{n})$ over all the directions of the half-space, $1/3$, we obtain the final result, the escape probability,

$$\boxed{\beta = \frac{1 - \exp(-3\tau_0)}{3\tau_0}.} \tag{3.66}$$

We notice that $\beta \to 1$ if $\tau_0 \to 0$ and that $\beta \to 1/3\tau_0$ if $\tau_0 \to \infty$. Remembering the statistical equilibrium equation (3.45), the expression for the brightness temperature in the line becomes, in the Rayleigh–Jeans approximation,

$$T_B = \frac{T_K}{1 + (kT_K/hv_0)\ln\{1 + (A_{ul}/3C_{ul}\tau_0)[1 - \exp(-3\tau_0)]\}}. \tag{3.67}$$

The term $[1 - \exp(-3\tau_0)]/3\tau_0$ allows us to estimate the *photon trapping* in the medium. If $\tau_0 \gg 1$, which is most often the case for the CO lines, this phenomenon is such that A_{ul} can be replaced by $A_{ul}/3\tau_0$.

In the spherical case, the escape probability, in the Eddington approximation (de Jong et al. [116]), is given by

$$\beta = \frac{1 - \exp(-\tau_0)}{\tau_0}.\qquad(3.68)$$

De Jong et al. [117] give the following expressions, valid in the plane-parallel case even if there is no velocity gradient, but which assume however that the radiation field at a given point depends only on local conditions as in the LVG approximation (the *on the spot* approximation)

$$\beta = \frac{1 - \exp(-2.34\tau_0)}{4.68\tau_0} \text{ for } \tau < 7,\qquad(3.69)$$

$$\beta = \frac{1}{4\tau_0[\ln(\tau_0/\sqrt{\pi})]^{1/2}} \text{ for } \tau \geq 7.\qquad(3.70)$$

In the more complex case of an inhomogeneous medium, or that of a medium with an arbitrary shape, the LVG approximation cannot be applied, and a Monte-Carlo method has to be used: photons are injected in all directions and their fate is examined statistically in a numerical model of the medium. Examples of this approach can be found in Spaans [484] or in Pagani [392].

3.3 The General Case; Masers

In some cases we cannot ignore the role of the population of energy levels other than those levels corresponding to the line under consideration. We must then solve all the statistical equilibrium equations for the populated levels, taking into account all important radiative and collisional transitions.

If we assume that all levels are in LTE the solution is simple. The *partition function* for the atom or molecule is defined as

$$Q = \sum_n g_n \exp\left(\frac{-E_n}{kT_K}\right),\qquad(3.71)$$

where E_n is the energy of level n, g_n its statistical weight, and the sum is over all levels with non-negligible population due to collisions. The population n_n of level n is simply given by

$$n_n = n_{tot} \frac{1}{Q} g_n \exp\left(\frac{-E_n}{kT_K}\right),\qquad(3.72)$$

where n_{tot} is the total density of the atoms or molecules.

It is very important to remark that the selection rules are in general not the same for radiative and collisional transitions. For example, only consecutive rotation levels are connected by allowed transitions in diatomic molecules such as CO: $\Delta J = \pm 1$, J being the total angular momentum. Conversely, collisions with neutral particles can yield transitions between any pair of levels, the transitions $\Delta J = \pm 2$ often being favored. If the temperature is high enough, the level $J = 3$ can be populated by collisions from the fundamental level $J = 1$, and the radiative cascade $J = 3 \rightarrow 2$ can in some cases overpopulate level 2 with respect to level 1. This creates a population inversion between levels 1 and 2, which an be expressed as a negative excitation temperature $T_{ex}(1\text{-}0)$ (Fig. 3.1). The medium then behaves as a weak maser in the CO(1-0) line. This is however an academic example because the maser effect arises only for an optically thin CO line while the CO lines are almost always optically thick in the interstellar medium.

The case of the interstellar masers is more interesting. Some masers like that of H_2O are mostly pumped by collisions as explained earlier. Others like the OH masers are pumped by radiation. The fundamental level of the OH molecule is split into two pairs of sublevels between which radiative transitions arise at wavelengths close to 18 cm (frequency 1 660 MHz, Fig. 3.2). These transitions have been long observed by radioastronomers: see e.g. Nguyen-Q-Rieu et al. [385]. The sublevels are radiatively connected with sublevels of higher rotational levels as shown in

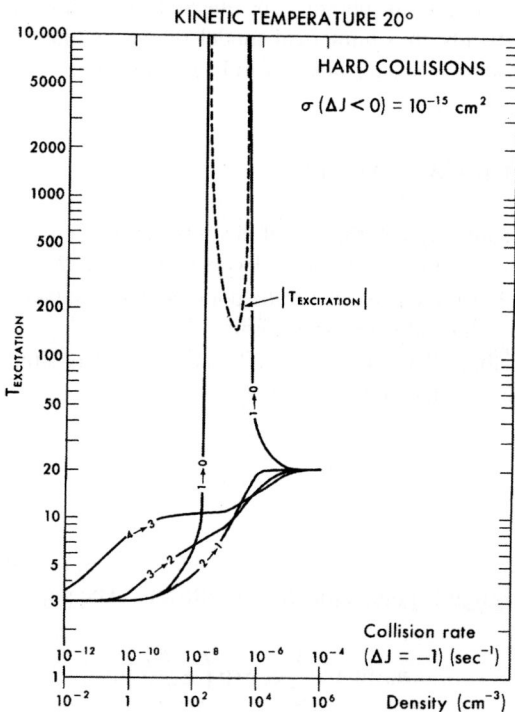

Fig. 3.1. Excitation temperatures of the 4 lowest rotational transitions of the CO molecule as a function of molecular hydrogen density in an interstellar cloud. The excitation temperatures have been calculated using a simple model for the $CO\text{-}H_2$ collisions. The kinetic temperature is 20 K. For densities close to 10^3 molecules cm^{-3} the excitation temperature is negative for the CO(1-0) line. $|T_{ex}|$ is then plotted as a dashed line. Reproduced from Goldsmith [203], with the permission of the AAS.

4 The Neutral Interstellar Gas

We begin with this chapter an investigation of the different components of the interstellar medium. We will first study the physics of the atomic gas and then that of the molecular gas. The chemistry will be discussed in Chap. 9. Chapter 5 will treat of the ionized gas, and the interfaces between neutral and ionized media will be examined in Chap. 8. High-energy phenomena will be the subject of Chap. 6, and the dust, which is intimately mixed with the gas, will be dealt with in Chap. 7.

We should mention at the beginning that the distinction between the so-called neutral medium and the ionized medium is somewhat arbitrary. The neutral medium is best defined by the absence of hydrogen Lyman continuum photons, so that hydrogen is neutral. However, interstellar extinction is considerably smaller at wavelengths longward of the Lyman discontinuity (911.7 Å, see Fig. 2.5), so that some elements like carbon or the metals as well as the dust grains can be ionized by UV radiation. Cosmic rays ionize a small fraction of all elements even deep inside molecular clouds (see Sect. 8.1), and X-rays, when present, can also weakly ionize all elements including hydrogen. The neutral medium has, in fact, a non-zero degree of ionization, which plays a very important role in its physics. Also, the so-called atomic medium contains a non-negligible quantity of molecules (cf. further Fig. 4.10), while the so-called molecular medium contains a small fraction of atoms.

4.1 The Atomic Neutral Gas

This component contains most of the mass of the interstellar medium. There are three main observables to study the neutral gas: the 21-cm line of atomic hydrogen which traces the main constituent and allows us to measure its temperature, the fine-structure lines in the far-IR which are the main cooling source for the medium, and the interstellar absorption lines that give the chemical composition and some physical parameters.

4.1.1 The 21-cm Line of Atomic Hydrogen

In this section we will examine successively the measurements of this line, first in emission, then in absorption, and we will summarize the main results obtained for interstellar atomic hydrogen.

Generalities, Emission Measurements

The 21-cm line, for which the exact frequency is 1.420 405 751 786(30) GHz ($\lambda = 21.106\,114$ cm), corresponds to the transition between the two hyperfine sublevels of the fundamental state of H I. The energy of the atom is larger by 6×10^{-6} eV when the spins of the electron and of the proton are parallel compared to when they are antiparallel. The total angular momentum is $F = 0$ in the latter case, with a statistical weight $g_l = 1$, while for the upper level $F = 1$ and $g_u = 3$. The excitation temperature of the line is often called the *spin temperature*, T_{spin}. The transition is strongly forbidden, the spontaneous emission probability being as small as $A_{ul} = 2.87 \times 10^{-15}$ s^{-1}, hence a radiative lifetime of the upper sublevel of $1/A_{ul} = 1.1 \times 10^7$ years, considerably larger than the time between H atom collisions even at the low densities of the interstellar medium, which is of the order of hours. Radiation becomes important only for densities lower than a critical density $n_{\text{crit}} < 10^{-2}\, T_K^{-1/2}$ cm^{-3}, where T_K is the kinetic temperature of the gas. The levels are thus essentially always in collisional equilibrium, hence at LTE, and $T_{\text{spin}} \simeq T_K$. Most of the hydrogen atoms are in the fundamental level, hence in the two hyperfine sublevels, because the level immediately above is at an energy of 10 eV and is not metastable. Then (3.37) shows that the total H I density is

$$n_{\text{H}} = n_u + n_l = 4n_l, \tag{4.1}$$

since $h\nu_0/kT_{\text{spin}} \ll 1$, the Rayleigh–Jeans approximation being valid at 21 cm.

If the optical thickness in the line is small, we can write

$$T_B(\nu) = T_K \tau(\nu), \tag{4.2}$$

and from (3.33) we have

$$\boxed{\tau(\nu) = 2.597 \times 10^{-15} \frac{N(\text{H\,I})\phi(\nu)}{T_K}.} \tag{4.3}$$

$N(\text{H\,I})$ is the column density of hydrogen atoms (the number of atoms in a column with unit cross-section, expressed in atoms cm^{-2}). We see that in the optically thin case the line intensity does not depend on the temperature since $\phi(\nu) \propto T_K$, but only on the total column density $N(\text{H\,I})$. Radioastronomers express the line profile, $\phi(\nu)$, as a function of radial velocity rather than of frequency. This is logical because line broadening is only caused by the Doppler effect, its natural width being extremely narrow since the lifetime of the upper level is only limited by collisions which are rare in the diffuse medium. We thus have

$$v - v_0 = c \frac{\nu - \nu_0}{\nu_0}. \tag{4.4}$$

We can then write

$$N(\text{H\,I}) \simeq 1.8224(3) \times 10^{18} \int \Delta T_B(v)\,dv \text{ atom cm}^{-2}(\text{K km s}^{-1})^{-1}, \tag{4.5}$$

where ΔT_B is the brightness temperature above the background continuum (which is at least equal to the blackbody radiation of the Universe). If the line profile is simple, for example gaussian, we have

$$N(\text{H\,{\sc i}}) \simeq 1.822 \times 10^{18} \Delta T_B \Delta v \text{ atom cm}^{-2}, \quad (4.6)$$

where Δv is the line full width at half maximum (FWHM) in km s^{-1}.

If the optical depth is large we must use the complete expression (3.18) with $T_B(0) = 0$, and

$$N(\text{H\,{\sc i}}) = 1.822 \times 10^{18} T_K \int \ln\left[\frac{T_K}{T_K - \Delta T_B(v)}\right] dv \text{ atom cm}^{-2}(\text{K km s}^{-1})^{-1}, \quad (4.7)$$

which requires a knowledge of T_K. Of course, the determination of $N(\text{H\,{\sc i}})$ is imprecise if $\tau > 1$.

For observations in the Galaxy, or in nearby galaxies like M 31, the radial velocity is generally given with respect to the local standard of rest (LSR). This velocity v_{LSR} is related to the heliocentric velocity v_{hel} (with respect to the centre of gravity of the Solar system) by

$$v_{\text{LSR}} = v_{\text{hel}} + 19.5(\cos\alpha_\odot \cos\delta_\odot \cos\alpha \cos\delta$$
$$+ \sin\alpha_\odot \cos\delta_\odot \sin\alpha \cos\delta + \sin\delta_\odot \sin\delta) \text{ km s}^{-1}, \quad (4.8)$$

where 19.5 km s^{-1} is the standard velocity of the Sun with respect to the LSR, α and δ are the equatorial coordinates of the line of sight at the same equinox as the equatorial coordinates $\alpha_\odot$ and $\delta_\odot$ of the solar apex. At equinox 1900 (not 2000 !), the standard apex is at $\alpha_\odot(1900) = 18$h, $\delta_\odot(1900) = +30°$, and we have simply

$$v_{\text{LSR}} = v_{\text{hel}} + 19.5(-0.866\,03 \sin\alpha \cos\delta + 0.5 \sin\delta) \text{ km s}^{-1}. \quad (4.9)$$

Absorption Measurements

The 21-cm line can be seen in absorption in front of a continuum radiosource or in front of the 21-cm emission of warmer H\,{\sc i}. In the case of absorption in front of a continuum varying slowly with frequency, we can directly measure the optical depth $\tau(v)$ of the intervening H\,{\sc i} cloud as follows. The solution of the transfer equation is then, assuming an uniform cloud,

$$T_B(v) = T_K(1 - \exp[-\tau(v)]) + T_C \exp[-\tau(v)], \quad (4.10)$$

where T_C is the brightness temperature of the continuum. In general, absorption measurements are made with interferometers with a high angular resolution, which are insensitive to the extended emission from the cloud. In this case the first term of the preceding equation vanishes, and we simply have

$$T_{abs}(v) = T_C \exp[-\tau(v)]. \tag{4.11}$$

Remember (3.33) that τ is inversely proportional to the excitation temperature (which is itself identical to the kinetic temperature), so that the column density of H I cannot be derived from an absorption observation alone.

If we observe the 21-cm emission in the directions immediately adjacent to that of a source with a large single antenna or with a low angular resolution interferometer, by interpolation we obtain the *expected emission* that would be observed in the direction of the source in the absence of this source. This supplies the quantity

$$\Delta T_{em}(v) = T_K(1 - \exp[-\tau(v)]). \tag{4.12}$$

From the two preceding equations we can derive the kinetic temperature and the optical depth, hence the column density in the cloud.

This method is practically the only one which can give, rather directly, the temperature of atomic interstellar clouds. It suffers however from several difficulties. One is the difficulty in determining the expected emission if the emission observations do not have sufficient angular resolution with respect to the spatial fluctuations in the emission. This difficulty was major for early observations but tends to disappear for more recent interferometer observations. Another difficulty is the presence of warm and cold H I phases along the line of sight. This is generally the case. Although there is no really safe method to separate their respective contributions, a method due to Mebold et al. [362] gives a substantial improvement and deserves some explanation. Let us assume that there is a mixture of warm and cold H I along the line of sight (a cold cloud immersed in a warm intercloud medium) but that only the cold H I contributes appreciably to absorption. Let q be the fraction of emission due to the warm gas located in front of the cold gas, hence unaffected by absorption. The brightness temperature $T(v)$ of the emission observed at velocity v is

$$T(v) = qT_{\text{warm}}(v) + T_{\text{cold}}(v) + (1-q)T_{\text{warm}}(v)e^{-\tau(v)}, \tag{4.13}$$

where the first term is the fraction of the emission of the warm gas unaffected by absorption, the second term is the emission of the cold gas and the third term is the rest of emission of the warm gas behind the cold gas affected by its absorption with an optical depth $\tau(v)$. Since $T_{\text{cold}} = T_{K,\text{cold}}[1 - e^{-\tau(v)}]$, $T_{K,\text{cold}}$ being the kinetic temperature of the cold gas, the preceding expression can be rearranged to give

$$T(v) = [1 - e^{-\tau(v)}][T_{K,\text{cold}} - (1-q)T_{\text{warm}}(v)] + T_{\text{warm}}(v). \tag{4.14}$$

Figure 4.1 shows an example of emission and absorption spectrum in the direction of the continuum source N 157b located in the Large Magellanic Cloud. In Fig. 4.2 $T(v)$ is plotted as a function of $1 - e^{-\tau(v)}$ and the points representative of the different spectral channels are connected to each other. Near the absorption peaks we can measure a slope $m(v)$ which is equal to

$$m(v) = dT(v)/d[1 - e^{-\tau(v)}] = T_{K,\text{cold}} - (1-q)T_{\text{warm}}(v). \tag{4.15}$$

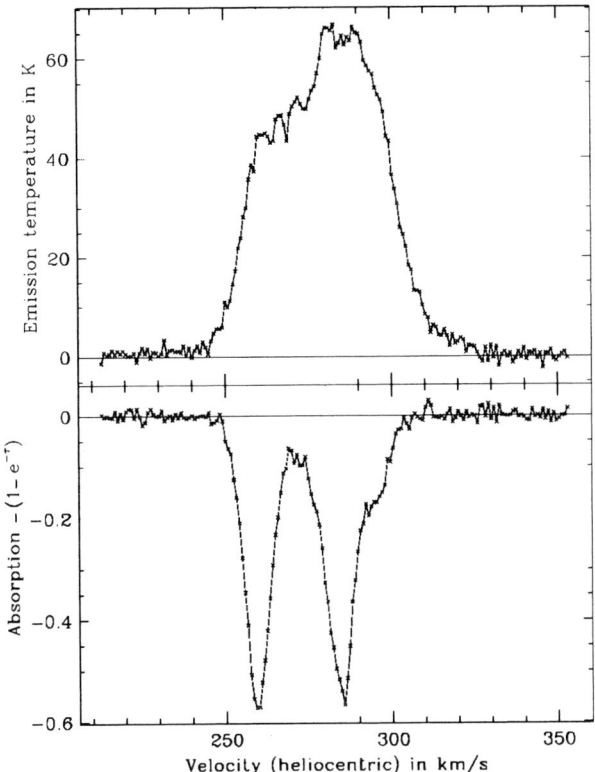

Fig. 4.1. The 21-cm emission spectrum in the vicinity of the continuum source N 157b in the Large Magellanic Cloud (top), and the absorption spectrum (optical depth) in the direction of the same source (bottom). Reproduced from Mebold et al. [362], with the permission of the AAS.

This equation allows us to derive $T_{K,\mathrm{cold}}$ for a given absorption component from $m(v)$ and T_{warm} if some hypothesis is made about the value of q. For the simple case in which T_{warm} is assumed to be constant over the velocity range of the absorption line, we can directly read the slope m and the intercept T_{warm} from Fig. 4.2. For example, the component centreed at 285 km s^{-1} has $m(v) \simeq 0$ and $T_{\mathrm{warm}} \simeq 65$ K, so that $T_{K,\mathrm{cold}} \simeq 33$ K if $q = 0.5$. For the other component at 260 km s^{-1}, $m(v) \simeq 25$ K (obtained as a mean of the two slopes on each side of the central velocity) and $T_{\mathrm{warm}} \simeq 45$ K, hence $T_{K,\mathrm{cold}} \simeq 39$ K if $q = 0.5$. Unfortunately there is no way to determine q, but this method is nevertheless better than the classical method which attempts to find in the emission profile those gaussian components which are seen in absorption.

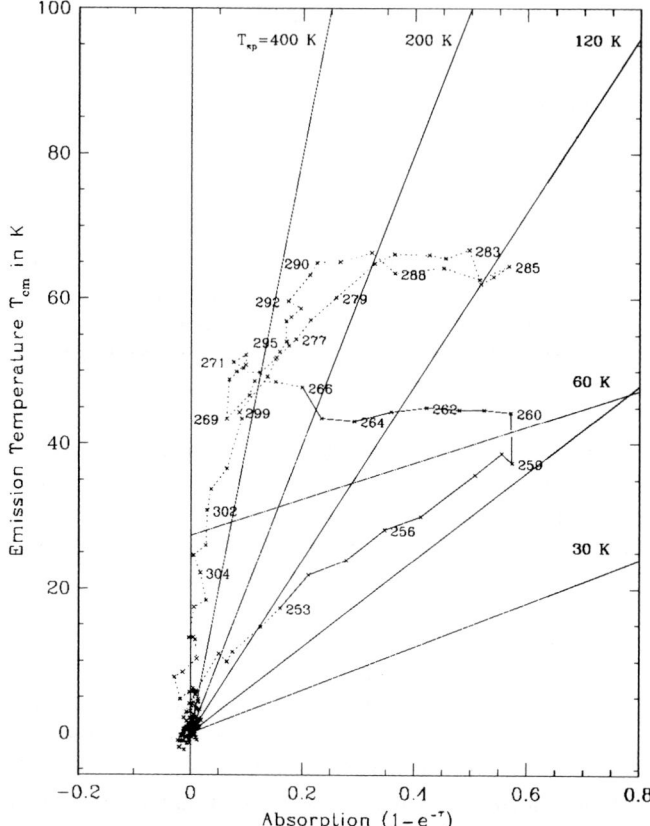

Fig. 4.2. Emission–absorption diagram for N 157b. The figures on the curves are the heliocentric radial velocity in km s^{-1}. For explanations see the text. A cloud with a given spin temperature T_{sp} (equal to the kinetic temperature) would plot on the straight line going through the origin, which is labelled by this temperature. Reproduced from Mebold et al. [362], with the permission of the AAS.

Results

The main application of the 21-cm line is the measurement of the mass, the distribution and the kinematics of atomic gas in our Galaxy and in external galaxies. For this, it is generally assumed that the line is optically thin, so that the column density does not depend upon the physical temperature of the gas (4.6). This can be unavoidably misleading because there is in general no way to determine the optical depth, but this depth can be large and we must keep in mind that the H I masses obtained in this way are lower limits. It is impossible here to give a complete overview of the results obtained from 21-cm line observations. We simply remind the reader that the atomic phase is almost always the dominant mass component of the interstellar medium of galaxies. Emission and absorption measurements have yielded very important results in spite of the difficulties in their interpretation. The main results can be summarized as follows:

- The atomic interstellar medium is extremely inhomogeneous. The 21-cm emission is dominated by filaments, sheets and shells (Plates 3, 4, 5, 6 and 32). This structure has a fractal appearance and could be due to turbulence.
- There are two phases in the atomic interstellar medium. One is warm (several thousands degrees), of low density (0.1–0.3 cm^{-3}), and is barely seen in absorption with some exceptions (Mebold & Hills [363]). The other is cold (60–100 K), denser (a few tens of atoms cm^{-3}), and dominates absorption. It is this component which is the main contributor to the complex structures we have just mentioned. It can also be found in the envelopes of molecular clouds. The warm component contains about as much matter as the cold component in the Galaxy, but it forms a thicker disk. Their respective mean half-thicknesses $\langle |z| \rangle$ are about 186 and 105 pc near the Sun (Falgarone & Lequeux [169], Malhotra [346], Dickey & Lockman [124]). The heating of the warm component implies partial ionization (see Sect. 8.1). Moreover, there seems to exist some neutral gas in the halo, with a high velocity dispersion (60 km s^{-1}) and a scale height of the order of 4.4 kpc (Kalberla et al. [276]). This gas might be in hydrostatic equilibrium, and its column density at the galactic poles is about 1.4×10^{19} atom cm^{-2}, quite low compared to the column densities of the cold and the warm components, which each have $\sim 1.5 \times 10^{20}$ atom cm^{-2}.
- H I structures are often called "clouds", a term which does not imply a more or less spherical shape. The distribution of cloud column densities is $\phi(N_{HI}) \propto (N_{HI})^{-1.3}$, which would correspond to a size distribution $\propto r^{-3.3}$ if they were spherical. The velocity dispersion between clouds along a galactic line of sight is approximately 9 km s^{-1}, after correction for differential rotation.
- There exists, at high galactic latitudes neutral gas that falls onto the galactic plane with velocities ranging from a few km s^{-1} to several hundreds of km s^{-1} (Plate 3, top). These *high-velocity clouds* might either be of extragalactic origin, or more probably originate in the hot ionized gas ejected by supernovae and bubbles from the galactic disk which then falls back onto the disk while cooling and recombining. We will come back to this point in Chap. 15.

4.1.2 Fine-Structure Lines in the Far-Infrared

Most important atoms and ions in the interstellar medium have their energy levels, in particular the fundamental level, split by *fine-structure interaction* between the total orbital momentum of the electrons and their total spin. The total orbital momentum is represented by the quantum number **L** which is the vector sum of the orbital moments of all the electrons, and the total spin **S** is the vector sum of all the electron spins. The total angular momentum is $\mathbf{J} = \mathbf{L} + \mathbf{S}$ in the case of the Russell–Saunders coupling (LS coupling), which generally applies to relatively small atoms. If this coupling applies, we note the terms of the fundamental state $n = 1$ as ^xY_J, with

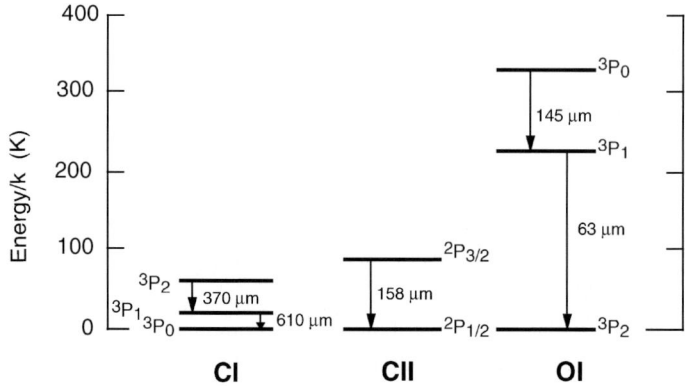

Fig. 4.3. Energy level diagram (Grotrian diagram) for C, C$^+$ et O, illustrating the fine-structure transitions. The wavelengths of the observable transitions are indicated.

$x = 2S + 1 = 1, 2, 3$ or 4, which correspond respectively to singlet, doublet, triplet and quadruplet, and $Y = S, P, D \ldots$ for $L = 0, 1, 2 \ldots$. The selection rules for electric dipole transitions are $\Delta S = 0$, $\Delta L = \pm 1$ and $\Delta J = 0, \pm 1$; transitions with $J = 0 \rightarrow J = 0$ are forbidden. Figure 4.3 shows an example of the fine structure sublevels of C I, C II and O I. Transitions between these sublevels are forbidden (magnetic dipole only); the electric dipole transitions are forbidden because they violate the selection rule $\Delta L = \pm 1$. The $^3P_2 \leftrightarrow {}^3P_0$ transitions, that also violate the $\Delta J = 0, \pm 1$ rule, are even weaker (electric quadrupole).

Table 4.1 gives a list of the most important forbidden transitions in the neutral interstellar medium. In order to avoid repetition, it also contains ionic lines that are found only in the ionized interstellar medium, which will be discussed in Sect. 5.1. The values of the atomic parameters should be considered as indicative and should be checked against the most recent sources. Here is a list of the URLs of Internet sites containing atomic and molecular data bases. These are generally regularly updated, with links between them or to other data bases:

Center for Astrophysics: *http://cfa-www.harvard.edu/amdata/ampdata/*
Cloudy (University of Kentucky): *http://www.pa.uky.edu/~verner/atom.html*
NIST: *http://physics.nist.gov/PhysRefData/*
OPACITY Project: *http://vizier.u-strasbg.fr/OP.html*
IRON Project (references only): *http://www.am.qub.ac.uk*
CHIANTI: *http://wwwsolar.nrl.navy.mil/chianti.html*.

The fine-structure transitions being forbidden, absorption of radiation at these wavelengths cannot appreciably populate the upper energy level[1]. For the same reason, we can generally neglect stimulated emission. If the line is optically thin, the upper level is only populated by collisions. De-excitation of the upper level

---

[1] There are however some cases (e.g. [O I]λ63 µm in photodissociation regions, see Chap. 10) in which the column density of the atom or ion is so large that the line becomes optically thick: absorption cannot be neglected in such cases.

4.1 The Atomic Neutral Gas

Table 4.1. The most important forbidden lines in the interstellar medium. Only C I, C II, O I, Si II, S II, and Fe II are present in the neutral medium. They are also present in the ionized medium, but generally in smaller amounts than more ionized species. Wavelengths λ are given in air for the visible transitions and in vacuum for the infrared/submillimetre ones. The collision strengths Ω_{ul} (cf. (4.17)) are for collisions with electrons at a temperature of 10^4 K. The critical densities $n_{crit} = A_{ul}/\langle \sigma_{ul} v \rangle$ (cf. (3.44)) correspond to collisions either with electrons (for $T_e \simeq 10^4$ K), or with H$_2$ molecules when between round brackets (for $T_K \simeq 100$ K). The values of the atomic parameters are only indicative and should be checked against the most recent sources. Some unobservable radiative transitions are not indicated, but the corresponding collisional transitions cannot be neglected. This is the case for the transition $^3P_2$–$^3P_0$ of O III, with $\Omega_{ul} = 0.21$, and for similar transitions in 3-level ions.

Ion	Transition l–u	λ μm	A_{ul} s^{-1}	Ω_{ul}	n_{crit} cm^{-3}
C I	$^3P_0$–$^3P_1$	609.1354	7.93×10^{-8}	–	(500)
	$^3P_1$–$^3P_2$	370.4151	2.65×10^{-7}	–	(3000)
C II	$^2P_{1/2}$–$^2P_{3/2}$	157.741	2.4×10^{-6}	1.80	47 (3000)
N II	$^3P_0$–$^3P_1$	205.3	2.07×10^{-6}	0.41	41
	$^3P_1$–$^3P_2$	121.889	7.46×10^{-6}	1.38	256
	$^3P_2$–$^1D_2$	0.65834	2.73×10^{-3}	2.99	7700
	$^3P_1$–$^1D_2$	0.65481	9.20×10^{-4}	2.99	7700
N III	$^2P_{1/2}$–$^2P_{3/2}$	57.317	4.8×10^{-5}	1.2	1880
O I	$^3P_2$–$^3P_1$	63.184	8.95×10^{-5}	–	2.3×10^4 (5×10^5)
	$^3P_1$–$^3P_0$	145.525	1.7×10^{-5}	–	3400 (1×10^5)
	$^3P_2$–$^1D_2$	0.63003	6.3×10^{-3}	–	1.8×10^6
O II	$^4S_{3/2}$–$^2D_{5/2}$	0.37288	3.6×10^{-5}	0.88	1160
	$^4S_{3/2}$–$^2D_{3/2}$	0.37260	1.8×10^{-4}	0.59	3890
O III	$^3P_0$–$^3P_1$	88.356	2.62×10^{-5}	0.39	461
	$^3P_1$–$^3P_2$	51.815	9.76×10^{-5}	0.95	3250
	$^3P_2$–$^1D_2$	0.50069	1.81×10^{-2}	2.50	6.4×10^5
	$^3P_1$–$^1D_2$	0.49589	6.21×10^{-3}	2.50	6.4×10^5
	$^1D_2$–$^1S_0$	0.43632	1.70	0.40	2.4×10^7
Ne II	$^2P_{1/2}$–$^2P_{3/2}$	12.8136	8.6×10^{-3}	0.37	5.9×10^5
Ne III	$^3P_2$–$^3P_1$	15.5551	3.1×10^{-2}	0.60	1.27×10^5
	$^3P_1$–$^3P_0$	36.0135	5.2×10^{-3}	0.21	1.82×10^4
Si II	$^2P_{1/2}$–$^2P_{3/2}$	34.8152	2.17×10^{-4}	7.7	(3.4×10^5)
S II	$^4S_{3/2}$–$^2D_{5/2}$	0.67164	2.60×10^{-4}	4.7	1240
	$^4S_{3/2}$–$^2D_{3/2}$	0.67308	8.82×10^{-4}	3.1	3270
S III	$^3P_0$–$^3P_1$	33.4810	4.72×10^{-4}	4.0	1780
	$^3P_1$–$^3P_2$	18.7130	2.07×10^{-3}	7.9	1.4×10^4
S IV	$^2P_{1/2}$–$^2P_{3/2}$	10.5105	7.1×10^{-3}	8.5	5.0×10^4
Ar II	$^2P_{1/2}$–$^2P_{3/2}$	6.9853	5.3×10^{-2}	2.9	1.72×10^6
Ar III	$^3P_2$–$^3P_1$	8.9914	3.08×10^{-2}	3.1	2.75×10^5
	$^3P_1$–$^3P_0$	21.8293	5.17×10^{-3}	1.3	3.0×10^4
Fe II	$^6D_{7/2}$–$^6D_{5/2}$	35.3491	1.57×10^{-3}	–	(3.3×10^6)
	$^6D_{9/2}$–$^6D_{7/2}$	25.9882	2.13×10^{-3}	–	(2.2×10^6)

occurs via collisions and emission of radiation, one process or the other dominating depending upon the density. For an optically thin line, the statistical equilibrium equation is thus simply

$$n_l C_{lu} = n_u (A_{ul} + C_{ul}), \quad (4.16)$$

with $C_{ul} = n \langle \sigma_{ul} v \rangle$ (3.42), n the density of those particles which are effective collision partners, in general free electrons, and $C_{lu} = C_{ul} \frac{g_u}{g_l} \exp(-h\nu/kT_K)$ (3.38). When electrons dominate the collisions, the collisional de-excitation probability is often written as

$$C_{ul} = \frac{8.63 \times 10^{-6}}{g_u T_e^{1/2}} n_e \Omega_{ul}, \quad (4.17)$$

where n_e and T_e are respectively the electron density and the electron temperature. This defines the *collision strength* Ω_{ul}, which with the chosen numerical coefficient is a parameter of order unity, and depends little upon the electron temperature.

The density in the upper level is given by

$$\frac{n_u}{n_l} = \frac{g_u}{g_l} \exp(-h\nu/kT_K) \frac{C_{ul}}{A_{ul} + C_{ul}}, \quad (4.18)$$

neglecting implicitly the stimulated emission, assuming a low intensity. The complete expression is given (3.39). Notice that, the upper level of fine-structure transitions being generally poorly populated, n_l is in practice the total density of the considered atom or ion. The *critical density*, n_{crit}, above which collisions dominate the de-excitation of the upper level, is such that $A_{ul} = C_{ul}$, so that $A_{ul}/C_{ul} = n_{crit}/n$, where n is the density of the particles responsible for the collisions. n_{crit} is temperature dependent. Approximate values are given in Table 4.1.

From (3.3), we see that the integrated intensity of an optically thin line at small densities is simply given by

$$I_{ul} = \frac{h\nu}{4\pi} n_u A_{ul} \text{ erg s}^{-1} \text{ sterad}^{-1}. \quad (4.19)$$

The exact expression, useful for high densities, is easily obtained from (4.16) to (4.18):

$$I_{ul} = \frac{h\nu}{4\pi} n_{ion} A_{ul} \frac{(g_u/g_l) \exp(-h\nu/kT_K)}{1 + (g_u/g_l) \exp(-h\nu/kT_K) + n/n_{crit}}, \quad (4.20)$$

where n_{ion} is the density of the emitting ion. If the density is much smaller than the critical density, the line intensity is proportional to nn_{ion}, and since the density of the ion is proportional to n the line intensity varies as the square of the density. Conversely, if $n \gg n_{crit}$, the levels are in thermal equilibrium (LTE) and the line intensity varies simply with n_{ion}.

Fine-structure lines are of major importance in the physics of the interstellar medium. In effect the [C II]λ157.7 µm line is very easily excited in the diffuse interstellar medium (see column 6 of Table 4.1), and is the main coolant of this medium as long as its temperature is smaller than about 100 K. At higher temperatures, the [O I]λ63 μm line also takes part in the cooling. In photodissociation regions and in shocks, emission lines of molecules such as H_2, CO and H_2O are also major coolants. All these lines have been observed in many sources with the Kuiper Airborne Observatory, the COBE satellite, stratospheric balloons and several other satellites, in particular the ISO and SWAS satellites. ISO has also observed the [O I] line in absorption in several directions with high column densities. This line is then probably optically thick, which prevents a good determination of the column density of O I.

4.1.3 Interstellar Absorption Lines

Generalities

Many interstellar absorption lines have been observed in the spectra of stars. They differ from stellar lines by being much narrower and also by having a fixed wavelength (when the target star is a close binary, the stellar lines vary periodically in wavelength due to the Doppler effect of the orbital motion). In the visible and near UV, we observe lines from atoms (Na, K, Ca), ions (Ca^+, Ti^+) and molecules (CN, CH, CH^+, C_2 at 10 140 Å and OH near 3 080 Å). Thanks to the Copernicus satellite and more recently to other space observatories such as the Hubble Space Telescope and the FUSE satellite, a very large number of atomic, ionic and molecular lines have been observed in the far-UV. Among the atomic lines the Lyman series of atomic hydrogen are conspicuous, and among the molecular lines the many lines of H_2. These lines supply important information on the chemical composition and the physical conditions in the diffuse interstellar medium. Let us now see how we can derive column densities from such observations.

The *equivalent width* of a line is defined as

$$W = \int_{line} \frac{I_0 - I_\lambda}{I_0} d\lambda, \quad (4.21)$$

where I_0 is the intensity of the stellar continuum on each side of the line and I_λ the intensity in the line at wavelength λ. Equivalent widths are given in wavelength units, generally in mÅ. The upper level of the transition being little populated because it lies at high energy, the spontaneous and stimulated de-excitations from this level can be neglected, and we have simply

$$W = \int (1 - e^{-\tau}) d\lambda, \quad (4.22)$$

where τ is the optical depth. For low column density N the line is optically thin ($\tau \ll 1$) and we have

$$W = \int \tau_\nu d\lambda = \int \tau_\nu \lambda^2 d\nu/c. \quad (4.23)$$

We have from (3.32) (for $h\nu_0/kT_{ex} \gg 1$)

$$\tau(\nu) = \frac{c^2 N_l g_u}{8\pi\nu^2 g_l} A_{ul} \phi_{ul}(\nu), \qquad (4.24)$$

the equivalent width for an optically thin line is then

$$W = \frac{\lambda^2 N_l g_u}{8\pi g_l} A_{ul} \frac{\lambda^2}{c}, \qquad (4.25)$$

since $\int \phi_{ul}(\nu) d\nu = 1$.

We often express the spontaneous emission probability A_{ul} as a function of the *oscillator strength*, a quantity equal to 1 for a harmonic oscillator and not very different from 1 for the strong resonance lines which are the main lines observed in interstellar absorption:

$$f = \frac{m_e c^3}{8\pi \nu^2 e^2} A_{ul} \frac{g_u}{g_l}, \qquad (4.26)$$

where m_e and e are the mass and the charge of the electron. From this we derive

$$\boxed{W = \frac{\pi e^2}{m_e c^2} N_l \lambda^2 f,} \qquad (4.27)$$

or numerically

$$\boxed{\left(\frac{W}{\text{Å}}\right) = 8.85 \times 10^{-13} \left(\frac{\lambda}{\mu\text{m}}\right)^2 \left(\frac{N}{\text{cm}^{-2}}\right) f.} \qquad (4.28)$$

If the line is not optically thin, we must consider its profile. The line profile is determined by Doppler broadening and damping broadening mechanisms. The latter is also called natural broadening. Collisional broadening is generally negligible for interstellar absorption lines. The optical depth is then given as

$$\tau(\nu) = N \frac{\pi e^2 f}{m_e c} \frac{\lambda e^{-(v/b)^2}}{b\sqrt{\pi}} \otimes \frac{\gamma}{\Delta\omega^2 + (\gamma/2)^2}, \qquad (4.29)$$

where $\otimes$ designates the convolution product between the first term which corresponds to the Doppler broadening and the second term which corresponds to the damping. b is $\sqrt{2}$ times the velocity dispersion σ_v along the line of sight, including the thermal velocity dispersion. $\omega = 2\pi\nu$ is the angular frequency. For a medium with kinetic temperature T_K and random motions (microturbulence) represented by ζ^2 we have

$$b^2 = b_{th}^2 + \zeta^2, \quad \text{with} \qquad (4.30)$$

$$b_{th} = \sqrt{\frac{2kT_K}{\mathcal{A} m_p}} = 12.9\sqrt{(T_K/10^4\,\text{K})/\mathcal{A}}\ \text{km s}^{-1}, \qquad (4.31)$$

where $\mathcal{A}$ is the mass number of the element and m_p the proton mass. γ is the damping parameter, the inverse of the radiative lifetime of the excited level.

4.1 The Atomic Neutral Gas   57

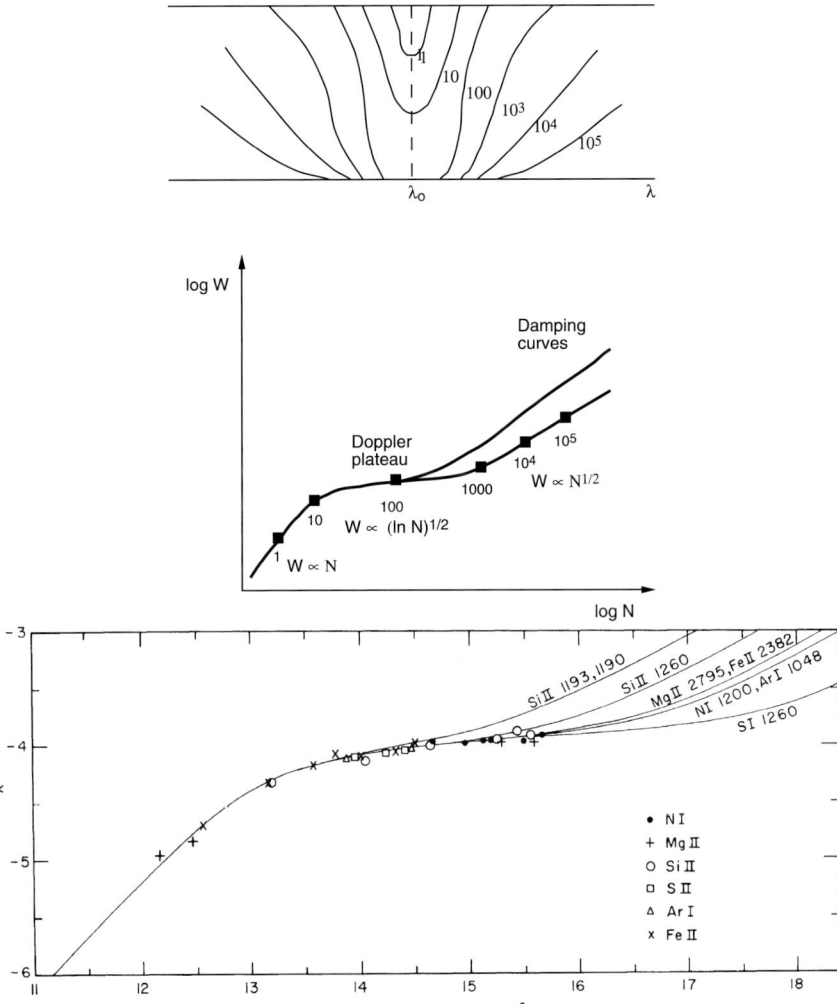

Fig. 4.4. Profiles of interstellar absorption lines and curves of growth. Top: schematic profiles of an absorption line as a function of the column density of the absorbing atoms (in arbitrary units). The very broad damping wings appear for high column densities. Middle: schematic curve of growth for the same line, where its equivalent width is plotted as a function of the column density. The two curves to the right correspond to two different values of the damping constant. The figures along the curve correspond to the column densities of the profiles above. Bottom: observed curve of growth for the interstellar medium in front of the star ζ Oph, reproduced from Morton [375] with the permission of the AAS. Here $\log(Nf\lambda)$ is given in abscissae instead of $\log N$, and $\log(W/\lambda)$ in ordinates instead of $\log W$. In this case, the curves of growth for the different atoms coincide except on the damping part, this because the damping constants are different for the different lines.

The convolution gives a *Voigt profile* that describes the line shape, so that

$$\frac{\tau(\nu)}{\tau_0} = \frac{a}{\pi} \int_{-\infty}^{+\infty} \frac{e^{-x^2} dx}{a^2 + (\nu/b - x)^2}, \quad (4.32)$$

with $a = \gamma\lambda/(4\pi b)$.

Calculating the integral of $\exp -\tau(\nu)$ over the line profile, which we do not show here, we can obtain its equivalent width (4.22). For increasing optical depths, we first find that the equivalent width is proportional to the column density N (4.5), then to $(\log N)^{1/2}$ (the *Doppler plateau* where the equivalent width depends very little upon the column density), and finally to $N^{1/2}$ (the *damping part* where the line profile is dominated by the damping wings and where the equivalent width is again sensitive to the column density). This is illustrated by Fig. 4.4. Unfortunately most observed absorption lines lie on the Doppler plateau.

Let us now consider several absorption lines from different elements, at different wavelengths, in the same cloud. We assume that they are broadened in the same way by the Doppler effect, with the same velocity dispersion: this is the case if the Doppler width is dominated by turbulence rather than by thermal motions, frequently the case. The Doppler width $\Delta\lambda_D$ is proportional to λ and for unsaturated lines $W \propto \Delta\lambda_D \propto \lambda$. As a result, if we plot the *curve of growth* (Fig. 4.4) which gives $\log(W/\lambda)$ as a function of $\log(\lambda N f)$, the representative points for the different lines are located on a single curve, even if these lines are saturated. This is no longer the case for larger column densities, in which case the line width is dominated by the damping wings, because the damping constant is not the same for all lines.

A better reduction technique than the use of curves of growth consists of fitting the line profiles by a model in which we introduce already known parameters for the given line of sight. This technique is the only one that can be used for complex line profiles.

Results: 1. Applications to the Physics of the Interstellar Medium

We observe in the visible spectrum of many stars a substantial number of atomic interstellar absorption lines. They are all resonance lines starting from the fundamental level of the atom. The principal lines are:

- The D_1 and D_2 doublet of neutral sodium (5 889-5 895 Å). These lines are almost always saturated and it is difficult to derive the abundance of Na from them. For this purpose, it is better to use the weak UV lines at 3 302.4 and 3 303.0 Å.
- The line of neutral potassium at 7 699 Å.
- The H and K doublet of ionized calcium at 3 933 and 3 968 Å.
- The line of neutral calcium at 4 226 Å.
- Weaker lines of Li, Ti$^+$ etc.

These lines are often multiple, indicating the presence of several interstellar clouds along the line of sight. The information obtained from the line profiles (internal velocity dispersion in each cloud, random motions between clouds) complements what can be extracted from the 21-cm hydrogen line.

The coexistence of Ca^0 and Ca^+ lines also allows us to obtain the degree of ionization of the corresponding cloud. We observe that Ca^+ is much more abundant than Ca^0. This is due to ionization by UV radiation from stars at energies between 6.11 eV (203.8 nm), the ionization potential of Ca^0, and 13.6 eV (91.1 nm), the ionization potential of hydrogen. The neutral interstellar medium is opaque at shorter wavelengths due to the very strong absorption by H atoms. The ionization equilibrium of calcium can be written as

$$n(Ca^0)\Gamma = n(Ca^+)n_e\alpha, \qquad (4.33)$$

where Γ is the ionization probability per second of neutral calcium, that can be estimated if we know the UV radiation field, and α the recombination probability per second. n_e is the electron density. α depends on the electron temperature T_e and varies as $T_e^{-0.7}$. T_e is practically equal to the kinetic temperature which can be estimated from 21-cm line observations. From the measured column densities of neutral and ionized calcium, which are assumed to be proportional to the volume densities, we can estimate the electron density. n_e is generally found to be lower than 1 electron cm^{-3}. Most of these free electrons are in interstellar clouds that are relatively transparent to the UV, and come from the ionization of carbon, the most abundant element with an ionization potential (11.260 eV) lower than that of hydrogen.

The UV range is much richer in absorption lines than the visible, because most atoms and ions have lines in the UV, with the notable exception of helium. The difficulties in interpretation are the same as for the optical range: strong lines are saturated and cannot yield accurate column densities. Fortunately, several very abundant elements such as oxygen and carbon also have very weak UV lines (see e.g. Sofia et al. [481]). In several cases the same element can be observed as a neutral atom and as an ion, allowing us to derive the electron density in the same way as for calcium.

We can also sometimes observe several absorption lines with wavelengths close to each other, which come from different fine-structure levels of the fundamental state of the same atom or ion. We can then directly obtain the respective populations of these levels. This gives valuable information on those physical parameters that determine their excitation, essentially the electron density. For example, ionized carbon has two lines, C IIλ1 334.57 and C II*λ1 335.70 Å, which unfortunaly are often saturated. They come respectively from the two fine-structure sublevels $^2P_{1/2}$ and $^2P_{3/2}$ of the fundamental (cf. Fig. 4.3). The fine-structure transition at 157.7 μm that was discussed in the previous section connects these sublevels. We can then in principle determine the cooling rate of the diffuse medium which is dominated by this 157.7 μm line, simply by observing the C II* absorption line, provided the difficulties due to its saturation can be resolved (Pottasch et al. [412]; Gry et al. [210]). The latter obtain a cooling rate of $3.5^{+5.4}_{-2.1} \times 10^{-26}$ erg s^{-1} per hydrogen nucleus, in agreement with the more direct determination of Bennett et al. [31] which is based on observations with COBE. The significance of these results will be discussed in Chap. 8.

Neutral carbon has a rather large number of UV absorption lines including the three sublevels of the fundamental $^3P_0$, $^3P_1$ and $^3P_2$ (cf. Fig. 4.3), designated as C I, C I* and C I** (Jenkins et al. [261]). As these numerous transitions have a large range of oscillator strengths, it is easy to solve the saturation problems and to obtain accurate column densities. The population ratios between the three fine-structure levels of C I are determined principally by the density of atomic hydrogen, the dominant collision partner, and by the kinetic temperature. They are good indicators of the pressure $P = nkT_K$, as shown by Jenkins & Shaya [260][2]. From these measurements Jenkins et al. [261] find that most of the diffuse interstellar medium is at pressures such that $10^3 < P/k < 10^4$ cm^{-3} K, but that there are regions with larger or smaller pressures. It is thus difficult to define a mean pressure for the interstellar medium. The large pressure variations from place to place are due to the constant agitation of the interstellar medium, and, in particular, to the shocks crossing it that arise from supernova explosions, stellar winds and the expansion of H II regions.

Results: 2. Determination of Elemental Abundances in the Interstellar Gas

The most important use of the UV absorption lines is the determination of the abundance of elements in the gaseous phase of the interstellar medium. The column density of atomic hydrogen is obtained, in principle, rather straightforwardly from the shape of the Lyman α line at 1 215.67 Å, since this line is almost always in the damping part of the curve of growth. For such lines the column density is related to the equivalent width by

$$\frac{N}{\text{cm}^{-2}} = \frac{m_e c^3}{e^2 \lambda^4} \frac{W^2}{f\gamma} = 1.07 \times 10^{33} \frac{1}{f\gamma} \left(\frac{W}{\text{mÅ}}\right)^2 \left(\frac{\lambda}{\text{Å}}\right)^{-4}, \qquad (4.34)$$

where γ is the damping constant. However, these lines are very broad. Due to difficulties in determining the stellar continuum it is better to fit a theoretical profile to the observed one. For the Lyman α line we can use the following approximate relation, which is not valid very close to the line centre (Bohlin [47]):

$$\boxed{\tau(\lambda) \approx 4.26 \times 10^{-20} \frac{N_\text{H}}{(\lambda - \lambda_0)^2},} \qquad (4.35)$$

where N_H is in atom cm^{-2} and λ_0 is the central wavelength of the line, λ being expressed in Å.

Of course, normalization of the abundances to hydrogen requires adding to the column density of H I, N(H I), twice the column density of molecular hydrogen N(H$_2$), which is obtained from its absorption lines at $\lambda < 115$ nm.

---

[2] We can also use the C II and C II* lines for a determination of the pressure (Kulkarni & Heiles [292]). Unfortunately these lines are usually very saturated and the derived column densities are accordingly uncertain.

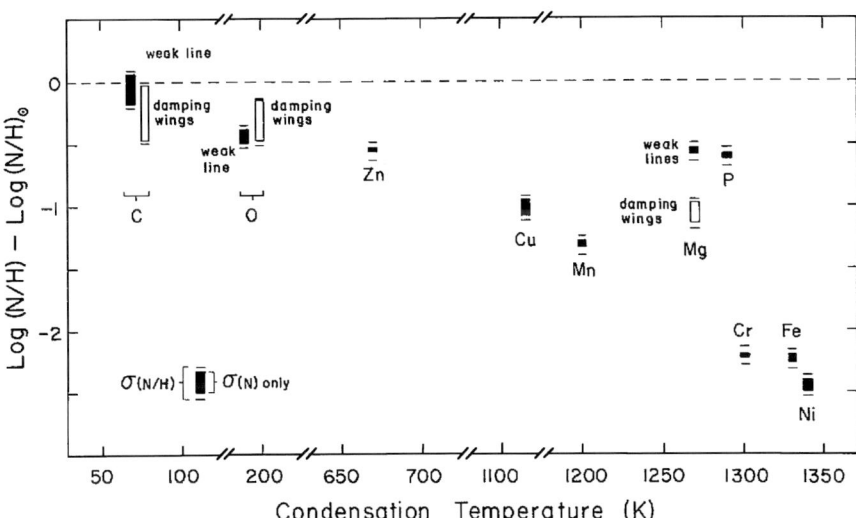

Fig. 4.5. The abundances of some elements in the gas phase measured with the GHRS UV spectrograph of the Hubble Space Telescope towards the star ζ Per. They are given relative to the Solar system abundances and expressed as $[X/H] = \log(X/H) - \log(X/H)_\odot$, and are plotted as a function of the condensation temperature of the elements. Full symbols correspond to abundances obtained from faint lines, and their heights correspond to an error of $\pm 1\sigma$ in the column density, the small horizontal lines including the uncertainty on the column density of hydrogen. The open symbols correspond to determinations using a fit to the damping wings of very saturated lines. The agreement between the two determinations is rather good for carbon and oxygen, but bad for magnesium. This points to errors in the adopted oscillator strengths for some Mg^+ lines. Reproduced from Cardelli *et al.* [78], with the permission of the AAS.

For other elements, it can be difficult to estimate their total abundance if only one ionization state can be observed. For O, N and the noble gases this problem does not arise because they are not ionized in the "neutral" medium. There is no difficulty either for most of the metals and carbon which essentially exist as singly ionized ions in this medium, provided the ion can be observed. Otherwise, the ionization equilibrium should be solved using knowledge of electron density obtained from an ion/neutral pair, calcium for example.

Table 4.2 summarizes the abundances obtained along two characteristic lines of sight, one in the cold diffuse medium and the other in the warm diffuse medium. They are compared to abundances measured in the Solar system, in young B stars, near the Sun, and in H II regions. Figure 4.5 shows an example of some interstellar abundances, relative to the Solar system abundances.

One point to note is that the Solar system is richer in heavy elements than the younger stars of the Solar neighbourhood. This poses a problem for studies of the chemical evolution of the Galaxy, because the heavy element abundances are expected to increase with time. It could be that the Solar system is anomalously overabundant in heavy elements. The abundances in H II regions are close to the

Table 4.2. The gas phase abundances of selected elements along two lines of sight, compared to abundances in the Solar system, nearby B, F and G stars and H II regions. Abundances are given as $12 + \log(X/H)$, X being the chemical symbol for the element and H that of hydrogen. The deficiencies given in columns 6 and 7 are expressed, with respect to Solar system abundances, as $[X/H] = \log(X/H) - \log(X/H)_\odot$. The data come mainly from Savage & Sembach [451] and from Snow & Witt [479]. Along the line of sight to μ Colombae, for which deficiencies are smaller by about 0.3 dex compared to those in the warm medium in front of ζ Ophiuchi, we observe $[X/H] = +0.05$ for P and $+0.10$ for S (Howk et al. [250]).

Element	Solar system $12 + \log(X/H)$	Stars	H II	T_c^1 K	ζ Oph cold [X/H]	ζ Oph warm
H	12.00	12.00	12.00	–	–	–
D	7.53	–	–	–	-0.33:[2]	–
He	10.99	–	10.95	–	–	–
Li	3.31	–	–	1 225	-1.58	–
B	2.88	–	–	650	-0.93	–
C	8.55	8.33	8.60	75	-0.41	–
N	7.97	7.82	7.89	120	-0.07	–
O	8.87	8.66	8.77	180	-0.39	0.00
Ne	–	–	8.03	–	–	–
Na	6.31	–	–	970	-0.95	–
Mg	7.58	7.40	–	1 340	-1.55	-0.89
Si	7.55	7.27	–	1 311	-1.31	-0.53
P	5.57	–	–	1 151	-0.50	-0.23
S	7.27	7.09	7.31	648	$+0.18$	–
Ar	6.56	–	–	25	-0.48	–
K	5.13	–	–	1 000	-1.09	–
Ca	6.34	6.20	–	1 518	-3.73	–
Ti	4.93	4.81	–	1 549	-3.02	-1.31
Fe	7.50	7.43	6.59	1 336	-2.27	-1.25

[1] Condensation temperature at thermal and chemical equilibrium, appropriate for the Solar nebula with an initial gas pressure of 10^{-4} bar.
[2] For lines of sight other than that of ζ Oph: Linsky et al. [325].

Solar system abundances, but they are somewhat uncertain except for oxygen. Iron is underabundant in H II regions, probably due to depletion into grains. We notice that most elements are underabundant in the diffuse interstellar medium with respect to the Solar system, young stars and H II regions. Exceptions are S, Zn and P (Howk et al. [250]), which do have *solar* abundances in the warm neutral medium! This shows that the elemental abundances are not precisely known in the interstellar medium.

The missing elements in the interstellar medium are believed to be in dust grains. The underabundance of an element is larger if its condensation temperature is higher, confirming this idea. Condensation must have taken place in circumstellar envelopes, and also the interstellar medium itself. The underabundances are smaller in the warm medium, suggesting that the evaporation of grains, probably as a consequence of shocks, has returned of a fraction of the elements to the gas phase. These phenomena will be discussed in Chap. 15.

Finally we must mention the discovery, through their absorption lines, of highly ionized elements like C IV, N V and O VI. C IV is even observed through emission lines. We discuss the interpretation of these observations in Sect. 5.3.

4.2 The Molecular Component

4.2.1 Introduction

Before 1965, only three molecules were known in the interstellar medium: CH, CH$^+$ and CN. They were discovered, through their absorption lines, in the direction of bright stars and therefore exist in the diffuse interstellar medium. The real harvest of interstellar molecule discoveries only began in 1965, thanks to radio observations of their emission lines. In this year OH, NH$_3$ and H$_2$O were discovered. After 1970, many other molecules were discovered through millimetre–wave observations. Today more than 120 different molecules are known in the interstellar medium and in the envelopes of cool stars (asymptotic giant branch, or AGB stars). A significant number of molecules are also found in comets. Many molecules have been discovered in external galaxies, in particular CO which has been observed in galaxies at very large redshifts.

Table 4.3 gives a list of the interstellar molecules known at the end of 2003, with indications as to where they are found. Most of them were discovered through their rotational transitions at centimetre and, more generally, at millimetre wavelengths. Figure 4.6 shows a portion of the submillimetre spectrum of the Orion molecular cloud. The confusion limit is reached in such spectra!

Contrary to atoms, for which only the energy of the electrons intervene in the transitions, molecules have three types of transitions: electronic, vibrational and rotational. We will examine each in turn, with examples of applications to the interstellar medium. For more details, the interested reader is referred to specialized treatises, e.g. those of Herzberg ([235], [236], [237]). First we give some general information.

Some molecules that have two identical nuclei with non-zero spin exhibit two forms between which radiative and non-reactive, hence non-destructive collisional transitions are forbidden. In the *ortho* form, the two nuclear spins are parallel and in the *para* form the spins are antiparallel. This is the case for H$_2$, H$_2$O and H$_2$CO. Once the molecule is formed in one of these forms, it remains practically frozen in this state and can stay there indefinitely. For example, the probability for the radiative transition ortho $\rightarrow$ para of H$_2$ at 84.392 μm is only 7×10^{-21} s^{-1}! The ortho/para ratio would witness to the conditions of formation of the molecule if it were not efficiently modified by reactions on the surfaces of grains (Le Bourlot [307]).

The electric dipole moment of symmetric molecules is zero at rest. For such molecules electronic transitions are permitted, but not rotational nor vibrational

Table 4.3. List of the interstellar and circumstellar molecules known at the end of 2003. Molecules *also* seen in comets are in boldface. Those seen *only* in comets are between round brackets. Circumstellar-only molecules are in italics. c- designates a cyclic molecule, and l- a linear molecule in cases of possible ambiguity. This list is regularly updated by Alan Wootten: see *http://www.cv.nrao.edu/~awootten*

2 atoms	3 atoms	4 atoms	5 atoms	6 atoms	7 atoms	8 atoms	≥9 atoms
Hydrogen family							
H_2	H_3^+						
Carbon family							
C_2	C_3	c-C_3H	C_5	C_5H	C_6H	C_7H	C_8H
CH	C_2H	l-C_3H	C_4H	l-C_4H_2	CH_3C_2H	C_6H_2	CH_3C_4H
CH^+	CH_2	$C_2\mathbf{H_2}$	l-C_3H_2	C_2H_4		(C_2H_6)	c-C_6H_6
		$CH_2D^+?$	c-C_3H_2				
			$\mathbf{CH_4}$				
Oxygen + hydrogen and/or carbon							
CO	C_2O	C_3O	**HCOOH**	**CH_3OH**	$HCOCH_3$	**$HCOOCH_3$**	$(CH_3)_2O$
CO^+	HCO	$HOCO^+$	H_2C_2O	HC_2CHO	c-C_2H_4O	CH_3COOH	CH_3CH_2OH
OH	$\mathbf{HCO^+}$	$\mathbf{H_2CO}$	H_2COH^+			CH_2OHCHO	$(CH_3)_2CO$
	HOC^+	H_3O^+			CH_2CHOH		
	H_2O						
	(H_2O^+)						
	CO_2						
Nitrogen + hydrogen and/or carbon							
CN	HCN	C_3N	$\mathbf{HC_3N}$	$\mathbf{CH_3CN}$	HC_5N	CH_3C_3N	HC_7N
NH	HNC	HCCN	CH_2CN	CH_3NC	CH_2CHCN		HC_9N
	N_2H^+	$HCNH^+$	HC_2NC	HC_3NH^+	NH_2CH_3		$HC_{11}N$
	NH_2	H_2CN	H_2CHN	C_5N			CH_3CH_2CN
	$\mathbf{NH_3}$		H_2NCN				$CH_3C_5N?$
			HNC_3				

4.2 The Molecular Component

Table 4.3. (continued)

2 atoms	3 atoms	4 atoms	5 atoms	6 atoms	7 atoms	8 atoms	≥ 9 atoms
Nitrogen + oxygen + hydrogen and/or carbon							
NO	HNO	HNCO		NH$_2$CHO			NH$_2$CH$_2$COOH?
	N$_2$O						
Molecules with sulphur							
CS	C$_2$S	C$_3$S		CH$_3$SH			
SO	HCS$^+$	HNCS					
SO$^+$	H$_2$S	**H$_2$CS**					
NS	**OCS**						
	SO$_2$						
Miscellaneous molecules							
SiO	SiCN	*SiC$_3$*	*SiC$_4$*				
SiS	*c-SiC$_2$*		*SiH$_4$*				
HF	*MgCN*						
SiC	*MgNC*						
CP	*NaCN*						
HCl	*AlNC*						
KCl							
NaCl							
PN							
SiN							
AlF							
AlCl							
SH							
FeO?							

transitions. Rotational dipole transitions, which correspond classically to the emission of a rotating electric dipole, are forbidden. They can only be quadrupolar and hence extremely weak. Symmetrical molecules such as H_2, O_2, $HC\equiv CH$, CH_4, etc., are thus essentially unobservable in rotation (not the case for O_2 which has a magnetic dipole, but this molecule has never been convincingly observed in the interstellar medium, probably because its abundance is too small). H_2 was only recently observed in the mid-IR through its rotation lines, thanks to its very large abundance. The vibration lines of symmetric molecules are also weak, but they are easily observed for H_2 and a few other abundant molecules. Due to this difficulty, the list of interstellar molecules presented in Table 4.3 is certainly incomplete with regard to symmetric molecules.

4.2.2 Electronic Transitions

The electronic transitions of molecules are the equivalent of the atomic transitions. Their energy is of the order of several eV and they are generally in the far-UV. For example, all the electronic transitions of H_2, the most abundant interstellar molecule by far, are at $\lambda < 115$ nm. Other molecules detected through their electronic transitions (almost always in absorption) are CO and OH in the far-UV, CH, CH^+ and CN in the near-UV accessible from the ground and C_2 and CN in the near-IR. Very few simple molecules have transitions in the optical range visible from the

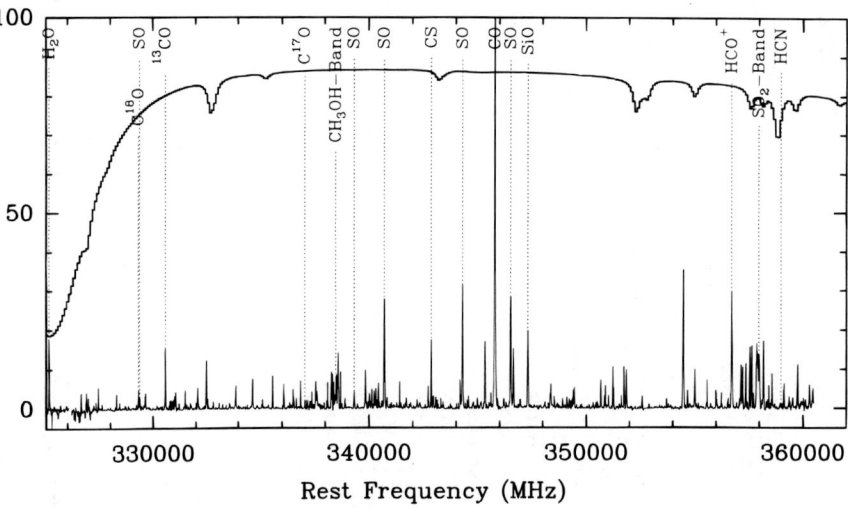

Fig. 4.6. A portion of the submillimetre spectrum of the Orion molecular cloud. The strongest lines are marked, and the atmospheric transmission is plotted above the spectrum. The spectrum is extremely complex due to the relatively high temperature of the cloud (about 60 K), such that highly excited rotation levels, and even vibration levels, are populated for some molecules. Some 6 per cent of the lines are not identified. Reproduced from Schilke et al. [457], with the permission of the AAS.

ground, and this is the reason why the discovery of interstellar molecules has been relatively late. Nevertheless, the many diffuse interstellar absorption bands seen in the optical spectra of bright stars are almost certainly electronic transitions of rather complex molecules, probably carbonaceous, which have not yet been identified (cf. Sect. 4.2).

When observed at high resolution, each electronic transition of a molecule is resolved into a series of different lines, each one coming from a transition between the vibrational levels of the two electronic states. Each one of these lines can itself be decomposed into several lines corresponding to the rotational sublevels. An electronic spectrum can thus be very complex. At lower resolution we only observe each electronic transition as a broad band resulting from the superimposition of all the ro-vibrational lines.

Figure 4.7 shows a fraction of the electronic spectrum of the H_2 and CO molecules. Figure 4.8 shows the energy of some levels of the H_2 molecule given as a function of the distance between the two H atoms (*potential curves*).

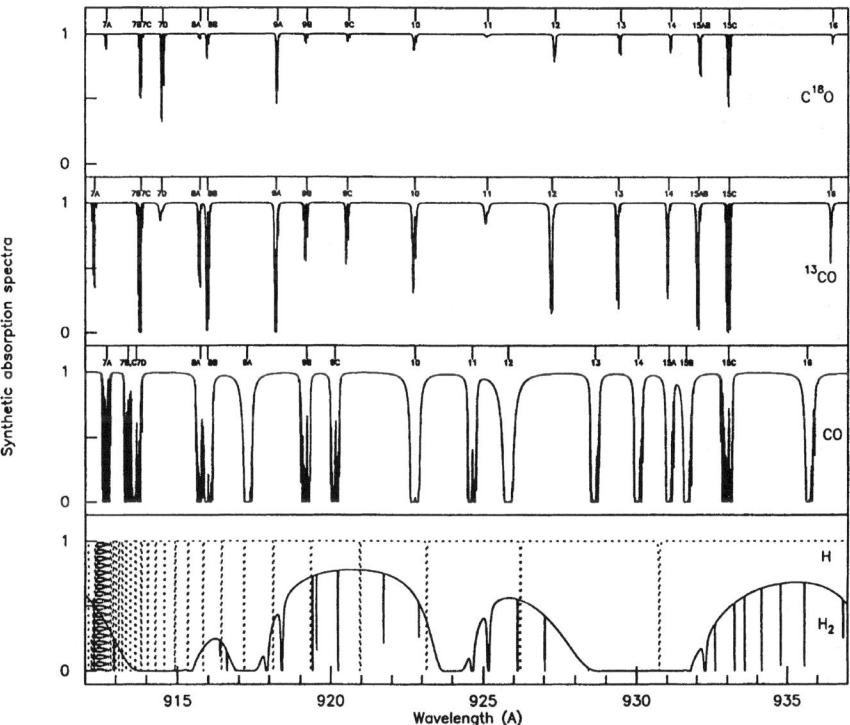

Fig. 4.7. A portion of simulated electronic absorption spectra of H_2 and of the ^{12}CO, ^{13}CO and C^{18}O molecules. The spectrum of the Lyman lines of atomic hydrogen has been added. These spectra result from calculations for a cloud of density $n = 10^3$ cm^{-3}, temperature $T = 25$ K and optical depth $A_V = 2$ mag. in the visible. From Warin et al. [541], with the permission of ESO.

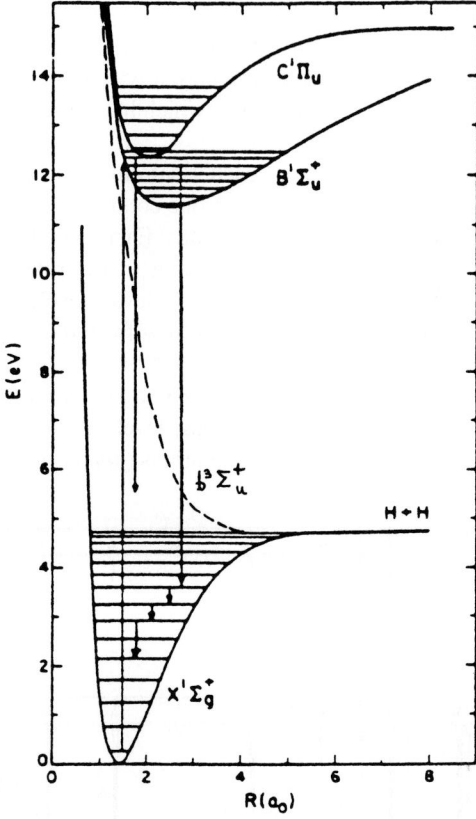

Fig. 4.8. Simplified energy level diagram for H_2. The energy of a number of different levels is plotted as a function of distance R between the H atoms, expressed in atomic units $a_0 = 0.529$ Å (radius of the first Bohr orbit for hydrogen). These are the *potential curves*. The lowest potential curve corresponds to the fundamental electronic state. The energy at its minimum, which is taken here as the zero energy, is smaller by 4.7 eV than the sum of the energy of two isolated H atoms in their fundamental state. The formation of H_2 from two H atoms thus liberates 4.7 eV of energy. The horizontal segments correspond to the vibrational levels in the fundamental electronic state. Each of these levels is decomposed into tight rotational levels not represented here. Two potential curves corresponding to excited electronic states are also shown, with their respective vibrational levels. The upwards arrow corresponds to absorption of one far-UV photon, and the downwards arrows either to de-excitation on one of the vibrational levels of the electronic fundamental state followed by a vibrational cascade (fluorescence), or to de-excitation on a continuum level higher than 4.7 eV which leads to the dissociation of the molecule. Another potential curve is indicated as a dashed line, it corresponds to a repulsive state: a molecule formed by two approaching H atoms is in this state, and because the transition with the lower-energy fundamental state is forbidden the molecule can only dissociate. It is thus impossible to form directly H_2 from two H atoms with a low energy, although the reaction is highly exothermic.

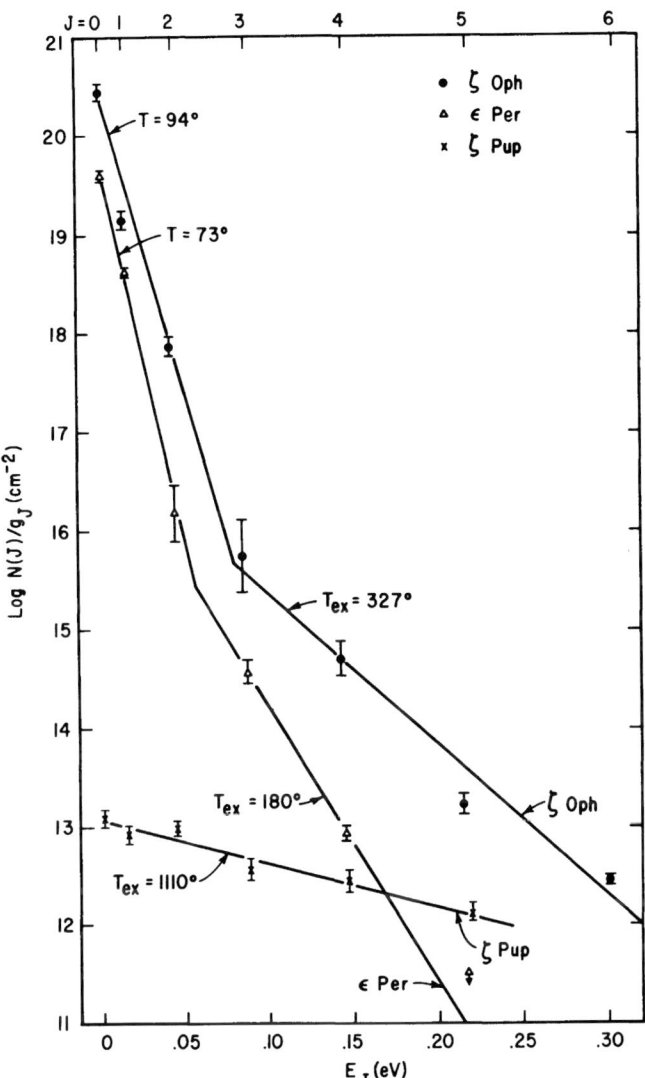

Fig. 4.9. Population of the different rotational J levels of H_2 determined from observation of electronic transitions starting from these levels. The observations have been made towards the stars ζ Ophiuchi, ζ Puppis and ϵ Persei. For each level we give the logarithm of the column density $N(J)$ divided by the statistical weight g_J, as a function of the excitation energy E_J of the level. At LTE the points corresponding to a given temperatures would then align on a straight line. Towards two of the three stars, there is apparently gas at LTE with two different temperatures. The colder temperature is probably the kinetic temperature of the medium. The warmer temperature may be due to a shock, to radiative cascades following a transition to a high excited level due to absorption of an UV photon, or perhaps to formation of H_2 on dust grains and expulsion from their surface in an excited state. Reproduced from Spitzer & Cochran [489], with the permission of AAS.

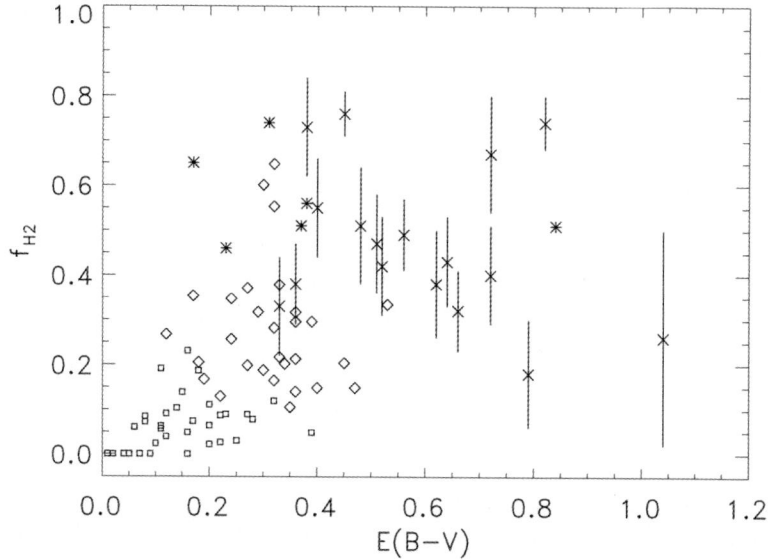

Fig. 4.10. Fraction $f = 2N(H_2)/[N(H\,{\sc i}) + 2N(H_2)]$ of molecular hydrogen as a function of colour excess. Crosses: data from FUSE; asteriks: data from FUSE with no independent measurement of $N(H\,{\sc i})$; diamonds: data from COPERNICUS with $N(H_2) > 10^{20}$ cm^{-2}; squares: data from COPERNICUS with $N(H_2) < 10^{20}$ cm^{-2}. The color excess is proportional to the total gas column density $N(H\,{\sc i} + H_2) = N(H\,{\sc i}) + 2N(H_2)$ with a good accuracy, as shown by COPERNICUS and FUSE. Reproduced from Rachford et al. [421], with the permission of the AAS.

The electronic transitions of H_2 in the far-UV have been observed by several rockets and satellites. The abundance of H_2 is so large, even in the diffuse interstellar medium, that its strongest lines are dominated by the damping wings and are thus on the damping portion of the curve of growth (see Fig. 4.7). It is then possible to obtain column densities through (4.35). If the observations have a sufficient wavelength resolution, it is possible to separate the different rotation levels of the fundamental electronic and vibrational state (the upper states are generally little populated in the diffuse medium). We can then obtain the populations of the rotational levels, that we can express in terms of excitation temperatures. An example is presented in Fig. 4.9. We can also determine the ortho/para abundance ratio.

The COPERNICUS and the FUSE satellites, and also the shuttle-borne telescope ORPHEUS, also allowed us to determine the H_2/H abundance ratio through simultaneous observations of the H_2 lines and of the Lyman α line in absorption towards stars (Savage et al. [449]; Rachford et al. [421]; Dixon et al. [127]). The result is presented in Fig. 4.10 for many lines of sight. We see that the gas is always partly molecular as soon as the extinction E(B−V) is larger than about 0.2 magnitude, corresponding to a total column density of $N(H\,{\sc i} + 2H_2) \simeq 10^{21}$ atom cm^{-2}. Since the observed stars are at distances from the Sun not very different from each other, the lines of sight for which $N(H\,{\sc i} + 2H_2)$ is large probably correspond to the crossing of rather high density clouds, which are expected to contain more molecular hydrogen. The COPERNICUS observations also show that molecular hydrogen is colder at larger densities. Its temperature is of the order of 80 K in clouds with $10^{19} < N(H) < 2 \times 10^{21}$ atom cm^{-2}, similar to the H I temperature.

4.2.3 Vibrational Transitions

The vibrational transitions of molecules occur between energy levels that result from the quantization of the possible modes of vibration. Most often these are *stretching modes* corresponding to variations in interatomic distances, the only possible vibration for diatomic molecules. More complex molecules also have *bending* and *deformation* modes. Vibrational energies are typically a fraction of an eV, and the corresponding wavelengths are in the near-infrared. These transitions are seen in emission, or in absorption if the background is bright enough. Each vibrational transition can be decomposed into rotational lines, forming a ro-vibrational band. Figure 4.11 shows an astrophysical example of a vibrational spectrum.

We can rather easily obtain the wavelengths of vibrational transitions for a diatomic molecule, such as H_2, from the potential curve of the corresponding electronic state (Fig. 4.8). In the vicinity of the minimum of the potential well V, where the interatomic distance is close to the equilibrium value R_e, this curve is parabolic and we can write the potential as

$$V(R) = V(R_e) + \frac{1}{2}(\partial^2 V/\partial R^2)_{R=R_e}(R - R_e)^2. \tag{4.36}$$

The solution of the vibration hamiltonian in this case is well known: it is the harmonic oscillator. The vibrational levels are quantized and given by

$$E_v = hc\omega\left(v + \frac{1}{2}\right), \quad \omega = (1/2\pi c)\sqrt{k/m_r}, \quad k = (\partial^2 V/\partial R^2)_{R=R_e}, \tag{4.37}$$

where $m_r = m_A m_B/(m_A + m_B)$ is the reduced mass of the system of the two atoms A and B, ω is the circular frequency of the transition and v the vibration quantum number. This simple formula gives a good approximation to the observed levels and is useful for predicting the vibrational frequencies of isotopically substituted molecules. In actual fact, the potential is not harmonic and for better accuracy we should use the development

$$E_v = hc\left[\omega_e\left(v + \frac{1}{2}\right) - \omega_e x_e\left(v + \frac{1}{2}\right)^2 + \omega_e y_e\left(v + \frac{1}{2}\right)^3 + \ldots\right]. \tag{4.38}$$

For the harmonic oscillator the selection rule for vibrational transitions is $\Delta v = \pm 1$, transitions with $\Delta v = \pm 2$ or 3 being much fainter.

These considerations can be extended to polyatomic molecules. These have several modes of vibration that can be considered as independent to a first approximation. Unfortunately we do not have space to discuss them further.

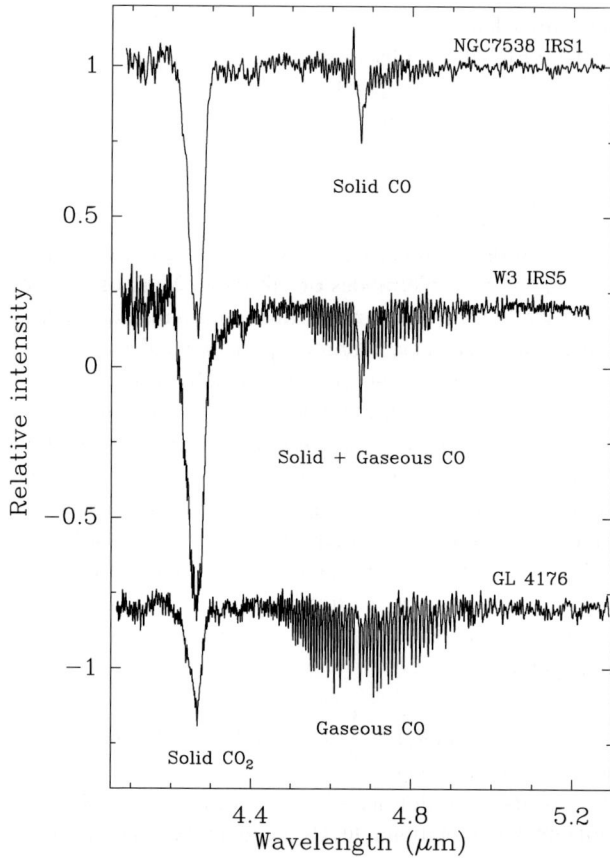

Fig. 4.11. The ro-vibrational band of the CO molecule at 4.7 μm observed in absorption in front of three very young stars embedded in a molecular cloud. Many rotation lines are located symmetrically with respect to the band centre. We also see in this spectrum broad bands of CO and CO_2 solid "ices" deposited on dust grains. We see that in this sequence there is less gaseous CO and more solid CO in going from the bottom to the top spectra, showing that gaseous CO was frozen onto the grains. From van Dishoeck et al. [531], with the permission of the authors.

The vibrational transitions of H_2 have been much observed from the ground and from space (Plate 17), in addition to those of CO and of H_2O. In shocks and photodissociation regions, excited vibrational levels can be populated either by collisions or by fluorescence (radiative cascades from electronic levels populated by absorption of ultraviolet photons). The ro-vibrational bands are then seen in emission. The required temperature for collisional excitation is at least 2 000 K, because the energy of the first excited vibrational transition of molecular hydrogen corresponds to $\sim$ 6 700 K. These bands can also be observed in absorption in front of an intense infrared continuum source and can be used, as per UV absorption lines, to determine the populations of the rotational levels of the electronic and vibrational ground state.

4.2.4 Rotational Transitions

The rotational transitions correspond to a quantization of the rotation of molecules. Rotation can be a global rotation around the principal axes of inertia, or some internal rotation for complex molecules. The energies associated with rotation are of the order of meV and the wavelengths are in the submillimetre to centimetre range, except for H_2 and HD. Other types of transitions to rotational transitions can occur at similar energies, for example the ammonia molecule NH_3, which is tetrahedral, can flip inside out like a glove (this is an *inversion* transition). In this case the energy is slightly different in the two configurations due to the antisymmetrical part of the wave function and the rotational levels are split. In the OH molecule, which possesses a single unpaired electron, and in similar molecules such as CH, the interaction between the orbital motion of this electron and the rotation of the molecule produces a splitting of the rotational energy levels called the Λ *doubling*. The OH lines, like those of many other molecules such as CN or HCN, are further split by hyperfine interactions (see Fig. 3.2).

Interstellar molecules can be extensively studied through their rotational transitions. We will first examine the diatomic and the linear molecules, the simplest case, and then the non-linear polyatomic molecules.

Rotation Spectra of Diatomic or Linear Molecules

For a linear molecule assumed to be rigid for the moment, the rotational energy is

$$H_{rot} = \frac{1}{2} I\omega^2 = \mathbf{J}^2/2I, \tag{4.39}$$

where I is the moment of inertia and $\mathbf{J}$ the angular momentum. For a diatomic molecule with a distance R_e between the two atoms A and B, the moment of inertia is

$$I = m_r R_e^2, \tag{4.40}$$

where $m_r = m_A m_B/(m_A + m_B)$ is the reduced mass. The solution of the Schrödinger equation for this system gives the values of the energy:

$$E_r/h = B_0 J(J+1), \tag{4.41}$$

with $B_0 = h/(8\pi^2 I)$. The quantum number J represents the angular momentum. This equation is strictly correct only for a rigid molecule. Actually molecules are deformable and rotation modifies the moment of inertia. The correct expression for the energy is

$$E_r/h = B_0 J(J+1) - D[J(J+1)]^2, \tag{4.42}$$

where the second term is due to centrifugal distorsion. It is 10^4 to 10^6 times smaller than the first term.

The selection rule for radiative dipolar transitions is $\Delta J = \pm 1$. The Einstein probability of spontaneous emission is

$$A_{ul} = \frac{64\pi^4}{3hc^3} v^3 |\mu_{ul}|^2, \tag{4.43}$$

where μ_{ul} is the element of the dipole moment matrix corresponding to the transition. For a transition $J + 1 \to J$ this element is

$$|\mu_{ul}|^2 = \mu^2 \frac{J+1}{2J+3}, \tag{4.44}$$

μ being the permanent electric dipole moment of the molecule. This yields the expression of $A_{J+1 \to J}$:

$$A_{J+1 \to J} = 1.165 \times 10^{-11} \mu^2 v^3 \frac{J+1}{2J+3}, \tag{4.45}$$

where A is in s^{-1}, v in GHz and μ in debye, the usual unit of dipole moment equal to 10^{-18} c.g.s. electrostatic unit. A most remarkable feature is the rapid increase of the spontaneous emission probability with the energy level, due to the factor v^3.

From (3.32) we directly obtain the general relation between the column density of the lower level, the optical depth and the excitation temperature of the transition:

$$N_l = 93.5 \frac{g_l v^3}{g_u A_{ul}} \frac{1}{[1 - \exp(-0.0480 v/T_{ex})]} \int \tau \, dv \text{ mol cm}^{-2}, \tag{4.46}$$

where v is the velocity in km s^{-1} and ν the frequency in GHz.

In the Rayleigh–Jeans approximation (only valid for centimetre and decimetre waves) this equation simplifies as

$$N_l = 1950 \frac{g_l v^2 T_{ex}}{g_u A_{ul}} \int \tau \, dv \text{ mol cm}^{-2}. \tag{4.47}$$

It is in general rather difficult to obtain the excitation temperature T_{ex}. As a first approximation when the density is large, thus assuming LTE, we can take $T_{ex} = T_K$.

The CO Molecule

CO is by far the most observed molecule in the interstellar medium (Plates 2, 5, 6, 7 et 8). In the case of CO the rotation lines are almost always optically thick and T_{ex} is close to the brightness temperature at the centre of the line. If we observe at the centre of the 2.3 mm (115 GHz) CO(1-0) line a "Rayleigh-Jeans" antenna temperature T_A^*

(cf. Sect. 3.1) above the blackbody radiation of the Universe ($T_{BB} = 2.73$ K), the brightness temperature is (for an optically thick line)

$$T_B^* = T_A^*/\eta_{mb} = \frac{h\nu}{k}\left[\frac{1}{\exp(h\nu/kT_{ex}) - 1} - \frac{1}{\exp(h\nu/kT_{BB}) - 1}\right]. \quad (4.48)$$

A useful expression for T_{ex} can be derived from this equation. For the CO(1-0) line we have numerically, without the Rayleigh–Jeans approximation:

$$T_{ex} = 5.53\left[\ln\left(\frac{1}{T_B^*/5.53 + 0.151} + 1\right)\right]^{-1} \text{ K.} \quad (4.49)$$

Let us now show how we can obtain the column density, and hence the abundance, of CO. A simple but disputable way is to assume that the important rotational energy levels of CO and of all its isotopic varieties, such as ^{13}CO, are in LTE at the temperature $T_K = T_{ex}$ determined for CO. Let us consider ^{13}CO, assuming further that its lines are optically thin because it is much less abundant than CO. We can then obtain the column density $N(0)$ of the fundamental level of ^{13}CO from a measurement of its brightness temperature at the centre of its (1-0) line ((4.46) in which T_{ex} is replaced by T_K). The total column density N_{tot} of ^{13}CO can then be obtained using the partition function $Q(T_K)$ (3.71). For a diatomic molecule at LTE, and in particular for CO, we have

$$Q(T_K) = N_{tot}/N(0) = \sum_n g_n \exp\left(\frac{-E_n}{kT_K}\right) \simeq 2\frac{kT}{h\nu_{1-0}} \simeq \frac{T_K}{2.8 \text{ K}} \text{ for CO.} \quad (4.50)$$

Since in the interstellar medium near the Sun (Wilson & Rood [551])

$$N(^{12}\text{CO}) \simeq 76\, N(^{13}\text{CO}), \quad (4.51)$$

we can estimate $N_{tot}(^{12}\text{CO})$. Then, assuming that about 20% of the interstellar carbon is in CO we can derive the column density of H_2, $N(H_2) \simeq 10^5 N(\text{CO})$. If the lines of ^{13}CO are suspected to be optically thick, it is preferable to use a rarer molecule like $C^{18}O$, with $N(^{12}\text{CO}) = 560\, N(C^{18}O)$.

Although the preceding method is currently used to estimate the column density of H_2 in dense regions, where extinction prevents observation of this molecule in the UV, it is quite uncertain and can only give order of magnitude estimates of the column densities of CO and H_2. In order to do better, a LVG analysis is recommended. Figure 4.12 shows two examples of diagrams corresponding to different kinetic temperatures, which show how this analysis can be performed in practice. Similar diagrams for CS can be found in Rohlfs & Wilson [439], p. 406. The difficulty with the LVG method is that it is necessary to make hypotheses about the density and/or the kinetic temperature that are not always easy to check. Another method consists of comparing observations with the predictions of a model for photodissociation regions. This makes sense because such models show that the emission of the ^{12}CO lines is dominated by the region where CO starts to be photodissociated (see Chap. 10).

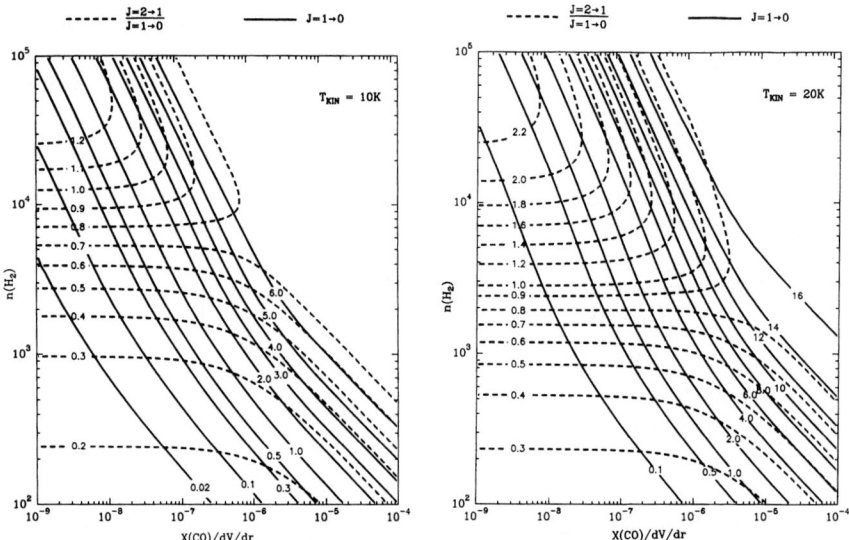

Fig. 4.12. Large velocity gradient (LVG) analysis of the two first rotational transitions of CO. The ratios of $X(CO) = n(CO)/n(H_2)$ to the velocity gradient dV/dR in km s^{-1}pc^{-1} are given in the abscissae, and the density of molecular hydrogen in mol. cm^{-3} as the ordinates. The kinetic temperature is fixed at 10 K in the left diagram, and at 20 K in the diagram to the right. The full curves give the brightness temperature T_B^* of the CO(1-0) line and the dashed curves give the brightness temperatures ratio between the CO(2-1) and the CO(1-0) lines. The line optical depth increases towards the top and right of the diagrams. From Castets et al. [80], with the permission of ESO.

Many other diatomic or linear molecules can be treated in the same way as CO. In order to guess if a given transition of a molecule can be considered in LTE, it is useful to know the corresponding critical density. Table 4.4 gives this, and other, parameters for some transitions important in the interstellar medium. As expected, the critical density increases with increasing dipole moment of the molecule, so that molecules like CS or HCN with their higher dipole moments allow us to explore higher density ranges than for CO. The critical density also increases strongly with J, due to the strong increase in $A_{J+1 \to J}$.

The H$_2$ Molecule

Rotational lines of H$_2$ have been observed with the ISO satellite in relatively warm regions of the interstellar medium. The wavelengths and spontaneous emission probabilities for the most important lines are listed in Table 4.5. The levels of para–H$_2$ have $J = 0, 2, 4...$ and those of ortho–H$_2$ have $J = 1, 3, 5....$ The S(0) line at 28.2 μm connects levels 0 and 2, the S(2) line at 12.3 μm levels 2 and 4, etc., while the S(1) line at 17.0 μm connects levels 1 and 3, the S(3) line at 9.6 μm levels 3 and 5, etc. The selection rules are different for these quadrupole transitions with respect

Table 4.4. Parameters for the rotational lines of diatomic or linear molecules. Only lines observable from the ground are indicated. μ is the dipole moment of the molecule and E_u/k the energy of the upper level of the transition, in K. n_{crit} is the critical density of hydrogen molecules such that the collisional and radiative de-excitation of the upper level are comparable, the radiation field in the line being assumed to be small. This critical density is $A_{ul}/\langle\sigma_{ul}v\rangle$ (cf. (3.44)). n_{crit} is given for a temperature of 100 K for CO and CS, and of 30 K for HCO$^+$, HCN and HNC. Its value must only be considered as an order of magnitude estimate.

Molecule	μ debye	Transition	Frequency GHz	E_u/k K	A_{ul} s^{-1}	n_{crit} cm^{-3}
CO	0.112	1-0	115.271 203	5.5	7.4×10^{-8}	3×10^3
		2-1	230.538 001	16.6	7.1×10^{-7}	1×10^4
		3-2	345.795 975	33.2	2.6×10^{-6}	5×10^4
CS	1.95	1-0	48.990 964	2.4	1.8×10^{-6}	1×10^5
		2-1	97.980 968	7.1	2.2×10^{-5}	7×10^5
		3-2	146.969 049	14.1	6.1×10^{-5}	2×10^6
		5-4	243.935 606	35.2	2.9×10^{-4}	8×10^6
HCO$^+$	4.07	1-0	89.188 518	4.3	3.0×10^{-5}	1.5×10^5
		3-2	267.557 625	25.7	1.0×10^{-3}	3×10^6
HCN	2.98	1-0 F(2-1)	88.631 847	4.3	2.4×10^{-5}	4×10^6
		3-2	265.886 432	25.5	8.5×10^{-4}	1×10^7
HNC	3.05	1-0 F(2-1)	90.663 574	4.3	2.7×10^{-5}	4×10^6
		3-2	271.981 067	26.1	9.2×10^{-4}	1×10^7

to the dipole ones. We can use the ratios between these lines, which are generally optically thin, to obtain the kinetic temperature and the ortho/para abundance ratio. LTE can be assumed as a first approximation. This allows us to easily calculate the line intensities, which are given by

$$I_{ul} = \frac{h\nu}{4\pi} x(u) N(\text{H}_2) A_{ul}, \quad (4.52)$$

where $x(u)$ is the fraction of molecules in the level u. At LTE, this fraction is given by

$$x(J) = \frac{(2J+1)g_J \exp(-h\nu/kT_K)}{Q(T_K)}. \quad (4.53)$$

The statistical weight g_J is 1 for the para species (even J) and 3 for the ortho species (odd J). $Q(T_K)$ is the partition function

$$Q(T_K) = \sum_J (2J+1)g_J \exp(-h\nu_J/kT_K). \quad (4.54)$$

This function is 1.00 at $T_K = 10$ K, 1.53 at 20 K, 2.67 at 100 K and 12.51 at 500 K. The three preceding equations allow us to calculate the line intensities, which are very temperature-dependent. For example, for the 28 μm line we have

Table 4.5. Vacuum wavelengths and spontaneous emission probabilities of some rotational transitions of the H$_2$ molecule.

	ortho-H$_2$			para-H$_2$	
Transition	λ μm	A_{ul} s^{-1}	Transition	λ μm	A_{ul} s^{-1}
v=0-0, S(0)	28.21883	2.95×10^{-11}	v=0-0, S(1)	17.03483	4.77×10^{-10}
v=0-0, S(2)	12.27861	2.76×10^{-9}	v=0-0, S(3)	9.66491	9.86×10^{-9}
v=0-0, S(4)	8.02505	2.65×10^{-8}	v=0-0, S(5)	6.90952	5.89×10^{-8}
v=0-0, S(6)	6.10856	1.14×10^{-7}	v=0-0, S(7)	5.51116	2.01×10^{-7}

$$I(28\mu m) = 6.56 \times 10^{-4} A_V \exp(-510/T_K)/Q(T_K) \text{ erg s}^{-1} \text{ cm}^{-2} \text{ sterad}^{-1}, \quad (4.55)$$

where A_V is the visual extinction related to the column density of H$_2$ by the standard galactic relation $2N(H_2) = 1.59 \times 10^{21} A_V$ cm^{-2} mag^{-1} (see later Sect. 7.1). Even with 30 mag. of extinction, corresponding to a very optically thick molecular cloud, this line is only detectable in emission if the temperature is larger than 60–80 K.

Rotation Spectra of Polyatomic Molecules

A non-linear molecule can be considered as a first approximation as a rigid rotator defined by three principal moments of inertia $I_{\alpha\alpha}$, with $\alpha = x, y, z$. The classical expression for the kinetic energy of rotation is

$$E_K = \frac{1}{2}\sum_{\alpha=1}^{3} I_{\alpha\alpha}\omega_\alpha^2 = \frac{1}{2}\sum_{\alpha=1}^{3} J_\alpha^2/I_{\alpha\alpha}, \quad (4.56)$$

where ω_α is the angular velocity around axis α and J_α is the component of the angular momentum on this axis. The quantum expression for the hamiltonian is very similar:

$$H = \frac{1}{2}\sum_{\alpha=1}^{3} J_\alpha^2/I_{\alpha\alpha}, \quad (4.57)$$

where the J_α are now the components of the operator for the angular momentum $-i\hbar\nabla$ in the rotating reference frame of the molecule. These components obey commutation rules inverse to the classical ones:

$$[J_x, J_y] = -i\hbar J_z. \quad (4.58)$$

On the other hand they commute with the components of the angular momentum J_X, J_Y et J_Z in a fixed frame with axes $OXYZ$. These considerations allow us to obtain the rotation energy in different cases.

- Spherical rotator: $I_{xx} = I_{yy} = I_{zz} = I$.

$$H = J^2/(2I). \tag{4.59}$$

$$E = hcB_e J(J+1), \text{ with} \tag{4.60}$$

$$B_e = \frac{h}{8\pi^2 I}. \tag{4.61}$$

Since J^2 commutes with J_z and J_Z, the degeneracy (statistical weight) of level J is $(2J+1)^2$.
- Symmetrical rotator: $I_{xx} = I_{yy}$ (examples: NH_3, CH_3CN).

$$H = \frac{J_x^2 + J_y^2}{2I_{xx}} + \frac{J_z^2}{2I_{zz}} = \frac{J^2 - J_z^2}{2I_{xx}} + \frac{J_z^2}{2I_{zz}}. \tag{4.62}$$

We can then find the eigenvalues of H, J^2 et J_z. The rotational state is defined by two quantum numbers, J and K, the latter being the quantum number of the projection of the angular momentum on the rotating axis Oz. We then obtain the proper values of the energy

$$F(J, K) = B_{xx}\left[J(J+1) - K^2\right] + B_{zz}K^2. \tag{4.63}$$

The corresponding degeneracy is $(2 - \delta_{0,K})(2J+1)$.
- Linear rotator.
This case has been studied in the previous section. It can be derived from the case of the spherical rotator by setting $K = 0$.
- Asymmetrical rotator. This is the general case (example: H_2O).
The three moments of inertia are different and the hamiltonian commutes neither with J^2 nor with J_z. The expressions for the rotation terms are complex and there are three rotational quantum numbers.

If we want more precision in the energies, we must develop the terms in series as follows (Dunham development):

$$T_{vj} = \sum_{l,m} Y_{lm}\left(v + \frac{1}{2}\right)^l J^m(J+1)^m, \tag{4.64}$$

where the Y_{lm} are correcting factors which depend on the vibrational quantum number v and on the rotational constants.

Let us now examine a few examples of non-linear molecules and their applications to the interstellar medium.

The NH_3 Molecule

Radio observations of several molecules with favorable energy levels can be used in order to determine the density and the temperature. Examples are CH_3CN, HC_3N,

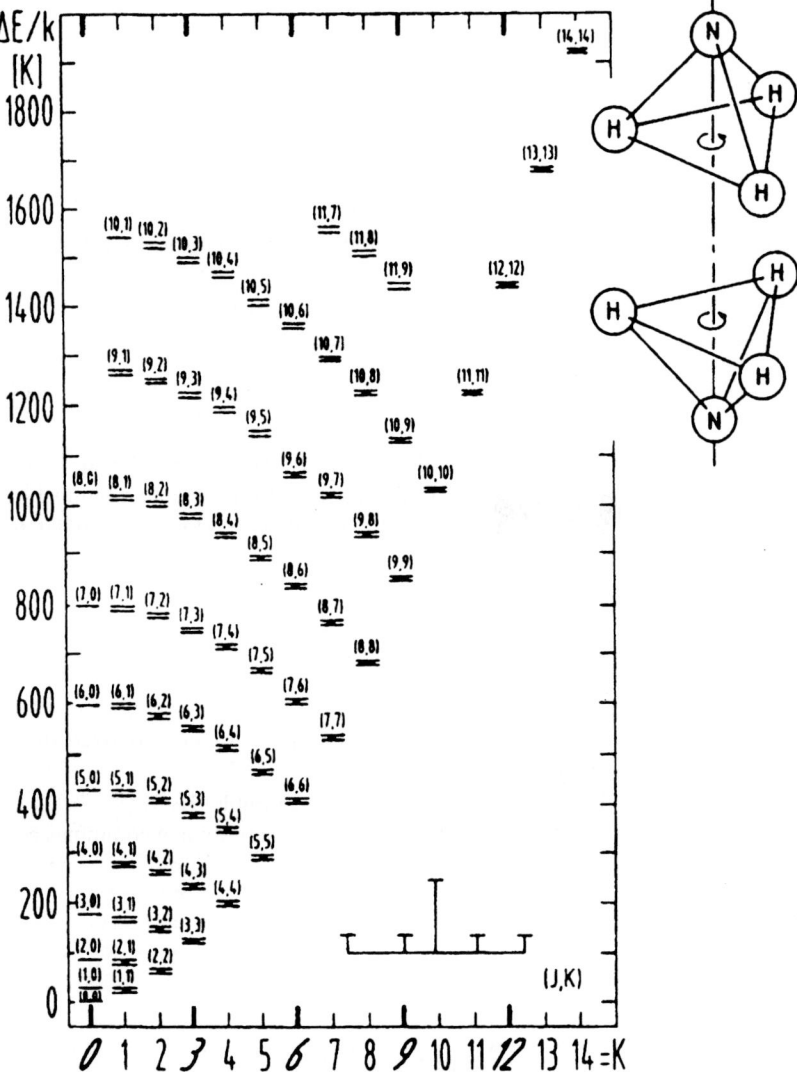

Fig. 4.13. Rotational energy level diagram for ammonia NH_3 in the ground electronic and vibrational states. The quantum number K is the abscissa and the energy expressed in units of degrees K is the ordinate. The levels are labelled by the values of the quantum numbers J and K. Each level is split by the inversion transition between the two configurations of the molecule represented at the top right. Moreover, each transition has a hyperfine structure due to the interaction between the spin of the nitrogen nucleus and the electrons. This structure is shown schematically at the bottom of the figure. The separation between the hyperfine components is of the order of 1 MHz. The ortho–NH_3 has $K = 0, 3, 6...$ (figures in italics) and K takes the other values for the para–NH_3. From Rohlfs & Wilson [439].

CH$_3$OH, and NH$_3$. We will only discuss the last molecule. NH$_3$ is a symmetrical rotator for which the rotation energy is thus determined by two quantum numbers J and K. Figure 4.13 illustrates the structure of the energy levels. The selection rules for electric dipole transitions are $\Delta K = 0$, and $\Delta J = 0, \pm 1$. As a consequence, transitions between the different K scales are forbidden, except for weak transitions with $\Delta K = 3$. The lowest levels of each scale (those with $J = K$) are thus metastable, while the upper levels ($J > K$) are not metastable and de-excite rapidly to levels with lower values of J. The corresponding lines are in the far infrared. Given the two possible orientations of the spins of the H atoms there are two different species of NH$_3$: ortho–NH$_3$ for which $K = 3n$, n being an integral number, with all H spins parallel, and para–NH$_3$ with $K = 3n \pm 1$, the H spins being not all parallel. As always, the ortho–para transitions are highly forbidden, even collisionally. Collisional transitions are possible between any other values of J and K, transitions with $\Delta K = 3$ being favored.

All rotational levels except those with $K = 0$ are split by the inversion. All the (allowed) transitions between inversion sub-levels of the same rotational state lie in the frequency range 21-25 GHz. The dipole moment of an inversion transition defined by the values of J and K is given by

$$|\mu_{JK}|^2 = \mu^2 \frac{K^2}{J(J+1)}, \tag{4.65}$$

with $\mu = 1.476$ debye. The spontaneous emission probability can be derived from this dipole moment using (4.43).

Observations of these inversion transitions, provided that they are in LTE (a frequent case) or at least that the density is known, allow us to obtain the population of the corresponding J, K level. Their optical thickness can be checked through observation of the hyperfine structure which is easily resolved. These very favorable circumstances permit an accurate determination of the populations of the different J and K levels, from which the temperature and the density can be derived as follows.

Since the population of the metastable levels is entirely defined by collisions, the ratios between their populations is almost only a function of temperature. On the other hand, the population of the levels J of a given K scale is a function of density, their critical densities being of the order of 10^7 H$_2$ molecules cm^{-3}. The critical densities for the inversion transitions are $10^3 - 10^4$ cm^{-3}, so that with NH$_3$ it is possible to explore this range of densities. The problem is that calculations of the collisional cross-sections between NH$_3$ and H$_2$ are difficult and that the cross-sections are different for ortho– and para–H$_2$, the ratio of which must be known (or rather guessed in practice).

The H$_2$CO Molecule

Formaldehyde H$_2$CO is an asymmetric rotator, but its asymmetry is small so that its energy diagram is not very different from that of a symmetric rotator (Fig. 4.14). It is defined by three quantum numbers: J, K_a (principal) and K_c (secondary); the

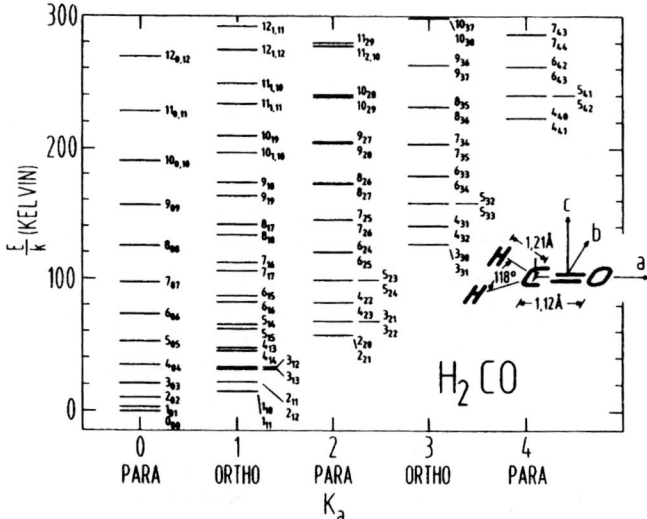

Fig. 4.14. Rotational energy level diagram for formaldehyde H_2CO in the ground electronic and vibrational state. The quantum number K_a is the abscissa and the energy expressed in units of degrees K is the ordinate. The levels are labelled by the values of the quantum numbers J, K_a and K_c. The configuration of the molecule is drawn on the right, with the axes **a** and **c** on which K_a and K_c are projected. From Rohlfs & Wilson [439].

direction **a** is along the axis C=O on which the dipole moment of the molecule is aligned, and **c** is perpendicular to the plane of the molecule. With its two H atoms, H_2CO has ortho and para forms that correspond respectively to odd and even values of K_a. The first transition observed historically is the 3_{12} - 3_{13} line of ortho-formaldehyde at 6 cm. Its physics is complex as it can be guessed by looking at Fig. 4.14. This line is often an anti-maser pumped by collisions. Its lower level is then overpopulated and leads to an anormalously enhanced absorption. We can also observe transitions of the $K_a = 0$ scale (para–H_2CO). They are easier to interpret but their critical density is much higher than that of the 6-cm line.

The H_2O Molecule

In spite of its apparent simplicity, the water molecule has an extremely complex rotation spectrum. This is due to the fact that it is a true asymmetrical rotator, the three quantum numbers J, K_a and K_c being of comparable importance. Moreover its dipole moment is along axis **b** (Fig. 4.15) implying that both K_a et K_c must change in a permitted transition. Like the previous molecules, H_2O exists in ortho and para forms. Observationally H_2O is a very difficult molecule because the Earth's atmosphere contains abundant water vapor and is totally opaque except for the 6_{16} - 5_{23} transition at 22 GHz and to a lesser extent the 3_{13} - 2_{20} transition at 180 GHz, which is more difficult to observe. Both transitions are maser, so that it is hopeless

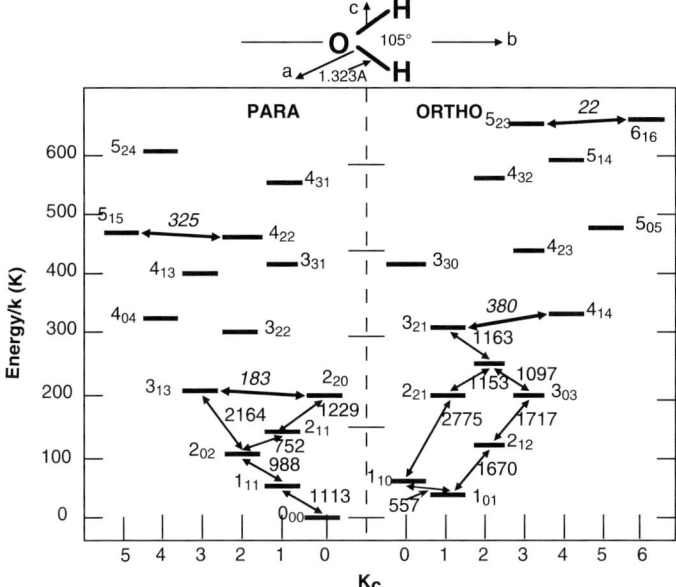

Fig. 4.15. Rotational energy level diagram for water H_2O in the ground electronic and vibrational state. The quantum number K_c is the abscissa and the energy expressed in units of degrees K is the ordinate. The levels are labelled by the values of the quantum numbers J, K_a and K_c. The configuration of the molecule is drawn above, with the axes **a** and **c** on which K_a and K_c are projected. Due to the two identical H nuclei, the diagram is split into two parts, ortho and para, between which there are no allowed radiative and collisional transitions under interstellar conditions. The frequencies of the most interesting transitions are indicated in GHz. All lower transitions are in the submillimetre range. The transitions with bold arrows are maser.

to attempt to obtain accurate column densities from them. The most interesting observations of H_2O are from satellites. ISO has observed a number of transitions, but did not cover wavelengths longer than 200 μm. The fundamental transition $1_{11} - 0_{00}$ of para–H_2O at 269.5 μm (1 113 GHz) has not yet been observed. The fundamental transition $1_{10} - 1_{01}$ of ortho–H_2O at 538.6 μm (557 GHz) was observed recently with the SWAS and ODIN satellites. These observations show that H_2O is not very abundant in molecular clouds, probably because it condenses easily as ice on the dust grains.

4.2.5 The Diffuse Interstellar Bands

Aside from the absorption lines discussed in Sect. 4.1, which are always very narrow except the highly saturated H and H_2, lines, we observe in the spectra of stars about 200 absorption bands, generally weak, with widths between 0.1 and a few Å. They are called the *Diffuse Interstellar Bands* (DIBs). They are found essentially in the visible, with a few in the near-IR: see Herbig [232] for a general review of the diffuse bands. Though the diffuse interstellar bands were discovered as early as 1934, their origin is still a mystery. It has been long considered that they come from absorption by impurities in dust grains or ice mantles on these grains, but the present opinion is that they are rather molecular absorption bands because we observe, in some bands, a characteristic rotation–vibration molecular structure (Fig. 4.16). The relative intensities of many diffuse bands vary from line of sight to line of sight. This has allowed to class them in families, the members of a given family varying together in the different directions.

There is a clear correlation between band intensity and extinction A_V (see Sect. 7.1 for a discussion of extinction). This correlation is not surprising because it is solely the effect of the amount of intervening interstellar matter in front of the different observed stars. In order to constrain the nature of the absorbers, observers have searched for correlations between DIBs and the parameters that define the shape of the extinction curve in the ultraviolet (see Sect. 7.1), unfortunately with rather disappointing results. There is however a loose correlation between the absorption band at 2 175 Å and the extinction excess in the far-UV, which suggests that the absorbers might be carbonaceous molecules, for example polycyclic aromatic hydrocarbons or PAHs (Sect. 7.2; Désert et al. [121]), particularly ionized species. However, the lack of laboratory data on isolated gas–phase PAH^+ prevents firm identifications. Other identifications have been proposed, for example fullerenes[3]. None of these identifi-

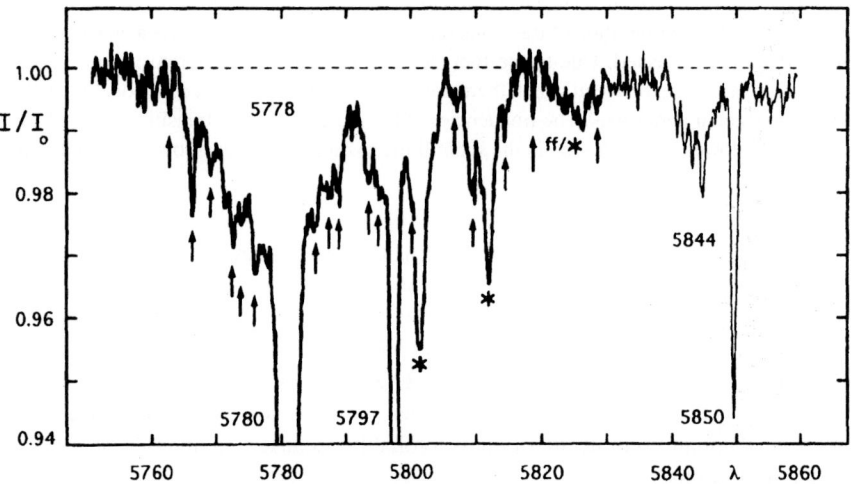

Fig. 4.16. Profiles of diffuse interstellar bands near 5780 Å. This spectrum is the average of the spectra towards five different stars observed with a high signal to noise ratio. The arrows point to absorption structures that may be ro-vibrational bands of molecules. The structures indicated by a star symbol are produced in the terrestrial atmosphere. From Jenniskens et al. [264], with the permission of ESO.

---

[3] Fullerenes are spherical or ellipsoidal shells formed with hexagonal and pentagonal aromatic cycles, the simplest of which is C_{60} with the structure of a soccer ball.

cations is fully convincing. Still, the absorbing molecules should be rather abundant, as we will now show. The strongest band at 4430 Å has an equivalent width W_{4430} related to visual absorption A_V by the relation $W_{4430}/A_V \simeq 0.8$ Å mag^{-1}. Since the absorber cannot have an oscillator strength f much larger than 1 we see from (4.28) that its column density must be such that $N_{mol} \geq 4.6 \times 10^{12}$ cm^{-2} mag^{-1}. Using (7.5), which relates the total column density of hydrogen N_H to $E(B-V)$ and taking $A_V/E(B-V) = 3.1$, we obtain

$$N_{mol}/N_H > 2.5 \times 10^{-9}, \qquad (4.66)$$

a relation that is true whatever the band carrier. If we assume as an example that there are 20 atoms other than H per molecule, the total number of absorbing atoms is larger than 5×10^{-8} times the number of interstellar H atoms. The same reasoning, applied to all the diffuse bands that have a combined equivalent width of about 5.2 Å per magnitude, leads to a total abundance of absorbing atoms larger than about 3×10^{-7} that of H. This is more than the atoms contained in most interstellar molecules, except the abundant species such as CO or OH, but if the atoms are carbon atoms the DIBs require only a few 10^{-3} of the total abundance of carbon. This does not provide a strong constraint for models that attempt to precise the origin of the diffuse interstellar bands.

5 The Ionized Interstellar Gas

The interstellar gas can be ionized by the far-UV radiation of hot stars or by other mechanisms such as collisional ionization in shocks, X-ray ionization or ionization by high-energy charged particles. In general we define three kinds of ionized interstellar medium, between which the distinction is not always completely clear: *HII regions* or *gaseous nebulae* (Plates 1, 9, 10, 14, 16, 17, 18, 30 and 31) which are in principle well-defined entities surrounding one or several hot stars; the *diffuse ionized medium*; and the *hot interstellar medium* which originates mainly from supernova remnants and bubbles that permeate the general medium. We will discuss these three components in turn. As far as H II regions are concerned, we do not intend to reach the same degree of detail as the specialized book of Osterbrock [389]. *Planetary nebulae* are formed from material ejected by intermediate-mass stars (about 1.5 to 8 M$_\odot$) at the end of their lifes. This material is ionized by the very hot remnant of the star. These nebulae are often considered as part of the interstellar medium. We will not discuss them here. Their physics is rather similar to that of H II regions and is treated by Osterbrock. We will examine supernova remnants and bubbles in Chap. 12, as well as the dynamics of H II regions.

5.1 H II Regions

Ideally, an H II region is an ionized sphere surrounding a hot star or a cluster of hot stars. This sphere will be considered in stationary state in the present chapter. Its dynamics will be examined in Sect. 12.3. Although this simple case is rarely encountered in nature it is interesting to discuss it because it is the only one that can be easily modelled. We will first study the photoionization of hydrogen and helium, then the different kinds of emission from H II regions: continuum, optical and radio recombination lines, and fine-structure forbidden lines.

5.1.1 Theory of Photoionization: the Strömgren Sphere

Considering atoms of a given species we designate their discrete energy levels by index j, and the ensemble of continuous levels resulting from the ionisation by index k (obviously levels of higher energy than the j levels). In the presence of ionizing radiation, let us call P_{jk} the probability of photoionization per second from level j, and n_j the population of level j. n_k is the population of level k and P_{kj} the probability of recombination with a free electron to level j. Assuming ionization equilibrium, we can write

$$n_k \sum_j P_{kj} = \sum_j n_j P_{jk}. \tag{5.1}$$

Letting σ_j be the cross-section for capture of an electron to level j, we have

$$P_{kj} = n_e \langle v\sigma_j \rangle = n_e a_j, \tag{5.2}$$

where n_e is the density of free electrons and v their mean velocity. We assume a maxwellian distribution of electron velocities with a temperature T_e.

Similarly, if $s_j(\nu)$ is the photoionization cross-section from level j by photons with frequency ν, $n_{phot}(\nu) = u_\nu/h\nu$ being the density of these photons (u_ν is the radiation energy density), then

$$P_{jk} = \int_\nu cn_{phot}(\nu)s_j(\nu)\,d\nu = b_j. \tag{5.3}$$

We can now write the ionization equilibrium equation for the two consecutive ionic states of the considered element, with ionizations r and $r+1$:

$$n_{r+1} n_e \sum_j a_j = \sum_j n_{r,j} b_j. \tag{5.4}$$

Radiative cascades from levels j to the fundamental level of the atom or ion being generally very fast, of the order of a few 10^{-9} second, and collisional excitations from the fundamental to excited j levels being generally negligible, we can consider that ionizations are only from the fundamental level, so that

$$n_{r+1} n_e \sum_j a_j \simeq n_r b_1. \tag{5.5}$$

In the case of hydrogen there is a further simplification: because hydrogen supplies most of free electrons, $n_e \simeq n_i$, n_i being the density of the hydrogen ions. If n_0 is the density of hydrogen atoms and x its degree of ionization,

$$n_H = n_0 + n_i, \tag{5.6}$$

$$x = n_i/n_H = n_e/n_H, \tag{5.7}$$

$$a = \sum_j a_j, \tag{5.8}$$

the equation of ionization equilibrium is then

$$x^2 n_H^2 a = (1-x) n_H b_1, \tag{5.9}$$

hence

$$\frac{1-x}{x^2} = \frac{n_H a}{b_1}. \tag{5.10}$$

Let us now consider a hydrogen cloud with uniform density n_H surrounding an ionizing star at the centre. The absorption coefficient of hydrogen in the Lyman continuum is

$$\kappa_\nu = s(\nu) n_0 = 6.6 \times 10^{-18} (\nu_1/\nu)^3 n_0, \qquad (5.11)$$

ν_1 being the frequency of the Lyman discontinuity:

$$\nu_1 = 3.288\,051 \times 10^{15} \text{ Hz}, \quad \lambda_1 = 911.763 \text{ Å}, \quad h\nu_1 = 13.605\,7 \text{ eV}. \qquad (5.12)$$

At a distance r from the star, the optical depth in the Lyman continuum is, neglecting absorption by dust

$$\tau_\nu = \int \kappa_\nu \, d\nu = 6.6 \times 10^{-18} (\nu_1/\nu)^3 \int_0^r n_0 \, dr. \qquad (5.13)$$

If $n_0 = 1\text{cm}^{-3}$, $\tau_\nu(\nu_1) = 1$ at only 0.05 pc from the star. The absorption is thus very abrupt if the hydrogen is not totally ionized, except where the density is extremely small. We can see that the ionization is almost complete out to a radius r_S, beyond which there are no ionizing photons, those inside this radius being absorbed by the few neutral atoms resulting from recombinations. The gas is neutral outside this radius r_S and ionized inside, forming the *Strömgren sphere*. Let us now calculate the radius r_S for the total density n_H and the ionizing flux $S(0)$ emitted by the star over 4π steradians. The ionizing flux $S(r_S)$ drops to zero at the surface of the Strömgren sphere. The number of ionizations per second and per cm³ is from (5.9)

$$(1-x) n_H b_1 = x^2 n_H^2 a \simeq n_H^2 a, \qquad (5.14)$$

since hydrogen is almost totally ionized in the H II region ($x \simeq 1$).

We have to take into account the fact that any recombination to the fundamental level yields a photon which will ionize the next neutral atom. Such recombinations must not be accounted for. Only recombinations to higher levels ($j \geq 2$) need be considered, and the recombination coefficient a must then be replaced by (Hummer & Seaton [251])

$$a^{(2)} = \sum_{j>1} a_{kj} = 1.627 \times 10^{-13} T_4^{-1/2} \left(1 - 1.657 \log T_4 + 0.584 T_4^{1/3}\right) \text{ cm}^3 \text{ s}^{-1}, \qquad (5.15)$$

with $T_4 = T/10^4 \text{K}$.

The radius r_S is such that the number of recombinations inside the volume is equal to the number of ionizations, so that

$$r_S = \left[\frac{3 S(0)}{4\pi n_H^2 a^{(2)}}\right]^{1/3}. \qquad (5.16)$$

The quantity

$$U = r_S n_H^{2/3} = \left[\frac{3 S(0)}{4\pi a^{(2)}}\right]^{1/3}, \qquad (5.17)$$

expressed in pc cm$^{-2/3}$, is called the *ionizing power* of the star and is often tabulated.

Table 5.1 gives the number $S(0)$ of photons in the Lyman continuum of hydrogen and $S(1)$ in the Lyman continuum of helium (see later) emitted by different types of hot stars, calculated from stellar atmosphere models by Schaerer & de Koter [456]. We can readily derive U from this table. For hydrogen, the results of these models are not much different from those of Panagia [395] which are widely used, but they are very different for helium.

Table 5.1. Fluxes of ionizing photons $S_0 = N_{LyC}(\text{H\,\sc{i}})$ and $S_1 = N_{LyC}(\text{He\,\sc{i}})$ in the hydrogen and helium Lyman continua for various types of hot stars with solar abundances, from Schaerer & de Koter [456].

Sp. type	V(dwarf)			III(giant)			I(supergiant)		
	$\log T_{eff}$ K	$\log S_0$ s^{-1}	$\log S_1$ s^{-1}	$\log T_{eff}$ K	$\log S_0$ s^{-1}	$\log S_1$ s^{-1}	$\log T_{eff}$ K	$\log S_0$ s^{-1}	$\log S_1$ s^{-1}
O3	4.710	49.85	49.42	4.707	49.97	49.52	4.705	50.09	49.63
O4	4.687	49.68	49.23	4.683	49.84	49.38	4.678	50.02	49.56
O4.5	4.676	49.58	49.12	4.670	49.78	49.32	4.665	49.98	49.53
O5	4.664	49.48	49.01	4.657	49.71	49.25	4.650	49.94	49.47
O5.5	4.652	49.38	48.86	4.644	49.64	49.16	4.636	49.88	49.35
O6	4.639	49.28	48.75	4.630	49.56	49.05	4.620	49.81	49.24
O6.5	4.626	49.17	48.62	4.615	49.47	48.91	4.604	49.73	49.12
O7	4.613	49.05	48.44	4.601	49.36	48.75	4.588	49.64	48.91
O7.5	4.599	48.93	48.25	4.585	49.24	48.53	4.571	49.53	48.65
O8	4.585	48.80	48.05	4.569	40.09	48.14	4.553	49.42	48.37
O8.5	4.570	48.64	47.74	4.553	48.94	47.80	4.534	49.29	48.05
O9	4.555	48.46	47.37	4.536	48.76	47.40	4.515	49.12	47.67
O9.5	4.539	48.25	46.92	4.518	48.56	46.95	4.495	48.90	47.21
B0	4.523	48.02	46.41	4.499	48.33	46.47	–	–	–
B0.5	4.506	47.77	45.86	4.479	48.11	46.03	–	–	–

Photons with energy only slightly higher than 13.6 eV are mostly absorbed by hydrogen. Those with higher energies can be absorbed by helium, nitrogen, etc. In practice, the most efficient element at a given photon energy is the one which has an ionization threshold immediately smaller than this energy. This results from the fast variation in ν^{-3} of the photoionization cross-sections. A consequence of this is the formation of an ionization structure in relatively uniform gaseous nebulae.

Photons with an energy larger than 24.6 eV, the ionization potential of helium, produce a region of ionized helium in the inner zone of the H\,\sc{ii} region. Table 5.1 gives the number of helium-ionizing photons as a function of the spectral type of the central star. If this star is hot enough, the He\,\sc{ii} zone is co-extensive with the H\,\sc{ii} one. This occurs when $S_1/S_0 > 0.1$, i.e. for stars hotter than O8. He\,\sc{iii} (54.4 eV) is only visible around the very hottest stars (some Wolf–Rayet stars). Oxygen having an ionization potential very close to that of hydrogen, the O\,\sc{ii} zone is co-extensive with the H\,\sc{ii} zone. O\,\sc{iii} (35.1 eV) is only found in the central regions if the star is very hot.

The preceding description is very idealistic. It is rare to find an isolated, spherical H II region with a true ionization structure. In fact the ionized medium has generally an heterogeneous structure. Moreover, massive stars tend to form near the surface of molecular clouds, so that the H II region tends rapidly to pierce the cloud and to expand outside it due to its high pressure. $P = nkT$ is about 1 000 times larger in the H II region, because of the large temperature difference (20 K in the placental cloud compared to about 10 000 K in the H II region), while n is a factor of 2 higher due to ionization. This is the *champagne effect* (Yorke et al. [565], see Sect. 12.3). The ionization structure of such a nebula is difficult to model. Recent models of H II regions are those of Stasinska & Schaerer [494] and the CLOUDY model of Gary Ferland et al. [176] *(http://www.nublado.org)*. These models assume a uniform density, or a uniform filling factor for the ionized gas, or a density varying in a regular way. They are thus quite idealistic and must be used with caution when comparing the results with observations. Moreover, a part of the ionizing photons may escape from the H II region into the surrounding, lower density medium. We then say that the H II region is *density-bounded*, while an ideal H II region in a uniform medium, where all Lyman photons are used to ionize the medium, is *ionization-bounded*.

We now examine the various types of emission that characterize H II regions.

5.1.2 Continuous Emission

Gaseous nebulae as well as planetary nebulae emit a continuum at all wavelengths from the ultraviolet to radio. This continuum is produced by a variety of mechanisms that we will now describe.

Free–Free Emission (Thermal Bremsstrahlung)

The *free–free* continuum is produced by the braking of free electrons in the electric field of an ion, but without capture. It is also called *Bremsstrahlung* from the german words for brake (Brems) and radiation (Strahlung). When the velocity distribution of the electrons is maxwellian, as it is the case here, the emission is said to be thermal. At any wavelength the free–free *emissivity* (the energy emitted by unit volume and unit frequency) of a plasma, where the electrons have a maxwellian distribution of velocities corresponding to a temperature T_e, is given by Lang [299] (see e.g. for a demonstration Rohlfs & Wilson [439] Sect. 9.4):

$$\epsilon_\nu = \frac{8}{3}\left(\frac{2\pi}{3}\right)^{1/2} \frac{n_\nu Z^2 e^6}{m_e^2 c^3} \left(\frac{m_e}{kT_e}\right)^{1/2} n_i n_e g_{ff}(\nu, T_e) \exp(-h\nu/kT_e), \quad (5.18)$$

where n_ν is the refraction index of the plasma (very close to unity in H II regions), Z the charge of the ions whose density is n_i (in practice $Z \simeq 1$ and $n_i \simeq n_e$). $g_{ff}(\nu, T_e)$ is the *Gaunt factor* which corresponds to the finite integration limits on the impact parameter of the electron approaching the ion. Following appropriate substitution, (5.18) can be written as

$$\epsilon_\nu = 5.4 \times 10^{-39} n_\nu Z^2 \frac{n_i n_e}{T_e^{1/2}} g_{ff}(\nu, T_e) \exp(-h\nu/kT_e) \text{ erg s}^{-1} \text{ cm}^{-3} \text{ Hz}^{-1} \text{ ster}^{-1}.$$

(5.19)

The Gaunt factor is given by

$$g_{ff}(\nu, T_e) = \frac{\sqrt{3}}{\pi} \ln \Lambda = 0.55 \ln \Lambda, \tag{5.20}$$

with

$$\Lambda \simeq 5.0 \times 10^7 \frac{T_e^{3/2}}{Z\nu}, \tag{5.21}$$

for typical conditions in H II regions. For the general case, the Gaunt factor multiplied by $\exp(-h\nu/kT_e)$ is shown (Fig. 5.1) as a function of frequency.

The absorption coefficient is

$$K_\nu = \frac{\epsilon(\nu)}{n_\nu^2 B_\nu(T_e)}, \tag{5.22}$$

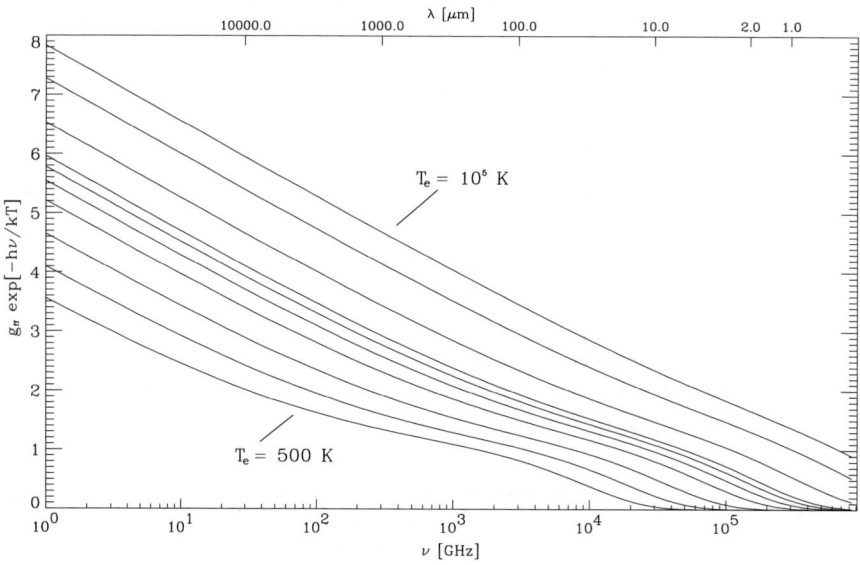

Fig. 5.1. Frequency- and temperature-dependence of the free–free Gaunt factor g_{ff} multiplied by $\exp(-h\nu/kT_e)$; this product is proportional to the emissivity ϵ_ν (cf. (5.19)). The different curves correspond to the temperatures 500, 10^3, 2×10^3, 4×10^3, 6×10^3, 8×10^3, 10^4, 2×10^4, 5×10^4 and 10^5 K. From Beckert et al. [26], with the permission of ESO.

where $B_\nu(T_e)$ is the Planck function and $n_\nu \approx 1$ is the refraction index of the medium at frequency ν.

In the radio wavelength case (Rayleigh–Jeans approximation), taking $n_i = n_e$, $Z = 1$, $n_\nu = 1$, the optical thickness is

$$\tau_\nu = 3.014 \times 10^{-2} \left(\frac{T_e}{K}\right)^{-3/2} \left(\frac{\nu}{\text{GHz}}\right)^{-2} \left(\frac{EM}{\text{pc cm}^{-6}}\right) g_{ff}, \quad (5.23)$$

where EM is the *emission measure*

$$EM = \int_0^s n_e^2 \, ds. \quad (5.24)$$

The expression for the Gaunt factor g_{ff} at radio wavelengths is

$$g_{ff} = \ln\left[4.955 \times 10^{-2} \left(\frac{\nu}{\text{GHz}}\right)^{-1}\right] + 1.5 \ln\left(\frac{T_e}{K}\right), \quad (5.25)$$

and $g_{ff} \simeq 1$ if $\nu(\text{MHz}) \gg (T_e/K)^{3/2}$.

Using this expression we obtain the simple formula

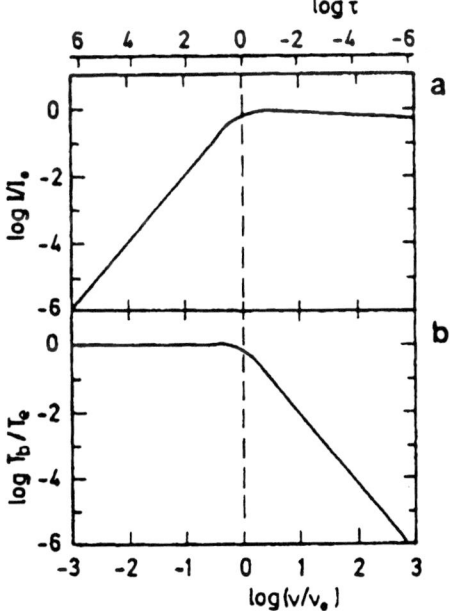

Fig. 5.2. Thermal radiation from an H II region at radio wavelengths. (a) Spectral energy distribution of the intensity as a function of frequency. (b) Brightness temperature as a function of frequency. The optical thickness τ is plotted as abscissae at the top of the figure. The frequency ν_0 is that for which $\tau = 1$. From Rohlfs & Wilson [439].

$$\tau_\nu = 8.235 \times 10^{-2} \left(\frac{T_e}{\text{K}}\right)^{-1.35} \left(\frac{\nu}{\text{GHz}}\right)^{-2.1} \left(\frac{EM}{\text{pc cm}^{-6}}\right) a(\nu, T_e). \quad (5.26)$$

$a(\nu, T_e)$ is a correction factor very close to 1, tabulated by Mezger & Henderson [369].

At high frequencies where the medium is optically thin, the intensity I_ν emitted per unit frequency by a H II region changes with frequency as $\nu^{-0.1}$, and is thus almost independent of frequency. The brightness temperature $T_B \propto I_\nu \nu^{-2}$ varies as $\nu^{-2.1}$. At low frequencies, where the medium is optically thick, it behaves like a blackbody at temperature T_e (Fig. 5.2).

Free–Bound Radiation

The recombination of free electrons with ions produces continuum radiation. The recombination to the different energy levels of hydrogen gives rise to discontinuities

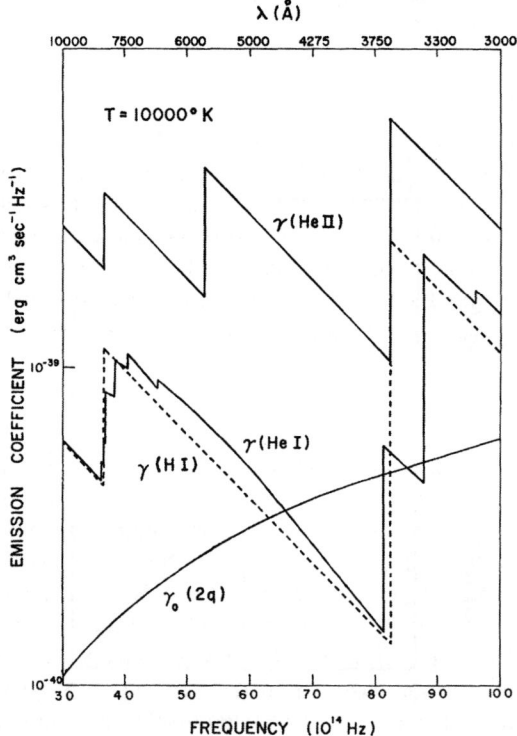

Fig. 5.3. Variation with frequency of the continuous emission coefficients of H I, He I and He II at a temperature of 10^4 K. The 2-photon continuous emission coefficient for hydrogen $\gamma_0(2q)$ is also given. Reproduced from Brown & Mathews [71], with the permission of the AAS.

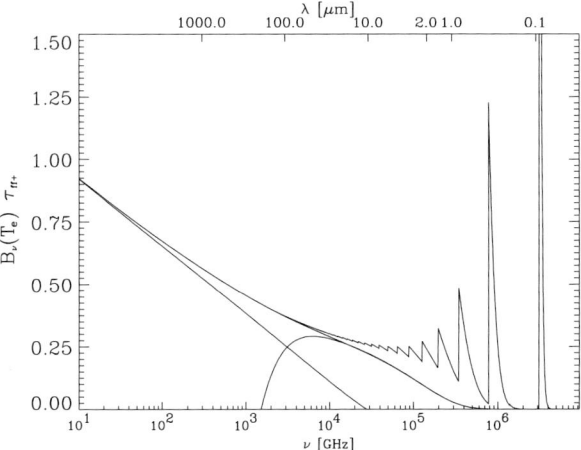

Fig. 5.4. The total continuous spectrum (free–free + free–bound) for a hydrogen plasma at 7×10^3 K, normalized at 5 GHz. The dashed line is the free–free contribution. The contribution of the 2-photon radiation, which is only important in the UV, is neglected here. From Beckert et al. [26], with the permission of ESO.

in the spectrum. The Balmer discontinuity corresponds to recombination to level $n = 2$, the Paschen discontinuity to level $n = 3$, etc. The expressions for the corresponding emissivities can be found in Lang [299], Sect. 1.31 or in Beckert et al. [26]. Figure 5.3 gives the coefficients for continuous emission $\gamma_\nu = \epsilon_\nu/(n_i n_e)$ free–free + free–bound, as well as for the two-photon radiation of hydrogen (see the next paragraph) as a function of wavelength in the visible range. Figure 5.4 shows the free–free + free–bound spectrum at all frequencies for a hydrogen plasma at $T_e = 7 \times 10^3$ K, which corresponds to a typical H II region.

Two-Photon Radiation

The emission of radiation by the radiative de-excitation of an atomic level can arise either directly by a single photon, or through the intermediate of a virtual non-quantified level. In the latter case, two photons are emitted, the sum of their energies being equal to the energy of the transition. The probability for this 2-photon emission is small, but it can become the main channel for the de-excitation of a metastable level if collisions are negligible. This is the case for neutral helium which possesses two series of singlet and triplet states between which direct radiative transitions are forbidden. Recombinations produce comparable amounts of singlet and triplet He I, and radiative cascades populate the lowest level of each configuration, the 2s ^{3}S level of the triplet configuration being 19.81 eV above the ground level 1s^2 ^{1}S of the singlet configuration. The triplet level can only de-excite radiatively by two-photon emission, the energy of each photon being between 0 and 9.9 eV. This radiation is however weak compared to that of other continuous radiations in H II regions.

Similarly, hydrogen can recombine to the metastable level $2^2S_{1/2}$ which can only de-excite radiatively by two-photon emission, producing a continuum which rises towards the ultraviolet. This continuum is more intense than the free–free and free–bound continua near 4 000 Å, as shown by Fig. 5.3 for hydrogen.

Scattered Stellar Light and Thermal Emission of Dust

H II regions contain dust which scatters the light of the exciting stars and causes some diffuse brightness in the visible and especially in the ultraviolet. Dust grains also absorb some of the photons emitted by the stars and some of the Lyman α emission that fills the H II region. They re-emit the absorbed energy in the mid- and far-infrared, producing a thermal continuum. This emission are quantitatively very important and will be discussed in Chap. 7: cf. for example Fig. 7.9.

5.1.3 The Recombination Lines

These are the permitted lines of H, He, O, C, etc., that are emitted by radiative de-excitation cascades following recombination to the higher levels of these atoms. Images of H II regions in the Hα recombination line of hydrogen are shown on Plates 1, 9, 10, 16, 18 and 29. Plate 17 shows an image in the infrared hydrogen recombination line Brackett γ. Direct population of the corresponding atomic levels by collisions from the fundamental level is negligible except for helium (Benjamin et al. [30]). We will only study the hydrogen recombination lines. For the other lines see Osterbrock [389].

Figure 5.5 shows the energy level diagram (*Grotrian diagram*) for hydrogen. It is interesting to note that, if the nebula is optically thick in the Lyman lines (Case B of Baker & Menzel), every recombination of hydrogen ultimately yields a Balmer line or continuum photon, as shown by Fig. 5.6. This is a very interesting property because it allows us to estimate the number of recombinations, hence the number of ionizations, from observations of any Balmer line or of the Balmer continuum (Balmer discontinuity, cf. Fig. 5.3 and 5.4). This is called the *Zanstra method*.

On the other hand, any Lyman α photon emitted in a nebula optically thick in the Lyman lines is soon reabsorbed by the first hydrogen atom it encounters. This atom immediately re-emits another Lyman α photon in an arbitrary direction. The Lyman α photons thus propagate by a random walk (*resonant scattering*) until they escape from the nebula or are absorbed by dust inside the H II region[1]. The $n = 2$ level can also de-excite by 2-photon emission, or ionization can occur from this level.

Let us now calculate the intensity of the recombination lines. Since these lines are always optically thin, it is sufficient to calculate the population of the levels.

---

[1] Resonant scattering of Lyman α photons also exists in the neutral medium, where these photons propagate by a random walk until they are absorbed by dust. For a study of this scattering see Neufeld [384] and references herein, and for an application to the escape of Lyman α photons from a galaxy see Lequeux et al. [318].

5.1 H II Regions

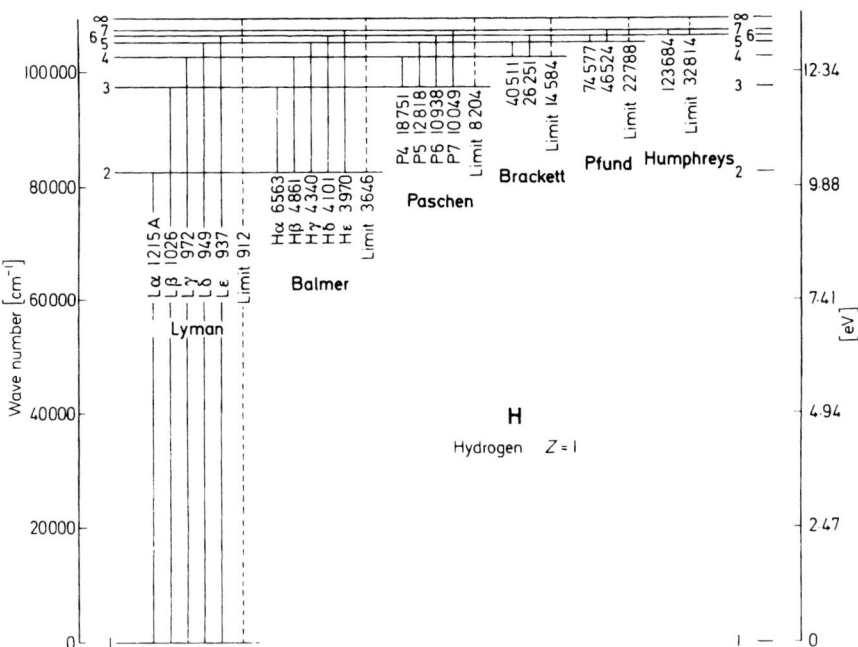

Fig. 5.5. Energy diagram for the hydrogen atom, with the different series designated. The principal quantum number n is indicated to the left of each level. The scale of the ordinate is in wave numbers, $1/\lambda$, and should be multiplied by hc in order to obtain the energies. From Lang [299].

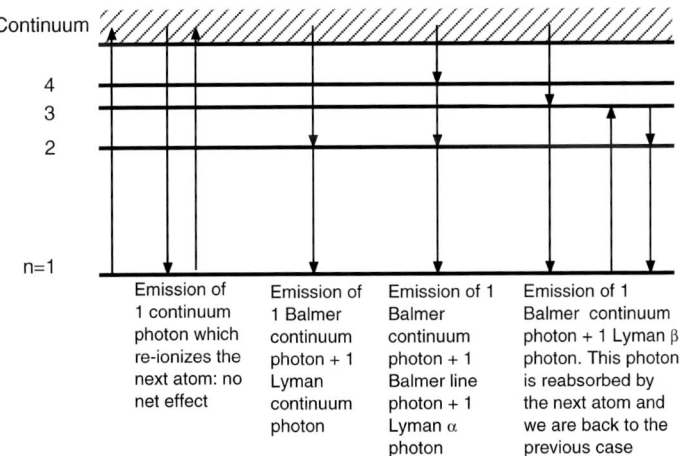

Fig. 5.6. Schematic showing that every recombination of hydrogen in an H II region produces a Balmer photon in case B.

5 The Ionized Interstellar Gas

The emitted intensity per cm^3 is immediately obtained from $n_u A_{ul} h \nu_{ul}$, stimulated emission being negligible. The line frequencies for a hydrogenoid atom are such that

$$\nu_{ul} = Z^2 R_M \left(\frac{1}{n_l^2} - \frac{1}{n_u^2} \right), \tag{5.27}$$

where n_u and n_l are here, respectively, the principal quantum numbers for the upper and the lower level of the transition. $R_M = R_\infty/(1+m_e/M)$ is the Rydberg constant for an atom of mass M and Z is its effective charge (m_e is the mass of the electron). Z is very close to unity for large values of n (≥ 100), because then the transitions concern a single electron very far from the rest of the atom, i.e. the nucleus with the remaining surrounding electrons. The atom is then hydrogenoid. Table 5.2 gives M and R_M for the most important atoms, as well as the velocity shift for the lines of high n of these atoms with respect to the corresponding lines of hydrogen.

Table 5.2. Atomic mass, Rydberg constant and velocity shift for high-n (radio) recombination lines with respect to the hydrogen recombination lines for the most abundant atoms.

Atom	Atomic mass	$R_M (10^{15}\,\text{Hz})$	Δv (km s^{-1})
^{1}H	1.007 825	3.288 051 29(25)	–
^{2}He	4.002 603	3.289 391 18	-122.166
^{12}C	12.000 000	3.289 691 63	-149.560
^{14}N	14.003 074	3.289 713 14	-151.521
^{16}O	15.994 915	3.289 729 19	-152.985
∞		3.289 842 02	-163.272

The oscillator strengths for the lower Balmer lines ($l = 2$) are $f_\alpha = 0.641 (u = 3)$, $f_\beta = 0.119 (u = 4)$, $f_\gamma = 0.044 (u = 5)$, $f_\delta = 0.021 (u = 6)$ and $f_\epsilon = 0.012 (u = 3)$. We can then calculate A_{u2} using (4.26). For the general expressions see Lang [299], Sect. 2.12. For radio recombination lines with high values of n, we have for all atoms:

$$A_{n+1,n} = \frac{5.36 \times 10^9}{n^5} \text{ s}^{-1}. \tag{5.28}$$

We usually express the level populations through the coefficients of departure from LTE, b_n, which are the ratios of the actual population of level n to the population it would have at LTE, given by the *Saha law*, which is demonstrated in many astrophysics or statistical mechanics textbooks:

$$\boxed{n_n = b_n n_i n_e \frac{g_n}{g_e} \frac{h^3}{(2\pi m_e k T_e)^{3/2}} \exp\left(\frac{hR}{n^2 k T_e}\right),} \tag{5.29}$$

the statistical weights being $g_n = 2n^2$ and $g_e = 2$. It is thus necessary to calculate the b_n's for the levels of interest by solving the equations for statistical equilibrium, which involve the following quantities:

- The rates $\mathcal{N}_{lu}$ of radiative transitions, in cm^{-3} s^{-1}. Collisional transitions can be neglected and we have, using (3.1), (3.3), (3.12), (3.13) and (4.26),

$$\mathcal{N}_{lu} = R_{lu} n_l = I_\nu B_{lu} n_l = I_\nu \frac{e^2}{h\nu m_e c} f n_l, \qquad (5.30)$$

f being the oscillator strength of the transition. I_ν is often taken as the blackbody intensity at the effective temperature of the exciting star, diluted by the factor $W = r_*^2/r^2$, r_* being the radius of the star and r the distance to it. However, because the far-UV spectral energy distribution of the stellar radiation is very different from that of a blackbody, it is far better to use stellar spectra given by models like those of Schaerer & de Koter [456].

- The rate $\mathcal{N}_{ul}$ of radiative transitions from u to l. Neglecting stimulated emission, an approximation not justified in radio as we will see in the next section, we have

$$\mathcal{N}_{ul} = R_{ul} n_u \simeq A_{ul} n_u = \frac{8\pi e^2 \nu^2}{m_e c^3} f \frac{g_l}{g_u} n_u. \qquad (5.31)$$

- The ionization rates: for example, from level l

$$P_{lk} n_l = \int \mathcal{N}_{lk} \, d\nu = \int \frac{4\pi I_\nu}{ch\nu} a_l n_l \, d\nu. \qquad (5.32)$$

- The recombination rates: for example, from level l

$$P_{kl} n_e = \int \mathcal{N}_{kl} \, d\nu = \int n_i n_e \frac{K}{T_e^{3/2}} \frac{g_l}{l^3} \exp(\chi_l) \exp(-h\nu/kT_e) \frac{d\nu}{\nu}, \qquad (5.33)$$

where K is a constant and χ_l is the ionization potential from level l. For the details of the derivation of this equation, see Lang [299] Sect. 2.10 and 1.31.

Solving this system of equations is a rather complex problem, that can be simplified in two cases:

- *Case A* of Baker & Menzel: the star does not emit in the Lyman lines (this is true for "normal" hot stars), and the nebula is optically thin in these lines so that there are no line absorptions from level 1. The equilibrium equation for level n is in this case:

$$\sum_{n''>n} \mathcal{N}_{n''n} + \int \mathcal{N}_{kn} \, d\nu = \sum_{n'=1}^{n-1} \mathcal{N}_{nn'} + \int \mathcal{N}_{nk} \, d\nu, \qquad (5.34)$$

in which the successive terms represent, respectively, the cascades from the higher levels to n, the recombinations to n, the cascades from n to lower levels and the photoionizations from n.

- *Case B* of Baker & Menzel: the star does not emit in the Lyman lines and the nebula is optically thick in the Lyman lines. This is generally the case for gaseous nebulae, and corresponds to Fig. 5.6. We now have $\mathcal{N}_{1n} = \mathcal{N}_{n1}$ and the equilibrium of level n is written:

$$\sum_{n''>n} \mathcal{N}_{n''n} + \int \mathcal{N}_{kn}\, dv = \sum_{n'=2}^{n-1} \mathcal{N}_{nn'} + \int \mathcal{N}_{nk}\, dv. \quad (5.35)$$

Photoionizations from levels $n \geq 2$ are generally neglected because these levels are much less populated than level $n = 1$.

In this way we arrive at an infinite (in principle) system of linear equations in n_n and thus in b_n. We find a solution in Hummer & Storey [252], Storey & Hummer [497] and references cited herein. These articles give the intensities of the Balmer lines and also the *Balmer decrement* which is the intensity ratio of the different Balmer lines to Hβ (the 4-2 line). This ratio is almost the same in cases A and B and depends very little on the physical conditions (n_e and T_e). Table 5.3 gives the wavelength and the intensity relative to Hβ of the first lines of the Balmer (n-2), Paschen (n-3), Brackett (n-4) and Pfund (n-5) series for $n_e = 10^4$ cm^{-3} and $T_e = 10^4$ K, for case B. See Hummer & Storey [252] and Brocklehurst [69] for higher lines, for the lines of He II and for other values of the physical parameters.

Table 5.3. Wavelengths in μm and intensities relative to Hβ for the most important recombination lines of hydrogen in the visible and in the infrared, for $n_e = 10^4$ cm^{-3} and $T_e = 10^4$ K, in case B. From Hummer & Storey [252].

	Balmer series			Brackett series	
Hα	0.6563	2.85	Brα	4.05	7.77×10^{-2}
Hβ	0.4861	1.00	Brβ	2.63	4.47×10^{-2}
Hγ	0.4340	0.469	Brγ	2.17	2.75×10^{-2}
Hδ	0.4101	0.260	Brδ	1.94	1.81×10^{-2}
Hϵ	0.3970	0.159	Brϵ	1.82	1.26×10^{-2}
	Paschen series			Pfund series	
Pα	1.875	0.332	Pfα	7.46	2.45×10^{-2}
Pβ	1.282	0.162	Pfβ	4.65	1.58×10^{-2}
Pγ	1.094	9.01×10^{-2}	Pfγ	3.74	1.04×10^{-2}
Pδ	1.005	5.53×10^{-2}	Pfδ	3.30	7.25×10^{-3}
Pϵ	0.9546	3.65×10^{-2}	Pfϵ	3.04	5.24×10^{-3}

The interest of the intensity ratios between recombination lines, and in particular the Balmer decrement, is that a comparison between the observed and the theoretical ratios (Table 5.3) can be used to estimate the reddening and hence the extinction by dust in the direction of the H II region (cf. Caplan & Deharveng [76]). In particular, the visual extinction A_V is related to the colour excess $E_{\beta-\alpha} = 2.5 \log[I_{theor}(H\beta)/I_{obs}(H\beta)]/[I_{theor}(H\alpha)/I_{obs}(H\alpha)]$ by the relation

$$A_V = 2.60 \, E_{\beta-\alpha}. \tag{5.36}$$

The intensity of recombination lines depends on the temperature and is proportional to $n_e n_i$, i.e., in fact to n_e^2. For example, the intensity of the Hβ line in Case B is $I(H\beta) = 15.410^{-26} \, n_e^2$ cgs units for $T_e = 5 \times 10^3$ K, $8.3 \times 10^{-26} n_e^2$ cgs units for $T_e = 10^4$ K and $4.2 \times 10^{-26} n_e^2$ cgs units for $T_e = 2 \times 10^4$ K. The ratio between a recombination line and the free–free radio continuum gives another determination of the extinction. This determination does not always coincide with that based on the Balmer decrement. This problem is discussed in detail by Caplan & Deharveng [76]. A useful relation is:

$$\frac{I(H\beta)}{I_{cont}(\nu)} = 3.01 \times 10^{-10} \frac{N(H^+)}{N(H^+) + N(He^+)} \left(\frac{T}{10^4 K}\right)^{-0.52} \left(\frac{\nu}{10^9 Hz}\right)^{0.1}, \tag{5.37}$$

where $I(H\beta)$ is in erg cm^{-2} s^{-1} and $I_{cont}(\nu)$ is in erg cm^{-2} s^{-1} Hz^{-1} or in 10^{23} Jy. A comparison of the observed ratio with the theoretical one is useful in determining the extinction. Since there is no extinction at radio wavelengths, we directly obtain the extinction at Hβ, $A(\beta) = 1.19 A_V$. Observation shows that the extinction determined in this way is generally larger than that determined from the Balmer decrement. This is mainly a selection effect due to inhomogeneities in the extinction: the extinction measured from the Balmer decrement corresponds to relatively transparent regions where the Hβ and Hα lines are both seen, while at radio wavelengths we see all the emission from the H II region.

The ratio of the nebular continuum discontinuity at 3 646 Å (the Balmer jump) to the intensity of a high Balmer line, chosen as close as possible to the discontinuity in order to avoid problems with the extinction, can yield a measure of the electron temperature thanks to the different variation with temperature of the two quantities. For an example, see Tsamis et al. [519].

Finally, it is worth mentioning that the intensity ratio between the helium and hydrogen recombination lines allows us to obtain the abundance of helium. This is probably its best determination, but it does have its own problems which are discussed in detail in the book by Pagel [394].

5.1.4 The Radio Recombination Lines

These lines correspond to transitions between levels with very high quantum numbers n ($n \geq 30$ roughly). We define α $(n+1 \to n)$ lines, β $(n+2 \to n)$ lines, γ $(n+3 \to n)$ lines, etc. Many lines of H and He have been observed, as well as lines of C and S. However the latter originate mainly from photodissociation regions rather than from H II regions, because these photodissociation regions contain much C II and S II while the H II regions contain little of these ions: carbon and nitrogen are mostly doubly ionized in H II regions.

For these high energy levels (*Rydberg states*), atoms behave in a quasi-classical way: they can be considered as a point charge around which a single electron orbits at a large distance. Since the oscillator strength of recombination lines decreases as

$1/n^2$ while the cross-section for collisional transitions with electrons is only weakly dependent on n, collisional transitions dominate over radiative transitions, contrary to the situation with optical recombination lines. We can therefore assume LTE as a first approximation. Then the level population obeys the Saha law (cf. (5.29) with $b_n = 1$). In the Rayleigh–Jeans approximation the population n_n of level n is given by

$$n_n = 4.2 \times 10^{-6} \frac{n_e n_i n^2}{T_e^{3/2}}. \tag{5.38}$$

Assuming that the line is only broadened by the Doppler effect, the absorption coefficient for an α line is then independent of the atom,

$$\kappa_\nu = \frac{\sqrt{\pi} e^2}{m_e c} \frac{n_n}{\Delta \nu} f_{n+1 \to n} e^{-[(\nu-\nu_0)/\Delta \nu]^2} \frac{h\nu}{kT_e}, \tag{5.39}$$

where ν_0 is the central frequency of the line and $\Delta \nu$ is the line width at half maximum. Since $f \propto 1/n^2$ and $n_n \propto n^2$, κ does not depend much on n. If the line is optically thin, which is most generally the case, the brightness temperature at the line centre is

$$T_{line} = 1.92 \times 10^3 \left(\frac{T_e}{K}\right)^{-3/2} \left(\frac{EM}{cm^{-6} pc}\right) \left(\frac{\Delta \nu}{kHz}\right)^{-1}. \tag{5.40}$$

Radio recombination lines are always superimposed on the thermal continuum of the H II region, which is generally optically thin for $\nu > 1$ GHz. The line to continuum intensity ratio is obtained by combining (5.40) and (5.26):

$$\boxed{\frac{T_{line}}{T_{cont}} \left(\frac{\Delta \nu}{km\, s^{-1}}\right) = \frac{6.99 \times 10^3}{a(\nu, T_e)} \left(\frac{\nu}{GHz}\right)^{1.1} \left(\frac{T_e}{K}\right)^{-1.15} \frac{1}{1 + n(He^+)/n(H^+)}.} \tag{5.41}$$

The last factor comes from the fact that both He II and H II contribute to n_e. This relation can be used to determine the electron temperature, as long as LTE is valid, which is true at high frequencies. For example we have $T_{66\alpha}/T_{cont} = 0.245$ at 22.364 GHz for the Orion nebula (Orion A). With $n(\text{He II})/n(\text{H II}) = 0.08$ and $\Delta \nu_{1/2} = 25.7$ km s^{-1}, we get $T_e = 8\,200$ K, a reasonable number at first glance. We often note as T_e^* the LTE temperatures determined in this way.

However, we caution the use of the LTE hypothesis. Although the level populations are almost at LTE, small deviations can have a large effect as we will see. Introducing the coefficients b_n of departure from LTE, we can write the absorption coefficient as

$$\kappa_n \propto 1 - \frac{b_n}{b_{n'}} \left(1 - \frac{h\nu}{kT_e}\right) = \frac{h\nu}{kT_e} \frac{b_n}{b_{n'}} \left[1 - \frac{kT_e}{h\nu} \left(1 - \frac{b_{n'}}{b_n}\right)\right]. \tag{5.42}$$

Even when b_n and $b_{n'}$ have close values, the term between the square brackets can be very different from unity and even negative when $kT_e/h\nu > 1$: in this case there is a maser effect. It is thus necessary to calculate b_n accurately. The dependence of b_n on n is shown in Fig. 5.7 together with the differential variation $d\ln b_n/dn$. In

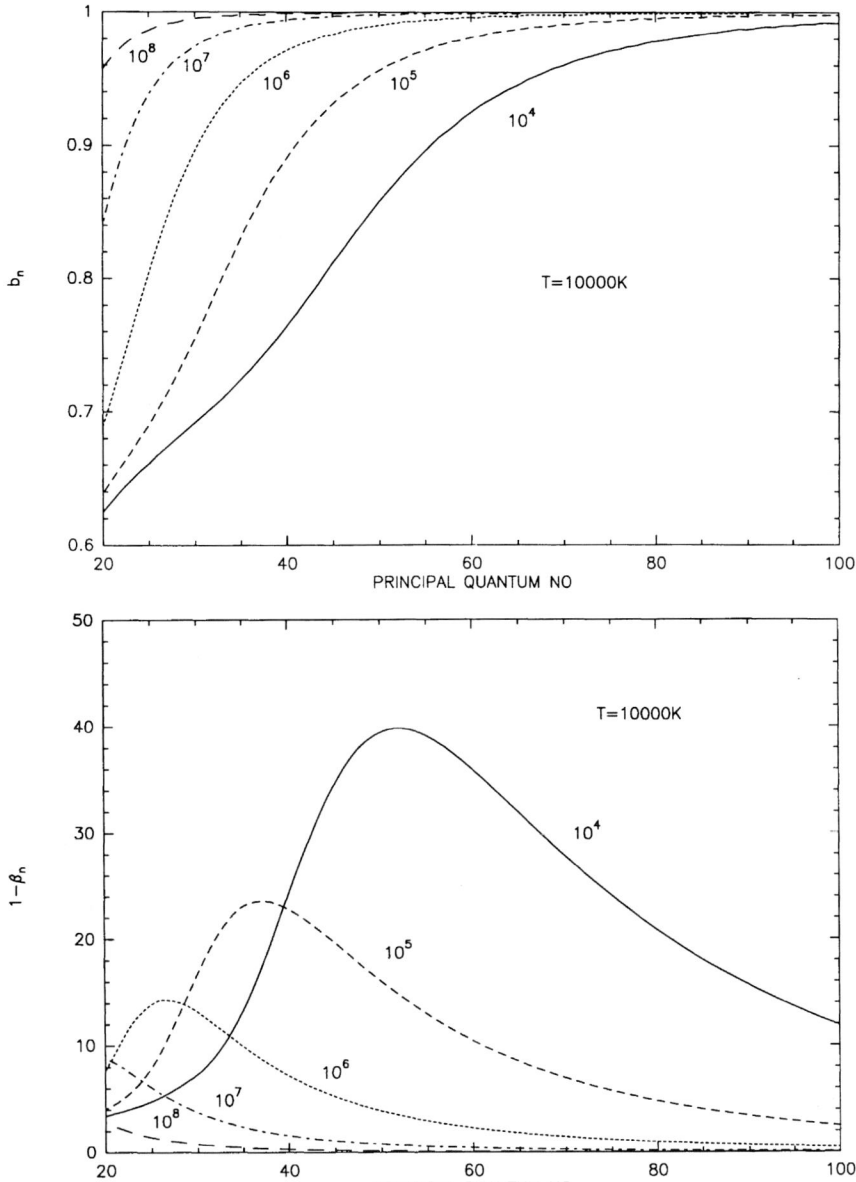

Fig. 5.7. b_n coefficients of departure from LTE for radio recombination lines (top) and their differential variation $d \ln b_n / dn$ (bottom), as a function of quantum number n, for a temperature $T_e = 10^4$ K and different values of the electron density n_e. From Walmsley [540], with the permission of ESO.

5 The Ionized Interstellar Gas

order to correctly calculate the line to continuum ratio, we must solve the transfer equation for the mixture line + free–free continuum. The transfer equation is[2]

$$\frac{dI(\nu)}{d\tau(\nu)} = \frac{j(\nu)}{\kappa_\nu} - I(\nu). \tag{5.43}$$

The total absorption coefficient is thus

$$\kappa_\nu = \kappa_{cont}(\nu) + \kappa_{line}(\nu). \tag{5.44}$$

Similarly:

$$j(\nu) = j_{cont}(\nu) + j_{line}(\nu). \tag{5.45}$$

Since the continuum is at LTE, we have

$$\frac{j_{cont}(\nu)}{\kappa_{cont}(\nu)} = \frac{2kT_e\nu^2}{c^2}, \tag{5.46}$$

and for the line

$$\frac{j_{line}(\nu)}{\kappa_{line}(\nu)} = \frac{2kT_{ex}\nu^2}{c^2} \text{ with } 1 - \frac{h\nu}{kT_{ex}} = \frac{b_n}{b_{n'}}\left(1 - \frac{h\nu}{kT_e}\right). \tag{5.47}$$

Combining these equations, we obtain the source function (cf. (3.7))

$$S(\nu) = \frac{j(\nu)}{\kappa(\nu)} = B_\nu(T_e) \frac{1 + b_n \kappa^0_{line}(\nu)/\kappa_{cont}(\nu)}{1 + (b_n/b_{n'})[\kappa^0_{line}(\nu)/\kappa_{cont}(\nu)]} \left(1 - \frac{d \ln b_n}{dn} \frac{kT_e}{h\nu} \Delta n\right), \tag{5.48}$$

$\kappa^0_{line}(\nu)$ being the absorption coefficient that the line would have if it were emitted at LTE. This coefficient is given by (5.39). The line/continuum ratio can be easily obtained by solving the transfer equation. From the preceding equations we obtain the relation between the true electron temperature T_e and the temperature T_e^* determined from (5.41) assuming LTE (see Rohlfs & Wilson [439] Sect. 13.5 and 13.6):

$$\boxed{T_e = T_e^* \left[b_n\left(1 - \frac{1}{2}\beta\tau_{cont}\right)\right]^{0.87},} \tag{5.49}$$

with

$$\beta = 1 - 20.836 \left(\frac{T_e}{K}\right)\left(\frac{\nu}{\text{GHz}}\right)^{-1} \frac{d \ln b_n}{dn} \Delta n. \tag{5.50}$$

In H II regions, this ratio is larger than the LTE ratio because of a maser effect, but the departure is large only at low frequencies, say for $\nu < 10$ GHz. For example, in the case of the H 66α line in the Orion nebula, $T_e = 1.07\, T_e^*$. The electron temperature obtained from the recombination lines is generally in good agreement with that derived from the forbidden lines in the visible, which will be discussed in the next section. It is easier to obtain, in particular if there is strong extinction,

---

[2] $j(\nu)$ is the emissivity per steradian, that is $1/4\pi$ times the emissivity ϵ_ν in (5.18) and (5.19).

because radio waves are not absorbed in the interstellar medium. The measurement of the frequency of a recombination line allows us to obtain the radial velocity of the H II region, even if it is invisible optically. This is of great interest for the determination of the structure of the Galaxy. Figure 1.2 rests heavily on observations of recombination lines.

Recombination lines of elements other than hydrogen have also been observed. The lines of helium can be used to obtain the He/H abundance ratio, but the result is less accurate than when using optical recombination lines. The lines of carbon and sulphur are mainly emitted in photodissociation regions (Chap. 10). Carbon recombination lines are also emitted in the diffuse atomic gas : for a discussion see Payne et al. [400]. We will only mention here that these lines are enhanced by *dielectronic recombination* (Goldberg & Dupree [200]): for a singly-ionized, complex ion, recombination with an electron can produce a doubly excited atom (i.e. with two electrons in an excited state) that de-excites on a singly excited state (see Lang [299] Sect. 2.10.g). This phenomenon is important for some elements and must be accounted for in a complete theory of ionized gases.

5.1.5 The Forbidden Lines

Many very strong lines emitted by H II regions were identified by Bowen in 1926 as forbidden transitions of common ions. Some of these transitions have already been presented Sect. 4.1. Figure 5.8 shows a typical optical spectrum of an H II region where the recombination lines and the forbidden lines can both be seen. An infrared spectrum is presented in Fig. 7.9. Monochromatic images of H II regions in some of these lines are presented in Plates 9, 16 and 25. Plate 22 shows an image in a forbidden line in the mid-infrared.

As an illustration, we present in Fig. 5.9 the energy diagram for the lowest levels of O III. The forbidden transitions of this ion, whose characteristics are given in Table 4.1, are of three types.

1. The far-infrared transitions between the sub-levels of the fine-structure ^{3}P triplet, $^3P_1 \to {}^3P_0$ at 88.4 μm, and $^3P_2 \to {}^3P_1$ at 51.8 μm. The $^3P_2 \to {}^3P_0$ transition is very strongly forbidden and cannot be observed as explained Sect. 4.1. We can easily calculate the population ratio of the levels 2 ($^3P_2$) and 1 ($^3P_0$), by writing the statistical equilibrium equations. These equations are simple in this case because both absorption and stimulated emission can be neglected, the lines being optically thin and their frequencies being high:

$$n_i \sum_{j<i}(C_{ij} + A_{ij}) = \sum_{j>i} n_j(C_{ji} + A_{ji}), \qquad (5.51)$$

with $C_{ji} = n_e \langle \sigma_{ji} v \rangle$ (cf. (3.42)) and $C_{ij} = C_{ji}(g_j/g_i)\exp(-h\nu_{ji}/kT_e)$ (cf. (3.38)). For a maxwellian distribution of the velocities of the free electrons, we have

$$C_{ji} = \frac{8.63 \times 10^{-6}}{g_j T_e^{1/2}} n_e \Omega_{ji}. \qquad (5.52)$$

106   5 The Ionized Interstellar Gas

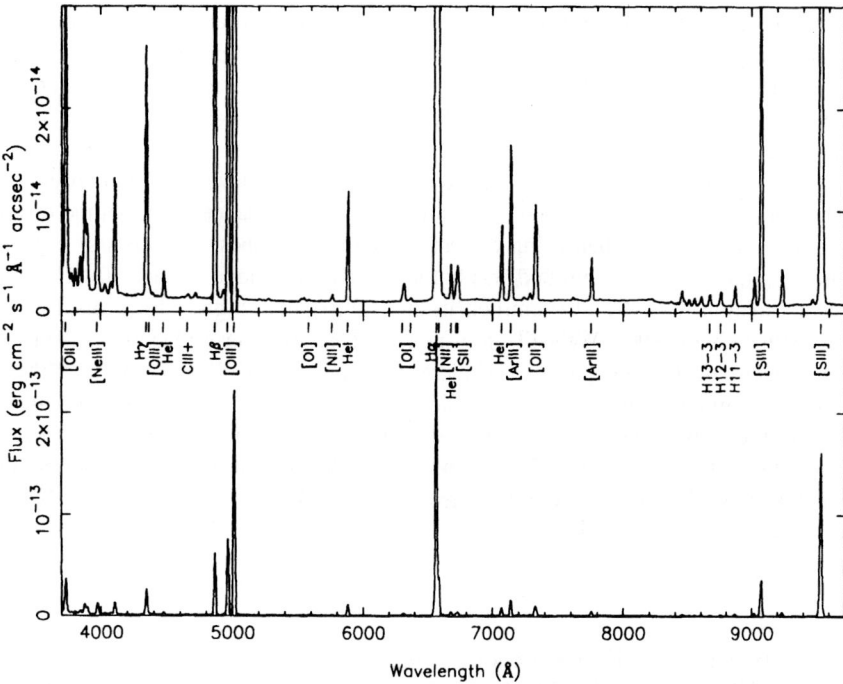

Fig. 5.8. The optical spectrum of the Orion nebula. The lower figure displays the strong lines and the upper figure the weak lines on an expanded intensity scale, the strong lines being cut. The main lines are identified. The N II lines are not separated from Hα at the wavelength resolution of the spectrum, as well as the [O III]λ4 363 line from Hγ. Note the high-n Paschen lines ($n-3$) on the right of the figure. Reproduced from Baldwin et al. [16], with the permission of the AAS.

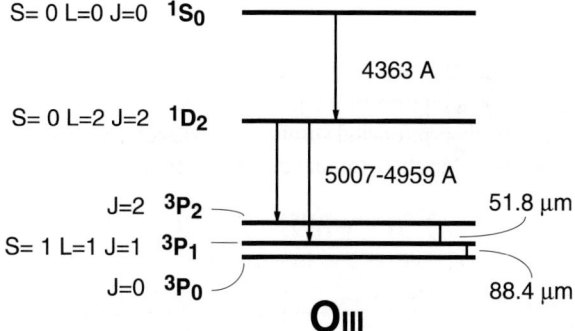

Fig. 5.9. Energy diagram for the lowest levels of O III. The values of the quantum numbers and the level designations are given to the left. All the transitions, which are indicated with their wavelengths, are forbidden. Nevertheless, they are very intense in H II regions.

If only three levels have to be considered, as in the present case, the solution is

$$\frac{n_2}{n_1} = \frac{C_{12}(C_{31} + A_{31}) + (C_{12} + C_{13})(C_{32} + A_{32})}{C_{12}(C_{12} + C_{13}) + C_{13}(C_{21} + A_{21})}. \quad (5.53)$$

Furthermore we have $A_{31} \simeq 0$. A knowledge of n_2/n_1 allows to obtain immediately the intensity ratio between the two lines:

$$\frac{I_{21}}{I_{10}} = \frac{n_2 A_{21} h\nu_{21}}{n_1 A_{10} h\nu_{10}}. \quad (5.54)$$

This ratio increases with density until it reaches the LTE ratio at high densities. It is almost independent of temperature because the energies of the transitions are small with respect to kT_e. This is illustrated in Fig. 5.10. The intensity ratio between the 88.4 and 51.8 μm lines is thus an excellent indicator of the electron density as long as it is not too high, which is generally the case in H II regions, and the intensity of one of these lines give a good measure of the abundance of O III. It is interesting to note that the determinations of the density, based on these lines, invariably shows that the H II regions are not homogeneous but exhibit large density fluctuations. The forbidden lines, as well as the recombination lines and the free–free and free–bound continua, preferentially originate in high-density regions because their emissivity is proportionnal to n_e^2.

2. the optical transitions $^1D_2 \rightarrow {}^3P_2$ at 5 007 Å and $^1D_2 \rightarrow {}^3P_1$ at 4 959 Å, are called the *nebular lines*. These two lines start from the same upper level and are always in the same ratio, equal to 2.9. The intensity of any of these lines can be used to determine the abundance of O III provided that both the density and the temperature are known.

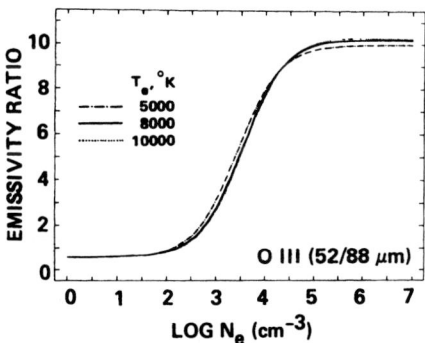

Fig. 5.10. Intensity ratio between the two forbidden lines of O III $^3P_2 \rightarrow {}^3P_1$ at 52 μm and $^3P_1 \rightarrow {}^3P_0$ at 88 μm, as a function of electron density n_e, for three values of the electron temperature T_e. The dotted-dashed curve is for $T_e = 5\,000$ K, the full curve is for 8 000 K and the dotted line, almost superposed on the previous one, is for 10 000 K. We see that the line ratio is almost independent of temperature. Reproduced from Rubin [440], with the permission of the AAS.

3. the optical transition $^1S_0 \to {}^1D_2$ at 4363 Å, is called the *auroral line*. The intensity ratio between this line and one of the nebular lines (or the ratio R_O with the sum of the two nebular lines) can be calculated from (5.53) and (5.54). With the numerical values of the C and A, we obtain the relations

$$R_O = \frac{I(\lambda 4959) + I(\lambda 5007)}{I(\lambda 4363)} = \frac{7.73 \exp[3.29 \times 10^4/T_e]}{1 + 4.5 \times 10^{-4}(n_e/T_e^{1/2})}, \quad (5.55)$$

and

$$T_e[\text{O\,{\sc iii}}] = \frac{14\,320\,\text{K}}{\log R_O - 0.890 + \log(1 + 4.5 \times 10^{-4} n_e/T_e^{1/2})}. \quad (5.56)$$

The value obtained for T_e is obviously that which corresponds to the region where oxygen exists as O III. Unfortunately the auroral line at 4363 Å can only be observed when the temperature is sufficiently high. Otherwise, it is difficult to obtain the electron temperature.

We can obtain similar relations for the optical lines of N II:

$$R_N = \frac{I(\lambda 6548) + I(\lambda 6583)}{I(\lambda 5755)} = \frac{6.91 \exp[(2.50 \times 10^4)/T]}{1 + 2.5 \times 10^{-3}(n_e/T_e^{1/2})}, \quad (5.57)$$

and

$$T_e[\text{N\,{\sc ii}}] = \frac{10\,860\,\text{K}}{\log R_N - 0.841 + \log(1 + 2.5 \times 10^{-3}(n_e/T_e^{1/2}))}. \quad (5.58)$$

In this case the temperature corresponds to the region where nitrogen is mostly in the form of N II. $T_e[\text{N\,{\sc ii}}]$ is lower than $T_e[\text{O\,{\sc iii}}]$.

5.1.6 Abundance Determinations in H II Regions

H II regions are ideal places to determine the abundances of the elements that are responsible for recombination and fine-structure lines. The list of these elements is generally limited to He, C, N, O, Ne, S, Cl, Ar and Fe at the present time. The lines of many more elements are observable in several planetary nebulae, such as NGC 7027, where the lines are very intense (cf. Baluteau et al. [19]).

However, there are problems in the determination of abundances in H II regions. These have been discussed in an abundant literature on this subject and are described in detail, for example in the book of Pagel [394] and more recently by Tsamis et al. [518]. Lack of space does not permit an extensive discussion, even a simplified one, but we do give a short overview of the problems.

- The abundances are given relative to hydrogen, which is observed by its recombination lines. Only a few abundant elements give observable recombination lines with similar physics: helium, and also carbon, nitrogen and oxygen whose lines are very weak. Other, even fainter, recombination lines have been detected for example by Esteban et al. [168], and can be used to obtain the abundances of rarer elements. There may be problems with fluorescence excitation effects for recombination lines (Tsamis et al. [518]).

C and O also have semi-forbidden lines in the far-UV ([C III] $\lambda 1\,909$ Å and [O III] $\lambda 1\,666$ Å: cf. for example Garnett et al. [192]), but their physics is complicated. In this case (as always) it is necessary to correct the observed line intensities for interstellar extinction, which fortunately can be derived to some extent from the Balmer decrement.

- The abundances derived from the fine-structure lines in the visible are sensitive to both temperature and density, and the interpretation of the line intensities is a delicate problem. In some cases, like those of O III and of N II which were discussed in the previous paragraph, the temperature of the emitting zone can be obtained and the abundance determination is safer. However, even in this case temperature fluctuations, if they exist, can yield systematic errors in the abundances (Peimbert [401]). However, it seems that these fluctuations are not very important in general.
- The abundances derived from the mid- and far-infrared fine structure lines are not sensitive to electron temperature and are little affected by extinction. These are considerable advantages with respect to the optical lines. There is now a large set of observations of infrared lines from H II regions, thanks to observations with the Kuiper Airborne Observatory and especially with ISO. They involve the lines of [N II], [N III], [O III], [Ne II], [Ne III], [S III], [S IV], [Ar II] and [Ar III]: see Table 4.1. More observations will come from the Spitzer satellite and the SOFIA airborne observatory. There are important differences between the abundances derived from infrared and from optical lines. These differences may originate in temperature fluctuations or in errors in some atomic parameters, but it should be remembered that the critical density for infrared lines is generally much smaller than for visible lines (see Table 4.1), so that the abundances derived from the infrared lines are underestimated if the density is high. This effect can be very large for planetary nebulae (Tsamis et al. [519])
- All elements with the obvious exception of hydrogen exist in several ionization states in H II regions. However, only the abundances of those ions that emit observable lines can be determined. If such ions are minor species, they yield no useful information because the physical parameters of H II regions are most often too uncertain to allow an accurate solution of the ionization equilibrium. This is for example the case for O I, C II, S II, or Si II. If moreover lines from these ions are emitted strongly by photodissociation regions at the interface of the ionized gas and the surrounding neutral medium, then there are problems with their observation in the H II region itself. If the observed ion is a major species the situation is more favorable since we can calculate, more or less accurately, the abundances of the unobserved ions of the same element. However, uncertainties remain if a high precision is required, as in the case of helium because of its cosmological importance. Some fraction of helium can exist as He I which is

essentially unobservable. The most favorable case is that of oxygen whose major ionization states, O II and O III, are observable optically and for which the electron temperature T_e(O III) can be determined. For this reason, oxygen is, after helium, the element whose abundance is best determined, at least if the temperature is large enough for the 4 363 Å line to be measured. If this is not the case, we may use the empirical relations between the oxygen abundance and the intensity of the [O II]$\lambda\lambda$3 726,3 729 and [O III]$\lambda\lambda$4 959,5 007 lines relative to Hβ (Pagel [394]), but this does not give accurate results.

5.2 The Diffuse Ionized Gas

Outside well-defined H II regions, the interstellar medium contains diffuse ionized gas which can originate either from leaks of ionized gas out of H II regions due to the champagne effect, or from ionization by the UV radiation of isolated hot stars, and perhaps from other mechanisms. Plates 1 and (more clearly) 10 display the emission of the Hα line by the diffuse ionized gas in two external galaxies. In our Galaxy, the diffuse ionized gas contains much more mass than the H II regions. Its total mass is of the order of 1/3 of that of H I. This component is observed by several methods.

– The dispersion of the radio radiation of pulsars yields an average electronic density $\langle n_e \rangle \simeq 0.03$ cm^{-3} in the diffuse interstellar medium. The Faraday rotation of the linear polarization of synchrotron radiosources gives the product $\int n_e B_{\parallel}\,ds$ (see Sect. 2.2), but it is rather used to determine the longitudinal component of the magnetic field. The *interstellar scintillation* which affects the radio radiation of pulsars and of small-diameter radiosources is another evidence for the existence of a diffuse ionized medium, but it cannot give the density because it depends also on the structure of this medium. Interstellar scintillation will not be discussed in this book. For a review, see Narayan [382]. Taylor & Cordes [506] have used most of the available information to build a model of the distribution of free electrons in the Galaxy.
– Fine-structure lines of N II at 122 and 205 μm originating from the diffuse ionized medium have been observed for the first time with the COBE satellite (Wright et al. [558]). There are also ultraviolet absorption lines coming from the three corresponding fine-structure levels of this ion (Gry et al. [210]). Since nitrogen has an ionization potential of 14.534 eV, larger than that of hydrogen (13.598 eV), the presence of ionized nitrogen implies that hydrogen is also ionized in the same regions.
– Free–free emission of a radio continuum, and free–free absorption of the synchrotron continuum of the Galaxy and extragalactic radiosources at very low frequencies (2-10 MHz), are not limited to H II regions but are more diffuse, showing the existence of a diffuse ionized gas.
– This property is also true for the emission of the radio recombination lines of hydrogen at relatively low frequencies, for example the H166α line at 1.44 GHz

(see e.g. Lockman [333]). Non-LTE effects are such that these lines are relatively more intense than the high-frequency lines in the diffuse medium compared to the H II regions.
- The best method for studying the diffuse ionized gas, which gives insight on its physics, is the observation of the optical recombination and fine-structure lines. The existence of a diffuse Hα emission from outside the H II regions has long been known (Sivan [475]). Although they are very weak, the Hα, [N II]$\lambda 6\,583$ and [S II]$\lambda 6\,716$ lines have been observed at high spectral resolution in large parts of the sky (Haffner et al. [218]). We will discuss some of the results obtained by this method.

At high galactic latitudes, the intensity of the Hα lines decreases with increasing latitude like (Reynolds [429])

$$\langle I_\alpha(b) \rangle \simeq 2.9 \times 10^{-7} \mathrm{cosec}|b| \; \mathrm{erg\,cm^{-2}\,s^{-1}\,sterad^{-1}}. \tag{5.59}$$

The emission measure EM (cf. (5.24)) of the diffuse ionized gas can be obtained through the relation

$$\boxed{I_\alpha = 8.7 \times 10^{-8} (T/10^4 \mathrm{K})^{-0.92} EM \; \mathrm{erg\,cm^{-2}\,s^{-1}\,sterad^{-1}},} \tag{5.60}$$

where the emission measure is expressed in cm^{-6} pc. Equations (5.59) and (5.60) yield an emission measure perpendicular to the galactic plane of 2.7 cm^{-6} pc. The scale height of the medium being probably of the order of 1 kpc, $\langle n_e^2 \rangle^{1/2} \simeq 0.05$ cm^{-3}. Since we know independently that $\langle n_e \rangle \simeq 0.03$ cm^{-3} (see above), we see that the medium is rather uniform.

The temperature of the diffuse ionized gas can be obtained by comparing the widths of the Hα and [S II] lines. These lines are broadened by thermal Doppler broadening which is proportional to $\mathcal{A}^{-1/2}$, $\mathcal{A}$ being the atomic mass number (~ 1 for H and 32 for S), and by turbulent Doppler broadening which does not depend on $\mathcal{A}$ (cf. (4.30) and (4.31)). We find in this way a temperature of the order of 8 000 K. The non-detection of the [O III]$\lambda 5\,007$ line and the relatively small abundance of He$^+$ (He$^+$/He^0 = 0.3-0.6) show that excitation (i.e. the fraction of very high-energy far-UV photons) is relatively weak in this medium. Nitrogen is then expected to exist mostly as N II, rather than N III which would require an energy of 29.6 eV, the ionization potential of N II. Haffner et al. [218] show that in these conditions the intensity of the N II$\lambda 6\,583$ line is

$$I_{6\,583}(R) = 5.95 \times 10^4 \left(\frac{\mathrm{H}^+}{\mathrm{H_{tot}}}\right)^{-1} \left(\frac{\mathrm{N_{tot}}}{\mathrm{H_{tot}}}\right) \left(\frac{\mathrm{N}^+}{\mathrm{N_{tot}}}\right) T_4^{-0.474} e^{-2.18/T_4} EM, \tag{5.61}$$

where the intensity $I(R)$ is in rayleigh (1R = 2.4 10^{-7} erg cm^{-2} s^{-1} sterad^{-1} at Hα)[3]. T_4 it the temperature in units of 10^4 K and EM is the emission measure in

---
[3] The rayleigh is either a unit of monochromatic brightness of the sky which is $10^6/4\pi$ photons cm^{-2} s^{-1} sterad^{-1} Å^{-1} (cf. Leinert et al. [313]), or a unit of integrated brightness in a line which is then $10^6/4\pi$ photons cm^{-2} s^{-1} sterad^{-1}.

cm^{-6} pc. Reynolds et al. [431] deduce from their observations of the [O I]λ6 300 line that hydrogen is almost entirely ionized in the diffuse ionized medium, so that $n(H^+)/n(H_{tot}) \simeq 1$. The reason is that the charge-exchange reaction $H^+ + O^0 \longleftrightarrow H^0 + O^+$, which is a fast reaction because the ionization potentials of H and O are very similar (13.598 et 13.618 eV respectively), produces a substantial quantity of neutral oxygen from O II, so that

$$\frac{I_{6\,300}}{I_{H\alpha}} \propto \frac{n(H^0)}{n(H^+)} \frac{n(O_{tot})}{n(H_{tot})} f(T_e), \tag{5.62}$$

$f(T_e)$ being a function of temperature given by Reynolds et al. [431]. From a measurement of the $I_{6\,300}/I_{H\alpha}$ ratio, O/H and T_e being known, we can derive H^0/H^+ which is small with respect to 1, so that H is almost fully ionized. There is a flaw in this reasoning, however: if a fraction of the [O I]λ6 300 line is produced in shocks, the degree of ionization of H is smaller. If we assume however that the medium is fully ionized, including for the elements with an ionization potential larger than that of hydrogen, there is no neutral nitrogen and $N^+/N_{tot} \simeq 1$. We can then simply write, using the expression giving the intensity of the Hα line $[I_{H\alpha}(R) = 0.364\,T_4^{-0.9} EM]$:

$$\frac{I_{6\,583}}{I_{H\alpha}} \simeq 1.63 \times 10^5 \left(\frac{N_{tot}}{H_{tot}}\right) T_4^{0.426} e^{-2.18/T_4}. \tag{5.63}$$

This ratio can be used to obtain the temperature. We verify that it is of the order of 8 000 K, but we also find that it is larger in regions of small density and that it increases with increasing distance to the galactic plane. High values of the temperature have also been derived more directly from the ratio of the auroral [N II]λ5 755 line and the nebular [N II]λ6 854 one (Reynolds et al. [433]). They are in part responsible for the enhancement in the [N II]/Hα and [S II]/Hα line ratios. They also suggest the presence of another source of heating than stellar radiation, which is proportional to n_e rather than to n_e^2. This might be the dissipation of plasma turbulence (Reynolds et al. [432]).

There are also variations in the intensity ratio of the [S II]/[N II] lines with distance to the galactic plane, that can be attributed to a partial ionization of S II into S III. This ionization is easier than for N II because the ionization potential of S II is only 23.33 eV.

5.3 The Hot Gas

As early as 1956 Spitzer [486] suggested the existence of a hot gas ($T > 10^6$ K) in the Galaxy, whose pressure would maintain in equilibrium the high-velocity H I clouds in the galactic halo. In 1968 Bowyer et al. [64] discovered a diffuse emission in soft X-rays (< 1 keV) that they attributed to this gas. In 1974 Jenkins & Meloy [259] and York [562] observed with the COPERNICUS satellite absorption lines of O VI at 1 032 and 1 038 Å in many lines of sight. These lines cannot be due to circumstellar envelopes because their radial velocities bear no relation with the radial velocity of the hot stars in front of which they are observed. Interstellar absorption lines of N v and C IV were also detected. Finally, Inoue et al. [256], Schnopper et al. [459] and more recently Sanders et al. [448] have observed the emission of interstellar X-ray lines from O VII, O VIII and other ions. All these observations demonstrate as we will see the existence of a hot diffuse gas at temperatures of a few 10^5 K to 10^7 K. This gas comes from supernova remnants and bubbles (see later Chap. 12): these objects contain hot gas which eventually fills up a portion of the interstellar medium. The existence of this component which adds to the components we have described previously lead McKee & Ostriker [359] to propose their 3-component model of the interstellar medium, that we will discuss Sect. 15.2.

5.3.1 Collisional Ionization by Electrons at High Temperatures

It would be very difficult to ionize oxygen radiatively to reach O VI, because this would require photons of 114 eV. On the other hand, collisional ionization of a gas is possible at high temperatures: a hot gas ionizes spontaneously through atom–atom collisions. The electrons liberated in this way thermalize very rapidly and are eventually responsible for most of the ionization, which leads to multi-charged ions at the temperatures of the interstellar hot gas. The equations that govern ionization and recombination can easily be derived from the equations we have given for H II regions, although ionization was radiative in this case. Collisional ionization can be represented by a probability per atom and per unit time P^c_{nk}. This quantity is given for an hydrogen-like atom by the semi-empirical approximate expression

$$P^c_{nk} \simeq 7.8 \times 10^{-11} T_e^{1/2} n^3 \exp(-\chi_n) n_e \text{ s}^{-1}, \tag{5.64}$$

where n is the quantum number of the level from which ionization occurs, k is the continuum level and χ_n is the ionization potential from level n, in units of kT_e.

Inversely, the recombination probability on level n writes

$$P^c_{kn} = \frac{n_e h^3}{(2\pi m_e k T_e)^{3/2}} n^2 \exp(\chi_n) P^c_{nk}, \tag{5.65}$$

which is obtained by applying the Saha equation (cf. (5.29) with $b_n = 1$).

The calculation and measurement of ionization and recombination cross-sections of the various important elements and ions are the subjects of considerable activities due to their importance for plasma physics. For a recent review see Böhringer [50]. Figure 5.11 shows the result of calculations of the ionization structure at equilibrium for multi-charged oxygen ions. We see that O VI (O^{5+}), which gives interstellar absorption lines, is abundant at 3×10^5 K, and that the O VII and O VIII ions whose X-ray lines are observed in emission indicate higher temperatures, of the order of 10^6 K. Iron is another very important element for the study of hot gas, but the atomic data for iron have been recently considerably modified and are perhaps still uncertain. The lines for multi-charged Fe ions have been well observed, in particular in the hot intergalactic medium of clusters of galaxies. It is certain that the recent

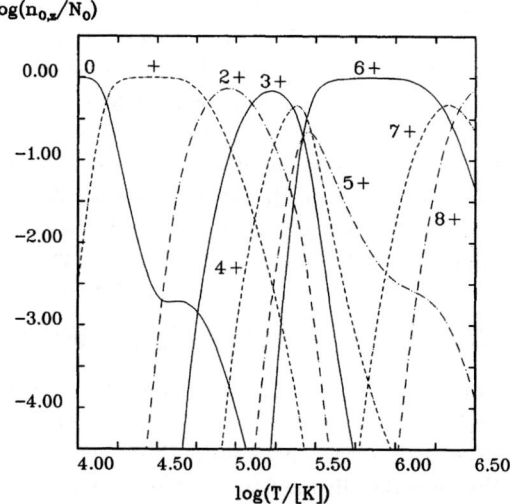

Fig. 5.11. Ionization structure of oxygen at equilibrium. The abundance of the different oxygen ions is given as a function of temperature. From Schmutzler & Tscharnuter [458], with the permission of ESO.

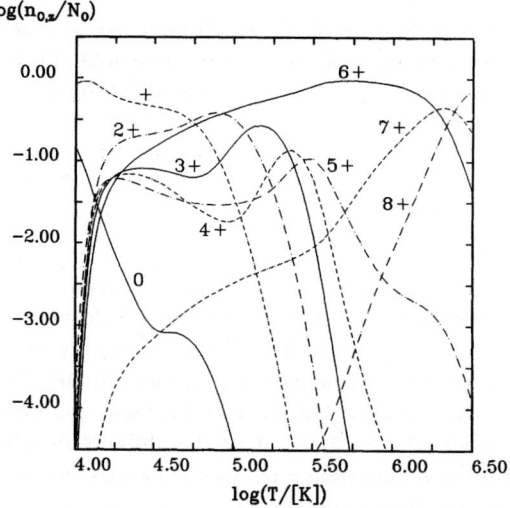

Fig. 5.12. Ionization structure of oxygen outside equilibrium in a cooling plasma. The abundance of the different oxygen ions is given as a function of temperature. Compare to Fig. 5.11. From Schmutzler & Tscharnuter [458], by permission of ESO.

X-ray satellites AXAF–CHANDRA and XMM–NEWTON will supply important data on the galactic hot gas thanks to observation of these lines and also of lines of Mg, Ne, S, Si, and of course O, for which the previous observations were somewhat marginal (cf. for exemple Sanders et al. [448]). Plate 11 shows a recent observation (imaging and spectroscopy) of the hot gas in the Large Magellanic Cloud obtained with XMM–NEWTON.

A problem in the interpretation of these observations is that it is not certain that the hot gas is in ionization equilibrium. Recombination is a very slow process at the very small densities of the medium ($n_e \ll 10^{-2}$ cm^{-3}). For example, the characteristic time for recombination of C v into C iv is of the order of 2×10^6 years for $n_e = 10^{-2}$ cm^{-3}. It might be necessary to use non-equilibrium models. Such models have been developed e.g. by Breitschwerdt & Schmutzler [68]. Figure 5.12 shows the model result for oxygen. Comparison to Fig. 5.11 shows very important differences. A diagnostic based on oxygen lines (or lines of other elements) would yield in this case temperatures much lower than with an equilibrium model. These temperatures would also be lower than that given by the observation of the X-ray continuum from the same gas. It is not yet possible with the existing data to make a choice between these models.

5.3.2 The Emission of X-Ray Lines

The interpretation of X-ray spectra was made as early as 1917. In an atom the electrons are arranged in layers designated as K, L, M, etc. starting from the deepest one that contains 2 electrons. An electron can be ejected from this K layer after absorption of a photon with sufficient energy, or by collision with a free electron. Of course the photon or electron energy must be higher than the ionization potential from level K. When illuminated by X-ray photons, atoms thus produce an X-ray absorption spectrum with a discontinuity corresponding to this ionization potential. But they do not produce X-ray absorption lines similar to optical absorption lines. The hole left in the K layer after an electron has been ejected can be filled by an electron belonging to an outer layer, with emission of a X-ray photon. $K\alpha$, $K\beta$, etc. lines correspond to these transitions from layers L, M, etc. respectively. The frequency of the $K\alpha$ line produced by an element with nuclear charge Z is given by the *Moseley law*

$$\nu = R(Z - a_s)^2 \left(\frac{1}{1^2} - \frac{1}{2^2}\right) = \frac{3}{4} R(Z - a_s)^2, \quad (5.66)$$

where R is the Rydberg constant and a_s, the *screening constant*, is close to 1. In reality level K is simple, level L is triple, level M is quintuple, etc. Figure 5.13 shows the disposition and designation of these levels and of the permitted lines which connect them. In particular the $K\alpha$ and $K\beta$ lines are double (and not respectively triple or quadruple), due to the selection rules.

Multi-charged atoms also emit optical forbidden lines, initially discovered in the solar corona. Some of these lines have been observed in the hot gas of interstellar shocks in supernova remnants and in the general medium (cf. Plate 11).

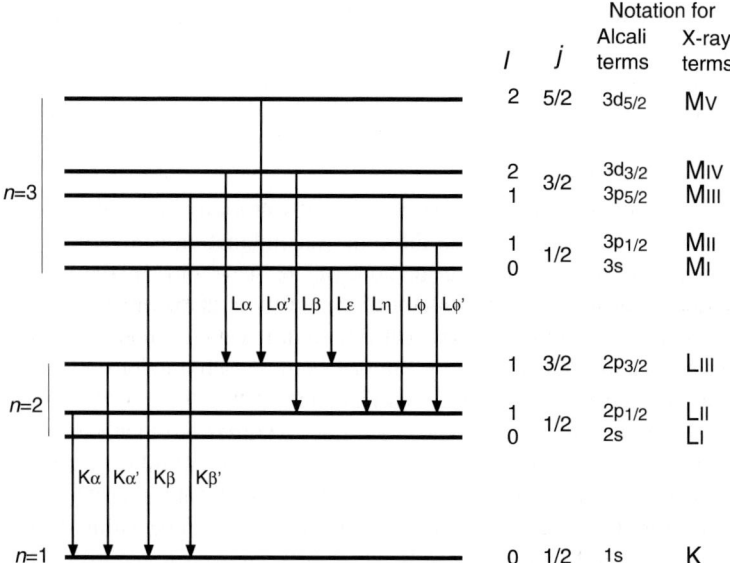

Fig. 5.13. Energy diagram showing the designation of X-ray lines and the correspondence of the levels to those of alcaline atoms, from which the selection rules can be derived. From Born [57].

5.3.3 The Thermal X-Ray Continuum

Hot plasmas emit a free–free emission, or *Bremsstrahlung*, and a free–bound emission similar to that of H II region plasmas at $\sim 10^4$ K.

The corresponding free–free emissivity is given by (5.18) or (5.19). The Gaunt factor $g_{ff}(\nu, T_e) = (\sqrt{3}/\pi) \ln \Lambda \simeq 0.55 \ln \Lambda$ (cf. (5.20)) at the energies of interest here is such that

$$\Lambda = \frac{4kT}{\gamma h\nu} \simeq 4.7 \times 10^{10} \, (T/\nu), \tag{5.67}$$

with $\gamma = 1.781$. The optical depth being always small in the interstellar case, the emissivity varies like $T^{1/2} \exp(-h\nu/kT)$. The shape of the observed spectrum allows to measure temperature.

The free–bound emission is also of importance. Tables giving the total continuum emission at soft X-ray energies (1 Å to 30 Å, or 12.4 keV to 0.41 keV) for temperatures between 8×10^5 K and 10^8 K are published by Culhane [109].

5.3.4 Results

In spite of an early start, our knowledge of the hot interstellar medium is fragmentary and uncertain, due to difficulties in observations and in their interpretation. The hypothesis of a pressure equilibrium between clouds and the hot medium, which triggered the first researches and is at the basis of models of the interstellar medium like that of McKee & Ostriker [359], is controversial. Also as we will see Sect. 15.2, thermal conduction between the hot gas and the colder clouds, that plays a major role in this model, is uncertain. As a consequence, the fraction of the volume of the galactic disk occupied by the hot gas is poorly known. From a numerical simulation of the evolution of the supernova remnants which are at the origin of the hot gas, Ferrière [178] estimates this fraction as about 30% in the plane of symmetry of the disk. It would increase fast with distance to the plane to reach some 80–90% at 2 kpc.

Along a given line of sight, we observe in the X-ray continuum a superimposition of emission by nearby regions with emission for more distant regions absorbed by the intervening interstellar clouds. However we can observe in some directions the emission in front of a cloud which can be considered as opaque to the more distant X-ray radiation. This is the case for the emission of the Local bubble in front of a cloud located at 55 pc (de Boer & Kerp [114]). The analysis of the X-ray continuous spectrum in this direction, after correction for the residual absorption by the local interstellar medium, allows to obtain with a good degree of certainty the temperature ($T_e = 10^{5.9}$ K), the density ($n_e = 0.004$ cm^{-3}) then the pressure P ($P/k = n_e T_e = 16\,500$ K cm^{-3}) in the bubble. The X-ray observations of the galactic halo give results difficult to interpret (Hurwitz & Bowyer [254]).

Much progress is coming from the FUSE satellite with which many observations of the O VI absorption lines have been made in the direction of extragalactic far-UV sources. We already know that the mean density of O VI in the galactic plane is about 2×10^{-8} ion cm^{-3} (Savage et al. [452]; Zsargó et al. [567]), which would correspond to a density of hot gas near 3×10^5 K of the order of 3×10^{-5} cm^{-3}. But there is also hot gas at other temperatures, and the actual density in the hot regions is certainly higher than 3×10^{-5} cm^{-3} since they fill only a fraction of the volume of the disk. The scale height of O VI above the disk is very high, 2.3 to 4 kpc. That of the ions with a lower ionization potential is still larger: 3.9 kpc for N V, 4.4 kpc for C IV, and 5.1 kpc for Si IV.

On the other hand, O VI associated with the high-velocity H I clouds has been observed with FUSE (Sembach et al. [465]), that could result from heating of their surface by conduction from the hot gas within which they are immersed.

5.4 The X-Ray Absorption

The X-ray absorption giving an information on the total column density of interstellar matter with little dependence on its physical state, we will examine it separately here.

As explained Sect. 5.3, any material not fully ionized in front of a X-ray source absorbs its radiation, with discontinuities characteristic of the absorbing elements. The Lyman continuum absorption of atomic hydrogen, with its discontinuity at 911.7 Å, is the absorption with the longest wavelength. Then comes the absorption by neutral helium with a discontinuity at 504 Å, etc. The free–bound absorption cross-section writes in a Coulomb electric field

$$\sigma_a(\nu) = \frac{32\pi^2 e^6 R Z^4}{3^{3/2} h^3 \nu^3} g_{fb}(T) = 2.8 \times 10^{29} \frac{Z^4}{\nu^3} g_{fb}(T) \text{ cm}^2, \qquad (5.68)$$

where g_{fb} is the free–bound Gaunt factor, which is not very different from 1 in the interstellar conditions (see complete tables in Karzas & Latter [279]). However, for multi-electron atoms or ions for which the electric field is not coulombian, we must introduce a screening factor. The absorption cross-section can then differ considerably from that given by (5.68). Molecular hydrogen has a cross-section per H atom larger by 40% than that of atomic hydrogen (Yan et al. [561]). These authors also give values for the X-ray absorption cross-sections of He I and He II. Of course, H II and He III being completely ionized do not absorb X-rays. If we take into account the most abundant elements with their usual cosmic abundances, we obtain the total absorption cross-section per H atom represented in Fig. 2.5. This cross-section is very large from 912 Å to about 0.1 keV (124 Å), but the radiation of a few nearby very hot stars could still be observed in this range, in particular with the EUVE satellite. This satellite has allowed to obtain the degree of ionization of helium in the very local interstellar medium from the ratio between the He II discontinuity at 228 Å and the He I one at 504 Å. This degree of ionization is rather variable from direction to direction, while being generally less than 60% (Bowyer [65]).

At those energies where it has initially been measured (0.5 to 2 keV), the X-ray absorption is dominated by heavy elements, in particular C, N and above all O. Ryter et al. [445] have shown that near the Sun the absorption measured in front of X-ray sources is proportional to the column density $N(H\,\text{I})$ of atomic hydrogen and also to the colour excess $E(B - V)$ hence to the visual extinction A_V, that are proportional to the amount of dust (see later Sect. 7.1). They give the relations

$$N_X/N(H\,\text{I}) = 1.2 \pm 0.6 \text{ equivalent H atom per H atom}, \qquad (5.69)$$

$$N_X/E(B - V) = (6.8 \pm 1.6) \times 10^{21} \text{ equivalent H atom cm}^{-2} \text{ mag}^{-1}. \qquad (5.70)$$

In these equations N_X is an equivalent column density of hydrogen obtained from the X-ray absorption assuming the solar abundances of Table 1 of Morrison & McCammon [374], which are very close to the solar abundances that we give Table 4.2, column 2. The first relation is uncertain because the 21-cm line is somewhat optically thick in several of the studied directions, so that we must assume a value for the gas temperature to derive $N(H\,\text{I})$ from the line intensity (here 100 K have been assumed). Moreover molecular hydrogen has been neglected. The second relation is safer. It differs little from the relation between colour excess and total hydrogen column density (including H_2), that is derived more directly from observations with the COPERNICUS satellite (see later (7.5)). This shows that the actual interstellar abundances of C, N and O are not very different from the solar abundances.

At large column densities of H I, $N(H\,\text{I}) > 6 \times 10^{20}$ atom cm^{-2}, N_X is considerably larger than $N(H\,\text{I})$, implying large amounts of molecular hydrogen. X-ray absorption in molecular clouds is discussed further in Sect. 6.3.

6 The Interstellar Medium at High Energies

In this chapter, we gather together the various subjects related to high-energy particles and radiation which are present in the interstellar medium: firstly cosmic rays (Sect. 6.1), then the gamma-ray continuum and its formation (Sect. 6.2), and finally the gamma-ray lines (Sect. 6.4). The intermediate Sect. 6.3 shows how observations of the gamma-ray continuum allows us to measure the mass of the interstellar medium and compares this with other methods.

6.1 Cosmic Rays

Cosmic rays consist of atomic nuclei, and approximately 1% electrons. They have relativistic energies that can be as high as 10^{20} eV. When arriving at the Earth, they interact with the high atmosphere and produce neutrons. These neutrons have been studied for a long time using neutron monitors. The cosmic particles with energies greater than 10^{12} eV also generate showers of secondary ionizing particles that spread over several thousands of square metres at ground level. The corresponding flux of these secondaries is about 1 particle cm^{-2} s^{-1}. The existence of ionizing particles able to discharge an electrometre was established by Hess [238] in 1912. The observation that the flux of the ionizing particles increases with altitude indicates their celestial origin.

The term "rays" is thus not appropriate to describe these particles, but it is anchored in history. Since their discovery the problem of the origin of cosmic rays has been actively debated, because their existence implies that nature is able to accelerate particles to extremely high energies with unexpected ease.

Those cosmic rays that reach the Earth should have kept some memory of their origin and of their propagation through the interstellar medium where they could have spent several millions years. To arrive at the Earth they had to interact with the solar environment (the solar wind and magnetic field) and finally had to penetrate the terrestrial magnetic field. An understanding of these various interactions is still the subject of active research.

We will show that cosmic rays are of galactic origin, and not universal or metagalactic except perhaps at very high energies, and we will present the effect of the solar and terrestrial environments on their propagation until they reach detectors. Then we will discuss the properties of cosmic rays and the abundances of the nuclei. The very-high energy particles will then be briefly discussed as well as the cosmic electrons. We will end with their confinement and lifetime in the Galaxy[1]. The acceleration of cosmic rays will be dealt with in Chap. 12.

6.1.1 The Origin of Cosmic Rays

The question of the origin of cosmic rays – galactic or universal – has been raised many times. It has recently been answered by the measurement of gamma-rays emitted by the interaction of intergalactic cosmic rays with the interstellar medium of a nearby galaxy. For this, we require a galaxy which cannot itself accelerate or confine a large cosmic ray flux. Spiral galaxies like ours are excluded because they contain abundant high-energy particles. A good choice is the Small Magellanic Cloud, a nearby low-mass galaxy which seems to be breaking up after a passage close to our Galaxy about 200 million years ago, and which is not gravitationally bound.

Indeed, as we will see later (Sect. 6.2), collisions of high-energy charged particles with interstellar nuclei produce a gamma-ray flux with energies of hundreds of MeV, with an intensity (given in photons cm^{-2} s^{-1} ster^{-1}) of

$$I_\gamma = N_H \epsilon_\gamma, \tag{6.1}$$

where ϵ_γ is the gamma-ray emissivity per hydrogen atom and per steradian given by (6.20), and N_H is the column density. If cosmic rays could fill up the whole Universe, or at least the Local group of galaxies, with the same flux as observed in the Galaxy, the gamma-ray flux at energies larger than 100 MeV received from the Small Magellanic Cloud would be 2.4×10^{-7} photon cm^{-2} s^{-1}. Observations give only 0.5×10^{-7} photon cm^{-2} s^{-1}, five times less than expected from interaction with a universal cosmic ray flux (Sreekumar et al. [493]). The observed flux is roughly what would be expected, given the structure and the mass of the Small Magellanic Cloud. Note also that the flux of relativistic electrons, as derived from the observation of their radio emission, is also five times smaller than the galactic electron flux, implying that the electron/nucleus flux ratio is the same in the two galaxies. Despite the difficulties with the measurements and their interpretation, these results are convincing enough to exclude a universal or metagalactic origin for cosmic rays.

6.1.2 Solar Cosmic Rays and Solar Modulation

The solar corona is not static, but evaporates at a rate of the order of 10^{-13} M$_\odot$/yr, producing an expanding ionized gas, i.e. the solar wind. Its intensity is modulated by solar activity. While there are sporadic high fluxes driven by the most active solar phenomena, the average intensity of the solar wind varies with the 11-year solar cycle. During quiet periods the density of solar wind particles at the Earth's orbit is

---

[1] A complete, but somewhat outdated, review will be found in the book of Ginzburg & Syrovatskii [198].

of the order of 5 cm^{-3}, their velocity 300 to 500 km s^{-1} and their temperature about 10^5 K. During an active period all these quantities can be one order of magnitude larger. The solar magnetic field is carried with, and frozen into, this expanding plasma. It forms magnetic tubes with the same sign as the magnetic field at the surface of the Sun. Due to the rotation of the Sun, these tubes deform as Archimedes spirals. However the irregularities in solar activity produce variations in the velocity and the density of the wind which affect this simple picture. The structure therefore reflects the history of solar activity during the last few weeks to months.

The intensity of the magnetic field is of the order of 10 µG near the terrestrial orbit (see e.g. Encrenaz et al. [166]). The shock wave which limits the solar cavity to where the solar wind encounters the interstellar medium is located at about 100 AU, assuming pressure equilibrium for the local interstellar medium of $T = 7\,000$ K and $n_e = 0.1$ cm^{-3} outside the solar cavity.

The high-energy particles entering the solar cavity from the interstellar medium can reach the Earth's orbit only if their gyration radius is large relative to the size of the interplanetary magnetic structures, typically a fraction of a AU. Their energy must then be larger than about 1 GeV/nucleon.[2] Particles with lower energies can only reach the terrestrial environment by diffusion, and only with increasing difficulty at lower energies and when the solar wind is strong. The cosmic-ray flux is strongly reduced at energies lower than 100 MeV/n. A clear anti-correlation can be observed between the cosmic-ray flux, at energies of 1–2 GeV/n, and solar activity as measured by the number of sunspots. This phenomenon is called the *solar modulation*. It is necessary to understand it well in order to derive the spectrum and composition of interstellar cosmic rays from satellite observations (see for example Fisk et al. [182]).

Solar bursts are most often accompanied by charged particles emitted by the Sun. Their energy can be as high as a few hundreds of MeV. They are carried by the magnetic field in the solar cavity and certainly reach the interstellar medium. When passing near the Earth they are detected by cosmic-ray detectors. These are the *solar cosmic rays*.

The terrestrial magnetic field also affects the propagation of charged particles but it is not very extended. It is sufficiently well-known for its effects to be predicted accurately and even exploited to determine the energy of the particles. Its structure is dipolar to a first approximation and so charged particles always reach the Earth's magnetic polar regions. This is, in particular, the case for electrons accelerated during solar bursts. They are guided to the polar regions and produce aurorae. At lower terrestrial latitudes, l, the particles must have an energy E which is larger than a limit called the geomagnetic cut-off in order to be detectable near the Earth's surface. Here are a few values of the geomagnetic cut-off, from Longair [336]:

$$l = 60° \quad E = 0.48 \text{ GeV/n}$$
$$l = 40° \quad E = 4.3 \text{ GeV/n}$$
$$l = 0° \quad E = 14.0 \text{ GeV/n}.$$

---

[2] The energy of cosmic-ray nuclei is generally given per nucleon, and is expressed in GeV/nucleon or MeV/nucleon, in short GeV/n or MeV/n.

6.1.3 Galactic Cosmic Rays

Traditionally, *galactic* or *primary cosmic rays* are the particles with energies between ~ 1 and $\sim 10^6$ GeV/n. Their energy is large enough for them to reach the solar environment but small enough to insure confinement in the Galaxy. They are easily detected by nuclear-physics instruments and have been studied by many space missions. The abundances of particles with atomic nuclei between 1 and 32 (from H to Ge) are shown in Fig. 6.1, compared to the standard (or "universal") element abundances. We see that the relative abundances of cosmic nuclei follows, more or less, that of the elements, with noticeable exceptions (Li, Be, B) that will be discussed later.

Figure 6.2 shows the observed energy spectrum for a few cosmic-ray nuclei between 0.01 et 100 GeV/n. The intensity variations at low energies show that the modulation depends upon the quiet or active state of the Sun. The rise at energies lower than a few tens of MeV/n corresponds to the average flux of solar cosmic rays and has nothing to do with galactic cosmic rays.

Various attempts have been made to estimate the flux of low-energy cosmic rays taking into account the solar modulation. Prantzos et al. [413] have calculated this flux from the abundances of the light elements Li, Be et B which are produced by the interaction of the low-energy cosmic rays with heavy interstellar nuclei (see next

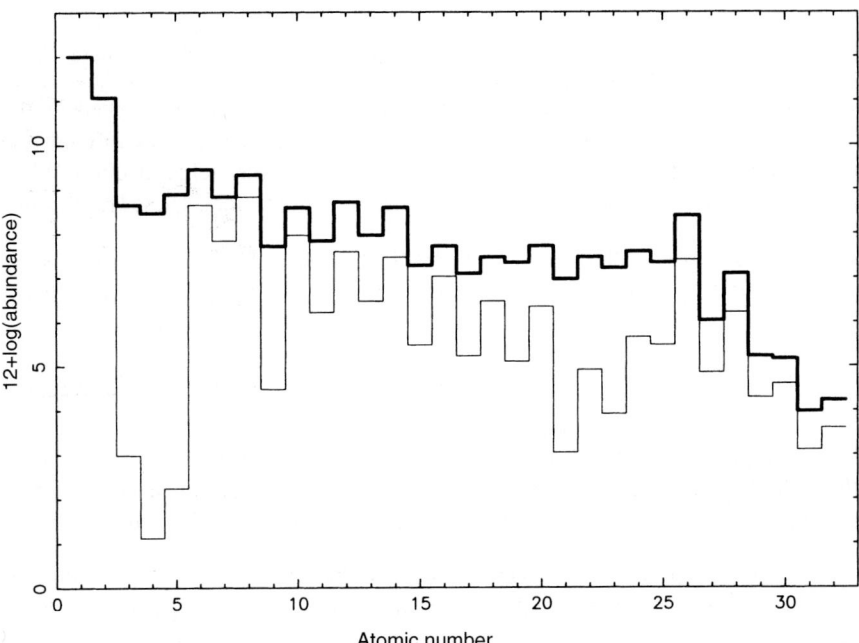

Fig. 6.1. Abundances of elements in the cosmic rays (thick line) compared to the Galactic abundances (thin line). They are both normalized to 10^{12} for hydrogen. Notice the considerable overabundance of Li, Be and B in cosmic rays. Adapted from Lund [342].

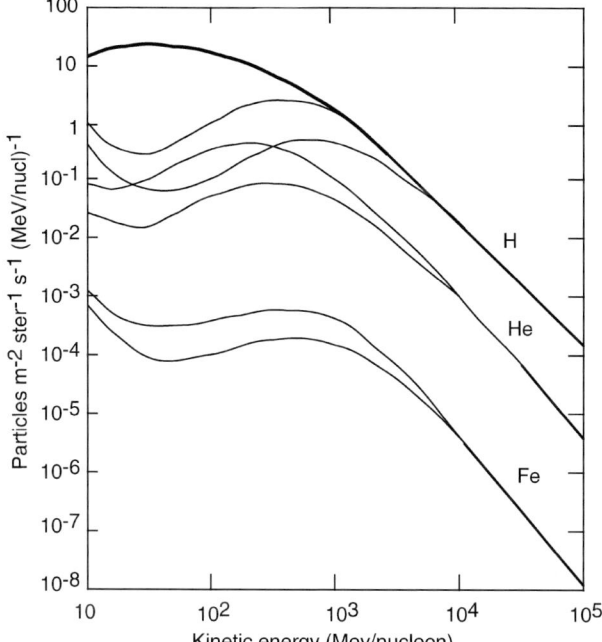

Fig. 6.2. Observed energy spectrum for cosmic-ray protons (H), α particles (He) and iron nuclei. There are two values below 10^4 MeV/nucleon. The higher one corresponds to the minimum of solar activity, the lower one to the maximum of activity: this is the solar modulation. At very low energies, the flux rise is due to the solar cosmic rays. The bold curve is an estimate of the proton spectrum outside the solar cavity, due to Prantzos et al. [413]. Adapted from Silberberg & Tsao [473].

section). This requires a model for the propagation of cosmic rays and a model of the chemical evolution of the Galaxy. The results are shown in Fig. 6.2 and are in good agreement with the calculations by Ip & Axford [257], based on a theory of cosmic-ray acceleration.

Between 3 and 300 GeV/n, the differential energy spectrum can be represented by a power law[3]:

$$I(E) = 1.34 \times 10^4 \left(\frac{E}{\text{GeV/n}}\right)^{-2.6} \text{m}^{-2}\text{s}^{-1}\text{ster}^{-1}(\text{GeV/n})^{-1}. \quad (6.2)$$

At higher energies the spectrum becomes steeper, the modulus of the exponent of the power law increasing by 0.5. We will discuss later the very high energy cosmic rays. The total pressure of cosmic rays is estimated to be 1.3×10^{-12} dyne cm^{-2}. Particles with less than 300 GeV/n contribute one quarter of the total pressure (Holzer [246]).

---

[3] Note that, according to specialist usage, the particle fluxes are given per square metre.

High-energy particles circulating in the interstellar medium experience various interactions corresponding to energy losses. The most important, which mainly affects the low-energy particles that cannot be observed inside the solar cavity, is ionization. This will be treated Sect. 8.1. Energy losses due to free–free interactions and synchrotron radiation are negligible except for electrons. Inelastic collisions with interstellar hydrogen nuclei are not important for the energetic balance but do produce π^0 mesons which desintegrate in a pair of gamma-ray photons. This phenomenon will be discussed Sect. 6.2, as well as the Bremsstrahlung and the inverse Compton effect.

Detailed studies of the abundances of cosmic-ray nuclei, as a function of atomic number, have shown that they can be explained by Galactic sources. We can express these abundances with reference to the standard elemental abundances (Fig. 6.1), even if there are some doubts about the exact values of these reference abundances as discussed Sect. 4.1. Three remarkable observations are:

- a high overabundance of the light elements Li, Be and B with respect to the standard abundances;
- a positive correlation of the abundances with the first ionization potential of the corresponding neutral atom;
- an unexpectedly high abundance of refractory atoms which are generally condensed in interstellar dust grains (see Chap. 7).

We will examine these three points in detail.

Light Elements

Figure 6.1 shows that the relative abundances of the L (for light) nuclei (Li, Be and B) in cosmic rays lie roughly on the interpolation of the abundances of elements with nearby atomic numbers, while their standard elemental abundances are lower by factors of 10^6 to 10^7 with respect to their neighbours.

These L elements are fragile. They are destroyed by capture of a low-energy proton in exothermal reactions like

^{6}Li + p → ^{4}He + ^{3}He
^{7}Li + p → 2 ^{4}He
^{9}Be + p → ^{6}Li + ^{4}He → 2 ^{4}He + ^{2}H
^{10}Be + p → ^{7}Be + ^{4}He
^{11}Be + p → 3 ^{4}He,

with reaction thresholds lower than 10 keV. It follows that when these elements are carried to some depth in the convective zone of stars they are destroyed in a few million years as soon as the temperature reaches some 10^7 K (cf. for example Reeves [427]). The L elements in the interstellar medium cannot come from stellar nucleosynthesis, like the other elements, but must have another origin. They originate in high-energy reactions in the interstellar medium.

Collisions of high-energy protons or alpha particles, circulating in the interstellar medium, with C, N or O nuclei produce *spallation* reactions that destroy these atoms

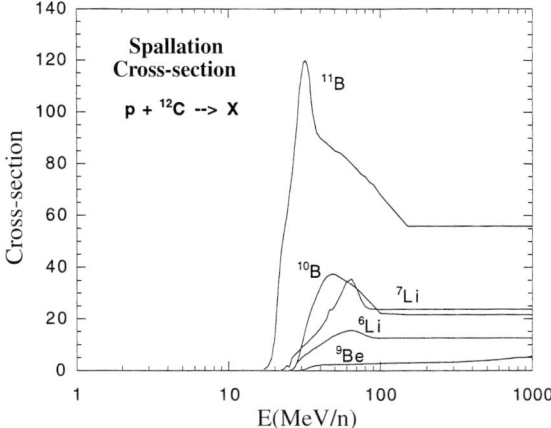

Fig. 6.3. Cross-section for production of Li, Be and B nuclei by spallation of a ^{12}C nucleus by a proton. The cross-sections for spallation of ^{16}O and ^{14}N are not very different. From Parizot (thesis, 1997, Université Paris VI), adapted from Read & Viola [426].

and produce smaller atoms including Li, Be and B. The inverse reactions for C, N and O nuclei in cosmic rays acting on atoms at rest in the interstellar medium also exist. However, their efficiency is of the order of 20% of that for the direct reactions. All of these reactions have a very sharp threshold near 50 MeV, which corresponds to the maximum cross-section. At higher energies the cross-sections decrease as shown in Fig. 6.3. The most efficient producers of the L elements are thus the low-energy cosmic-ray particles, whose flux cannot be measured directly because of the solar modulation. The abundance of the L elements is proportional to the product of the flux of cosmic rays near 50 MeV and the abundance of C, N and O, integrated over the life-time of the Galaxy. After the pioneering work of Meneguzzi et al. [365], a self-consistent treatment of the problem was performed by Prantzos et al. [413]. They included the temporal evolution of the abundances of C, N, O, the shape of the galactic halo, and the lifetime and energy losses of cosmic rays through ionization. They assumed an initial cosmic-ray spectrum that was flatter and more intense in the past due to the higher number of supernovae. The escape length $L = 6$–10 g cm^{-2} (see (6.5)) is derived from the present abundance ratio (Li,Be,B)/(C,N,O) in cosmic rays. The derived cosmic-ray spectrum is represented in Fig. 6.2 by a bold line. Unfortunately this method gives no information on cosmic rays at energies lower than 50 MeV. There is some hope to derive the low-energy flux using future observations of nuclear gamma-rays, as will be discussed in Sect. 6.3.

It is remarkable that these L elements, which contrary to all other elements are destroyed, rather than formed, in stellar interiors, are produced by spallation with abundances similar to those of the other elements. This is only one of the many curious coincidences in the study of the Universe.

Other Elements (A > 10)

It was noted long ago that the cosmic ray abundances, relative to the standard abundances, tend to decrease with the value of the first ionization potential of the element, as shown by Fig. 6.4. In particular, protons (hydrogen nuclei) and alpha particles (helium nuclei) are roughly 10 to 30 times less abundant than the heavier nuclei in comparison to the cosmic abundances. This suggested that the originating material from cosmic rays comes from stellar coronae, and that it is more easily accelerated when it is easier to ionize. Of course, when the energy of the accelerated particles increases they are stripped of electrons from deeper layers and end up as bare nuclei.

This reasoning cannot apply to the refractory elements which are generally locked into interstellar dust grains (Table 4.2). We know for example that iron is almost entirely in grains and that the interstellar gas contains no more than 1% of atomic or ionic iron (cf. Table 4.2). However, Fig. 6.5 shows that heavy and refractory elements like iron are amongst the most abundant in cosmic rays. The solution to this apparent contradiction must be searched for in a deeper study of the acceleration mechanisms. These will be discussed in Sect. 12.4. We just mention here that dust grains are in general electrically charged and that, despite their enormous gyration radii in the interstellar magnetic field, they can be accelerated as very heavy ions. When accelerating, they experience sputtering arising from the impact of the

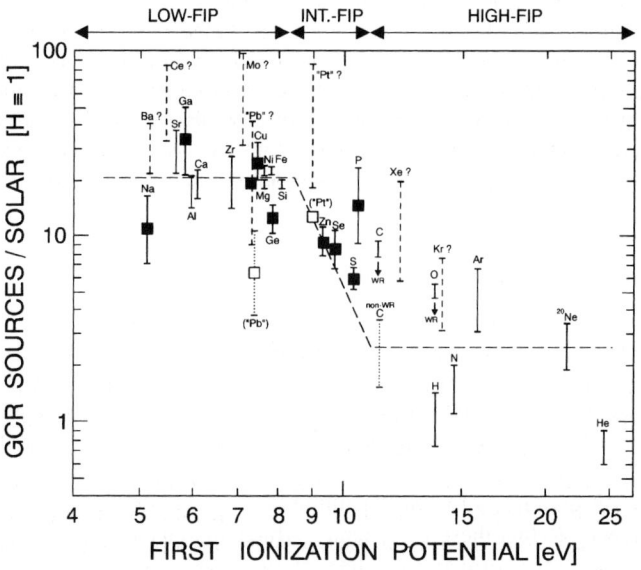

Fig. 6.4. Abundances of the elements in galactic cosmic rays (GCR) normalized to the standard abundances as a function of the first ionization potential (FIP). The elements represented by a filled square are refractory. Note that they are as abundant as the volatile elements. Reproduced from Meyer et al. [367], with the permission of the AAS.

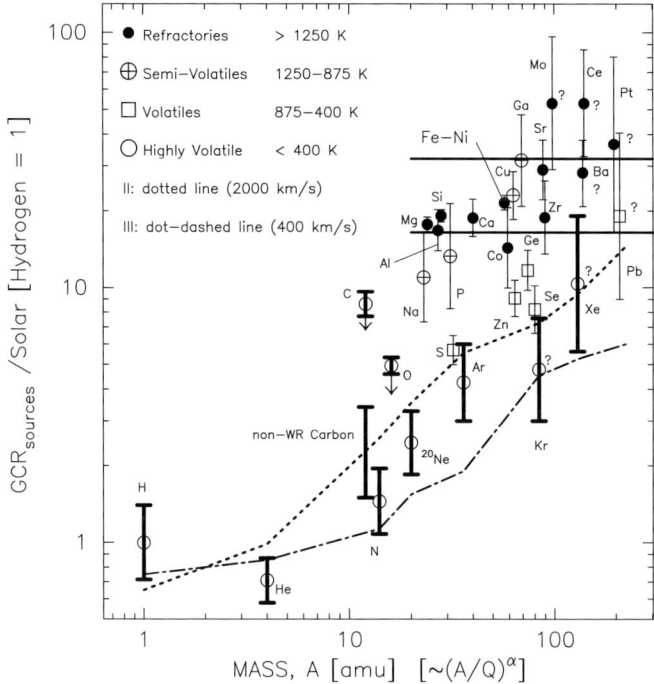

Fig. 6.5. Abundances of the elements in galactic cosmic rays (GCR) normalized to the standard abundances as a function of atomic number A. The normalization is to hydrogen. The symbols indicate the volatility of the element. Note that the abundances are larger for the heavy elements, and, amongst them, for the less volatile elements. The dashed and the dashed-dotted lines correspond to the predictions of an acceleration model which is described in Chap. 12. In this model, charged interstellar grains are accelerated in shock waves, the velocity of which is indicated in the figure. Reproduced from Ellison et al. [154], with the permission of the AAS.

atoms of gas atoms. During this process they release the elements of which they are composed. These elements, as ions, are then accelerated to form cosmic rays.

We will see later that supernova explosions are a sufficient energy source for the formation of cosmic radiation in a galaxy. However, the simple idea according to which the supernova explosions would inject material with a high energy into the surrounding medium is not tenable. This is because the abundances in cosmic rays do not correspond to the abundances of the elements created by explosive nucleosynthesis[4]. On the other hand, the shock waves generated by supernova explosions do offer an efficient acceleration mechanism: the energy in cosmic rays is about one tenth of that of supernova remnants. The acceleration mechanism will be discussed in detail in Sect. 12.4.

---

[4] Some of the overabundances of ^{22}Ne, ^{16}O and ^{12}C in cosmic rays might, however, suggest a supernova origin but these elements could also originate in Wolf–Rayet stars.

6.1.4 Very High-Energy Cosmic Rays

We have seen that galactic cosmic rays follow a power-law energy spectrum up to an energy of 10^6 GeV/n (10^{15} eV/n). At higher energies the modulus of the spectral index changes to 3.0, as shown by Fig. 6.6. A particle with energy $E > 10^6$ GeV/n, interacting with the high altitude terrestrial atmosphere, produces an avalanche of particles that are detected on the ground over an area of several thousands m^2. These secondary particules form an *extensive air shower*. We can also observe the Čerenkov light emitted by the relativistic particles whose velocity is larger than the velocity of light in the terrestrial atmosphere. The number of particle impacts at ground-based detectors allow us to derive the energy of the primary particle. The hadronic physics which occurs in the formation of the extensive air showers is not well known and the determinations of the atomic number or of the charge of the primary nucleus are still debated. The energy is probably better determined because it depends only upon the number of particles in the shower. The differences between the arrival times at the different detectors, which are scattered over a large area, allow us to determine the direction of the primary particle.

The existence of particles with energy as high as 10^{18} eV (10^9 GeV) is well established. Assuming that they are protons, the gyration radius of their trajectory

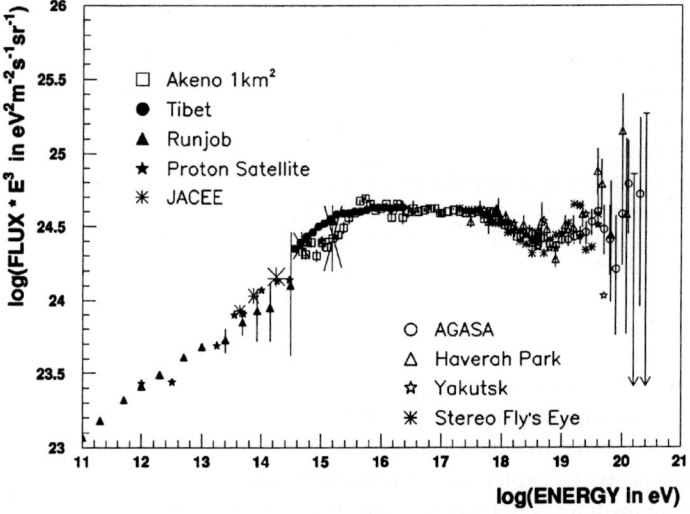

Fig. 6.6. Differential energy spectrum of the very high-energy cosmic rays. Here the abcissa gives the energy E per nucleus, and not per nucleon. The ordinate is the logarithm of the flux multiplied by E^3 for better presentation. The source of the data is indicated on the figure. Note that the plots above 10^{18} eV are from four different experiments which show reasonable compatibility, although the dispersion is large near the upper limit. There is no sign of the Greisen–Zatsepin–Kuz'min cutoff, which is expected to arise above 10^{19} eV if the cosmic rays come from very large distances. From Nagano & Watson [380], with the permission of Reviews of Modern Physics.

in a 5 μG magnetic field is 200 pc. This is at the limit of compatibility with the scale height of the Galaxy and an extrapolation of confinement models to these energies would lead to a lifetime of only 100 years. Particles with even higher energies, reaching as much as 10^{20} eV, do exist as shown by Fig. 6.6. The flux of these particles corresponds to one event per km^2 per century! An extragalactic origin has often been discussed, however this raises serious problems, in particular because of the energy loss of particles coming from high-redshift sources due to the expansion of the Universe. Even more serious, Greisen as well as Zatsepin & Kuz'min have shown that collisions of 10^{20} eV protons with the photons of the blackbody radiation of the Universe at 2.73 K should destroy them in $\sim 3 \times 10^8$ years. This would restrain the distance of possible sources to some 100 Mpc and would produce a cut-off in the energy spectrum near 5×10^{19} eV which is not observed: see Cesarsky [83] and the very complete review by Nagano & Watson [380]. Finally, no particular direction of arrival nor correlation with extragalactic active objects have been seen in spite of many searches.

The problem is less serious if the very high-energy cosmic rays are heavy nuclei. For example, an iron nucleus (mass 56) of 10^{20} eV would only contain 2×10^9 GeV/n, with a gyration radius of about 1 000 pc. This might still be compatible with a galactic origin. The problem then is that the number of events seems to be too high given the abundance of iron in lower-energy cosmic rays, if an extrapolation to very high energies is valid. We must acknowledge that the interpretation of the observations of very high-energy cosmic rays is still very speculative. The AUGER project, with a very large detector in each terrestrial hemisphere, might help in solving these problems.

6.1.5 Cosmic Electrons

Electrons with energies between 1 and at least 10^3 GeV are also detected near the Earth. Their flux at 10 GeV is 1/100 of that of protons. Their observed spectrum *in the solar environment*, affected by solar modulation, follows a power law (Longair [336]):

$$I(E) = 700 \left(\frac{E}{\text{GeV}} \right)^{-3.3} \text{m}^{-2}\text{s}^{-1}\text{ster}^{-1}\text{GeV}^{-1}. \tag{6.3}$$

This spectrum is steeper than for protons, essentially due to the synchrotron energy losses of the electrons[5].

In Sect. 2.2 we discussed, and gave some useful formulae for, the galactic synchrotron radiation. Cosmic electrons are relativistic and their gyration radius is approximately the same as for protons of the same energy. The probable sites of their acceleration are the shock waves associated with supernova remnants. The synchrotron radiation of these electrons traces the location of these shock waves, see Sect. 12.1 and Plates 26 and 28. The energy spectrum of cosmic ray electrons,

---

[5] Spectra published in the sixties are harder, probably due to a poor discrimination of parasitic events at energies larger than 10 GeV.

as derived from radioastronomy measurements of their synchrotron radiation, has a spectral index of about 2.2 in modulus (Webber [545]). The difference with the spectrum measured inside the solar cavity (6.3) comes from a modulation by the solar magnetic field and from synchrotron losses.

Some 1% of the high-energy electrons are in fact positrons. They come from high-energy reactions between cosmic rays and atomic nuclei.

6.1.6 Confinement of Cosmic Rays in the Galaxy

Although high-energy particles circulating in the interstellar medium experience collisions with atoms, molecules and dust grains, their trajectories are dominated by the effect of the magnetic field. The total energy of a particle with an energy E_n per nucleon is $\mathcal{A} E_n$, where $\mathcal{A}$ is the atomic number. The gyration radius R of this particle of mass $\mathcal{A} m_p$, with a relativistic velocity perpendicular to a magnetic field B, is

$$r = 3.33 \times 10^{12} \frac{\mathcal{A}}{Q} \left[\frac{E_n/(\text{GeV/n})}{B/\mu\text{G}} \right] \text{ cm.} \tag{6.4}$$

For example, protons with an energy $E \simeq 10^6$ GeV in a magnetic field of $B = 5$ µG have a gyration radius of $r \simeq 7 \times 10^{17}$ cm (0.2 pc), much smaller than the scale height of the Galaxy. These particles are therefore well-confined by the magnetic field. They can reside in the Galaxy for a long time before possibly escaping. For a general and historical review of the subject, see for example Cesarsky [82]. The length of the trajectory of the particles is generally expressed as

$$L(x) = \int_0^x \rho(s) \, ds, \tag{6.5}$$

where L is the total column density of matter along the trajectory, in g cm^{-2}. $\rho(s)$ is the mass density at point s. This quantity is called the *grammage* and is most representative and useful in studies of nuclear interactions. The grammage is related to the column density along the trajectory of the particle by the relation $L = N/\mu m_H$. We will see that it is generally in the range 6 to 10 g cm^{-2}.

The pressure exerted by cosmic rays is $1/3$ of their energy density, or about 4×10^{-12} dyne cm^{-2} (Holzer [246]). This pressure, in part, supports the magnetic field, which is tied to the gas, the gas being confined to the Galaxy by gravitation. If they were too many cosmic rays they would necessarily escape from the Galaxy, carrying with them the magnetic field and the gas (see Fig. 15.1). Thus, gravitation indirectly sets a limit to the density of cosmic rays.

At energies not affected by solar modulation the directions of arrival of cosmic rays are isotropic with respect of the Local standard of rest. The small anisotropy (streaming) which is observed is due to the motion of the solar system. No preferred direction, which would indicate the presence of a source, is detectable. This is *a priori* difficult to understand because intuitively the cosmic rays are expected to follow the magnetic field lines. However a strong lateral diffusion is possible due to irregularities in the magnetic field (see later Sect. 12.4), but this is difficult to quantify because the strength and spectrum of these irregularities are not well known. Many authors have discussed this question, introducing in particular the fact that cosmic rays can excite Alfvén waves in the interstellar medium and showing that the resulting magnetic irregularities can scatter particles. As a consequence, there is a tendency for the equipartition of energy between cosmic rays and the interstellar medium. Particles diffuse along large-scale magnetic structures and cross the Galaxy many times. Transverse diffusion can lift them in the galactic halo where they can stay for some time, then come back to the disk or escape to intergalactic space. This is the *leaky box* model. The solution to the diffusion equation indicates that the distribution of the path-lengths, before escape, is exponential, $L(x) = \exp(-x/x_0)$, where x_0 is a characteristic length which decreases with increasing energy: see for example Berezinskii [34]. The high-energy particles thus escape earlier than the low-energy ones. As a consequence, the modulus of the index of the energy spectrum of cosmic rays is larger by 0.5 to 0.6 in the interstellar medium compared to the source spectrum (see e.g. Longair [336]). The modulus of the source spectrum should be 2 from theory, (Sect. 12.4), while a value of 2.65 is observed near the Sun (Sect. 6.1).

Duric et al. [146] have checked this change of spectral index by observations of the radio radiation from the nearby galaxy M 33. Assuming that the spectral index of relativistic electrons is the same at the source as that of cosmic protons, and remembering that if this spectrum has an index γ, the spectral index of the synchrotron radiation is $\alpha = (\gamma - 1)/2$ (2.17). The observed average index of the radio spectrum of supernovae in this galaxy is $\alpha = 0.6$, while that in the interstellar medium is 0.8 to 0.9. The corresponding difference between the γ indices is thus of the order of 0.5 to 0.6 as expected.

The leaky box model is supported by three different arguments that we will evoque now.

The Energy Spectrum

Syrovatskii [502] presented the following heuristic argument which explains in a natural way, based on the stationnarity of the energy equipartition, the spectrum of the high-energy particles.

Let u_{cr} be the energy density of cosmic rays in the interstellar medium, u_{turb} that of its kinetic energy (dominated by random macroscopic, mostly turbulent, motions), and u_B that of the magnetic field. As discussed previously, the total energy is approximately distributed evenly between these three forms. The total energy of the system is $u_{tot} = u_{cr} + u_{turb} + u_B$. Let us assume that

$$u_{cr} = \delta(u_{turb} + u_B), \tag{6.6}$$

with $\delta \approx 1/2$. Then the total energy is

$$u_{tot} = \left(1 + \frac{1}{\delta}\right) u_{cr} = \left(1 + \frac{1}{\delta}\right) n E_0, \tag{6.7}$$

in which we have assumed that initially the particles have all the same energy E_0, with the density n. The total energy loss of the system is assumed to be due solely to the escape dn of these particles and is thus $E_0 dn$. However, in order to preserve the relation given by (6.7), the energy of the remaining particles has to be

$$du_{tot} = E_0\, dn = \left(1 + \frac{1}{\delta}\right) d(n E_0), \tag{6.8}$$

which gives by developping the right hand term

$$\frac{dn}{n} = -(1+\delta)\frac{dE_0}{E_0}. \tag{6.9}$$

This has the stationary solution

$$n \propto E_0^{-(1+\delta)}. \tag{6.10}$$

This equation is satisfied by the differential spectrum

$$\frac{dn}{dE} \propto E^{-(2+\delta)}. \tag{6.11}$$

Observation of this spectrum shows that δ is close to $1/2$, which confirms that equipartition is approximately satified, as postulated. The power-law spectrum can be explained by noting that cosmic rays have essentially no mutual interaction and do not tend to accumulate around an average energy. A more rigorous reasoning than the heuristic one we have given can be found in Ginzburg & Syrovatskii [198].

The Abundance of Light Elements and the Path Length

The production of secondary elements by spallation depends, of course, upon the interaction lengths of the primary particles before they escape, and upon the formation cross-sections. Modelling the path length distribution at every particle energy, and taking energy losses into account whenever necessary, it is possible to determine the mean grammage as a function of energy. This mean grammage is the only quantity that can be used in the interpretation. Taking into account not only the abundances of the spallation products of C, N and O (which are Li, Be and B), but also the abundance of the spallation products of iron (Sc, Ti, V, Cr and Mn), it is possible to confirm and to refine the exponential distribution of the path lengths of the parent atoms and the mean grammage as a function of energy. The results obtained by various authors differ slightly from one another due to different choices of the cross-sections and of the path-length distributions. This is still a subject under development. Here, we will only cite, as representative, the results of Duvernois et al. [147]. This article contains a summarized discussion of the method and gives all the necessary references. The best agreement is obtained by an exponential distribution of the path lengths

$$L(x) = \exp(-x/x_0), \tag{6.12}$$

with $x_0 \simeq 7$ g cm^{-2}, and an energy dependence exhibiting a peak of 9 g cm^{-2} near 300 MeV/n, which reflects the dependence of the formation cross-section upon energy. The path length corresponding to x_0 in the galactic disk is approximately 1 000 kpc, 50 times the diameter of the Galaxy. Note, however, that the fraction of the path length that the particles follow in the galactic halo, where collisions are negligible because of the low density, is not included in this number.

Cosmic Clocks

Amongst the elements formed by spallation, some are radioactive. When their formation cross-sections are known and when their lifetimes are of the order of a million years, a measurement of the abundance ratio between the parent element and its daughter allows us to determine the time interval between the formation of the radioactive element and its observation. The most useful element for this determination is ^{10}Be, which desintegrates by β radioactivity into ^{10}B with a half-life of 1.5 million years[6]. ^{26}Al, with a half-life of 0.85 million years, is another interesting candidate but it is less abundant than ^{10}Be. Simpson & Garcia-Munoz [474], in their very complete study using that method, find an average confinement time for cosmic rays of 11 to 22 million years. More recent studies tend to confirm this longer value. Combining this with the average grammage of 7 g cm^{-2} given earlier, we find an average density along the path of about 0.2 atom cm^{-3}. If the density of the interstellar medium in the disk is 1 atom cm^{-3}, this means that high-energy particles spend 4/5 of their lifetime in the halo. The presence of high-energy particles in the halo is confirmed by the synchrotron radiation from the halo, which indicates the presence of relativistic electrons.

6.2 The Gamma-Ray Continuum

The interaction of cosmic-ray particles with the interstellar medium produces gamma-ray photons by three mechanisms which we will examine in turn: nuclear interactions, Bremsstrahlung and inverse Compton scattering.

6.2.1 Gamma-Ray Production by Nuclear Interactions

Collisions between charged, high-energy cosmic-ray nuclei and interstellar nuclei produce gamma-ray photons through intermediate π^0 mesons. Protons being the most abundant cosmic particles, we can neglect p-α and α-p interactions to a first approximation. Proton–proton interactions produce charged mesons (pions) $\pi^{\pm}$ and neutral pions π^0. Charged pions decompose into $\mu^{\pm}$ muons and neutrinos. Muons finally transform into electrons and positrons, according to their charge, and into neutrinos. It is this process that is at the origin of the cosmic positrons. The neutral pions, that are of interest here, decompose into two gamma-ray photons with equal energies.

---

[6] The value of 2.5 million years found in old tables is obsolete.

The threshold for the p–p reaction is about 280 MeV, and the total cross-section for this reaction is about 2.7×10^{-26} cm^{-2} for proton energies larger than 2 GeV. The number of charged pions produced per reaction is $2(E_p/\text{GeV})^{0.25}$, E_p being the proton energy. The number of neutral pions is about half that of the charged pions. Each of these pions has an energy of roughly

$$E_\pi = 175 \left(\frac{E_p}{\text{GeV}}\right)^{3/4} \text{MeV}, \tag{6.13}$$

i.e., approximately 1/10 of the proton energy for $E_p = 10$ GeV. For more exact formulae, taking into account the energy distribution of pions, see Dermer [119].

Each neutral pion, π^0, is decomposed into a pair of equal-energy gamma-ray photons, emitted in the direction of propagation of the pion. In the reference system of the pion, each of the photons has an energy $1/2\, m_{\pi^0} c^2 = 68$ MeV, where m_{π^0} is the rest mass of the pion. We obtain the energy E_γ of the gamma-ray photons, in the reference system of the observer, by Lorentz transformation:

$$E_\gamma = \frac{1}{2}\gamma_{\pi^0} m_{\pi^0} c^2 (1 + \beta_{\pi^0} \cos\theta'), \tag{6.14}$$

where $\gamma_{\pi^0} = E_{\pi^0}/(m_{\pi^0} c^2)$ is the Lorentz factor for the pion; $\beta_{\pi^0} = v/c$, where v is the pion velocity (in fact not much different from c), and θ' its direction with respect to the line of sight.

Assuming an isotropic distribution of the pion velocities, we obtain by integrating over all directions the conditional probability for obtaining a pair of photons of energy E_γ from a pion of energy E_{π^0}:

$$f(E_{\pi^0}, E_\gamma) = 1 - 1/\gamma_{\pi^0}^2 \quad \text{if } \frac{1}{2}E_{\pi^0}(1-\beta_{\pi^0}) \le E_\gamma \le \frac{1}{2}E_{\pi^0}(1+\beta_{\pi^0})$$
$$= 0 \quad \text{otherwise.} \tag{6.15}$$

We now have to integrate the partial spectrum of gamma-rays emitted by pions with energy E_{π^0}, over the energy spectrum of these pions. If the spectrum of pions is $d\mathcal{N} = \mathcal{N}(E_p)\,dE_p$, and if n is the density of the interstellar medium, we obtain the production rate of gamma-rays per unit volume in all directions

$$Q_\gamma(E_\gamma) = \frac{d\mathcal{N}_\gamma}{dt} = \int_{E_{\pi^0,min}}^{E_{\pi^0,max}} \left[2f(E_{\pi^0}, E_\gamma)\int n\mathcal{N}(E_p)\sigma(E_p, E_{\pi^0}) c\, dE_p\right] dE_{\pi^0}, \tag{6.16}$$

where $\sigma(E_p, E_{\pi^0})$ is the differential cross-section for the production of a pion with energy E_{π^0} by collision of a proton with energy E_p with an interstellar proton. This cross-section is approximately $9 \times 10^{-26}\, (E_p/\text{GeV})$ cm^2 for $E_p \ge 1$ GeV, the relation between E_{π^0} and E_p being given by (6.13).

6.2 The Gamma Ray Continuum

We still have to fix the integration limits. In the highly relativistic case where $\beta_{\pi^0} \simeq 1$, the energy E_γ can be achieved even from pions with an infinite energy, as per (6.15), so that

$$E_{\pi^0,max} = \infty. \tag{6.17}$$

To calculate $E_{\pi^0,min}$ we note that $E_{\pi^0} = \frac{1}{2}E_{\pi^0}(1-\beta_{\pi^0}) + \frac{1}{2}E_{\pi^0}(1+\beta_{\pi^0}) = E_{\gamma,min} + E_{\gamma,max}$, and that $E_{\gamma,min}E_{\gamma,max} = \frac{1}{4}E_{\pi^0}^2(1-\beta_{\pi^0}^2) = \frac{1}{4}m_{\pi^0}^2 c^4$, so that

$$E_{\pi^0} = E_{\gamma,max} + \frac{m_{\pi^0}^2 c^4}{4E_{\gamma,max}}, \tag{6.18}$$

from which we derive

$$E_{\pi^0,min} = E_\gamma + \frac{m_{\pi^0}^2 c^4}{4E_\gamma}. \tag{6.19}$$

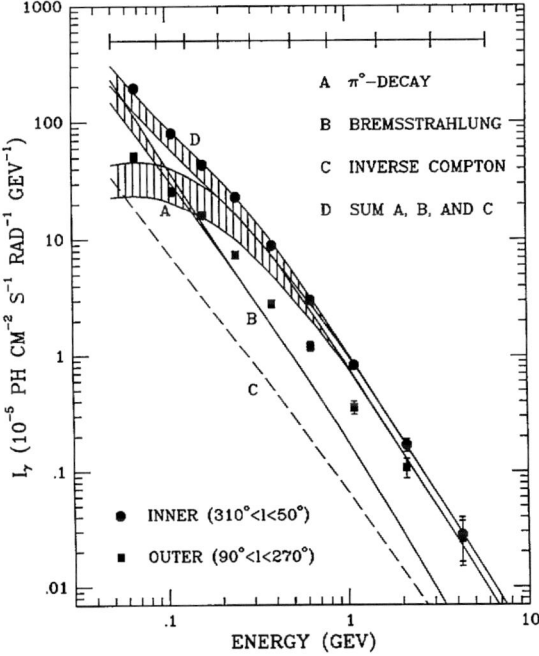

Fig. 6.7. The energy spectra of gamma-rays emitted in the inner part of the galactic disk (circles) and in the outer part (squares), as observed with the COS-B satellite. The spectra are compared to a theoretical spectrum derived from the local energy spectrum of cosmic-ray protons and electrons. The sum of the spectra for the three production mechanisms for gamma-ray photons described in the text is normalized at 1 GeV to the observed spectrum for the inner Galaxy. The hatched regions correspond to the uncertainties in the energy spectra of cosmic-rays, due mainly to the correction for solar modulation. From Bloemen [43], with the permission of Annual Reviews, www.AnnualReviews.org.

Equation (6.16) allows us to calculate the emitted gamma-ray spectrum (Fig. 6.7). If cosmic protons have a differential power spectrum $I_p(E_p) = K_p E_p^{-\Gamma}$, we find that the differential gamma-ray spectrum has almost the same slope: $d\mathcal{N}_\gamma/dt \propto n K_p E_\gamma^{-\Gamma}$, with $\Gamma \simeq 2.75$.

Actually, pions and, subsequently, gamma-rays are also produced by the interactions of cosmic rays with interstellar atoms heavier than hydrogen. A detailed study (Dermer [119]) shows that we must multiply the above production rate by 1.45 to take this into account.

We often use an emissivity per hydrogen atom and per steradian

$$\boxed{\epsilon_\gamma \equiv Q_\gamma/(4\pi n_\mathrm{H}) \text{ photon (H atom)}^{-1} \text{ s}^{-1} \text{ ster}^{-1}.} \qquad (6.20)$$

The intensity of the gamma-ray radiation, for a direction with a hydrogen atom column density N_H, is simply written $I_\gamma = N_\mathrm{H} \epsilon_\gamma$ photon cm^{-2} s^{-1} ster^{-1}, assuming a uniform flux of cosmic rays.

6.2.2 Gamma-Ray Production by Bremsstrahlung

The electromagnetic (Coulomb) scattering of high-energy cosmic-ray electrons by interstellar nuclei and electrons produces a braking of the cosmic-ray electrons, the corresponding energy being emitted as *Bremsstrahlung* gamma-ray photons (Bremsstrahlung is the German for braking radiation). This is the high-energy version of the free–free radiation studied Sect. 5.1. The differential cross-section for the production of a gamma-ray photon of energy E_γ by a highly relativistic electron of energy E_e, interacting with the nucleus of charge Z of a neutral atom, is given by the simple formula, which takes into account the screening of the atom by its electrons:

$$\sigma_B(E_e, E_\gamma) dE_\gamma \simeq \frac{m_n}{L E_\gamma} dE_\gamma, \text{ with } E_e \gg \frac{m_e c^2}{\alpha Z^{1/3}}, \qquad (6.21)$$

where $\alpha = e^2/\hbar c = 1/137.036$ is the fine-structure constant, m_n is the mass of the nucleus and L is the *radiation length*. L defines the quantity of matter traversed such that the energy of the electron is decreased by a factor e. The radiation length is expressed as a grammage and is equal to $L = 62.8$ g cm^{-2} for neutral hydrogen and 93.1 g cm^{-2} for neutral helium. For a mixture of atoms with solar abundances we have $\langle L \rangle \simeq 66$ g cm^{-2} and $\langle m_n \rangle \simeq 2 \times 10^{-24}$ g. For energies smaller than the validity limit of this expression, i.e. ~ 70 MeV, see Blumenthal & Gould [45].

If the medium is completely ionized, the screening effect of the electrons bound to the atom on the electrostatic potential of its nucleus is reduced or vanishes, but we must then take into account the scattering by free electrons. This results in an increase of 30% in the value given by (6.21). This change has a practical importance for the calculation of the gamma-ray emissivity at high galactic latitudes because ionized gas dominates at large distances from the galactic plane (see Sect. 5.2).

At a given gamma-ray energy, Bremsstrahlung is dominated by electrons with an average energy $E_e \simeq 3 E_\gamma$. The emissivity $Q_\gamma(E_\gamma) \propto n I_e(E_e > E_\gamma)/E_\gamma$ so that for electrons, with an energy spectrum $I_e(E_e) = K_e E_e^{-\Gamma}$, the energy spectrum of the emitted gamma-ray photons has a similar power law: $Q_\gamma(E_\gamma) \propto n[K_e/(\Gamma-1)]E_\gamma^{-\Gamma}$. The calculated spectrum in the vicinity of the Sun is plotted in Fig. 6.7. Its intensity is about 1/4 of that of the gamma-rays originating from nuclear interaction, but it becomes progressively more important at lower energies.

6.2.3 Gamma-Ray Production by the Inverse Compton Effect

The Compton effect is the interaction between photons and free electrons. In its original form, it is the energy loss of X-ray photons scattered inelastically by low-energy electrons: hence, the name of *inverse Compton effect* is given to the inelastic scattering of high-energy electrons by low-energy photons which is of interest here.

The cross-section for this process is

$$\sigma_C \simeq \sigma_T \left(1 - \frac{2\gamma_e h\nu}{m_e c^2}\right) \text{ for } \gamma_e h\nu \ll m_e c^2, \tag{6.22}$$

where $\sigma_T = 8\pi e^4/(3m_e^2 c^4) = 6.65 \times 10^{-25}$ cm^2 is the Thomson scattering cross-section, $\gamma_e = E_e/m_e c^2$ is the Lorentz factor for the electron and $h\nu$ is the photon energy. Except close to intense radiation sources, those photons that dominate in the interstellar medium come from the blackbody radiation of the Universe, i.e. $\langle h\nu \rangle \simeq 6 \times 10^{-4}$ eV.

The photon energy after scattering is given by the simple expression

$$E_\gamma \simeq \frac{4}{3}\gamma_e^2 h\nu, \tag{6.23}$$

so that electrons of approximately 200 GeV ($\gamma_e \simeq 4 \times 10^5$) are responsible for the emission of 100 MeV gamma-ray photons by interaction with the cosmological blackbody photons.

If the energy spectrum of the electrons is $I_e(E_e) = K_e E_e^{-\Gamma}$, the inverse Compton spectrum has approximately the shape $Q_\gamma(E_\gamma) \propto n_{ph}(h\nu)K_e E_\gamma^{-(\Gamma+1)/2}$. The spectrum calculated in the solar vicinity is shown in Fig. 6.7. We see that the inverse Compton effect contributes only 10% of the production of gamma-rays.

The *total* gamma-ray emission per hydrogen atom near the Sun is estimated by Dermer [119] to be $\epsilon_\gamma = 1.22 \times 10^{-26}$ photon atom^{-1} s^{-1} ster^{-1} for photons with energy larger than 100 MeV. Values for ϵ_γ over different energy intervals are given in Fig. 5 of Bloemen [43]. From this reference, $\epsilon_\gamma \approx 2.4 \times 10^{-26}$ photon (atom H)$^{-1}$ s^{-1} ster^{-1} for photons between 70 MeV and 5 GeV in the solar neighbourhood. This figure is slightly higher than that given by Dermer and is to be preferred .

6.3 The Mass of the Interstellar Medium

The gamma-ray emission of the interstellar medium is, as we have seen, proportional to the mass of this medium with little dependence on its physical state: neutral or ionized, or even solid. Gamma-ray observations offer an excellent method to determine this mass. We will examine it and compare the results with other methods. Note from the start that the mass of molecular gas obtained with these methods (with the exception of the last one) is only that inside those molecular clouds and complexes which also emit CO lines, but does not include the mass of the H_2 molecules in the diffuse H I medium.

6.3.1 The Use of Gamma-Ray Observations to Determine the Mass of the Interstellar Medium in the Galaxy

Galactic gamma-ray emission has been mapped by several satellites, first with the NASA satellite SAS-2, then with the european (ESA) satellite COS-B and more recently with the EGRET instrument on board the NASA satellite GRO. The angular resolution of the maps is approximately one degree (Plate 2). The spectrum has been measured above 30 MeV and confirms the production mechanisms just described (see Fig. 6.7). There is an extended emission on which is superimposed the emission of more or less point sources. Some of these sources coincide with active star formation regions. A good critical review of the SAS-2 and COS-B results can be found in Bloemen [43]. Figure 6.8 gives a recent version of the gamma-ray spectrum of the inner regions of the Galaxy.

The diffuse emission, which will be the only one discussed in this Section, is dominated, as can be seen in Fig. 6.7, by the contribution of nuclear interac-

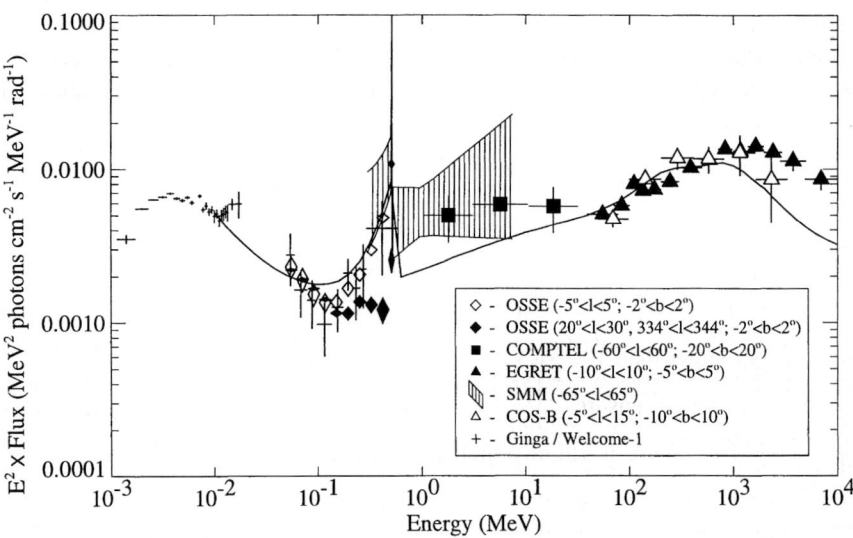

Fig. 6.8. Spectrum of the diffuse X-ray and gamma-ray emission of the central radian of the galactic plane. The scale of the ordinates has been multiplied by the square of the energy in MeV in order to make the figure more legible. From Purcell et al. [420], with the permission of ESO.

tions and Bremsstrahlung. Both emissivities are proportional to the product of the matter density and the cosmic-ray flux of nuclei and electrons, respectively. It is possible, to a certain extent, to discriminate between the two mechanisms by observing the gamma-ray radiation at relatively low energies (less than 70 MeV) where Bremsstrahlung dominates, and at high energies where nuclear interactions dominate. The contribution of inverse Compton scattering is a small correction which is easy to calculate. The great interest of gamma-ray emission is, as we have said, that it is sensitive to matter *whatever its form*, atomic, molecular, or ionized, provided that it is optically thin to gamma-ray photons and is traversed by cosmic-ray particles at the energies of interest (more than 100 MeV). It gives the best available way to determine the mass of the interstellar medium, in particular of the mass of the molecular component which is not well known. Of course, the flux of cosmic-ray particles must then be known. We can also attempt to determine this flux and its space variations in the Galaxy, which was the initial objective of the gamma-ray satellites. There are difficulties with both determinations, which have often been considered as controversial and have been discussed in an abundant literature. Here we will only deal with the first determination.

The gamma-ray intensity, that we define as the number of received photons per unit energy, per unit area and per unit time from a given direction, is

$$I_\gamma(E_\gamma) = N_H \epsilon_\gamma(E_\gamma), \quad (6.24)$$

where $\epsilon_\gamma(E_\gamma)$ is the emissivity per hydrogen nucleus and N_H the column density of hydrogen nuclei in all forms. Using a uniform emissivity over the line of sight assumes that the flux of cosmic rays is uniform. This is a realistic assumption for protons, which diffuse easily and without energy losses far from their sources, but a debatable one for electrons because they may experience important synchrotron losses. We must also assume that cosmic-ray particles penetrate all interstellar matter, which is not completely clear for the large molecular complexes, because of deviations by the magnetic field. Also, we must assume that the discrete gamma-ray sources have been well subtracted from the map in order to obtain the contribution of the interstellar medium alone. We will suppose for the moment that all these assumptions are correct.

On the other hand, it is clear that we do not observe the gamma-ray emission of regions with very large column densities which are optically thick to gamma-rays. This is the case, for example, for relatively large, dense objets like stars, planets, etc., which ensures that only the emission of the interstellar medium is observed. However, this could also be the case for very dense, small interstellar clouds should they exist. We showed in Sect. 6.2 that Bremsstrahlung is inefficient at grammages larger than about 66 g cm^{-2}, due to the energy losses of the electrons. This corresponds to a column density of approximately 3.3×10^{25} nuclei cm^{-2}. The propagation of high-energy photons is also limited by pion creation in the medium, with a similar penetration depth (Sect. 6.2). The gamma-ray photons produced deep in the medium are destroyed by electron–positron pair creation in the field of nuclei, with a characteristic interaction depth of 100 g cm^{-2} for hydrogen, and less for heavy elements (Chupp [100]). Consequently, matter will not be observed via gamma-rays at column densities larger than a few 10^{25} nuclei cm^{-2}. The existence of dense clouds with large column densities is one of the possibilities invoked to account for the baryonic dark matter in the Universe (Pfenniger et al. [408]).

Despite these difficulties, we will assume that we can use (6.24) for interstellar matter. The similarity between the gamma-ray maps and the H I and CO maps of the Galaxy gives some evidence in favor of its use. This equation is used in the following way in order to separate the respective contributions of the atomic and molecular components of the interstellar medium in the solar neighbourhood. As a first approximation we can neglect the contribution of the diffuse ionized gas if the study is limited to low galactic latitudes.

- We first assume that the intensity of the 2.63 mm CO(1-0) line is proportional to the column density of the molecular gas. This hypothesis rests on the idea that in an antenna beam we always see a number of similar molecular clouds, and that the antenna temperature depends mainly upon the number of clouds in the beam. The justification is that the (usually complex) profiles in the ^{12}CO(1-0) line and in the line of the isotopically substituted molecule ^{13}CO(1-0) are remarkably similar for most galactic lines of sight, although the optical depth is 76 times smaller in the latter line (cf. (4.51)). This is interpreted as indicating some degree of uniformity in the properties of the clouds[7]. The overlap of the different clouds along the line of sight is not a serious problem because the clouds have a substantial velocity dispersion between them and because galactic differential rotation spreads out the velocities: at a given velocity in a line we most often observe only one cloud, at least in well-chosen directions. There are however doubts about all these affirmations.
- We can then write (6.24) as

$$I_\gamma(E_\gamma) = [N(\text{H I}) + 2XW_{\text{CO}}]\epsilon_\gamma(E_\gamma), \qquad (6.25)$$

where $W_{\text{CO}} = \int T_B^* \, dv$ is the integrated intensity of the ^{12}CO(2-1) in K km s^{-1}, and

$$\boxed{X \equiv N(\text{H}_2)/W_{\text{CO}} \text{ mol. cm}^{-2} \, (\text{K km s}^{-1})^{-1}.} \qquad (6.26)$$

We then obtain ϵ_γ (see the end of Sect. 6.2), and also X, by a correlation analysis between the H I, CO and γ maps. The advantage of this method is that no hypothesis is made about the value of ϵ_γ. It is also possible to use a value for ϵ_γ derived from the measurement of the flux of cosmic-ray nuclei and electrons observed near the Earth (cf. Sect. 6.1) combined with nuclear parameters measured using accelerators (cf. Sect. 6.2). Then X can be determined without correlation analysis. This method is limited to high-energy gamma-rays

---

[7] Another, more convincing, qualitative argument is that the integrated intensity of the CO(1-0) line increases with increasing size of the cloud, due to an increase in the velocity dispersion, hence of the mass if the clouds are in virial equilibrium and have similar densities (Sect. 14.1).

since the cosmic-ray flux is poorly known at low energies due to the solar modulation (cf. Figs. 6.2 and 6.7).

The correlation analysis of the COS-B data by different groups give values for ϵ_γ in reasonable agreement with the theoretical predictions, and values for X between 2.3×10^{20} and 3×10^{20} mol. cm^{-2} (K km s^{-1})$^{-1}$, with some preference for the lower figure. The method using the knowledge of the local cosmic-ray flux gives a slightly lower value, $X \simeq 1.5 \times 10^{20}$ mol. cm^{-2} (K km s^{-1})$^{-1}$. More recent correlation analyses of the EGRET data from the GRO satellite give values for X between 0.9 and 1.6×10^{20} mol. cm^{-2} (K km s^{-1})$^{-1}$(Digel et al. [125], [126], Hunter et al. [253], etc.).

A further step would be to determine, for various parts of the Galaxy, both X and ϵ_γ, i.e. the flux of cosmic rays or, even better, the fluxes of cosmic-ray electrons and of cosmic-ray protons. A method that is currently used consists of fitting the observed gamma-ray distribution with a model of the form $I_{gamma}(E_\gamma) = \sum_i (N(\text{H\,\textsc{i}})_i + 2X_i W_{\text{CO},i}) \epsilon_{\gamma,i}(E_\gamma, R_i)$, where the sum is over concentric rings i with galactocentric radii R_i. The unknowns are $\epsilon_{\gamma,i}$ and X_i, which can be obtained by a maximum likehood analysis. The results, which will not be detailed here because they are somewhat controversial, are described by Bloemen [43] and by Hunter et al. [253].

Although there are some doubts about the hypothesis that the integrated intensity of the ^{12}CO(1-0) lines is really representative of the column density of the molecular component of the interstellar medium, the fact that consistent results are obtained in the correlation analysis for H\,\textsc{i}/CO/gamma is an empirical justification of this idea. It would probably be preferable to perform a similar analysis with a much less optically thick line like ^{12}C^{18}O(1-0), which is emitted in less superficial, more extended zones of molecular clouds (see Chap. 11). This is however a major observational project. For the moment, we find it useful to compare the values of X obtained by the correlation method with the results of other methods that will now been described briefly.

6.3.2 Use of the Virial Mass of Molecular Clouds

Assuming that for molecular clouds there is an equilibrium between their self-gravity and the dynamic pressure of the macroscopic random motions of the gas, which are often much larger than the thermal pressure, we can estimate their mass using the *virial theorem*. The demonstration and the validity of application of this theorem will be discussed in Chap. 14. For a spherical, uniform and isolated cloud with zero magnetic field and no external pressure, we have

$$\boxed{\frac{M}{M_\odot} = 210 \left(\frac{\Delta v_{1/2}}{\text{km s}^{-1}}\right)^2 \left(\frac{R}{\text{pc}}\right),} \quad (6.27)$$

where $\Delta v_{1/2}$ is the width of the CO line at half intensity. The profile of this line is assumed to faithfully reproduce the internal motions of the cloud, which are assumed to be gaussian. R is the external radius of the cloud. For non-uniform clouds with a radial density gradient, the numerical factor differs: it is 190 if $n \propto 1/r$, and 126 if $n \propto 1/r^2$; these are more realistic density distributions (see Sect. 14.1 and Fig. 14.2). This method yields values of X of the order of 1×10^{20} mol. cm^{-2} (K km s^{-1})$^{-1}$ (MacLaren et al. [345]), using the width of the ^{13}CO(1-0) line rather than that of the ^{12}CO(1-0) line. There are several difficulties with this method. One of them is the observational definition of the radius of a cloud, because its emission is generally superimposed on an irregular CO background emission.

6.3.3 A Comparison Between W_{CO} and Extinction

Assuming that the dust and gas are well mixed, it is possible to use the extinction of the light from background stars produced by dust in a molecular cloud to obtain the column density of dust, and then of the gas by assuming a gas/dust ratio. This method can be applied to nearby interstellar clouds for which contamination by foreground stars is small. We can either determine the reddening for individual stars, from which extinction is derived (see Sect. 7.1), or use star counts in well-defined magnitude intervals, compared to similar counts in a nearby reference field. If the *luminosity function*, which gives the number $n_r(m)$ of stars as a function of their apparent magnitude m, has the form $\log n_r(m) = a + bm$ in the reference field, it becomes $\log n_c(m) = a + b(m - A_\lambda)$ in the direction of the cloud where extinction is A_λ. A_λ is thus given by

$$A_\lambda = -\frac{1}{b} \log \left[\frac{n_c(m)}{n_r(m)} \right]. \tag{6.28}$$

See Thoraval et al. [510] for refinements of this method. If there are CO maps of the cloud, we can study the correlation between the extinction and W_{CO}. As expected, because of the saturation of CO lines, W_{CO} reaches a constant value as soon as A_V is of the order of 3 mag. As a consequence, X cannot be determined in this way. On the other hand there is a good correlation with the integrated intensity of the lines of isotopically substituted CO molecules up to rather large values of A_V, of the order of 10 magnitudes for $C^{18}O(1-0)$ (Fig. 6.9) or $C^{17}O(1-0)$. This can be explained by the much weaker optical depth for these lines compared to that of CO(1-0). The saturation at $A_V \simeq 10$ mag. is probably due to the condensation of CO on dust grains in the deeper regions of the cloud where the grains are very cold (Kramer *et al.* [288]). In any case, the large dynamical range of the correlation leads to some optimism about the possibility of performing a new analysis of the gamma-ray observations using the $C^{18}O(1-0)$ line when large maps in this line become available.

Studies of the distribution of dust and of W_{CO} in molecular clouds show a surprising result: the distribution of the dust is smooth at small scales while that of the CO emission is much more fragmented (Thoraval et al. [510], [511]). This is probably the main reason for the dispersion in the relation extinction/CO presented in Fig. 6.9. For a given value of extinction W_{CO} can vary by a factor 5 at small scales. Either there is a mechanism able to decouple dust from gas at small scales, or there

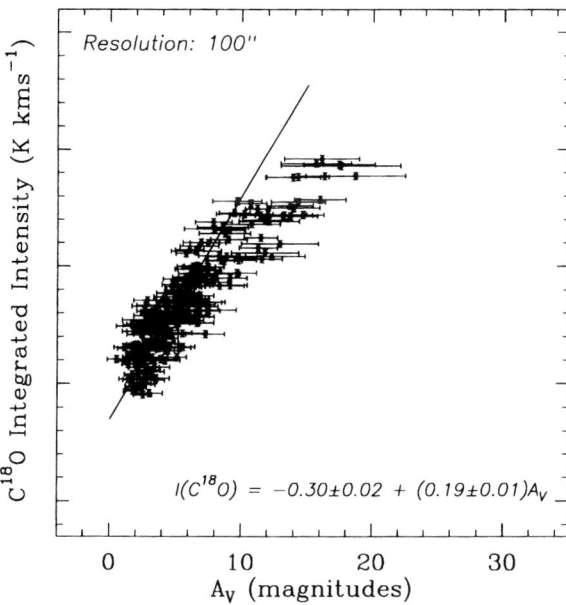

Fig. 6.9. Comparison between visual extinction and the integrated intensity of the $C^{18}O(1-0)$ line in the L 977 molecular cloud. The extinction is derived from star counts in the near infrared using a standard extinction curve. At extinctions larger than 10 magnitudes, the line intensity is no longer proportional to extinction, probably because CO condenses onto grains. Reproduced from Alves et al. [9], with the permission of the AAS.

are large fluctuations in the abundance and excitation of CO (this in regions of low density), or both.

6.3.4 A Comparison Between W_{CO} and Millimetre/Submillimetre Dust Emission

Another property of interstellar dust that will be studied in detail in Sect. 7.2 is its thermal emission. This emission peaks between 60 to 300 μm, depending upon the dust temperature; at much longer wavelengths, e.g. 1.2 mm, the emission is in the Rayleigh–Jeans regime. Its intensity is then proportional to temperature, while it varies roughly as T^5 at the maximum of the emission. As the optical depth is always small, the brightness of a dust cloud at millimetre wavelengths is

$$I_\nu = \sigma_\nu^H B_\nu(T_d) N_H, \qquad (6.29)$$

where $B_\nu(T_d)$ is the brightness of the blackbody at the dust temperature T_d, N_H is the column density of hydrogen nuclei in atomic and molecular form, and σ_ν^H is the absorption cross-section of dust per hydrogen nucleus. The gas/dust ratio and the dust properties both intervene in this cross-section. It can be calculated either from a dust model, which is rather uncertain, or calibrated by observation of a region where the gas is mostly atomic and where the dust temperature can be reasonably estimated. This gives (Neininger et al. [383])

$$\sigma^H_{1.2\,mm} = 5 \times 10^{-27} \text{ cm}^2 \text{ (H atom)}^{-1}. \qquad (6.30)$$

Equation (6.29) can then be used to determine N_H in other regions, in particular in molecular clouds. T_d is estimated and the properties of the dust are assumed to be the same as in the reference region. This is an excellent method because the estimation of T_d is not critical, but the absorption cross-section can be different in molecular clouds because of the possibility of condensing ice mantles on dust grains inside molecular clouds (see further Sect. 7.4, in particular Table 7.4). Subtracting the emission of dust corresponding to the atomic component, we obtain that of the molecular component, from which the column density can be derived, then X can be determined after the CO line is observed. This method gives values of X close to 1.0×10^{20} mol. cm^{-2} (K km s^{-1})$^{-1}$. Notice that if σ^H_ν is larger in molecular clouds, this decreases N_H then X. Progress in bolometre array development allows us to predict a great future for this method: see for example Johnstone et al. [267]. Observations at several submillimetre wavelengths, which are now available (see e.g. Lis et al. [327]), should better constrain σ^H_ν and T_d. See also Sect. 7.2 for another recent determination of the submillimetre emissivity of interstellar dust.

A method, similar in its principle, has been used by Dame et al. [113] in order to determine X at high galactic latitudes, hence near the Sun. They have compared their complete maps of the sky in H I and in the CO line with a map of the infrared dust emission at 100 μm. In the regions with no CO emission, the infrared brightness is well correlated to the H I column density ($N_H/I_{100\mu m} = (0.9 \pm 0.4)10^{20}$ atom cm^{-2} (MJy ster^{-1}). Subtracting from the infrared map the contribution of H I, they obtain the contribution of the molecular gas that they find well correlated to the CO emission. Assuming the same relation between $2N_{H_2}$ and $I_{100\,\mu m}$ as between N_H and $I_{100\,\mu m}$, they derive $X = (1.8 \pm 0.3)10^{20}$ mol. cm^{-2} (K km s^{-1})$^{-1}$.

6.3.5 X-Ray and Mid-Infrared Absorptions

We discussed X-ray absorption in Sect. 5.4 (see also Fig. 2.5) and showed that this absorption allows us to indirectly obtain the column density of H_2. Ryter et al. [445] showed that the relation between the column density N_X of equivalent H atoms and the colour excess $E(B-V)$ (cf. (5.70)) differs between the solar neighbourhood and the direction of the galactic centre. Observations in the near infrared yield $E(B-V) \simeq 9$ mag. and $N_X = (9 \pm 2) \times 10^{22}$ (equivalent H atoms) cm^{-2} mag^{-1} towards the galactic centre, a factor 1.5 times larger than that given by (5.70). This can be explained by a larger abundance of heavy elements in the inner regions of the Galaxy. On the other hand 21-cm observations yield only $N_H = 2 \times 10^{22}$ atoms H cm^{-2}, showing that a large part of the hydrogen is molecular.

X-ray absorption studies are currently experiencing a revival thanks to the data coming from the CHANDRA and XMM-NEWTON satellites. Very recently,

Vuong et al. [537] compared the X-ray absorption measured from these satellites with the near-IR J band extinction, for pre-main sequence stars in several nearby molecular clouds. Assuming standard solar abundances for the ISM, they convert the X-ray absorption into a column density of hydrogen atoms (or rather nuclei), $N_{H,X}$. They find for their best-studied case, the ρ Oph cloud, $N_{H,X}/A_J = 5.6 \ (\pm 0.4) \ 10^{21}$ cm^{-2} mag^{-1}. This is about 20% less than what can be derived from the comparison of UV absorption lines with the visual extinction (see later (7.5)). Their interpretation is that the standard solar abundances are overestimates for the interstellar ones. We have already alluded to this problem in Sect. 4.1.

Similarly, observations of mid-infrared absorption in front of sources embedded inside molecular clouds, or located in the background, will offer other, precious information on the column density of the absorbing dust (see Sect. 7.1). For this, a systematic exploitation of data from the ISO and Spitzer satellites is desirable.

To summarize this section, we have seen that the determinations of the mass of the molecular component in the interstellar medium of our Galaxy, and of galaxies in general, are still fraught with uncertainties but also that better determinations are possible. All the methods that we have described, which use in one way or the other the intensity of the CO lines, are intended to give the masses of the atomic and molecular component of the interstellar medium. However, they do not actually supply the respective masses of atomic and of molecular hydrogen. The diffuse "atomic" medium in fact contains important quantities of H_2 without CO, as revealed by observations with COPERNICUS and more recently with FUSE. This comes from the fact that H_2 is more resilient to photodissociation than CO (Sect. 9.1). For the same reason, there are in the envelopes of molecular clouds regions containing H_2 but no CO. The respective extents of these zones depend upon the metallicity and the UV radiation field. We will come back to this point in Chap. 10.

6.4 The Gamma-Ray Lines

The interaction of cosmic-ray particles with interstellar nuclei produces nuclear emission lines in the gamma-ray range that result from the de-excitation of nuclei. These nuclei can either be stationary interstellar nuclei or cosmic-ray nuclei. To these lines add the gamma-ray photons coming from the desintegration of radioactive nuclei, and the 511 keV photons created by positron–electron annihilation. Several of these lines have been observed in the interstellar medium, and many more in solar eruptions. The astronomy of gamma-ray lines will strongly develop with the observations with the INTEGRAL satellite. Good reviews of the subject can be found in Ramaty et al. [422] and Prantzos [414].

The following lines are expected to be observed in the interstellar medium.

1. Lines emitted following the excitation of nuclei by protons or α particles (helium nuclei):

examples of such lines are:
- the de-excitation line of ^{12}C* at 4.439 MeV;
- similar lines of ^{14}N* at 2.313 and 5.105 MeV;
- lines of ^{16}O* at 2.741, 6.129, 6.917 and 7.117 MeV;
- lines of ^{20}Ne* at 1.634, 2.613 and 3.34 MeV;
- lines of ^{24}Mg* at 1.369 and 2.754 MeV;
- lines of ^{28}Si* at 1.779 and 6.878 MeV;
- lines of ^{56}Fe* at 0.847, 1.238 and 1.811 MeV; etc.

A more complete list and relevant nuclear data will be found in Ramaty et al. [422], completed by Dyer et al. [150], [151], Lesko et al. [321] and Tatischeff [505]. The excitation cross-section peaks for incident particle energy of about 10 MeV/nucleon (MeV/n). The cross-section decreases rapidly with increasing energy until about 100 MeV/n, at which point the nuclei disintegrate. As a consequence, these lines are potentially good indicators of the flux of low-energy cosmic rays (say from 5 to 50 MeV/n). Presently, we know almost nothing of these particles due to the strong solar modulation.

2. Lines emitted following spallation reactions:
These are reactions which synthetize the light elements ^{6}Li, ^{7}Li (in part), ^{9}Be, ^{10}Be et ^{11}Be from ^{12}C, ^{14}N et ^{16}O, they were discussed Sect. 6.1. They also produce more common elements from nuclei heavier than C, N and O. The produced elements are often in a nuclear excited state. We thus expect to observe de-excitation lines like that of ^{12}C* at 4.439 MeV, and probably also of Li, Be and B, but the fraction of excited nuclei produced by the spallation reactions is unknown. We have already seen that the threshold for these reactions is higher than for direct impact excitation. It is of the order of 30 to 100 MeV/nucleon (cf. Fig. 6.3).

The lines produced after the impact of cosmic-ray particles with heavy interstellar nuclei are narrow, with a typical width $\Delta E \simeq 0.03E$, because the velocity of these nuclei remains small after the collision. Conversely, the lines produced by the de-excitation of heavy cosmic-ray nuclei are very broad due to their high velocities, which are not much modified by collisions with the light interstellar nuclei. The shape and width of these lines depend upon the direction of arrival of the high-energy particles, if they are anisotropic, and upon their spectrum. This problem has been treated by Kozlovsky et al. [287]. These authors explain a broad (3–7 MeV) emission observed with COMPTEL in the Orion region as due to the de-excitation of a large quantity of C and O nuclei with energies of some 10 MeV/n. The important flux of the particles with energies $\geq$ 30 MeV that is necessary to excite those nuclei might come from recent supernova explosions, an environment able to accelerate these particles (Parizot [398]).

3. Decay lines of radioactive nuclei:
Nuclear reactions in stars, in particular in supernova explosions, and spallation reactions also produce radioactive elements. The best-known example is that of ^{26}Al which decays to an excited nuclear level of ^{26}Mg with a lifetime of 1.1×10^6 years. De-excitation of ^{26}Mg produces a 1.809 MeV photon. The

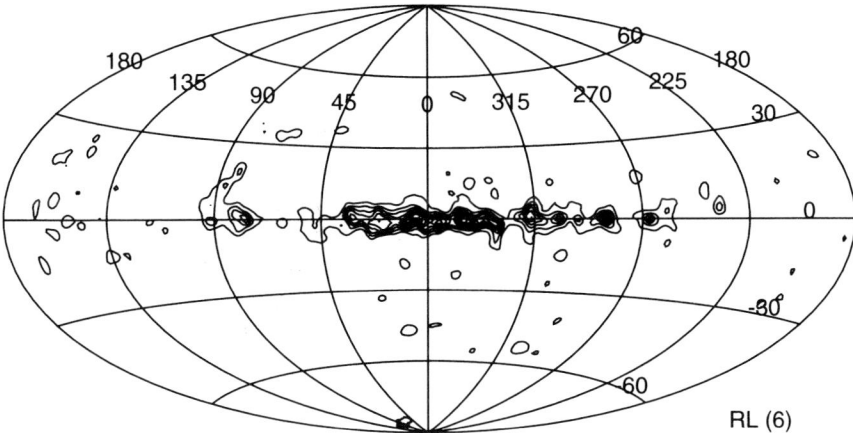

Fig. 6.10. Map of the distribution of the galactic emission in the 1.809 MeV line, in galactic coordinates. The data were obtained with the COMPTEL instrument on board the GRO satellite. From Knödlseder et al. [284], with the permission of ESO.

corresponding line, improperly designated as the "^{26}Al line", was observed as early as 1984 in the direction of the galactic centre. Observations with COMPTEL have shown that the source is diffuse and thus interstellar (Fig. 6.10), and that the mass of ^{26}Al present at a given time in the Galaxy is about 2 $M_\odot$. The sources of ^{26}Al are massive Type II supernovae and Wolf–Rayet stars (Signore & Dupraz [472], Meynet et al. [368]). The ^{26}Al nuclei produced by these stars are dispersed over large distances in the interstellar medium by the explosions or by the winds of these stars and form a diffuse source. Some meteorites contain a small quantity of ^{26}Mg which might originate from ^{26}Al present in the protosolar nebula, but there are doubts about the idea that this ^{26}Al could come from a nearby supernova explosion that occured just before the formation of the Solar system.

The review by Prantzos [414] gives a list of lines from radioactive nuclei. A few have already been observed in supernova remnants. Apart for the ^{26}Al line, examples are (the detected lines are underlined):

- lines from the decay chain ^{56}Ni → ^{56}Co → ^{56}Fe at <u>0.847</u>, 1.771, <u>1.238</u> and 2.598 MeV;
- lines from the chain ^{57}Co → ^{57}Fe at <u>0.122</u> and <u>0.136</u> MeV;
- lines from the chain ^{44}Ti → ^{44}Sc → ^{44}Ca at 0.068, 0.078 and <u>1.156</u> MeV.
- lines from the chain ^{60}Fe → ^{60}Co → ^{60}Ni at 1.173 and 1.173 MeV.

All these lines are narrow because the radioactive nuclei have time to thermalize before decaying. Their observation will be of great importance as a check on nucleosynthesis theories, and also to locate of the birth sites of massive stars in the Galaxy (Fig. 6.10).

4. The 511 keV electron–positron annihilation line:

Annihilation of a positron with an electron yields 2 gamma-ray photons of 511 keV, the mass energy of these particles. A diffuse emission has been observed in this line in the region of the galactic centre (cf. Prantzos [414]). Photons produced by the annihilation of high-energy cosmic-ray positrons are not observable because of the very large width of the line. The positrons that give the observed line are probably produced by the β^+ decay of the radioactive products of supernova explosions. The responsible nuclei are probably ^{56}Co and ^{44}Sc (see the list above), with a small contribution of ^{26}Al. The observed production rate of positrons is $1.5 \pm 0.5 10^{43}$ positrons s^{-1}. The annilation lifetime of these positrons being long in the interstellar medium, at least 10^5 years, they have time to diffuse far from their sources.

7 Interstellar Dust

Dust grains, with sizes ranging from nanometres to micrometres, are intimately mixed with the interstellar gas. They are formed in the atmospheres of evolved stars as well as in novae and supernovae, but they are also destroyed and re-formed in the interstellar medium (see Chap. 15). Although dust makes up only about one hundredth of the mass of the interstellar medium, it plays an extremely important role in the physics and chemistry of this medium, in the energy balance of the Galaxy, in phenomena that determine the evolution of interstellar clouds and the formation of stars, and even in the initial stages of the acceleration of cosmic rays. More precisely:

– Dust absorbs and scatters stellar light. Scattering being wavelength-dependent, the scattered light that forms *reflection nebulae*[1] is bluer than that of the illuminating star, while the light transmitted by a dust cloud is redder. If the dust grains have an anisotropic shape and are oriented, the scattered and transmitted lights are partly polarized (cf. Sect. 2.2). Close to an illuminating source, the scattered light is always radially polarized in the plane of the sky, due to the anisotropy of the radiation. Plates 1, 2, 12, 13, 14, 18, 20 and 24 illustrate extinction by interstellar dust, while scattering is visible on Plates 15 and 20.
– The energy absorbed by dust grains heats them and is re-emitted in the mid- and far-infrared. Almost half of all the energy emitted by stars in the Galaxy in the ultraviolet, the visible and the near infrared is absorbed by dust, then re-emitted at much longer wavelengths. The other half is either unaffected or scattered by dust. The thermal emission of dust is illustrated in Plates 2, 18, 19, 21, 22 and 23.
– Atoms or molecules striking dust grains can stick, forming a surface "mantle" of "ices" where chemical reactions can occur[2]. Some of these reactions are not possible in the gas phase, so that dust grains act as catalysts. Heating of the grain can evaporate the mantle and release new molecules in the interstellar medium. The H_2 molecule can only form on grains. These phenomena will be discussed in Chap. 9.

---

[1] Reflection nebulae are extended objects with, in general, a lower brightness than H II regions. They are always close to a bright star. They do not emit light by themselves, but are illuminated by the star as shown by their spectrum which exhibits the same lines as that of the star. Examples are shown by Plates 15 and 20.
[2] The term of "ice" designates the products condensed on grains at low temperatures: this includes the molecules H_2O, CO, CO_2, NH_3, CH_4, etc.

- The destruction or evaporation of dust releases heavy elements into the interstellar gas. Conversely, these elements can condense on grains to form refractory mantles. The grains can agglomerate together to form bigger grains. This will be discussed in Chap. 15.
- Dust grains irradiated by UV photons can release electrons via the photoelectric effect. These electrons play a major role in the heating of the interstellar medium. At densities larger than about 10^4 atoms or molecules per cm^3 collisions can efficiently transfer energy from the gas to dust and vice-versa. These phenomena will be examined in Chap. 8.
- Charged dust grains can be accelerated in shocks, in particular in supernova remnants. Collisions with the atoms or ions of the gas, and with other dust grains, release heavy elements into the gas that can themselves be accelerated and form the seeds of cosmic rays. This will be discussed in Sect. 12.4.

The book by E. Krügel [289] gives an excellent account of the physics of interstellar dust.

7.1 Interstellar Reddening and Extinction

7.1.1 General Ideas

The decrease in the luminosity of a star when seen through a dust cloud is due to two different physical phenomena: the absorption of photons by the material of the grains and the scattering of photons in directions other than the incident direction. The ensemble of these phenomena is called *extinction*. Extinction depends upon the grain composition, shape and size distribution and also upon the wavelength. The extinction A_λ is expressed as the ratio of the emerging flux $I(\lambda)$ and the incident flux $I_0(\lambda)$ such that:

$$I(\lambda) = I_0(\lambda) 10^{-(A_\lambda/2.5)} = I_0(\lambda) e^{-\tau_\lambda}, \tag{7.1}$$

so that the optical depth $\tau_\lambda = 0.921 A_\lambda$, where A_λ is expressed in magnitudes.

The *extinction curve* or *extinction law* is the curve in which the extinction is plotted as a function of wavelength, or more frequently as a function of the inverse of the wavelength. It is obtained by comparing, over a wavelength range as extended as possible, the spectral energy distributions of at least two stars of the same spectral type and luminosity class, which are assumed to be identical. One of these stars should be strongly affected by extinction, hence by the presence of a large amount of interstellar matter along the line of sight, and the other one should be little affected. This enables a *relative* extinction curve to be derived[3]. Usually the curve is normalized to the ratio of the extinctions in the Johnson *B* (blue, centered near

---
[3] It is very difficult to directly obtain the absolute value of the extinction. This would require a very good knowledge of the intrinsic luminosity of the star and of its spectral energy distribution. These quantities are rarely available with sufficient accuracy, even for hot stars.

4 400 Å) and V (visible, centered near 5 500 Å) broad bands. Defining the *colour excess* as

$$E(B-V) = A_B - A_V, \qquad (7.2)$$

the extinction curve is given as $A_\lambda/E(B-V)$ as a function of $1/\lambda$ (generally expressed in μm^{-1}). Other normalizations are possible, e.g. A_λ/A_V, also in frequent use. The average extinction curve from the near infrared to the ultraviolet (Fig. 2.5 and 7.1) is smooth, with a strong, broad band centered at 2 175 Å. For this average curve we have $R = A_V/E(B-V) = 3.1$ hence $A_B/E(B-V) = 4.1$. The extinction curve in the mid-infrared (Fig. 2.5, 7.1 and 7.2) exhibits broad bands at 9.7 and 18 μm, which are due to absorption by silicates. They correspond, respectively, to the stretching vibration mode of Si–O bonds and to the bending mode of the O–Si–O bonds. The 2 175 Å band is probably due to carbonaceous particles containing aromatic cycles which produce a characteristic absorption around this wavelength.

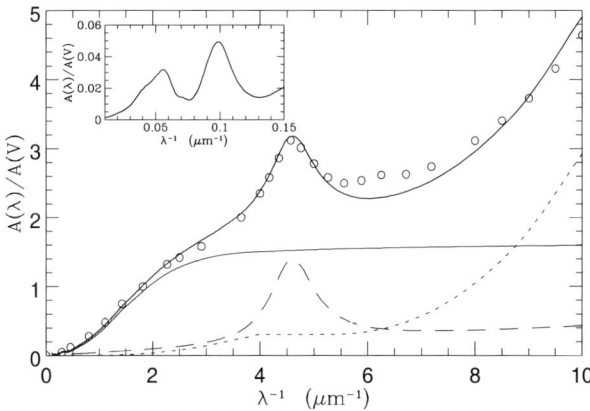

Fig. 7.1. The standard galactic extinction curve, normalized to visual extinction $A(V) \equiv A_V$. The circles correspond to the determination by Savage & Mathis [450]. The thick line is a fit by one of the possible three-component models, with the following contributions: *full thin line*: large cylindrical core-mantle grains; *dashed line*: very small graphitic grains; *dotted line*: PAHs (cf. Sect. 7.2). From Li & Greenberg [322], with the permission of ESO.

Table 7.1 gives the extinction curve in a numerical form, this is especially useful in the infrared[4].

There are large variations in the galactic extinction curves obtained along different directions. These correspond to variations in the grain properties (Fig. 7.3). The variations may correspond, for example, to the coagulation of small grains into larger grains (cf. Dominik & Tielens [128]), or conversely to the partial or or total

---

[4] We mention here that Lutz and collaborators [343] find a larger mid-infrared extinction in the direction of the galactic centre; their determination is based on the ratios of the hydrogen recombination lines (see Sect. 5.1 and Fig. 7.2). This difference might originate in the particular properties of the dust in the central regions of the Galaxy.

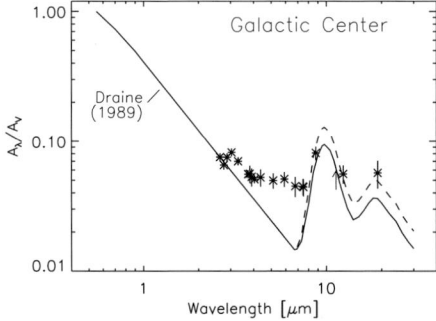

Fig. 7.2. Extinction curve in the mid infrared. The ratio of extinction A_λ to the visual extinction A_V is given in the ordinate. The full line is the extinction law for the dust model of Draine & Lee [134] and the dashed line that corresponding to observations by Rieke & Lebovsky [435]. Note the maxima at 9.7 and 18 μm due to silicate absorptions. The symbols correspond to the determination by Lutz [343] in the direction of the galactic centre, which is based on the intensity ratios between the hydrogen recombination lines observed with the ISO satellite. Notice the large difference in the wavelength range 4–8 μm. From Lutz [343], with the permission of the author.

Table 7.1. Extinction curve in the visible and the infrared. The standard spectral bands centered near the wavelengths of the first column are indicated. From Rieke & Lebofsky [435] and Draine & Lee [134].

λ (μm)	$\frac{E(\lambda-V)}{E(B-V)}$	A_λ/A_V	λ (μm)	$\frac{E(\lambda-V)}{E(B-V)}$	A_λ/A_V
0.34 (U)	1.64	1.531	9.7	−2.86	0.075
0.44 (B)	1.00	1.324	14.3	−3.03	0.0186
0.55 (V)	0.00	1.000	19.0	−3.00	0.0284
0.70 (R)	−0.78	0.748	30	−3.05	0.0130
0.90 (I)	−1.60	0.482	50	−3.07	0.0058
1.25 (J)	−2.22	0.282	100	−3.09	0.0017
1.65 (H)	−2.55	0.175	200	−3.09	0.00043
2.22 (K)	−2.74	0.112	350	−3.09	0.00014
3.5 (L)	−2.92	0.056	850	−3.09	0.000024
4.8 (M)	−3.02	0.023	1300	−3.09	0.000014
7.0	−3.04	0.015			

destruction of some types of grains (see later Sect. 11.3). Fitzpatrick & Massa [184] show that the ultraviolet extinction curve can be parametrized as follows[5]:

$$\frac{E(\lambda - V)}{E(B - V)} = c_1 + c_2 \lambda^{-1} + c_3 D(\lambda) + c_4 F(\lambda), \qquad (7.3)$$

---

[5] This parametrization is valid until about 1 200 Å. At shorter wavelengths, until the Lyman discontinuity at 911.7 Å, extinction keeps increasing considerably and seems very variable: cf. Hutchings & Giasson [255].

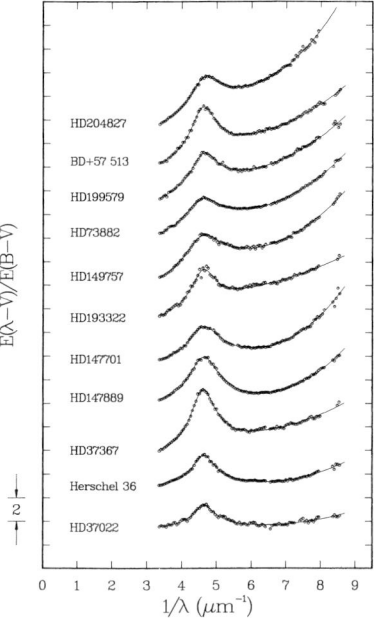

Fig. 7.3. Variations in the extinction curve in the direction of 11 different stars. The points correspond to observations and the lines to the parameter fit described in the text. Reproduced from Fitzpatrick & Massa [184], with the permission of the AAS.

where c_1, c_2, c_3 and c_4 are parameters depending of the line of sight. $D(\lambda)$ is the Drude function that describes the 2 175 Å band:

$$D(\lambda) = \frac{2}{\pi} \frac{\gamma \lambda_0}{(\lambda/\lambda_0 - \lambda_0/\lambda)^2 + \gamma^2}, \quad (7.4)$$

where λ_0 is the central wavelength (here 2 175 Å) and γ is a broadening parameter (here 0.217); the full width at half maximum of the Drude profile is $\gamma \lambda_0$.

The quantity $F(\lambda) = 0.539(\lambda^{-1} - 5.9)^2 + 0.0564(\lambda^{-1} - 5.9)^3$, with $F(\lambda) = 0$ for $\lambda^{-1} <$5.9 μm, gives the shape of the far-ultraviolet extinction excess whose amplitude is given by c_4.

Another presentation of the extinction curve consists of plotting the optical depth per unit column density, N_H, of hydrogen (atomic and molecular). This presentation has been adopted for Fig. 7.5 and 7.6. We can convert from one presentation to the other by using the relation between colour excess and total hydrogen column density derived from observations with the COPERNICUS satellite by Bohlin et al. [48], very well confirmed by the observations with FUSE up to $E(B-V) \simeq 1.0$ (Rachford et al. [421])

$$N_H/E(B-V) = 5.8 \times 10^{21} \text{ (H atom) cm}^{-2} \text{ mag.}^{-1}. \quad (7.5)$$

The absorption and scattering of electromagnetic waves by small particles has been treated in detail in the classical treatises of van de Hulst [524] and of Bohren & Huffman [49]. Here, we only give some simple results. We are dealing with an interference phenomenon that can be described as follows. The electric field of the incoming radiation induces a motion in those electrons of the grain that are weakly bounded (the motion of the nuclei is much smaller due to their larger masses). These electrons in turn emit radiation with the same frequency as the incoming radiation and which is locally in phase with it. The ratio between the intensity absorbed by an electron and the emitted intensity is $(8\pi/3)(e^2/m_e c^2)^2$ where e and m_e are respectively the charge and mass of the electron. The amplitude of the radiation emitted per unit volume of the grain depends upon the polarizability α of its material at the frequency of the radiation: this polarisability is such that an electric field $\mathbf{E}$ produces a dipole $\mathbf{p} = \alpha \mathbf{E}$ per unit volume. This is the phenomenon that causes refraction (see standard physics textbooks). The emitted radiation leaves the grain with a phase difference relative to the incoming radiation, that depends upon the shape, dimensions and refractive index of the grain. The interference between the radiations coming from different parts of the grain between them and with the incident radiation produces the scattering. For particles which are very small compared to the wavelength we can, to a first approximation, replace the secondary radiation by the emission from a single dipole slightly out of phase with the incoming radiation. If the grain is extremely small, the phase change is negligible, there is no scattering and extinction is only due to absorption.

Let us consider a spherical grain of radius a, made of an homogeneous material with a complex refractive index $n = m - ik$ (the imaginary part of the refractive index corresponds to absorption). We can define an extinction cross-section $\sigma_e = \sigma_a + \sigma_s$, σ_a and σ_s corresponding respectively to the absorption and scattering cross-sections. We can write these cross-sections as:

$$\sigma_a = \pi a^2 Q_a, \text{ and } \sigma_s = \pi a^2 Q_s, \tag{7.6}$$

Q_a and Q_s being, respectively, the *absorption efficiency* and the *scattering efficiency*. The *extinction efficiency* is $Q_e = Q_a + Q_s$. For spherical grains that are small compared to the wavelength, the optical calculation (*Mie theory*, see e.g. Lang [299] Sect. 1.40) yields, defining $x = 2\pi a/\lambda \ll 1$,

$$\boxed{Q_a = -4x \operatorname{Im}\left(\frac{n^2-1}{n^2+2}\right) \text{ and } Q_s = \frac{8}{3} x^4 \operatorname{Re}\left(\frac{n^2-1}{n^2+2}\right)^2.} \tag{7.7}$$

From (7.6) and (7.7) we see that for grains small with respect to wavelength the absorption cross-section σ_a is proportional to their volume, and the scattering cross-section σ_s is proportional to the square of their volume. Conversely, big absorbing grains with sizes much larger than the wavelength act as an opaque screen ($Q_a \simeq 1$). Their edges diffract the incoming radiation and it can be shown that $Q_s \simeq 1$ also, at least if the grain is not extremely large, so that $Q_e \simeq 2$ independent of wavelength and grain shape.

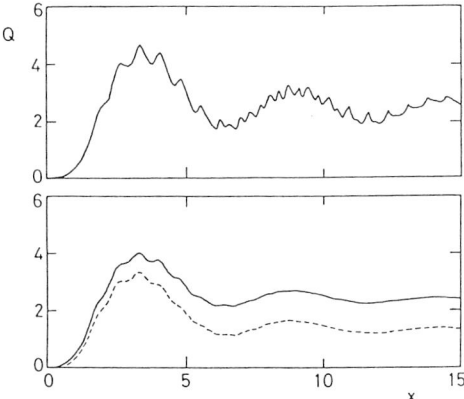

Fig. 7.4. Variation of the extinction efficiency Q_e and the scattering efficiency Q_s as a function of the inverse of wavelength, for a non-absorbing spherical grain with a refractive index 1.6 (top, here $Q_e = Q_s$), and for an absorbing spherical grain with a refractive index $1.6 - 0.05i$ (bottom, here Q_e is plotted as a full line and Q_s as a dashed line). The dimensionless quantity $x = 2\pi a/\lambda$ is given in the abscissae, a being the radius of the grain and λ the wavelength. From Whittet [546], with the permission of IOP Publishing Ltd.

The general behaviour of the variations of the efficiencies Q as a function of grain size is displayed in Fig. 7.4.

The *albedo* is the ratio between the scattering cross-section and the extinction cross-section σ_s/σ_e, or equivalently between the efficiencies Q_s/Q_e. The *phase function*, $g(\theta)$, gives the angular distribution of the scattered light, θ being the angle with the direction of the incoming radiation. Scattering is strongly forward-directed, so that $g(\theta)$ is maximum for $\theta = 0$. These quantites can be calculated for grain models using the Mie theory (cf. van de Hulst [524] and Bohren & Huffman [49]); a useful approximation for the phase function is that of Henyey & Greenstein [231]. The diffusion also produces polarization that we will not discuss here (see however Sect. 2.2 and for an interesting study and references Martin et al.[349]). Observationally, the determination of the albedo and phase function is difficult (cf. Mathis [353]). They can in principle be obtained from photometry of reflection nebulae illuminated by a star with well-determined properties, in cases where the geometric distribution of dust with respect to the star is known (for an example see Calzetti et al. [75]). The albedo is close to 0.6 in the visible and UV, except near the 2 175 Å band where it is much smaller. This shows that the 2 175 Å band is due to absorption and hence to particles that are very small compared to the wavelength.

7.1.2 Extinction and Dust Models

For an ensemble of grains the extinction is obtained by integrating Q_e over the grain size distribution. Several types of grains contribute to extinction in the interstellar medium including, at least, amorphous silicate grains and carbonaceous grains (in part composed of graphite or aromatic components). Grains with a nucleus of one material and a mantle made of a different material, ices or refractory products, may also exist. We often uses the distribution of grain sizes derived by Mathis et al. [351], the so-called *MRN distribution* :

$$dn_i = A_i n_H a^{-3.5} da, \quad a_{min} < a < a_{max}, \tag{7.8}$$

where A_i is a normalization constant, dn_i is the number of grains of species i with radii between a and $a + da$, and n_H is the density of hydrogen nuclei that is taken as the abundance reference. Mathis et al. [351] estimate $a_{min} \simeq 0.005$ μm and $a_{max} \simeq 0.25$ μm. These limits are not well-determined and this distribution should therefore be used with some care.

The optical depth can be written as

$$\tau_\lambda = \sum_i \int_{a_{min}}^{a_{max}} Q_e(a_i, \lambda) \pi a_i^2 N_i(a_i) \, da_i, \tag{7.9}$$

where $N_i(a_i)$ is the column density of grains of species i with radius a_i.

If we are only interested in absorption, we can write an equation similar to (7.9), replacing Q_e by Q_a. We will then see, that as long as grains are small with respect to λ, τ_{abs} is proportional to their total volume and hence to their mass in a column with unit cross-section along the line of sight (7.7).

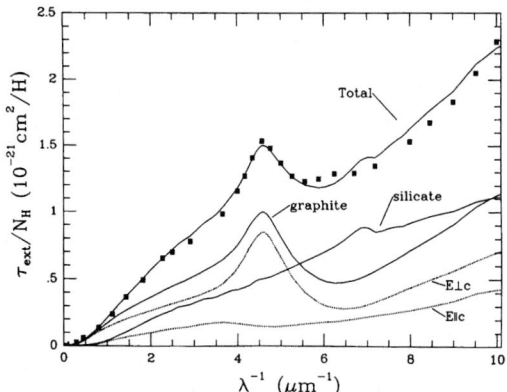

Fig. 7.5. Extinction per hydrogen atom calculated according to the dust model of Draine & Lee [134], compared to observations of the average galactic extinction curve (squares). This extinction is the sum of the contributions of amorphous silicate and of graphite grains. Due to the anisotropy of graphite crystals, their extinction is decomposed into components parallel and perpendicular to the unique crystal axis (orthogonal to the graphite basal plane, the crystallographic c-axis). In this model small graphite particles produce the absorption band at 2 175 Å. Reproduced from Draine & Lee [134], with the permission of the AAS.

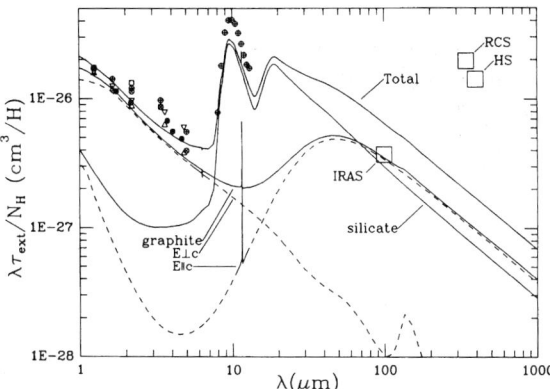

Fig. 7.6. Infrared extinction curve per hydrogen atom (multiplied by λ for better presentation) as calculated according to the model of Draine & Lee [134], compared to observations. As in Fig. 7.5, the respective contributions of silicates and of graphite (total, parallel and perpendicular components) are given. The symbols labelled IRAS, RCS and HS correspond to old measurements of the far-infrared opacity. The recent determination by Boulanger et al. [61], which is based on observations with the COBE satellite, is not plotted here but is in very good agreement with the model results. Reproduced from Draine & Lee [134], with the permission of the AAS.

In practice, we try to reproduce the observed extinction law with a dust model and a grain size distribution. The inverse problem is very difficult to treat due to the large number of free parameters. There is an abundant literature on the subject. We will only cite here the development by Draine & Lee [134] of the model of Mathis et al. [351]. A recent version of this work, due to Weingartner & Draine [544], differs mainly by the addition of very small carbon grains, which intervene essentially in the far-UV extinction, and also by their infrared emission that we will discuss Sect. 7.2. Draine & Lee [134] fit the average extinction curve from the UV to the mid infrared by a mixture of spherical particles of graphite and silicate particles with the MRN size distribution of (7.8). This fit, presented in Figs. 7.5 and 7.6, is excellent. It is obtained by assuming the following grain abundance factors in (7.8): $A_{sil} = 10^{-25.11}$ cm$^{2.5}$/H and $A_C = 10^{-25.16}$ cm$^{2.5}$/H. The silicate grains are supposed to have an olivine-type chemical composition $Mg_{1.1}Fe_{0.9}SiO_4$. In this model 90%, 95%, 94% and 16% of the total abundances of Mg, Fe, Si and O, respectively, are used. This is in general agreement with what is known of the depletion of these elements in the interstellar gas (cf. Table 4.2). All the carbon not in the gas phase is used by the model. Some authors consider that there is a possible abundance problem for carbon, the available carbon abundance being perhaps insufficient (Snow & Witt [479]). But we should remember that the interstellar abundances themselves are somewhat problematic (Sect. 4.1), and we will refrain from discussing this point. We will, however, highlight the fact that fluffy, porous dust grains have larger scattering efficiencies, and hence extinction efficiencies, than compact grains with the same mass. This might help in solving the "abundance crisis" if it is indeed real. However, observations do not favor a large abundance of fluffy grains in the ISM (Boulanger et al. [61]; Smith & Dwek [478]).

The model of Draine & Lee, as any model, is only a representation of reality. For exemple, only spherical particles are considered while observations of interstellar polarization show that elongated grains are present. The optical properties of interstellar silicates are poorly determined, because of the wide variety of possible materials, and are in any case adjusted to fit the observations. Other models also give a good match with the observed extinction curve, for example that of Li & Greenberg [322] which involves silicate grains with a refractory mantle of carbonaceous materials (Fig. 7.1). However this model tends to use too much carbon. Finally, in view of the uncertainties on the interstellar abundances of elements, it does not seem necessary to abandon the simple model of Draine & Lee which is a good starting point for studies of the interstellar dust.

It is interesting to mention that the Kramers–Kronig relations, which link the real and imaginary parts of the dielectric constant of any material, can yield a very general lower limit for the amount of dust necessary to produce extinction (Purcell [419]). They also yield a lower limit for the temperature of grains in a given radiation field. We can obtain in this way a strict lower limit of the dust-to-gas mass ratio in the galactic interstellar medium, that is

$$\rho_d/\rho_H > 0.5 \times 10^{-2} \rho_g, \qquad (7.10)$$

where ρ_d and ρ_H are respectively the density of matter in grains and in gas, and ρ_g is the density of the matter that form the grains, in g cm^{-3} (Aannestad & Purcell [1]; for a recent detailed discussion see Kim & Martin [282]). This limit is just compatible with the ratio $\rho_d/\rho_H \simeq 0.6 \times 10^{-2}$ obtained for the grain model of Weingartner & Draine [544].

7.1.3 X-Ray Scattering by Dust

Interstellar dust grains scatter X-rays in directions close to that of the incident radiation. While the scattering of optical light is not very anisotropic, due to the small average size of dust grains (grains with sizes comparable to, or larger than, the wavelength are rare), most grains have sizes larger than X-ray wavelengths, so that scattering is at small angles[6]. For a dust size distribution the superimposition of the scattering produced by the different grain sizes gives rise to a halo around X-ray sources. As we will see, the size of this halo depends upon the size distribution of the largest grains that are the most efficient scatterers. It also depends upon the composition and density of the grains. The observation of X-ray halos can thus supply information on these parameters and, in particular, constrain the large grain sizes in grain models. Even if this method has not yet given many results, it has an important future and it seems worthwhile to give some elements of the analysis, following Smith & Dwek [478]. More details can be found in their paper.

---

[6] Remember that scattering is actually a diffraction phenomenon, and that diffraction occurs in directions closer to the incident radiation for larger diffracting targets.

The mean scattering angle is less than 10′ for a grain with radius 0.1 μm, or than 5′ for a 0.25 μm grain, for 2-keV X-rays. The refractive index for X-rays is very different from the optical index. The real part m is only slightly less than unity[7]. X-ray photons thus penetrate the entire grain, surface reflection being negligible. Their energy being considerably larger than that of the electronic bands of the solid which forms the grain, all electrons can be considered as active in the process, so that refraction and scattering depend upon the total electron density n_e in the grain (conversely, only weakly bound electrons take part in optical processes). Since m is close to 1, the phase shifts between the incident radiation and the emitted radiation are small and can often be neglected (the *Rayleigh–Gans approximation*). These phase shifts must however be taken into account for big grains at energies smaller than about 1 keV, and the Mie theory must then be used. The absorption of X-ray photons by the grain material is generally small above 1 keV but it is important at lower energies (cf. Fig. 2.5).

As an example, we show in Fig. 7.7 the intensity of the scattering halo around an X-ray point source, as a function of the angle with the direction of the source. This is calculated using an MRN size distribution of grains (7.8). For comparison with observations, see Fig. 7 of Smith & Dwek [478]. Another example is described by Draine & Tan [142].

7.2 Interstellar Dust Emission

7.2.1 Grains in Thermal Equilibrium

Interstellar grains are in general heated by the absorption of UV and visible radiation, and cool by the thermal emission of infrared photons. Other heating and cooling mechanisms can be efficient in particular cases: these are molecule–grain collisions deep inside molecular clouds, which cool the grains because they are generally warmer than the gas, and electron–grain collisions in the hot gas of supernova remnants. We examine each of these mechanisms in Sect. 8.1.

A spherical grain with radius a subjected to a radiation field of density u_ν absorbs a total energy

$$E_{abs} = \int_0^\infty 4\pi a^2 Q_a(\nu) \pi \frac{cu_\nu}{4\pi} d\nu, \qquad (7.11)$$

$4\pi a^2$ being the area of the grain. Indeed, per unit area a grain absorbs an energy $Q_a(\nu)\pi I_\nu$, $I_\nu = cu_\nu/4\pi$ being the flux per steradian. The factor $\pi = \int_0^{\pi/2} 2\pi \cos\theta \sin\theta \, d\theta$ (θ being the angle with the normal to the surface) comes from integration over the half-space. Since $Q_a(\nu)$ increases in the UV, the energy is mostly absorbed in the UV and the visible in the general interstellar radiation field

---

[7] An approximate expression for the real part of the refraction index for X-rays is $m = 1 - n_e r_e \lambda^2/2\pi$, n_e being the density of all electrons in the material and $r_e = 2.82 \times 10^{-13}$ cm being the classical radius of the electron.

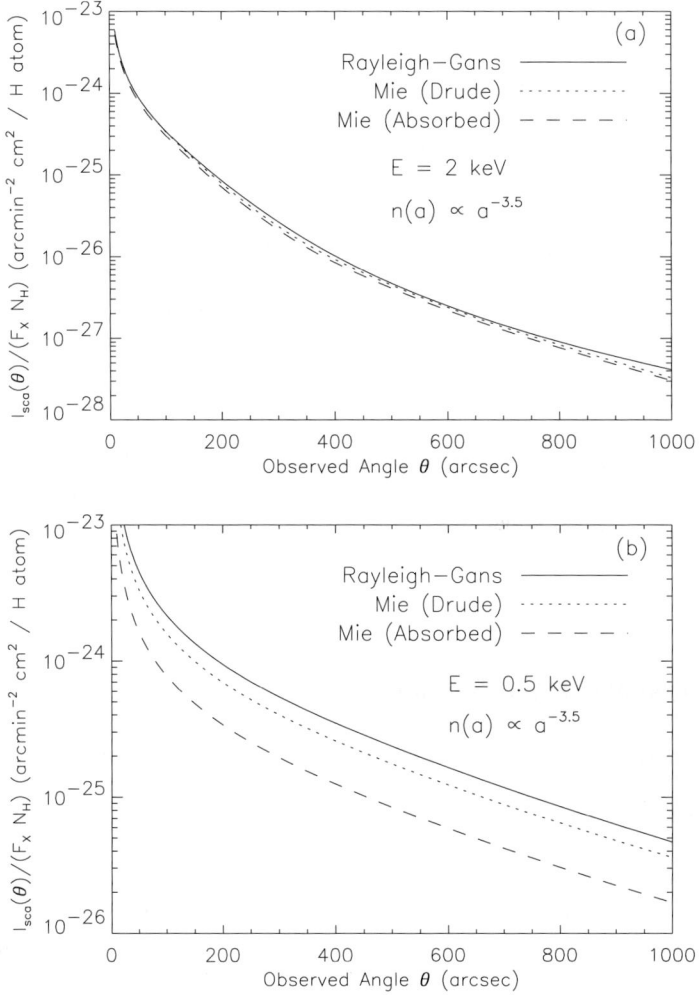

Fig. 7.7. Intensity of the X-ray scattering halo as a function of the angular distance to a point source. It is calculated for a mixture of silicate and graphite grains with a composition similar to that of Draine & Lee [134] and the MRN size distribution of Mathis et al. [351] (cf. (7.8)). The intensity relative to that of the source F_X is given per square arc minute and normalized to the column density N_H of hydrogen atoms. The grains are assumed to be distributed uniformly between the source and the observer. The results are given (a) for 2 keV photons and (b) for 0.5 keV photons. The three curves in each figure correspond respectively to the simplified Rayleigh–Gans theory where all electrons in a grain are assumed to emit in phase, to the full theory without absorption [Mie (Drude)], and to the same theory including absorption [Mie (Absorbed)]. The differences are large at small energies. Reproduced from Smith & Dwek [478], with the permission of the AAS.

7.2 Interstellar Dust Emission

or near a sufficiently hot star. On the other hand, the same grain at a temperature T emits a total energy

$$E_{em} = \int_0^\infty 4\pi a^2 Q_a(\nu) \pi B_\nu(T) \, d\nu, \qquad (7.12)$$

where $B_\nu(T)$ is the Planck function. $4\pi a^2$ is the emitting surface, Q_a is the emission efficiency and $\pi B_\nu(T)$ is the power emitted in the half space per unit frequency by a blackbody at temperature T. Now, the grain temperature being low as we will see later, the emission is in the mid- or far-infrared. If the grain is in thermal equilibrium, we have

$$W(T) = \int_0^\infty Q_a(\nu) \frac{c u_\nu}{4\pi} \, d\nu = \int_0^\infty Q_a(\nu) B_\nu(T) \, d\nu, \qquad (7.13)$$

from which we can obtain the temperature T of the grain.

We can calculate analytically this temperature with a simple grain model, if the absorption efficiency in the mid-infrared is approximately represented by

$$Q_a(\nu) = Q_0 \left(\frac{\nu}{\nu_0}\right)^\beta \frac{a}{a_0}, \qquad (7.14)$$

with $\beta \simeq 2$.[8] According to Draine & Lee [134], $\lambda Q_a / a \simeq 1$ at 100 μm, allowing us to determine Q_0 once the reference frequency ν_0 is chosen. The factor a in (7.14) comes from the assumption that the grain is supposed to be much smaller than the wavelength (cf. (7.7)). Putting $y = h\nu/kT$ and expanding the Planck function, we obtain the expression for the right-hand part of (7.13):

$$W(T) = \frac{2h}{c^2} \left(\frac{kT}{h}\right)^{4+\beta} \frac{Q_0}{\nu_0^\beta} \frac{a}{a_0} \int_0^\infty \frac{y^{3+\beta}}{e^y - 1} \, dy, \qquad (7.15)$$

or numerically, taking $\beta = 2$,

$$W(T) = 4.6 \times 10^{-11} \left(\frac{a}{0.1 \, \mu m}\right) T^6 \, \text{erg cm}^{-2} \, \text{s}^{-1}. \qquad (7.16)$$

The left-hand term of the equation for thermal equilibrium (7.13) can also be evaluated in a simplified way assuming that the emitting dust is highly absorbing in the UV and the visible, so that $Q_a \simeq 1$. This is not a bad approximation for the big grains that dominate the far-infrared emission, and which have sizes of the same order as UV wavelengths. We then find a temperature of the order of 20 K for grains with radii $a = 0.1$ μm in the interstellar radiation field of the solar neighbourhood.

---

[8] The value of the exponent depends upon the nature of the grain and is not well determined by laboratory measurements, which are difficult to perform. An observational determination can be made using the spectral energy distribution of the emitted radiation: see for example Lagache et al. [295], (7.18) and Fig. 7.8.

This temperature varies as $(a/a_0)^{-1/6}$. A more exact calculation gives 18.8 K for graphite and 15.4 K for silicate (Draine & Lee [134], Table 3). The temperature obviously depends upon the ratio Q_a(UV, visible)$/Q_a$(far-infrared).

In order to obtain the spectrum emitted by an ensemble of grains we must, of course, integrate over their size distribution. A rough but often sufficient approximation is given by Désert et al. [120].

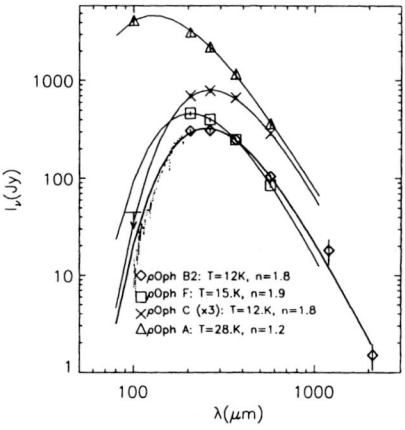

Fig. 7.8. Dust emission spectra in different parts of the molecular cloud associated with the star ρ Oph. The spectra are fitted by the emission of dust at a single temperature T, whose submillimetre emissivity varies as ν^{-n}. The values of T and n which result from these fits are indicated. From Ristorcelli et al. [436], with the permission of EDP Sciences.

We have seen that the spectrum emitted by grains depends upon both the radiation field and the grain properties. As an example, Fig. 7.8 shows submillimetre spectra of cold dust observed in different directions in the molecular cloud of ρ Ophiuchi, and Fig. 7.9 the infrared spectrum of warm dust in the direction of the Orion H II region. Lagache et al. [295] derived, from observations with the COBE satellite, the intensity emitted at submillimetre wavelengths by dust heated by the local interstellar radiation field near the Sun, which can be well-represented by

$$I_\nu = \tau B_\nu(17.5\,\text{K}), \tag{7.17}$$

where $B_\nu(T)$ is the Planck function, τ being given by

$$\tau/N_\text{H} = (8.7 \pm 0.9)10^{-26}(\lambda/250\,\mu\text{m})^{-2}\ (\text{H atom})^{-1}. \tag{7.18}$$

7.2.2 Small Grains out of Thermal Equilibrium

For a very small dust grain, which has an accordingly small heat capacity, the energy input from the absorption of a single UV or visible photon can lead to a strong,

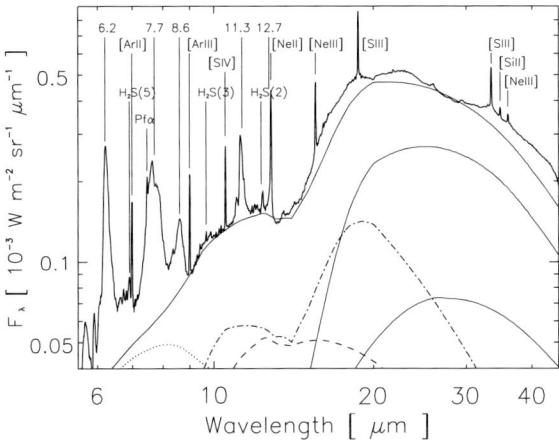

Fig. 7.9. The mid-infrared spectrum of the Orion nebula (in log scale). We see many fine-structure lines identified on the figure: the Pfund α hydrogen recombination line, rotation lines of H_2 emitted by the photodissociation region located on the back of the H II region, and the classical aromatic bands coming from this region. In the wavelength range displayed in this figure, the continuum is dominated by the emission of dust inside the H II region and in the photodissociation region. The full curve is a fit of this continuum, which is the sum of the following contributions: emission by amorphous silicates at 80 K (thin dash-dotted line) and at 130 K (thick dash-dotted line), emission by amorphous carbon at 85 K (thin dashed line) and at 155 K (thick dashed line). These temperatures are not chosen arbitrarily, but are extreme temperatures estimated for grains of different sizes in the known radiation field at this location of the H II region. The dotted curve is the emission of very small grains of amorphous carbon at 300 K. The continuum excesses observed at several wavelengths with respect to the model might correspond to emission bands of crystalline silicates. From Cesarsky et al. [86], with the permission of ESO.

instantaneous increase in temperature followed by rapid cooling. The grain then spends some time at low temperature until another photon is absorbed. There is no thermal equilibrium since the grain experiences large temperature fluctuations. If $C(T)$ is the heat capacity of the grain at temperature T, the condition for strong temperature fluctuations can be written as

$$h\nu_m \geq \int_0^{T_{eq}} C(T)\, dT, \tag{7.19}$$

where ν_m is the mean energy of the absorbed photons and T_{eq} the equilibrium temperature that the grain would have if no quantum effect was present. This temperature is given by (7.13).

Absorption of a photon with frequency ν heats the grain to a temperature T such that

$$\boxed{h\nu = \int_{T_0}^{T} C(T)\, dT,} \tag{7.20}$$

where T_0 is the initial grain temperature. The determination of the heat capacity for very small grains (or large molecules) is a rather difficult problem, treated in a complete way by Draine & Li [141]. We will later give an approximate solution. As a first approximation, we may assume that the thermal energy of a grain composed of $\mathcal{N}$ atoms is about $3\mathcal{N}kT$, hence $C(T) \simeq 3\mathcal{N}k$. A photon with frequency ν thus very rapidly brings the grain temperature to $T \simeq h\nu/3\mathcal{N}k$, neglecting the initial thermal energy of the grain. For example, a grain made of 50 atoms absorbing a 1 000 Å photon is heated to some 1 000 K! It then emits in the near infrared with peak emission near 3 μm and rapidly cools according to

$$\frac{dT}{dt} = \frac{1}{C(T)} \int 4\pi a^2 Q_a(\nu) \pi B_\nu(T)\, d\nu, \qquad (7.21)$$

from (7.12). The grain obviously emits photons of increasingly longer wavelengths during cooling. From the Stefan-Boltzmann law, which says that the total power emitted by a blackbody is proportional to T^4, most of the energy is emitted at high temperatures, hence at wavelengths not much longer than 3 μm. Cooling occurs in a few seconds, while the interval between two successive photon absorptions is several months for a 50-atom grain in the local interstellar field. A very important consequence is that the shape of the emitted spectrum does not depend upon the intensity of the incident radiation, as long as the interval between successive photon absorptions is longer than the cooling time. The maximum temperature, and hence the shape of the spectrum, do however depend upon the particle size and the energy of the absorbed photon. Figure 7.10 illustrates schematically the time variation of temperature for grains of different sizes.

This mechanism was initially proposed by Andriesse [10] then later discussed by Sellgren [464] in order to account for the infrared emission from reflection nebulae. It has been modelled in detail by Draine & Anderson [135] and by Guhathakurta & Draine [213] for very small graphite and silicate grains, and by Léger & Puget [312] and Draine & Li [141], who treated the case of the polycyclic aromatic hydrogenated carbons which will be discussed later. Figure 7.11 illustrates the results of Draine & Anderson for graphite, as a temperature probability distribution for various grain sizes. Figure 7.12 shows the spectrum emitted by the silicate/graphite mixture of Draine & Lee [134], with a minimum radius of 3 Å.

A better approximation for the heat capacity of small grains of polycyclic aromatic hydrocarbons (or more generally of hydrogenated carbonaceous grains) can be obtained as follows. Their thermal energy content $E(T)$ at temperature T is given by the following formula:

$$E(T) = (3\mathcal{N} - 6)kT\eta(T), \qquad (7.22)$$

where $\mathcal{N}$ is the number of atoms in the grain and $\eta(T)$ is the ratio between the specific heat at temperature T and that at $T = \infty$. $\eta(T)$ is given with a reasonable accuracy for an isotropic crystal by the harmonic approximation of Einstein:

$$\eta(T) = \frac{C_V(T)}{C_V(\infty)} = \frac{\hbar\omega}{kT}\left(\frac{1}{2} + \frac{1}{e^{\hbar\omega/kT} - 1}\right), \qquad (7.23)$$

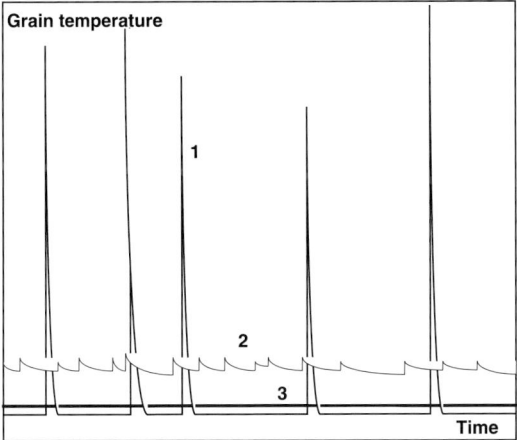

Fig. 7.10. Scheme for the time evolution of the temperature, hence of the emitted intensity for grains of different sizes, submitted to the interstellar radiation field. The big grains (broad horizontal line, 3), are in thermal equilibrium and their temperature, hence the emitted intensity, does not vary. The smallest grains experience an immediate, very strong increase in temperature after absorption of a photon, and rapidly cool down to a low temperature (1). The temperature of intermediate-size grains increases slightly after each photon absorption, causing temperature fluctuations less pronounced than in the preceding case (thin line, 2). The scales are arbitrary and different for each case.

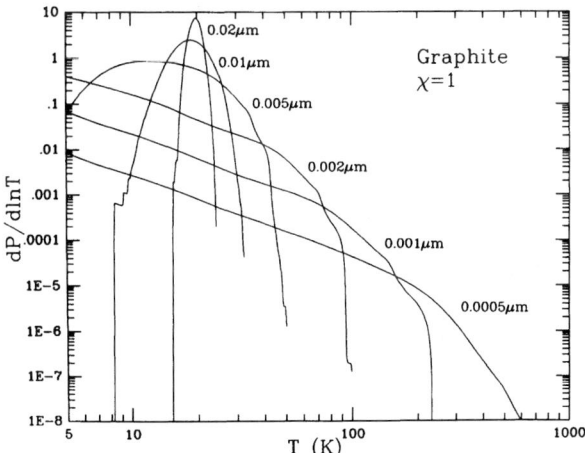

Fig. 7.11. Temperature distributions $dP/d\ln T$ for graphite grains of various radii a exposed to the interstellar radiation field near the Sun. $P(T)$ is the probability that a grain will have a temperature larger than T. Note that the big grains are approximately in equilibrium at a temperature of about 20 K. Reproduced from Draine & Anderson [135], with the permission of the AAS.

166   7 Interstellar Dust

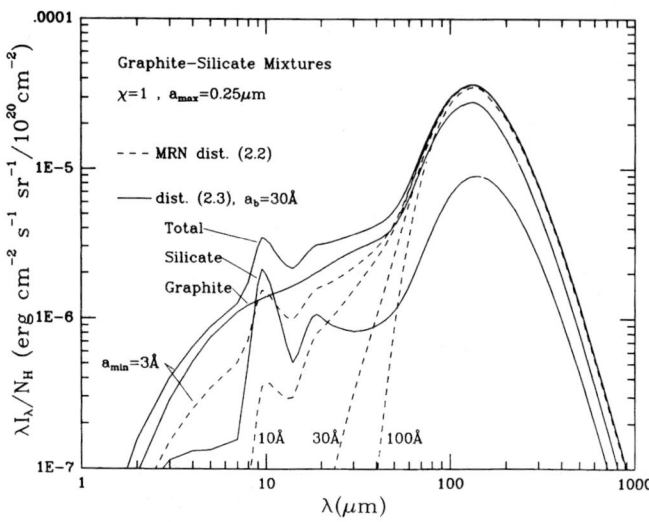

Fig. 7.12. The calculated emission spectrum for a mixture of silicate and graphite grains with different size distributions, heated by the mean interstellar radiation field near the Sun. The curves are labelled according to the minimum grain size a_{min}, the maximum size being fixed to $a_{max} = 0.25$ μm. The results are for an MRN size distribution with a slope of 2.2, and also for a size distribution with a slope of 2.1 where small grains are favored. This figure is only shown for pedagogical reasons because the size spectrum is unknown for very small silicate grains, and the total calculated spectrum does not well match the observations (cf. Sect. 7.2). Reproduced from Draine & Anderson [135], with the permission of the AAS.

ω being the angular vibrational frequency of the atoms in the crystal, which is assumed to be the same for all atoms in this approximation. For anisotropic crystals, see the textbooks of statistical physics, e.g. Reif [428]. $\eta(T)$ for graphite is plotted in Fig. 7.13 from d'Hendecourt et al. [123], as well as the same function for two light aromatic molecules from the *ab-initio* calculations of Cook & Saykally [104]. From an interpolation between these we can derive the number of atoms in a carbonaceous hydrogenated grain, such that it is brought to temperature T after the absorption of a photon of wavelength λ. This number is given in Table 7.2, which also gives the wavelength of the maximum of emission of a blackbody at temperature T. This wavelength is chosen to be consistent with the main mid-IR emission bands that will be discussed in the next section.

Let us also mention that rotating very small grains, when non-spherical and electrically charged, emit a dipole radiation at millimetre radio wavelength (Draine & Lazarian [139]). This mechanism is one of the possibilities that have been invoked to explain an excess radiation observed near 600 GHz by the COBE satellite.

7.2 Interstellar Dust Emission   167

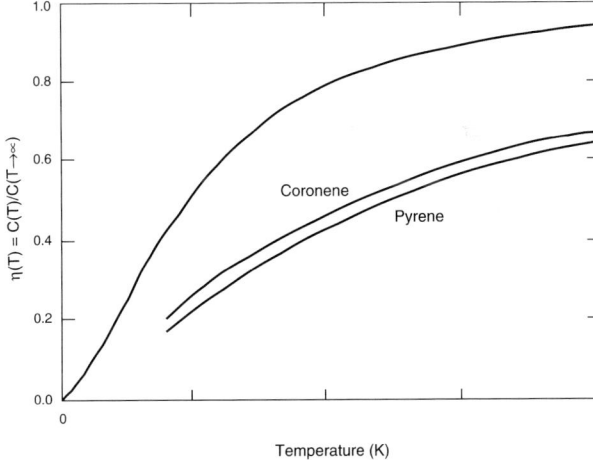

Fig. 7.13. Curves showing the specific heat capacity of big molecules and very small carbonaceous grains. See the text for the definition of η. The upper curve corresponds to the harmonic approximation for graphite, and the two lower curves to the results of *ab initio* calculations for two polycyclic aromatic hydrocarbons (PAHs), coronene and pyrene.

Table 7.2. Number of atoms in a small carbonaceous grain heated to a temperature T by the absorption of a photon of wavelength λ (columns 3, 4 and 5). The table also gives, in column 2, the wavelength of the peak of emission λ_{em} for a blackbody at a temperature T.

T K	λ_{em} μm	λ_{abs} 1 250 Å	λ_{abs} 2 500 Å	λ_{abs} 5 000 Å
402	12.7	231	137	80
455	11.3	189	122	70
662	7.7	129	78	41
822	6.2	99	58	29
1545	3.3	44	24	12

7.2.3 The Aromatic Emission Bands in the Mid-Infrared

The infrared-submillimetre emission spectrum of the diffuse interstellar medium is now well known (Fig. 7.14). No silicate band emission at 9.7 and 18 μm is visible, showing that very small silicate grains are absent or not abundant. But there are very intense emission bands at 3.3, 6.2, 7.7, 8.6, 11.3 and 12.7 μm, as well as weaker bands at other wavelengths. They are designated in the astronomical literature as UIBs (for *Unidentified Infrared Bands*) or IEFs (for *Infrared Emission Features*), which tells nothing about their origin, or as AIBs (for *Aromatic Infrared Bands*), our preferred designation because it is more precise, or, finally, as *PAH bands* (this term will been defined later, and may be too precise as we will see).

168   7 Interstellar Dust

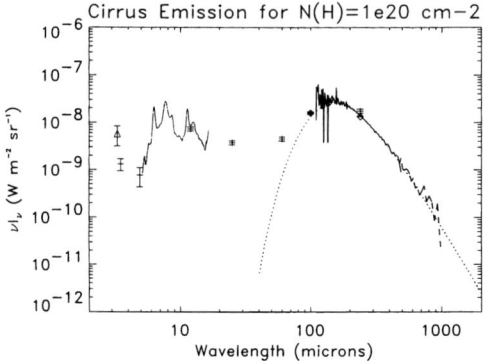

Fig. 7.14. The emission spectrum of interstellar dust from near-infrared to millimetre wavelengths, normalized to a column density of 10^{20} H atoms per cm^2. The dotted curve is the emission of a 17.5 K blackbody multiplied by an emissivity proportional to ν^2. Compare this figure to Fig. 2.4 which uses older data. From Boulanger et al. [63], with the permission of Springer Verlag and EDP Sciences. References to the relevant observations will be found in this article.

These aromatic bands contribute an important fraction of the interstellar dust emission, and are practically always found where UV or even visible photons can excite them, except if the UV radiation field is very intense, in which case the carriers are destroyed. The regions which show dominant aromatic band emission are the photodissociation regions at the interfaces between H II regions and neutral clouds. This can be explained by the fact that these regions contain large amounts of matter and are subjected to a strong flux of UV and visible photons (Plates 21 and 22). Photodissociation regions will be discussed in Chap. 10. Aromatic bands are also observed in the H I medium and in many external galaxies. The constancy of their spectrum under very different conditions is remarkable (Fig. 7.15). However the mid-infrared spectrum of the neutral interstellar medium can sometimes be very different, an extreme case being that of the Andromeda galaxy M 31 that will be briefly discussed in Sect. 15.4.

Properties of the Aromatic Bands

Despite many observations with the European ISO and the Japanese IRTS satellites, the exact nature of the emitters of the aromatic bands is still debated at the time of writing. Before discussing it, let us give a short summary of the present knowledge on these emitters.

1. The band carriers are certainly very small grains or big molecules containing hydrogenated aromatic rings. The different bands can be assigned to characteristic vibration modes of aromatic materials, i.e. Allamandola et al. [6]:
    - the 3.3 µm band to the stretching vibration mode of C–H attached to an aromatic ring (for the same group attached to an aliphatic skeleton, the band is near 3.4 µm);

7.2 Interstellar Dust Emission 169

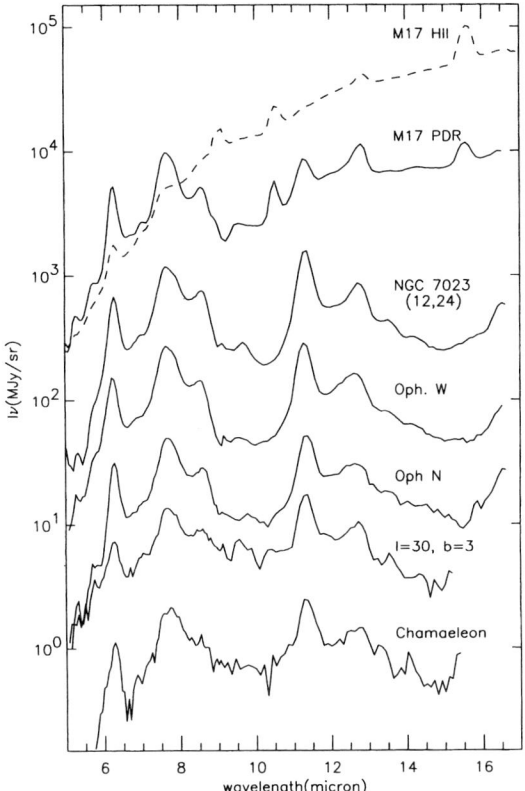

Fig. 7.15. The mid-infrared, low-resolution spectra of a variety of interstellar regions illuminated by radiation fields increasing in intensity from the bottom to the top of the figure. We see that the aromatic band spectra differ little in shape from region to region. The two spectra to the top, which correspond to particularly intense UV radiation environments, also show fine-structure lines at 7, 9, 10.5, 12.8 et 15.5 μm due respectively to Ar II, Ar III, S IV, Ne II and Ne III. The spectral resolution of these spectra obtained with the ISO satellite is only $\lambda/\Delta\lambda \simeq 40$, so that these lines appear as broad as the aromatic bands. For these two spectra the interstellar radiation field is more than 10^4 times the field in the solar neighbourhood and a continuum emission from very small grains is seen in addition to the aromatic bands. From Boulanger et al. [63], with the permission of Springer Verlag and EDP Sciences.

- the 6.2 and 7.7 μm to the C–C stretch in an aromatic solid or molecule;
- the 8.6 μm band to the bending of a C–H group attached to an aromatic ring and moving in the plane of this ring;
- the 11.3 and 12.7 μm and other bands in this rather complex spectral region to bending of C–H groups attached to an aromatic ring and moving perpendicularly to its plane. The exact wavelength depends upon the presence and number of the adjacent C–H groups.

2. The band shapes can be well-fitted by Lorentz functions or by combinations of Lorentz functions[9]

$$\phi(\nu) = \frac{\Delta\nu/2\pi}{(\nu - \nu_0)^2 + (\Delta\nu/2)^2}, \qquad (7.24)$$

where $\Delta\nu = (\pi\tau)^{-1}$, τ being the lifetime of the excited state and ν_0 the central frequency of the line (Boulanger et al. [62]). This suggests that the energy of the absorbed photon is very rapidly redistributed between the different vibration modes of the emitters with a characteristic time of the order of 10^{-13} s.

3. The emitters are heated by individual UV and visible photons (Uchida et al. [522]; Cesarsky et al. [85]; Pagani et al. [393]): see Plate 23 for the case of the Andromeda galaxy M 31. The contribution of visible photons is important in the atomic medium near the Sun (cf. Fig. 7.16). In this case, the emission of a band at a wavelength as short as 6.2 μm (the 3.3 μm band has not yet been clearly observed in the diffuse medium) implies that some of the emitting species do not contain more than about 30 atoms (cf. Table 7.2), which for compact grains corresponds to a radius of about a nanometre: these grains can thus be considered as big molecules.

4. The carbon abundance in these emitters is an important fraction of the total carbon abundance. It can be estimated, if we know the exciting radiation, the radiation field, the absorption cross-section of the emitters per carbon atom as a function of wavelength, the intensity of the bands and the column density of the interstellar medium. For example, for the high-latitude interstellar medium (the infrared "cirruses") a total emission of 1.6×10^{-24} erg s^{-1} per H atom between 2 and 15 μm is calculated. Assuming a specific absorption of the emitters of 2.3×10^{-20} erg s^{-1} per carbon atom for the interstellar radiation field near the Sun, from Joblin et al. [265], the carbon abundance in the emitters is found to be C/H = 7×10^{-5}, or about 20% of the total carbon abundance. In the reflection nebula Ced 201, Cesarsky et al. [87] find an abundance of the order of 15% of the total carbon abundance. These abundances are uncertain given the poor knowledge of the various parameters, but they are probably of the right order of magnitude.

5. Those particles which are responsible for the emission of infrared bands can be excited by absorption of visible photons; they produce a non-negligible part of the extinction in the V band (5 500 Å), of the order of 13% (an indicative figure only).

The Nature of the Aromatic Band Emitters

The aromatic bands were first attributed by Léger & Puget [312] and by Allamandola et al. [5] to very small grains or big molecules of polycyclic aromatic hydrocarbons,

---

[9] It is also possible to use Drude function fits (7.4) which are appropriate for describing bands of solids but are in any case not very different.

or PAH, heated by the absorption of single UV photons. There is some interesting evidence in favor of this identification, and also some contradictory aspects, some of which will be discussed here: the PAH hypothesis is still partly an hypothesis. This is mainly due to difficulties with producing and studying PAHs big enough to simulate interstellar particles in the laboratory.

In order to match the observed AIB emission spectra with PAHs, it is necessary to use a mixture of neutral and positively ionized PAHs (PAH$^+$). Neutral PAHs mainly emit the 3.3 µm band and the bands at 11.3–12.7 µm[10], while PAH$^+$ generally emit the 6.2–8.6 µm bands (Pauzat et al. [399]). Of course, the emitted spectrum depends much upon the maximum temperature experienced by the particle, because it is the product of the absorption cross-section and the spectrum of a blackbody at the given temperature. The shorter-wavelength bands are preferentially emitted at high

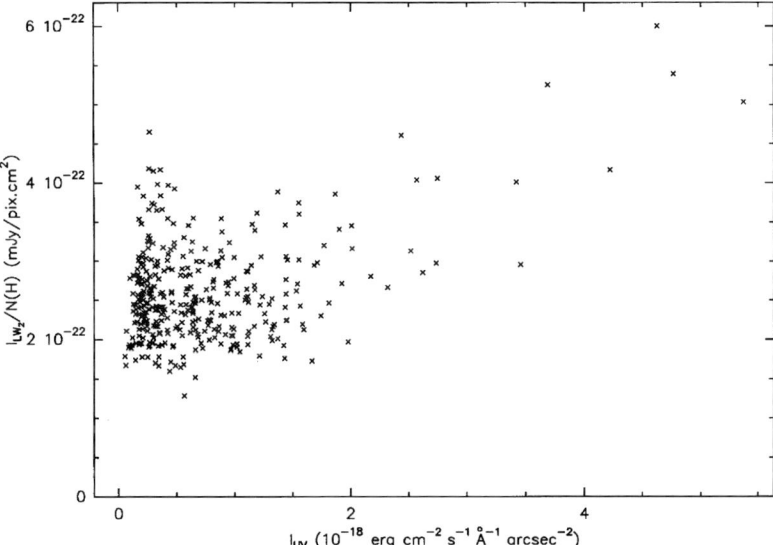

Fig. 7.16. Intensity per hydrogen atom for the 6.2 and 7.7 µm aromatic bands observed with ISO in the disk of the Andromeda galaxy M 31, as a function of the ultraviolet radiation field at 2 000 Å. See Pagani et al. [393], from which this figure originates, for the exact designation of the ordinate scale. Here it is sufficient to know that it is proportional to the band intensity and would be similar, to within 30%, for galactic cirrus. The studied regions are immersed in a visible radiation field approximately equal to that near the Sun. This visible radiation dominates the excitation for small UV intensities, until the observed UV brightness (abscissae) reaches about $I_{UV} = 2 \times 10^{-18}$ units. The UV intensity near the Sun would correspond to half this value. We conclude that the excitation of the emitters of the aromatic bands near the Sun is dominated by visible radiation. From Pagani et al. [393], with the permission of ESO.

---

[10] The spectra of PAH$^-$ which are sometimes invoked is rather similar to that of the neutral species.

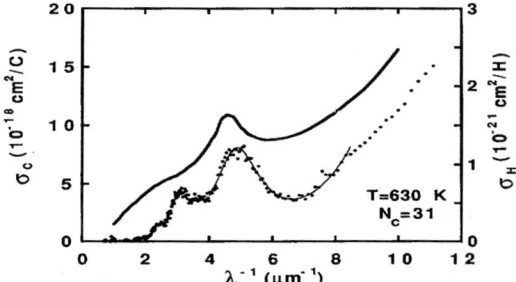

Fig. 7.17. Absorption cross-section per carbon atom for a mixture of neutral PAHs (points, left ordinate scale) compared to the interstellar extinction curve (full line, right ordinate scale). The mean number of carbon atoms in these PAHs is N_c and T is the temperature at which the mixture has been evaporated from the solid phase. Absorption is negligible for $\lambda^{-1} < 2$ µm^{-1} ($\lambda > 5\,000$ Å). These particles, but also other aromatic particles, might account for the extinction in the ultraviolet, which is rather well matched if we do not consider some bands between 2 and 4 µm^{-1} that do not exist in the interstellar medium. Reproduced from Joblin et al. [265], with the permission of the AAS.

temperatures and those at longer wavelength at lower temperatures (see examples in Cook & Saykally [104]).

The profile of the 3.3 µm band is well reproduced by very small PAHs (less than about 30 atoms), for which there are laboratory measurements that can directly be compared to the observations. These PAHs exhibit a broad variety of absorption spectra from 6 to 15 µm. The spectra of the bigger neutral and ionized PAHs, that are expected to dominate the interstellar emission, are not known so that we do not really know if it is possible to reproduce the interstellar spectra by a collection of PAHs et PAH$^+$s as Allamandola et al. [7] or Li & Draine [323] have attempted to do.

On the other hand, the predicted variability of the mid-IR spectrum of PAHs, which is related to their degree of ionization and hydrogenation, is not observed in the interstellar medium. This suggests that larger species, whose spectral properties are less sensitive to the physical conditions, are the carriers of the AIBs. By larger species, we mean aggregates of hydrogenated, partly aromatic, carbon. Diamond nanoparticles, whose surface appears to be covered with aromatic structures, are possible candidates amongst many other species (Jones & d'Hendecourt [271]).

7.2.4 The Very Small Grains

In order to account for the observed emission at wavelengths intermediate between those of the aromatic bands and those where the big grains emit in a moderate radiation field like that near the Sun, we are lead to postulate the existence of very small grains able to emit a continuum. Given the wavelength range in which they emit, these grains must be heated to several tens of degrees and cannot be in thermal equilibrium with the local radiation field. They are in an intermediate regime between thermal equilibrium, where grain cooling is negligible between the absorption of two consecutive photons, and the fully stochastic regime of the carriers of the infrared bands, where the grain has time to cool completely before the arrival of the next photon. They must therefore consist of several hundred atoms. Their thermal energy is approximately given by (7.22). In this intermediate situation the shape of the emitted spectrum is not independent of the intensity of the radiation field, contrary to the case with smaller particles. We know very little of the nature of these grains, except that the emission is presumably dominated by carbonaceous grains rather than by silicates: we have seen that the silicate bands are visible in emission only in very strong radiation fields, more 10^4 times stronger than the radiation field near the Sun (cf. Fig. 7.9): in this case the emission can be well accounted for by the usual big grains. However, small silicate grains may well exist but their infrared emission is difficult to distinguish from that of the other grains (Li & Draine [323]). It is empirically found that the maximum temperature reached by the very small grains becomes large enough for them to produce appreciable emission near 15 μm when the radiation field is greater than about 1 000 times that near the Sun (Contursi et al. [103], cf. Fig. 7.15).

7.2.5 The Big Grains

Grains bigger than those previously discussed are responsible for most of the extinction in the visible and in the infrared, and for most of the emission at wavelengths longer than about 60 μm. They also contain most of the solid matter in the interstellar medium. The exchange processes between these grains and the smaller ones play an important role in the energy balance of the interstellar medium. They will be examined in Chap. 15. The main problems raised by the big grains are their nature and their emissivity at submillimetre and millimetre wavelengths (cf. (7.14), (7.17) and (7.18)). These points are still somewhat controversial.

7.3 Global Dust Models

Astronomers have long ago attempted to build global interstellar dust models that would account for all the observed properties while still being compatible with what is known of the depletions of some elements from the gas phase of the interstellar medium (Table 4.2). The main difficulty is to know the origin of the different parts of the extinction curve. Extinction in the visible is not very variable from region to region and is due, as we have seen, to relatively big grains, the MRN size distribution (7.8) giving satisfactory results. We should, however, take into account the non-negligible but poorly known contribution of the carriers of the aromatic bands to the absorption (they do not contribute to scattering because of their extremely small size). Difficulties arise in the ultraviolet where important variations in the extinction curve are observed from region to region. There is a lack of correlation between the absorption band at 2 175 Å, extinction at longer wavelengths which looks like an extrapolation of the optical extinction and varies as $1/\lambda$, and extinction at shorter wavelengths which is particularly variable (Fig. 7.3). Désert et al. [120] and Li & Greenberg [322] assumed that the contribution of the very small carbonaceous grains is the absorption in the 2 175 Å band, and that PAHs dominate the far-UV excess of the extinction curve and give a part of the visible and UV extinction. In the model of Dwek et al. [149], the very small grains are made of graphite with a size distribution that is an extension of that of the big grains; they are responsible for a part of UV extinction and for the 2 175 Å band. The PAHs extend the size distribution to molecules with 20 carbon atoms. Weingartner & Draine [544] assume that the 2 175 Å band comes entirely from PAHs. All these hypotheses are rather arbitrary, and in fact the 2 175 Å band is rather anticorrelated to the emission at 12 µm which might be attributed to PAHs (Boulanger et al. [60]). Note however that all models, including that of Weingartner & Draine, are able to reproduce in a satisfactory way the observed variations of the extinction law, including the very different laws observed in the Magellanic Clouds, by appropriate modifications of the size distribution and of the abundances of the grains.

Another application of global grain models, that can possibly yield indirect information on the absorbing and emitting properties of the grains, consists of building a simplified model of the disk of a galaxy. We assume a vertical distribution of the hot stars, responsible for most of the dust heating, and also for the dust itself for which the scale height, assumed to be similar to that of the neutral gas, is somewhat larger (cf. Sect. 1.3). Multiple scattering of the stellar light by dust is taken into account (Witt & Gordon [552]). The model calculates the far-infrared emission of the disk, which is compared to observations. For an example see Xu & Helou [560]. For the propagation of radiation in an inhomogeneous dusty medium, see Boissé [51] where an analytic approximation to the solution can be found.

7.4 Infrared Absorptions and Ice Mantles

Where there is a large column density of dust in front of a sufficiently intense infrared source the dust is seen in absorption and bands characteristic of different solids are observed. Dust absorption studies are rapidly expanding thanks to observations with the ISO satellite and its successors. For example, the absorption bands of hydrocarbons have been observed in molecular clouds and even in the diffuse medium. The aromatic band at 6.2 µm has been seen in absorption towards a few sources (Schutte et al. [460]). It is probably produced by the same particles that emit the aromatic infrared bands. Another band with a complex structure centered at 3.4 µm has been observed along different directions, also originating in the diffuse medium (see e.g. Pendleton [402]). It can be identified with the C–H stretch vibration in –CH_2– et –CH_3 groups of aliphatic hydrocarbons (the C–H stretch of H atoms attached to aromatic cycles gives a band at 3.3 µm which is not seen in absorption). The 3.4 µm band provides evidence for the presence in the interstellar medium of carbonaceous compounds different from the aromatic particles seen in emission. For reasons not yet understood the 3.4 µm band is not observed in molecular clouds.

7.4 Infrared Absorptions and Ice Mantles   175

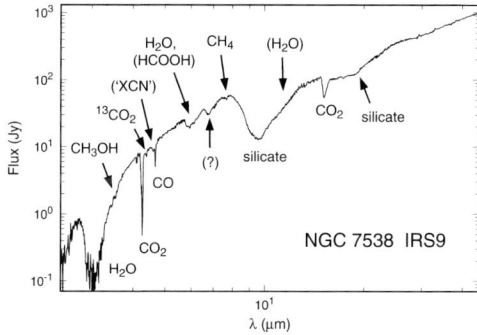

Fig. 7.18. Mid-infrared spectrum in the direction of the young massive stellar object NGC 7538 IRS9 embedded in a molecular cloud. This spectrum was obtained with ISO. The major absorption bands are identified. The band at 6.2 μm marked (?) is probably due in part to methyl alcohol. From Whittet et al. [547], with the permission of ESO.

The silicate absorption bands at 9.7 and 18 μm and some other absorption bands were discovered long ago, but most of the current knowledge on solid bands comes from observations with the ISO satellite, and in particular its SWS spectrograph. Figure 7.18 shows as an example the spectrum obtained towards a young star deeply embedded in a molecular cloud. The silicate bands are obvious, but there are also vibration bands of solid H_2O, CO, CO_2 and of other materials. These ices are formed by condensation of interstellar molecules onto sufficiently cold grains, and modified by chemical reactions. These phenomena will be studied Sect. 9.2.

It is interesting to note that the deposition of an ice mantle onto grains increases their size and changes the visible extinction curve in molecular clouds. This phenomenon can be ignored in the diffuse medium where grains are too warm to allow efficient ice condensation, but it might be of importance in molecular clouds. However, grain coagulation is more efficient than ice deposition in increasing the size of grains (see later Chap. 15). In any case the ratio $R = A_V/E(B-V)$, which usually has a value close to 3.1 in the diffuse medium, can reach 5 or 6 in some molecular clouds. This implies, on the average, bigger grains which give an extinction that is less dependent upon wavelength than for smaller grains.

We can derive the column density N of molecules in the solid from the intensity of an absorption band, if the integrated band absorbance A, expressed in cm per molecule, is known:

$$\boxed{\frac{N}{\text{mol. cm}^{-2}} = \left(\frac{A}{\text{cm mol.}^{-1}}\right)^{-1} \int \tau(\nu)\, d\nu \simeq \left(\frac{A}{\text{cm mol.}^{-1}}\right)^{-1} \tau_{max} \left(\frac{\Delta \nu}{\text{cm}^{-1}}\right),}$$
(7.25)

where τ_{max} is the optical depth at the band centre and $\Delta\nu$ its width in cm^{-1}. Table 7.3 gives the absorbances for the main bands of pure solids. For ice mixtures these absorbances can change by up to a factor 2.

Table 7.3. The main vibration bands of absorbing ices and absorbances in these bands. From Boulanger et al. [63]

Molecule	Mode	Frequency cm^{-1}	λ μm	Absorbance $cm\,mol^{-1}$
H_2O	O–H stretch	3 280	3.05	2.0×10^{-16}
	H–O–H bend	1 660	6.0	1.2×10^{-17}
	Libration	760	13.1	3.1×10^{-17}
CO	C≡O stretch	2 139	4.67	1.1×10^{-17}
^{13}CO	C≡O stretch	2 092	4.78	1.3×10^{-17}
CO_2	C=O stretch	2 343	4.27	7.6×10^{-17}
	O=C=O bend	660, 665	15.2	1.1×10^{-17}
$^{13}CO_2$	C=O stretch	2 283	4.38	7.8×10^{-17}
CH_4	C–H stretch	3 012	3.32	6.0×10^{-18}
	C–H deformation	1 304	7.69	6.4×10^{-18}
X–CN (OCN^-)	C≡N stretch	2 167	4.61	2.7×10^{-17}
NH_3	N–H stretch	3 208, 3 375	2.96	2.2×10^{-17}
	N–H bend	1 674	5.97	4.7×10^{-18}
	Umbrella	1 070	9.35	1.7×10^{-17}
H_2CO	C=O stretch	1 720	5.81	9.6×10^{-18}
CH_3OH	O–H stretch	3 251	3.07	1.3×10^{-16}
OCS	C=O stretch	2 040	4.90	1.5×10^{-16}

Table 7.4 gives the abundances of several ices towards three embedded young stellar objects and a star located behind a molecular cloud. The latter abundances are more characteristic of molecular clouds because the radiation from the embedded stars heats the dust and modifies the ices.

Table 7.4. The abundances of ices relative to the H_2O ice in the direction of three embedded young stellar objects (RAFGL 7009S, NGC 7538:IRS 9, IRAS 19110+1045) and of the star Elias 16 located behind a molecular cloud. For Elias 16 abundances are also given as number of molecules per H atom (isolated or in H_2). From Boulanger et al. [63] and Schutte [461].

Molecule	% relative to H_2O				Abund. rel. to H
	RAFGL 7009S	NGC 7538	IRAS 19110	Elias 16	Elias 16
H_2O	100	100	100	100	1.25×10^{-4}
CO	15	15	<1.5	33.5	4.1×10^{-5}
CO_2	21	12	5	14	2.1×10^{-5}
$^{13}CO_2$	0.33	–	–	–	–
CH_4	3.6	1.6	–	–	–
OCN^-	3.7	0.5	–	–	–
H_2CO	3.0	<3	–	–	–
CH_3OH	30	7	–	<5	$<4.3\times 10^{-6}$

There are also absorption bands in the far infrared. The most remarkable is a band centered at 44 μm due to amorphous H_2O ice. In the case of absorption in front of the protostellar object IRAS 19110+104, a narrow absorption at 43 μm, due to crystalline ice, is superimposed. This is an interesting example of the transformation of ice due to the radiation of the protostar. We will come back to such transformations in Chap. 15.

7.5 The Infrared Fluorescence

An emission band centered near 6 500 Å with a width of about 1 500 Å is observed in many interstellar objects, in particular in photodissociation regions, reflection nebulae and carbon–rich planetary nebulae. It is also seen in the diffuse interstellar medium (Gordon et al. [207]). It has been attributed for a long time to carbonaceous species illuminated by ultraviolet radiation. There are several difficulties with this idea, in particular the low efficiency for fluorescence of these species. The large band intensity implies that the emitter has a very high fluorescence efficiency, close to unity. Nanoparticles of amorphous silicon meet this criterion and are excellent candidates (Ledoux et al. [308], Witt et al. [553]). The agreement between observations and laboratory data is remarkable and the amount of Si required is only a small fraction of its total abundance. These nanocrystals are not necessarily isolated, but can be included in a matrix, for example as part of a SiC grain.

8 Heating and Cooling of the Interstellar Gas

With this chapter we begin the study of the physical processes in the interstellar medium. We will treat thermal exchange and thermal stability, describing in turn the different heating and cooling processes for the gas. We will also examine the static equilibrium between the different components of the interstellar medium. We will see, however, that the equilibrium condition is rarely met. It is, nevertheless, of interest to analyse it because it is to a zero-order approximation of reality.

By heating and cooling we mean the transfer of kinetic energy to or from atoms, molecules and ions of the interstellar gas. The principal heating processes begin with the removal of an electron from an interstellar species, gas or grain, by an energetic particle or photon. The suprathermal electron produced in this way heats the interstellar gas by thermalization through elastic collisions. The radiative excitation of atoms, molecules or ions does not correspond to a transfer of kinetic energy and does not directly produce any heating, although collisional de-excitation of the excited levels can transfer energy to the gas and heat it if the medium is dense.

The cooling processes mainly arise from inelastic collisions between the light particles of the gas, the colliders (electrons, H, H^+, etc.) and heavy targets (atoms, molecules, ions or grains). If these targets possess energy levels low enough to be excited by the collider, this collider loses kinetic energy and the gas cools through thermalization with the collider. The excitation energy of the target is then dissipated by the emission of radiation, in general in the infrared. This radiation escapes easily because of the small opacity of the interstellar medium in the infrared, except of course if it is produced deep inside molecular clouds.

The thermal aspects relating to dust grains were treated in Chap. 7, with the exception of thermal exchanges with the gas that will be discussed here. In the present chapter we will not discuss the heating and cooling by dynamical processes such as shock waves and turbulence, that will be dealt with in chapters 11 to 13. The hot interstellar medium described Sect. 5.3 can only be heated by such a process (Chevalier & Oegerle [92]).

Table 1 of the paper of Wolfire et al. [555] contains an extensive list of heating and cooling processes in the diffuse neutral interstellar medium. More processes intervene in other components of the medium. In this chapter, we will examine all of these processes, being careful to state to which component each process applies. Following the most frequent usage we will designate by Γ (for gain) the energy gain per unit volume and unit time, and by Λ (for loss), the energy loss per unit volume and unit time[1]. In thermal equilibrium obviously $\Gamma = \Lambda$, which allows us to determine the temperature after these functions, which are generally temperature-dependent, have been calculated.

8.1 Heating Processes

8.1.1 Generalities, Thermalization Time

Most heating mechanisms involve suprathermal particles, mainly electrons, which are rapidly thermalized by elastic Coulomb interactions with thermal electrons. The electron gas so heated equilibrates thermally with the ion gas. Then, if the gas is not completely ionized, elastic collisions between ions and neutral particles transfer energy more slowly to the neutral particles. It is interesting to give orders of magnitude for the characteristic times of these transfers. The thermalization of suprathermal particles has been dealt with in detail in Chap. 2 of the book by Spitzer [490], from which we borrow what follows.

Let us consider first the interaction between charged particles, which occurs via long-range electrostatic forces. Assume that a suprathermal electron enters a partly ionized gas with an initial velocity v_e, much larger than the random thermal velocity of the electrons of that gas (and *a fortiori* of heavier particles). A braking, or thermalization time t_t can be defined as

$$t_t = -\frac{v_e}{\langle \Delta v_{e\|} \rangle}, \tag{8.1}$$

where $\langle \Delta v_{e\|} \rangle$ is the mean change per second of the longitudinal velocity v_e of the suprathermal electron due to its interaction with the gas. Its sign is negative because velocity decreases.

Consider first the interaction with a heavy charged particle (ion). It only produces a deflection of the trajectory of the electron, with little change in its kinetic energy since the ion can be considered to be at rest, and is little affected by the collision to a first approximation. The trajectory of the electron is a hyperbola, and it is easy to see that the deflection is 90° when the impact parameter p, which is the closest distance of approach of the electron to the ion if there was no electrostatic force, is

$$p_0 = \frac{Z_i e^2}{m_e v_e^2}, \tag{8.2}$$

where $Z_i e$ is the charge of the ion and m_e is the mass of the electron. In this case $\langle \Delta v_{e\|} \rangle = -v_e$, hence an estimate of t_t which is

$$t_t \simeq \frac{1}{n_i v_e \pi p_0^2} = \frac{m_e^2 v_e^3}{\pi n_i Z_i^2 e^4} \text{ s}, \tag{8.3}$$

---

[1] The notation for Γ and Λ is sometimes ambiguous in the literature. We here always define them per unit volume, and not per atom or molecule.

where n_i is the ion density. This result is, however, only an approximation because distant collisions have been ignored, and their effect dominates over the close collisions[2]. The full result (Spitzer [487]) is

$$t_t(e, i) = \frac{m_e^2 v_e^3}{\pi n_i Z_i^2 e^4 \ln(\Lambda_D m_e v_e / 3kT_e)}, \qquad (8.4)$$

where Λ_D is the Debye screening factor that results from the screening of the electrostatic field of an ion by the surrounding plasma. This factor is

$$\Lambda_D = \frac{3}{2Z_i e^3} \left(\frac{k^3 T_e^3}{\pi n_e} \right)^{1/2}, \qquad (8.5)$$

where n_e and T_e are, respectively, the electron density and temperature. For example, for $n_e = 1$ cm^{-3} and $T_e = 10^4$ K, $\ln \Lambda_D = 23$ for hydrogen.

As we said before, the energy transfer form suprathermal electrons to the ions is negligible. The situation is different for electrons whose velocity is much affected by collisions with suprathermal electrons. If the velocity v_e of these suprathermal electrons is much higher than the random thermal velocity of the electrons in the gas, we can use (8.4), replacing n_i by n_e, so that

$$t_t(e, e) = \frac{m_e^2 v_e^3}{4\pi n_e e^4 \ln(\Lambda_D m_e v_e / 3kT_e)} = \frac{1.24 \times 10^{-18} v_e^3}{n_e \ln(\Lambda_D m_e v_e / 3kT_e)} \text{ s.} \qquad (8.6)$$

Beware! n_e and T_e refer to the gas and v_e to the suprathermal electrons. We can in the same way evaluate the thermalization time $t_t(i, e)$ for suprathermal ions of mass m_i (or $\mathcal{A}_i$ in atomic units of 1.66×10^{-24} g) and of charge $Z_i e$ moving in an ionized gas. We find

$$t_t(i, e) = \frac{3m_i (2\pi)^{1/2} (kT_e)^{3/2}}{8\pi m_e^{1/2} n_e Z_i^2 e^4 \ln \Lambda} = \frac{503 \mathcal{A}_i T_e^{3/2}}{n_e Z_i^2 \ln \Lambda} \text{ s.} \qquad (8.7)$$

The electrons in the gas, when heated by this process, rapidly thermalize. The ions thermalize more slowly with the electrons. Spitzer [490] showed that the thermalization rate between the electron gas, with temperature T_e, and the ion gas, with temperature T_i, is

$$\frac{dT_i}{dt} = -\frac{2(T_i - T_e)}{t_t(i, e)}. \qquad (8.8)$$

The characteristic time for energy equipartition (or temperature equipartition) of ions and electrons is then of the order of $t_t(i, e)/2$.

---

[2] The Bremsstrahlung energy losses have also been neglected. This is justified at the relatively low energies that we consider. The relative Bremsstrahlung energy loss during a single collision is only of importance for very high electron energies. It is of the order of 30% at very high energies, cf. Sect. 6.2.

We now have to evaluate the equipartition time for the kinetic energy between the neutral atoms and the electrons and ions. Here we are dealing with short-range forces and the cross-sections for elastic collisions are smaller than those between charged particles. As a first approximation they are close to the geometrical cross-sections, but a correct evaluation requires quantum mechanical calculations. In this case the transfer of momentum has to be considered. Collisions between electrons and neutral particles are much less efficient than collisions between ions and neutrals due to the small momentum of the electrons. We will thus only consider ion-neutral collisions, but it would be easy to extend the results to any type of collisions in which short-range forces intervene, in particular to neutral–neutral collisions.

Consider the relative velocity $\mathbf{v}_r$ between an ion with velocity $\mathbf{v}_i$ and a nearby neutral particle with velocity $\mathbf{v}_n$. Let $\sigma(v_r)$ be the collision cross-section. It depends, in general, upon the modulus v_r of the relative velocity and upon the geometrical circumstances of the collision, due to quantum effects. For example, the two particles can be temporally quasi-bound and orbit around each other due to the attraction of the ion by the electric dipole it induces in the neutral particle, if v_r is small and if the impact parameter has an appropriate value. Assuming first that $\mathbf{v}_r$ is the same for all neutral particles (this is the case if the ion has a relatively large energy), the collision rate for the ion is $v_r \sigma(v_r) n_n$, n_n being the density of the neutrals. $\sigma(v_r)$ is defined in such a way that the mean loss of momentum of the charged particle per unit time is equal to the product of this collision rate and the available momentum, which is $m_r v_r$, m_r being the reduced mass of the two particles, i.e.,

$$m_r = \frac{m_i m_n}{m_i + m_n}. \tag{8.9}$$

With respect to the mass centre, the momentum of either particle is $m_r v_r$. The mean change $\langle \Delta \mathbf{v}_i \rangle$ of the ion velocity per unit time is then such that

$$m_i \langle \Delta \mathbf{v}_i \rangle = -n_n v_r \sigma(v_r) m_r \mathbf{v_r}. \tag{8.10}$$

We now have to average this expression over the directions and velocities of the neutrals, assuming a maxwellian distribution. Assuming that the velocity of the charged particle is much higher than that of the neutrals we have, approximately,

$$\langle \mathbf{v_r} v_r \sigma(v_r) \rangle \simeq \mathbf{v}_i \langle v_r \sigma(v_r) \rangle. \tag{8.11}$$

The average value of $v_r \sigma(v_r)$ is called the *slowing coefficient*. If it is constant, the preceding equation is exact since the velocities $\mathbf{v}_i$ are isotropic. $\langle \Delta \mathbf{v}_i \rangle$ remains fixed in the incoming direction of $\mathbf{v}_i$. We can now estimate the slowing time of the ion $t_t(i,n) = -v_i / \Delta v_i$ from (8.9), (8.10) and (8.11):

$$\boxed{t_t(i,n) = \frac{m_i + m_n}{n_n m_n \langle v_r \sigma \rangle},} \tag{8.12}$$

where the argument of $\sigma(v_r)$ has been omitted. As before, the equipartition time between two interacting maxwellian gases with different temperatures is $t_t(i,n)/2$.

A similar reasoning can be applied to the collisions between dust grains and the atoms or molecules of the gas. The grain is practically at rest with respect to the gas particles and Spitzer [490] showed that t_t then has to be multiplied by 3/4. The slowing time of the grain through friction with the neutrals is

$$t_t(n, g) = \frac{3m_g}{4n_n m_n \sigma \langle v_n \rangle}, \tag{8.13}$$

where σ is the geometrical cross-section of the grain ($\sigma = \pi a^2$ for a spherical grain with radius a).

Let us come back to the ion–atom collisions. Quantum calculations of cross-sections have been performed for elastic collisions between C^+ and H. The cross-section is found to be proportional to v_r^{-1} so that $v_r \sigma$ is independent of v_r. $\langle v_r \sigma \rangle$ (C^+ – H) = 2.2×10^{-9} cm^3 s^{-1}. The cross-sections for other positive ions are similar. This is also the case for H-H$^+$ collisions at $T < 100$ K. For collisions between positive ions and He atoms, the cross-sections are approximately two times smaller.

Let us summarize using the example of the diffuse neutral gas, for which we will take as characteristic parameters $n_H \simeq 25$ cm^{-3}, $n_e/n_H = 3 \times 10^{-4}$ resulting from ionization of carbon, and $T = 100$ K. Assume that this gas is heated by electrons of energy $E = 1$ eV, which have a velocity $(2E/m_e)^{1/2} = 6 \times 10^7$ cm s^{-1}, far larger than the thermal velocity $(kT/m_e)^{1/2} = 4 \times 10^6$ cm s^{-1} of the electrons of the gas. Suprathermal electrons thermalize with the gas electrons within a characteristic time $t_t(e, e) = 1.5 \times 10^6$ s (8.6). Energy equipartition between electrons and carbon ions occurs with a characteristic time $t_t(i, e)/2 = 1.7 \times 10^7$ s (8.7). Finally, equipartition between carbon ions and hydrogen atoms takes place in a characteristic time $t_t(i, n)/2 = 1.2 \times 10^8$ s (8.12). All these times are of the order of a year (1 year $= 3.16 \times 10^7$ s). They are, thus, very short with respect to the characteristic times for the evolution of the interstellar medium, which are of the order of 10^6 years, and equipartition is therefore achieved in this case. Spitzer [490] also demonstrated that interstellar particles have velocity distributions close to maxwellian. However the equipartition of energy may not be achieved in some fast phenomena such as shocks (Chap. 10) or turbulent intermittency (Sect. 13.2).

8.1.2 Heating by Low-Energy Cosmic Rays

The question of the heating of the interstellar gas raised as soon as the first determinations of the temperature of the neutral interstellar medium were made by emission/absorption 21-cm line measurements (cf. Sect. 4.1). The first mechanism thought of was heating by low-energy cosmic rays (Goldsmith et al. [202]; we will follow closely this article). Cosmic rays transfer energy to the bound electrons of the atoms or molecules of the gas (ionization and excitation) and also to the free electrons (Coulomb interactions and excitation of plasma waves). We will see, however, that part of this energy is used for atom excitation. The heating of the gas by plasma waves will be briefly treated Sect. 8.1.8.

Cosmic particles have very high velocities compared to the gas particles, and the cross-sections σ for ionization, excitation and Coulomb interactions, as well as the corresponding collisions rates, respectively C_1, C_2 and C_3 ($C = nv\sigma$ where n is the particle density in the interstellar medium), are proportional to each other and depend only upon the velocity v of the cosmic particle. The kinetic energy E_1 of the electrons ejected by ionization and the energy E_3 acquired by Coulomb interactions do not depend upon v because v is very large ($v \gg \sqrt{2E_2/m_e}$). Similarly, the fraction of energy $f_1 E_1 + f_3 E_3$ that is used to heat the gas is independent of v. As a consequence, the heating rate $f_1 C_1 E_1 + f_3 C_3 E_3$ is proportional to C_1 whatever the value of v. Since the energy spectrum of cosmic rays decreases strongly at high energies (Fig. 6.2) it is clear that high-energy cosmic rays have only a marginal effect compared to those of a few MeV which are more numerous; however the flux of the latter particles is very poorly known due to the solar modulation (Sect. 6.1).

The heating rate, being proportional to C_1, is also proportional to the ionization rate ζ, which is usually considered as the characteristic parameter for the problem. This rate is, per interstellar nucleon,

$$\zeta = \sum_Z \int_0^\infty 4\pi I_Z(v) \sigma_{ion,Z}(v)\, dv \; \text{s}^{-1}, \qquad (8.14)$$

where Z is the charge number of the cosmic ray nuclei whose flux is $I_Z(v)$. $\sigma_{ion,Z}(v)$ is the ionization cross-section for a cosmic-ray nucleus of charge Z and velocity v. The kinetic energy E *per nucleon* for a cosmic-ray nucleus is

$$E = m_H c^2 [(1 - v^2/c^2)^{-1/2} - 1], \qquad (8.15)$$

and we can write the ionization rate as

$$\zeta = \sum_Z 4\pi \int_0^\infty I(E, Z) \sigma_{ion}(E, Z)\, dE, \qquad (8.16)$$

$I(E, Z)$ being expressed in cm^{-2} s^{-1} erg^{-1} ster^{-1}. This expression does not take into account the secondary ionizations of interstellar atoms and molecules by electrons liberated by cosmic ray ionizations. We will come back to this point later. For $E > 0.3$ MeV, $\sigma_{ion}(E, Z)$ is given by

$$\sigma_{ion}(E, Z) = \frac{1}{\beta^2}(1.23 \times 10^{-20} Z^2) \left[6.0 + \log\left(\frac{\beta^2}{1 - \beta^2}\right) - 0.43\beta^2\right] \text{cm}^2, \qquad (8.17)$$

with $\beta = v/c$. Then $\sigma_{ion}(E, Z) = Z^2 \sigma_{ion}(E, 1)$. Assuming that the abundances of heavy ions in cosmic rays, with respect to hydrogen, are independent of the energy and are the same at the energies of interest (a few MeV/nucleon) as at high energies where they are well measured, we have

$$\sum_Z Z^2 \frac{I(E, Z)\, dE}{I(E, 1)\, dE} \simeq 2.1. \qquad (8.18)$$

8.1 Heating Processes

Equation (8.17) can be written for cosmic protons as $\sigma_{ion}(E, 1) \simeq v^{-2}\Phi(E)$, neglecting the terms in β^4. This defines the function $\Phi(E)$. Equation (8.16) can then be re-written simply as

$$\zeta = 2.1 \langle \Phi \rangle \int_0^\infty \frac{1}{v^2} 4\pi I(E, 1)\, dE \; \text{s}^{-1}, \tag{8.19}$$

$\langle \Phi \rangle$ being the average value of Φ over the spectrum of cosmic protons.

Let us now calculate the rate of energy loss of a cosmic ray by the ionization and excitation of atoms and molecules of the gas. It is

$$-\left(\frac{dE_t}{dt}\right)_{ion,ex} = nv\langle \sigma \Delta E_t \rangle \; \text{erg s}^{-1}, \tag{8.20}$$

where E_t is the total energy of the particle and ΔE_t is the energy lost per interaction. This energy loss is given by the Bethe–Bloch formula

$$-\left(\frac{dE_t}{dt}\right)_{ion,ex} = -v\frac{dE_t}{dx}$$
$$= nz\frac{4\pi e^4 Z^2}{m_e v}\left\{\ln\left[\frac{2m_e v^2}{\Delta E_{eff}}(1+E/Mc^2)^2\right] - v^2/c^2\right\}, \tag{8.21}$$

where $\Delta E_{eff} \simeq 15$ eV is an effective value for ΔE_t (actually the mean ionization potential of the target weighted by the abundances of the elements) and M the rest mass of the cosmic particle. n is the density of the target particles (essentially H, H_2 and He) and z their mean charge. This formula is obtained after a long quantum-mechanical calculation that is rarely presented in a complete way (see e.g. Heitler [228] for references giving the details). A simplified, very physical derivation can be found in Longair [336], including the (small) corrections which arise for very high energies and dense media.

The expression for the energy loss of a cosmic particle by interaction with free electrons is obtained simply by replacing n in (8.21) by the electron density n_e and ΔE_{eff} by $\hbar \omega_p$, $\omega_p = (4\pi e^2 n_e/m_e)^{1/2}$ being the angular *plasma frequency* (Ginzburg & Syrovatskii [198]). This gives

$$-\left(\frac{dE_t}{dt}\right)_{el} = n_e \frac{4\pi e^4 Z^2}{m_e v}\left\{\ln\left[\frac{2m_e v^2}{\hbar \omega_p}(1+E/Mc^2)^2\right] - v^2/c^2\right\}. \tag{8.22}$$

Since the most effective cosmic particles are not relativistic, we may neglect the term v^2/c^2 in (8.21) and (8.22). Electrons resulting from the ionizations have an energy $\mathcal{E}$ of the order of 35 eV (Spitzer [490]). This is a favorable value for the efficient collisional excitation of atoms and molecules, so that only a fraction f of their energy leads to heating. On the other hand, these electrons also produce as we have said secondary ionizations of interstellar particles. The secondary electrons resulting from these ionizations lead to further heating via Coulomb interactions (Spitzer & Scott [488]), which is included in the factor f.

Let us define in (8.21) et (8.22) the respective functions

$$\phi_1 = \ln\left[\frac{2m_e v^2}{\Delta E_{eff}}(1 + E/Mc^2)^2\right], \tag{8.23}$$

$$\phi_2 = \ln\left[\frac{2m_e v^2}{\hbar\omega_p}(1 + E/Mc^2)^2\right]. \tag{8.24}$$

The ϕ_1/ϕ_2 ratio is of the order of 0.2, showing that the two corresponding processes are both important if the gas is partly ionized. The heating rate per unit volume is

$$\Gamma_{CR} = (nf\langle\phi_1\rangle + n_e\langle\phi_2\rangle)\int_0^\infty \frac{4\pi e^4}{m_e}\frac{1}{v^2} 4\pi I(E, 1)\, dE, \tag{8.25}$$

where the $\langle\ \rangle$ symbols mean an average over the energy spectrum of cosmic rays and over the different kinds of particles. Using (8.19) we obtain

$$\Gamma_{CR} = \frac{4\pi e^4}{m_e}(nf\langle\phi_1\rangle + n_e\langle\phi_2\rangle)\frac{\zeta}{2.1\langle\Phi\rangle}. \tag{8.26}$$

The electronic part can be evaluated immediately from the expressions for ϕ_2 and for Φ, for heating by protons of a given energy. For example, for 2 MeV protons we have $\Gamma_{CRe} \simeq 4.6 \times 10^{-10} n_e \zeta$ erg cm^{-3} s^{-1}. The other term is much more difficult to calculate, due to the difficulty in evaluating f.

Wolfire et al. [555] propose the following, directly usable, expression for the heating rate which takes secondary ionizations into account:

$$\boxed{\Gamma_{CR} = \zeta E_h(E, x)n,} \tag{8.27}$$

where $E_h(E, x)$ is the energy deposited as heat by a primary electron with energy E. Fig. 3 of Shull & van Steenberg [471] gives E_h/E as a function of E and x. For $E = 35$ eV, an appropriate value for the interstellar medium, E_h varies between 6 eV for a neutral gas to 35 eV for an ionized gas.

ζ depends directly on the flux of low-energy cosmic rays, which is much affected by solar modulation. Its determination is problematic. The estimated spectrum of low-energy cosmic rays, presented as a bold curve in Fig. 6.2, corresponds to $\zeta = 1.8 \times 10^{-17}$ s^{-1}. Then $\Gamma_{CR} \simeq 1.7 \times 10^{-28} n$ erg s^{-1} cm^{-3} for a neutral medium, from (8.27). Some authors have attempted to estimate ζ semi-empirically. They all agree on values of $\zeta \simeq 1-2 \times 10^{-17}$ s^{-1}. This ionization rate is far too small to correspond to an efficient heating of the diffuse interstellar medium to the observed temperatures (100 K or more). Goldsmith et al. [202] used a much larger value, $\zeta = 4 \times 10^{-16}$ s^{-1}, which is now considered as unrealistic for the Galaxy, except perhaps close to supernovae which are the main sources of cosmic rays. However, cosmic-ray heating is the only efficient mechanism in the depths of molecular clouds (however, if the density is large enough, the heating of the gas by heat exchange with the grains may also intervene: see later in this section).

8.1.3 Photoelectric Heating from Grains

This process is the most efficient at heating the cold diffuse interstellar medium[3]. This was not realized at the time of Goldsmith et al. [202]. This process was first proposed by Watson [542] in 1972, and developed in detail by Draine [132]. Ultraviolet radiation from hot stars remove electrons from interstellar dust grains. These electrons have an excess energy which is an important fraction of the energy of the ultraviolet photon. They thermalize with the free electrons of the gas, producing the heating. Their energy, being of the order of a few eV, is too small to ionize or to excite the atoms or molecules, so that we can consider that all this energy is used for heating. What remains of the photon energy heats the grain, but a part is lost to the photoelectron as it overcomes the potential energy barrier which may exist due to the possible positive charge of the grain, as will be discussed soon. The nature, dimension and to a lesser extent shape of the grain, and above all its charge, play an important role in the process.

It is of importance to notice that in the MRN size distribution of grains $n(a) \propto a^{-3.5}$ where a is the grain radius (7.8), the distribution of the grain areas $a^2 n(a) \propto a^{-1.5}$ is largely dominated by the smallest grains. As discussed Sect. 7.2, there may not be very small silicate grains in the interstellar medium, so that it is enough to consider carbonaceous grains only, as done by Bakes & Tielens [13]. We will closely follow their approach.

The photoelectric heating rate $H(\mathcal{N}, q)$ by a dust grain composed of $\mathcal{N}$ carbon atoms with a charge q is

$$H(\mathcal{N}, q) = \int_{\nu_q}^{\nu_H} \sigma_{abs}(\mathcal{N}) Y_{ion}(\mathcal{N}, \Phi_q) c u_\nu g(\mathcal{N}, \Psi_q) \, d\nu, \tag{8.28}$$

where u_ν is the energy density of the radiation. For isotropic radiation[4], $u_\nu = 4\pi I_\nu/c$, where I_ν is the intensity of the radiation field in erg cm^{-2} s^{-1} sterad^{-1}. This radiation field is usually normalized to that in the solar neighbourhood by setting $I_\nu = \chi I_{\nu,\odot}$ (Sect. 2.1). $\sigma_{abs} = Q_{abs} \pi a^2$, a being the grain radius, is the absorption cross-section (7.6). For the small grains of interest here, it is proportional to $\mathcal{N}$ because the absorption is due to the π electrons of carbon (1 electron per carbon atom), and because these grains are optically thin to UV radiation. Y_{ion} is the effective number of photoelectrons ejected per absorbed photon. It is the ratio of the photoionization cross-section to the absorption cross-section. An experimental determination of the photoionization cross-section per carbon atom for PAHs and for larger carbonaceous grains is shown in Fig. 8.1. Compare this figure to Fig. 7.17 which gives the absorption cross-section for PAHs. We see that only the hardest photons (roughly 912 to 1 600 Å) are efficient at heating the grains. For neutral grains Y is rather large, but it is reduced by the positive charge q, if any, and depends upon the electrostatic potential Ψ_q. $g(\mathcal{N}, \Psi_q)$ is the partition function of the kinetic

---
[3] X-ray heating is also very important for the warm diffuse medium, as we will see later.
[4] Bakes & Tielens [13] consider a directional radiation field originating from a single distant star and their expressions are different. However, the final results are the same.

energy of the photoelectrons. For a mass distribution of grains $n(\mathcal{N})d\mathcal{N}$, where $n(\mathcal{N})$ is the number of grains per unit volume with a number of carbon atoms between $\mathcal{N}$ and $\mathcal{N} + d\mathcal{N}$, the rate of photoelectric heating per unit volume is

$$\Gamma_{pe} = \int_{\mathcal{N}_{min}}^{\mathcal{N}_{max}} \sum_q H(\mathcal{N}, q) f(\mathcal{N}, q) n(\mathcal{N}) d\mathcal{N}, \tag{8.29}$$

where $f(\mathcal{N}, q)$ is the probability of finding a grain with $\mathcal{N}$ carbon atoms and a charge q.

However, we have to take into account the inverse process: the recombination of electrons and ions onto the grain corresponds to some energy loss for the gas with a rate $C(\mathcal{N}, q)$ (energy loss $(3/2)kT$ per recombination), with a net heating rate of

$$\Gamma_{net,pe} = \int_{\mathcal{N}_{min}}^{\mathcal{N}_{max}} \sum_q [H(\mathcal{N}, q) - C(\mathcal{N}, q)] f(\mathcal{N}, q) n(\mathcal{N}) d\mathcal{N}. \tag{8.30}$$

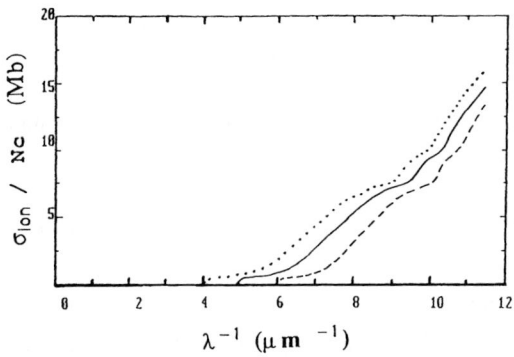

Fig. 8.1. Experimental measurement of the photoionization cross-section per carbon atom for coronene, a small PAH with 24 carbon atoms (dashed line) and extrapolation to a grain with 80 carbon atoms (full line) and to an infinite graphite plane (dotted line). The cross-section is given in Mbarn per carbon atom, i.e., 10^{-18} cm^2/C. The theoretical evaluations of Bakes & Tielens [13] for spherical carbon grains give results in reasonable agreement with those presented here. From Verstraete et al. [532], with the permission of ESO.

It is necessary to evaluate the charge on the grain. The problem is similar to that of the ionization equilibrium of atoms examined Sect. 5.1, with the complication that we have to consider recombination with two kinds of particles, electrons and ions. Let $f(q)$ be the probability that the grain has a charge q, and $\mathcal{J}_{pe}$, $\mathcal{J}_e$ and $\mathcal{J}_{ion}$, respectively, the emission rate of photoelectrons and the accretion rates of electrons and ions (mainly H$^+$ and C$^+$). The equation of charge equilibrium between two successive numbers of charges is

$$f(q)[\mathcal{J}_{pe}(q) + \mathcal{J}_{ion}(q)] = f(q+1)\mathcal{J}_e(q+1), \tag{8.31}$$

since ions and electrons have charges with opposite signs. If the radiation field is large enough ($\chi = 100 - 10^4$ as in a classical photodissociation region, see Chap. 10), $\mathcal{J}_{ion}$ is much smaller than $\mathcal{J}_{pe}$ and can be ignored. The grain charge can be positive, zero or even negative according to the circumstances. For example, small flat grains like PAHs are negatively charged in the diffuse interstellar medium. A full treatment of this complex problem can be found in Draine & Sutin [136], neglecting the photoelectric effect. It has been extended by Bakes & Tielens [13] and by Salama et al. [446] by the addition of the photoelectric effect. More revisions are probably necessary in order to take into account chemical reactions between PAHs and the ions of the gas, in particular the fast neutralization reaction PAH$^-$ + C$^+$ $\rightarrow$ PAH + C (Bakes & Tielens [15]). Revised values of the recombination cross-sections of ionized PAHs are also necessary. Here we will only give the result obtained by Bakes & Tielens [13]. The photoelectric heating rate (8.29) is given by

$$\boxed{\Gamma_{pe} = 10^{-24} \epsilon \chi n_H \text{ erg s}^{-1} \text{ cm}^{-3},} \quad (8.32)$$

where ϵ is a heating efficiency, i.e., the ratio between the energy of the radiation converted into heating of the gas and the energy absorbed by the carbonaceous grains. This expression assumes that these grains absorb an ultraviolet energy of 10^{-24} erg s^{-1} per hydrogen atom in the solar neighbourhood. This corresponds to the energy absorbed in the UV part of the extinction curve. Only the small grains ($a \leq 100$ Å) are efficient at heating. An approximate analytical expression for the heating efficiency is:

$$\epsilon = \frac{4.87 \times 10^{-2}}{1 + 4 \times 10^{-3} (\chi T^{1/2}/n_e)^{0.73}} + \frac{3.65 \times 10^{-2} (T/10^4 \text{K})}{1 + 2 \times 10^{-4} (\chi T^{1/2}/n_e)}, \quad (8.33)$$

where T is the gas temperature and n_e the electron density (electron cm^{-3}). ϵ is of the order of 0.05 for small values of the parameter $\chi T^{1/2}/n_e$, which is proportional to the ratio between the photoemission rate and the recombination rate of electrons and determines the grain charge. It decreases for values of this parameter larger than 10^3. Figure 8.2 shows the variation of the heating rate per hydrogen atom Γ_{pe}/n_H with $\chi T^{1/2}/n_e$, for three temperatures. Note that this rate does not depend upon the radiation field for large values of $\chi T^{1/2}/n_e$ and is then proportional to $n_e n_H$, because $\epsilon \propto \chi^{-1}$. This arises from the fact that the grains are strongly positively charged in this case, the heating rate being simply proportional to the recombination rate of electrons on the grains. The heating then becomes inefficient because it is almost compensated by the recombination cooling (compare Fig. 8.2 and 8.3, and see further Fig. 8.8).

Bakes & Tielens [13] have also evaluated the cooling rate Λ_{rec} by recombination of charged particles on grains. An analytical approximate expression is

$$\boxed{\Lambda_{rec} = 3.49 \times 10^{-30} T^{0.944} (\chi T^{1/2}/n_e)^{0.735 T^{-0.068}} n_e n_H \text{ erg s}^{-1} \text{ cm}^{-3}.} \quad (8.34)$$

Figure 8.3 shows this expression in graphical form. The net heating rate of the interstellar gas by the photoelectric effect of small grains is thus $\Gamma_{pe} - \Lambda_{rec}$.

190   8 Heating and Cooling of the Interstellar Gas

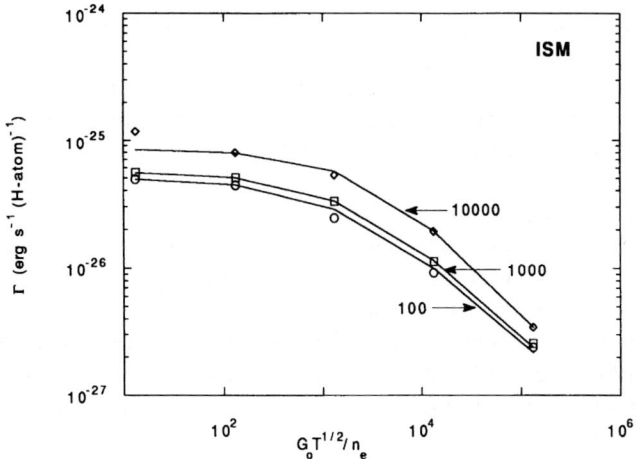

Fig. 8.2. Photoelectric heating rate Γ_{pe}/n_H *per hydrogen atom* as a function of the UV radiation field $G_0 \equiv \chi$, for a gas of temperature T and electron density n_e, for three values of the temperature. Reproduced from Bakes & Tielens [13], with the permission of the AAS.

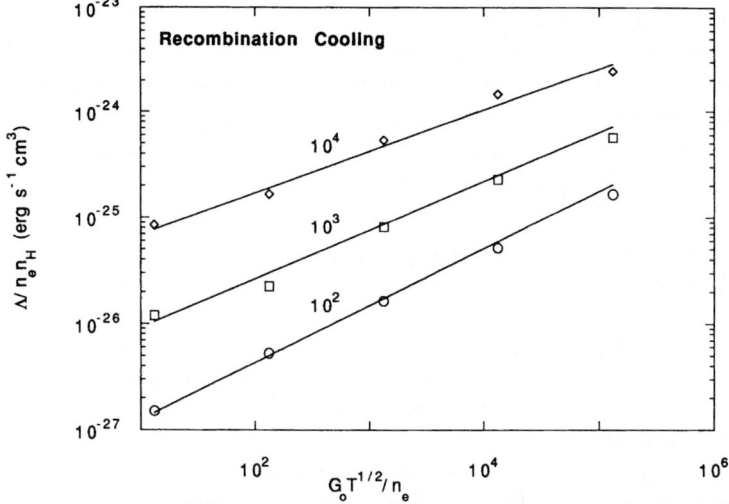

Fig. 8.3. Cooling rate for the recombination of electrons and ions on grains as a function of the UV radiation field $G_0 \equiv \chi$, for a gas of temperature T and electron density n_e, for three values of the temperature. Reproduced from Bakes & Tielens [13], with the permission of the AAS.

Considering Figs. 8.2 and 8.3 for the conditions in the cold, neutral interstellar medium (the H I "clouds"), i.e. roughly $\chi = 1$, $T = 100$ K, $n_H = 25$ cm^{-3} and $x = n_e/n_H = 3 \times 10^{-4}$, we see that recombination cooling is negligible with respect to photoelectric heating. The photoelectric heating through grains appears to dominate the heating of this medium. It yields some 3×10^{-26} erg s^{-1} per H atom from Fig. 8.2, a value close to the observed cooling rate by the [C II]λ158 µm line (Sect. 4.1). For the warm diffuse "intercloud" medium (Sect. 4.1 et 5.2) with $\chi = 1$, $T = 8\,000$ K, $n_H = 0.25$ cm^{-3} and $x = 10^{-2}$, the photoelectric heating rate is 10^{-26} erg s^{-1} per H atom from the model, a large enough value to explain the heating of the gas. However, the recombination cooling is important and reaches the same magnitude as the heating near 10^4 K, producing a thermostatic effect at this temperature. Photoelectric heating is very efficient at the photodissociation interfaces between H II regions and neutral clouds ($\chi \simeq 10^3 - 10^5$, $n_H = 10^3 - 10^5$ cm^{-3} and $x = 3 \times 10^{-4}$) and the gas can reach (as we will see) temperatures of several thousands degrees near the surface. Of course, this mechanism is totally inefficient in molecular clouds where UV radiation cannot penetrate. In H II regions, it is negligible with respect to the mechanism that will be discussed in the next section.

A few remarks are necessary at this point. The radiation field used by Bakes & Tielens comes from a rather old reference (Draine), but it is not much different from more recent determinations plotted in Fig. 2.3. Some quantities like the ionization efficiency of the grains and the value of the UV flux absorbed by carbonaceous grains near the Sun (10^{-24} erg s^{-1} per H atom) are more disputable. These parameters are in fact rather poorly known given our incomplete knowledge of the nature and of the absorption properties of the grains in the UV and, in particular, of the carriers of the infrared emission bands discussed Sect. 7.2.

8.1.4 Photoelectric Heating by the Photoionization of Atoms and Molecules

The ionization of atoms and molecules by UV photons liberates electrons whose energy is the difference between the photon energy and the ionization energy $h\nu_i$. The average kinetic energy of the photoelectrons is therefore given by

$$\langle E \rangle = \frac{\int_{\nu_i}^{\infty} h(\nu - \nu_i)\sigma_\nu u_\nu \, d\nu/\nu}{\int_{\nu_i}^{\infty} \sigma_\nu u_\nu \, d\nu/\nu}, \tag{8.35}$$

σ_ν being the photoionization cross-section and u_ν the energy density of the radiation.

This mechanism dominates the heating in H II regions. The estimation of u_ν is complicated in this case because we must solve the transfer equation for the lines and for the continuum, taking into account radiation scattering. This problem is treated in specialized books and solved in models of H II regions. Here we will only say that the mean energy of the photoelectrons in an H II region excited by a star with an effective temperature T_{eff} (defined in the far-UV) is of the order of $\langle E \rangle \simeq 1.4\,kT_{eff}$ (Spitzer [490] Sect. 6.1.a).

In the neutral interstellar medium the main source of photoelectrons is carbon with an ionization potential of 11.2 eV. As there is no photon with an energy larger than 13.6 eV, we see immediately that the mean photoelectron energy is of the order of (13.6–11.2)/2 eV, about 1 eV. Given the photoionization cross-section of carbon, the corresponding heating rate in the solar neighbourhood is approximately

$$\Gamma_C = 2 \times 10^{-22} n(C^0) \text{ erg s}^{-1} \text{ cm}^{-3}, \tag{8.36}$$

where $n(C^0)$ is the carbon atom density. This density is related to that of C^+ ions by the equation of ionization equilibrium (4.33). Using our example of cold diffuse interstellar medium with $n_e = 0.0075$ cm^{-3} and the local interstellar radiation field, $n(C^0)/n(C^+) = 2 \times 10^{-4}$, hence $n(C^0) = 2 \times 10^{-6}$ cm^{-3}. Then $\Gamma_C = 4 \times 10^{-28}$ erg s^{-1} cm^{-3}. Comparing with the results given previously, we see that this mechanism is negligible in the diffuse interstellar medium.

The ionization of molecular hydrogen by soft X-rays or far-UV can be an important source of heating in shocks (cf. Hollenbach & McKee [242]).

8.1.5 X-Ray Heating

X-ray emission is observed in the interstellar medium, coming mainly from the hot gas discussed in Chap. 5. It results in the heating of the interstellar gas. X-rays remove photoelectrons from the K shell of atoms and ions (Sect. 5.3). As in the case of cosmic-ray heating, the energetic primary photoelectrons produced in this way can provoke secondary ionization and liberate other electrons. All these electrons heat the gas as in the previous cases. As for the cosmic-ray heating, the ionization rate is chosen as the main determining parameter for the problem. Following Wolfire et al. [555], X-ray heating is only important up to some depth in atomic clouds, thereafter they can be considered as opaque to soft X-rays. The energy density of X-ray radiation to be taken into account at the cloud surface is thus only $2\pi I_\nu/c$, I_ν being the flux per steradian in free space. The ionization rate ζ_X^i of species i is

$$\zeta_X^i = \int 2\pi (I_\nu/h\nu) e^{-\sigma_\nu N_H} \sigma_\nu^i \, d\nu \text{ s}^{-1}, \tag{8.37}$$

where σ_ν is the total X-ray photoionization cross-section of the medium and σ_ν^i that of element i. The $e^{-\sigma_\nu N_H}$ term represents the absorption of X-rays for a column density N_H. The heating rate is obtained introducing the average energy of photoelectrons in the integral above. Figure 8.4 gives the ionization rate and the heating rate of a H atom by X-rays near the Sun (Sect. 5.4) as calculated by Wolfire et al. [555], as a function of the ionization degree and column density of the interstellar medium. At very low column densities this ionization rate is very much higher than the ionization rate by cosmic rays (about 1.8×10^{-17} s^{-1}, cf. Sect. 8.1), and X-ray heating is efficient. The heating rate reaches 10^{-26} erg s^{-1} per H atom for the interstellar medium not protected against X-ray radiation. This is not much smaller than the photoionization heating by grains. However the soft X-rays that are the most efficient at ionizing are rapidly absorbed by the interstellar medium (cf. Fig. 2.5). X-ray ionization and heating decrease quite rapidly when there is more interstellar matter between the X-ray source and the region of interest. Since the Sun is in a low density region which emits soft X-rays, the Local bubble, we can consider that the X radiation field used by Wolfire et al. [555], which results from direct satellite measurements, is the unattenuated source. In general, the energy density of X-rays is smaller in the interstellar medium. Note that the UV field responsible for the grain photoelectric heating is also attenuated by extinction, but the effect of this attenuation is much more limited than for the heating by X-rays. The intensity at 1 000 Å, hence the grain photoelectric heating by far-UV radiation, is reduced by only 2% by an hydrogen column density of 10^{19} atom cm^{-2} (Fig. 2.5), while the X-ray heating is reduced by an order of magnitude (Fig. 8.4).

To summarize, heating by X-ray radiation is in general not efficient for the cold atomic medium because the column densities of matter are relatively large, but can be very efficient for the warm, less dense atomic medium for which the column densities are small. It can be completely neglected in molecular clouds where only

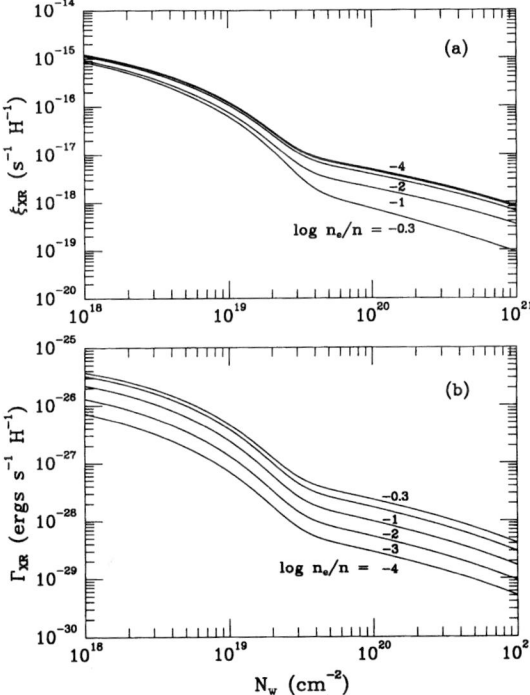

Fig. 8.4. The ionization rate per hydrogen atom by X-rays in the solar neighbourhood (a) and the heating rate by this radiation (b). The secondary ionization of H and He is included. The curves correspond to different degrees of ionization n_e/n. The results are given as a function of the hydrogen column density that attenuates the radiation. Reproduced from Wolfire et al. [555], with the permission of the AAS.

hard X-rays can penetrate. Their intensity is very small and their heating effect negligible. Of course, the situation can locally be very different near discrete X-ray sources, in particular near supernova remnants and pre main-sequence stars that are often strong X-ray emitters.

8.1.6 Chemical Heating

Chemical reactions that occur in the interstellar medium can release their products with some kinetic energy that can heat the gas. The most important process is undoubtely the formation of molecular hydrogen by the direct association of H atoms, which takes place on the surfaces of grains. This reaction will be studied in Chap. 9. It is very exothermal and yields an energy of 4.48 eV which is distributed between the rotational and vibrational excitations of H_2, grain heating and the kinetic energy of the molecule leaving the grain. The distribution between these different forms of energy is somewhat uncertain. The excitation energy might be 4.2 eV and the kinetic energy 0.2 eV while the grain heating would be negligible (cf. Hollenbach & McKee [242] Sect. VIc). The kinetic energy directly heats the gas, and a part of the excitation energy is also used: de-excitation of the H_2 molecule by inelastic collisions with the atoms or molecules of the gas transfers kinetic energy that takes part in the heating. The heating rate is thus

$$\Gamma_{H_2} = R_f n^2 x_H (0.2 + 4.2\eta) \text{ eV s}^{-1} \text{ cm}^{-3}, \tag{8.38}$$

where R_f is the formation rate of H_2 on grains and η is the fraction of the excitation energy of H_2 that is used for heating. These quantities are given by Hollenbach & McKee [242] (their (3.8), (6.43) and (6.45)). x_H is the fraction of the gas particles that are H atoms. Notice that the heating rate per unit volume is proportional to the square of density because it is a collisional process, contrary to the preceding processes which depend only linearly upon density. Chemical heating by H_2 is therefore efficient in shocks and dense photodissociation regions, but not under other circumstances.

H_2 molecules can also take part in heating in another way. The absorption of UV photons by H_2 excites rotation and vibration levels, and the de-excitation of these levels transfers energy to the gas. This mechanism is only of importance in regions exposed to strong UV radiation (Tielens & Hollenbach [513]). A similar mechanism which can be important in dense and warm neutral regions is the collisional de-excitation of H_2O excited by the infrared radiation from grains (Takahashi et al. [504]).

8.1.7 Heating by Grain-Gas Thermal Exchange

At relatively high densities, collisions between atoms or molecules and dust grains can be frequent enough for the energy transfer to be efficient with respect to other processes, at least in the neutral medium. In the neutral diffuse medium, the grains are always colder than the gas and can only cool it. However, this process is negligible because the density is low. Conversely, the grains are warmer than the gas in deep regions of the molecular clouds and can then heat the gas.

The heating of grains by the gas is unimportant in H II regions because it is much smaller than the heating by the ultraviolet radiation. It is also negligible in the diffuse ionized medium due to the low density, although the radiative heating of the grains is smaller than in H II regions (see Lagache et al. [295] for the thermal emission of dust in this medium). However, grain heating by thermal exchange with the gas is dominant in supernova remnants where the gas temperature is very high and the density is larger. This mechanism has been treated by Dwek [148] who considered not only the heating of big grains but also that of the very small grains out of thermal equilibrium.

In the deep regions of giant molecular cloud complexes, the grains are heated by the far-infrared radiation coming from outside. This radiation can easily penetrate to considerable depths since the optical depth at 100 μm is of the order of unity for a visual extinction of 300 magnitudes. Due to this the grain temperature is never smaller than 8 K (Falgarone & Puget [170]), and the grains can heat the gas if the density is large enough. The mean flux of kinetic energy from gas particles which strike the grain and are at least temporarily captured is $E = 2kT$ for a maxwellian distribution, T being the gas temperature[5]. The particles which struck the grain (or an equal number of other particles) leave the grain with a different mean kinetic energy which corresponds to a temperature T_2 intermediate between T_d, the grain temperature, and T. The *accomodation coefficient* of atoms or molecules on the grain is defined as

$$\alpha = \frac{T_2 - T}{T_d - T}, \quad (8.39)$$

so that each collision gives to the gas a mean energy $2\alpha k(T_d - T)$. The collision rate per unit volume is, assuming for the moment that the particles are H atoms,

$$\tau_C = n_H n_d \sigma_d v_{th}, \quad (8.40)$$

where n_d and $\sigma_d = \langle \pi a^2 \rangle$ are the number density and geometrical cross-section averaged over the size distribution of grains. v_{th} is the mean velocity of the atoms.

The heating rate per unit volume is

$$\Gamma_{gas,grains} \simeq n_H n_d \sigma_d \left(\frac{8kT}{\pi m_H}\right)^{1/2} \alpha 2k(T_d - T) \text{ erg s}^{-1} \text{ cm}^{-3}, \quad (8.41)$$

Measurements of the interstellar extinction suggest that

$$n_d \sigma_d \simeq 1.5 \times 10^{-21} n_H \text{ cm}^{-1}. \quad (8.42)$$

The accomodation coefficient is defined by Burke & Hollenbach [73] in such a way that the expression of $\Gamma_{gas,grains}$ given above is valid if n_H is the total number of hydrogen *nuclei* per unit volume. If hydrogen is fully molecular, they give $\alpha = 0.35$. The heating rate is then

---

[5] This flux is $\langle (1/2)mv^2 v \rangle = 2kT \langle v_{th} \rangle$, $\langle v_{th} \rangle = (8kT/\pi m)^{1/2}$ being the mean velocity.

$$\boxed{\Gamma_{gas,grains} \simeq 1.6 \times 10^{-33} n_H^2 T^{1/2}(T_d - T) \text{ erg s}^{-1} \text{ cm}^{-3},} \quad (8.43)$$

A more exact calculation has been performed by Falgarone & Puget [170] taking into account the variation of grain temperature with size. All this assumes, however, that the grains have the same properties in the molecular clouds and in the diffuse medium, which might not be true. Nevertheless, the order of magnitude should be correct. It is interesting to compare this process with other important heating processes. Combining (8.43) and (8.32) we find that, at $T = 10$ K and for the interstellar field near the Sun ($\chi = 1$):

$$\Gamma_{gas,grains}/\Gamma_{pe} = 1.3 \times 10^{-7} n_{H_2}(T - T_d), \quad (8.44)$$

and combining (8.43) and (8.27), still at 10 K:

$$\Gamma_{gas,grains}/\Gamma_{CR} = 4 \times 10^{-5} n_{H_2}(T - T_d). \quad (8.45)$$

Equation (8.44) shows that gas–grain collisions are unimportant if the UV radiation field is not very small. Equation (8.45) conversely shows that this process dominates in molecular clouds at densities higher than about 2×10^4 mol. cm^{-3}.

8.1.8 Hydrodynamic and Magnetohydrodynamic Heating

The heating mechanisms we have described up to now rest on microscopic processes. However, the macroscopic motions of the gas can also contribute to gas heating. For example, the gravitational collapse of a cloud due to its own gravity, a process that occurs during star formation, is a source of heat for the gas that must be dissipated in order not to stop the collapse. Its expression is given by the perfect gas law:

$$\boxed{\Gamma_{coll.} = \rho P \frac{d}{dt}\left(\frac{1}{\rho}\right) \simeq 2.6 \times 10^{-31} n_H T \text{ erg s}^{-1} \text{ cm}^{-3},} \quad (8.46)$$

for a molecular hydrogen gas, ρ and P being the density and the pressure, respectively.

Supernova explosions, stellar winds and to a lesser extent the expansion of H II regions produce large quantities of mechanical energy that is dissipated in the interstellar medium as radiation and as the kinetic energy of the atoms, ions and molecules. The principal intermediates as far as kinetic energy is concerned are shocks and turbulence. These processes will be examined in Chapters 11 to 13.

The magnetic field is also a heat source due to different processes. One of these processes is the viscous dissipation of Alfvén waves, which can be considered as the oscillations of magnetic field lines that are coupled to the gas because it is generally a good conductor of electricity. These waves are produced by cosmic ray propagation in the interstellar medium, and by other mechanisms such as the differential rotation or the shear of a magnetic gas which produces a torsion of the magnetic tubes (Hartquist [221]). The relative motions of the ionized and neutral components of a weakly ionized plasma (ion magnetic slip, most often designated as *ambipolar diffusion* which will be studied Sect. 14.1) also produce a viscous heating that was studied by Scalo [453]. In the present chapter we will not expand on these phenomena (which are actually rather poorly known), but simply mention that magnetohydrodynamic (MHD) waves created by supernova remnants can give a large contribution to the heating of the low-density interstellar medium (Ferrière et al. [179]). We will come back to these processes in the following chapters, for the regions in which they become important.

8.2 Cooling Processes

There are two categories of cooling processes in the interstellar gas: radiative processes in which radiation by atoms, ions and molecules excited by collisions transfers part of the kinetic energy into radiation, and processes that are the inverse of some of the heating processes we have just examined. Let us discuss them in turn.

8.2.1 Fine-Structure Line Cooling

This process dominates almost everywhere in the interstellar medium, with the exceptions of the hot gas and the regions deep within molecular clouds. Other mechanisms are also of importance in shocks and in H II regions. The atoms and ions that are the most efficient are those that are the most abundant and that have fine-structure levels close to the fundamental level: they are hence easy to excite. These are principally C II and O I in the neutral medium, and O II, O III, N II, N III, Ne II and Ne III in H II regions. Table 4.1 lists the main transitions of interest.

Consider first the neutral medium. Almost all the gaseous carbon is present as C II and almost all the oxygen in the form of O I. N I is also abundant but it has no transition of interest for cooling. Si II, S II and Fe II are abundant and have fine-structure transitions but the energy necessary for their excitation is only reached by collisions with electrons and neutrals at relatively high temperatures. At low temperatures only the upper fine-structure level of C II, with an energy corresponding to 91.2 K, can be excited. The first excited fine-structure level of O I is at 228 K and therefore requires higher temperatures than C II. Neglecting collisional de-excitation, a justified assumption at the low densities of the atomic medium (cf. Table 4.1), the statistical equilibrium equation for the two first levels of C II (4.16) reduces to $n_u = n_l C_{lu}/A_{ul}$. The intensity of the line emitted at 157.7 μm is $I_{ul} = n_u A_{ul} h\nu/4\pi$ erg s^{-1} cm^{-3} ster^{-1} (4.19). This assumes that the line is optically thin, which is generally true. The cooling rate is independent of the spontaneous emission probability A_{ul} because collisions determine the excitation rate to the upper level, so that at statistical equilibrium the radiative de-excitation rate for this level is equal to the excitation rate. The cooling rate per unit volume for ionized carbon excited by electrons is simply:

$$\Lambda_{e,\text{CII}} = n_C C_{lu} h\nu = n_C n_e \frac{8.63 \times 10^{-6} \Omega_{ul}}{g_l T^{1/2}} \exp(-h\nu/kT) h\nu, \qquad (8.47)$$

from (3.38) and (4.17). With the data of Table 4.1, assuming a reference abundance for carbon of $n_C/n_H = 3.55 \times 10^{-4}$ and a carbon depletion in the gas of d_C (cf. Table 4.2), and also assuming that carbon is almost entirely ionized (Sect. 8.1) and supplies all the free electrons so that $n_e \simeq d_C n_C$, we find numerically for densities smaller than the critical density for collisional de-excitation, $n_{crit} \sim 3\,000\ \mathrm{cm}^{-3}$ (Table 4.1):

$$\Lambda_{e,\mathrm{CII}} = 1.23 \times 10^{-27} n_H^2 d_C^2 e^{-91.2\,\mathrm{K}/T} \left(\frac{T}{100\,\mathrm{K}}\right)^{-1/2} \mathrm{erg\ s^{-1}\ cm^{-3}}. \quad (8.48)$$

Notice that the cooling rate is proportional to the square of the carbon depletion in the gas because carbon supplies the free electrons. This carbon depletion is unfortunately not very well known.

C II can also be excited by collisions with hydrogen atoms. This gives (Wolfire et al. [555], Appendix B)

$$\Lambda_{H,\mathrm{CII}} = 7.9 \times 10^{-27} n_H^2 d_C e^{-91.2\,\mathrm{K}/T}\ \mathrm{erg\ s^{-1}\ cm^{-3}}, \quad (8.49)$$

which is larger than the cooling rate via electronic excitation, at least as long as the degree of ionization remains small. At temperatures higher than about 100 K, we must also take into account the excitation, by electrons and mainly H atoms, of other atoms and ions, in particular O I which is very important at higher temperatures, i.e., in the warm interstellar medium (Sect. 4.1 and 5.2). The collisional excitation rates of O I by H and He are given by Péquignot [403]. The cooling rate is shown in Fig. 8.5, along with the results for higher temperatures. This gives the global cooling rate. One problem is that the O I lines are often optically thick, so that the result of the optically thin calculation can yield an overestimated cooling rate.

In H II regions electron excitation of the fine-structure lines is the main cooling mechanism. The calculation of cooling via these forbidden lines is similar to that that we presented for the electronic excitation of C II, with the difference that the collisional de-excitation cannot always be neglected. As a consequence, the complete expression given in (4.18) for the densities of ions in the upper levels of the fine-structure transitions must be used. Figure 8.6 shows a simplified example of the results for the cooling function in a low-density H II region where collisional de-excitation has been neglected. The full calculations are included in photoionization models (cf. Sect. 5.1) and generally take into account not only the cooling by fine-structure lines but also by ion recombination despite its weakness (see Sect. 8.2). The free–free continuum is important, and the free–bound and 2-photon continua are negligible (see Sect. 8.2 for the justification of the neglect of the free–bound continuum). A useful approximate expression for the free–free cooling is

$$\Lambda_{ff} = 1.42 \times 10^{-27} T^{1/2} n_e (n_{H^+} + n_{He^+} + n_{He^{++}})\ \mathrm{erg\ s^{-1}\ cm^{-3}}. \quad (8.50)$$

Everything else being equal, the temperature is higher if the abundances of the heavy elements are lower, because these elements take a part in the cooling while the heating is dominated by the most abundant elements, hydrogen and helium.

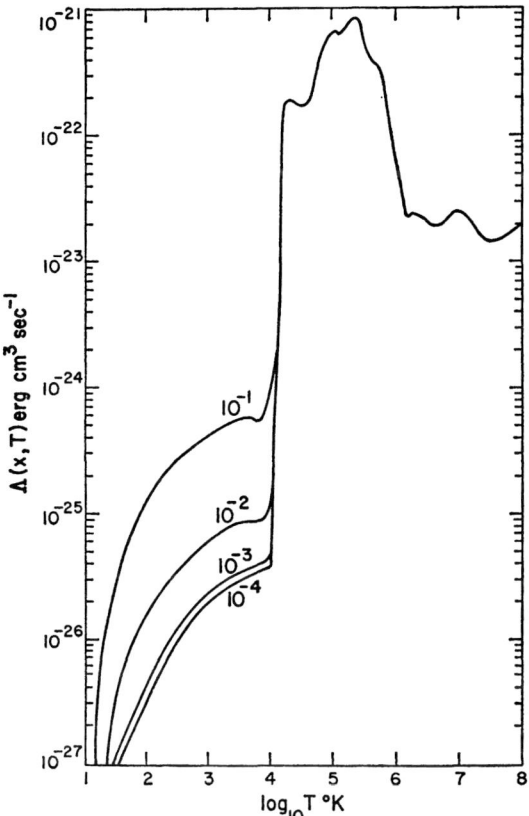

Fig. 8.5. The cooling rate for the interstellar gas. The values of $\Lambda(T)/n_H^2$ are plotted as a function of temperature. For $T < 10^4$ K the cooling is dominated by the excitation of C II and O I. The different curves correspond to different values of the ionization fraction n_e/n_H. The step at 10^4 K corresponds to cooling by the Lyman α line of hydrogen. At $T > 10^4$ K the cooling is due to the excitation of various lines and to Bremsstrahlung; the medium has been assumed to be fully ionized by collisions. Cooling by dust grains and by H_2 has been ignored and no account has be taken of the depletions of carbon and oxygen. The values of the atomic parameters are somewhat outdated and this figure gives only an order of magnitude estimate for the cooling rate, except at high temperatures. From Spitzer [490] (itself from Dalgarno & McCray [111]), with the permission of John Wiley & Sons, Inc.

8.2.2 Cooling by the Collisional Excitation of Permitted Lines

At high temperatures other levels than the fine-structure levels can be populated by collisions with electrons and can contribute to cooling. The most important phenomenon is the collisional excitation of the $n = 2$ level of hydrogen, whose de-excitation produces the Lyα line, but the excitation of other atomic and ionic levels can also contribute. In shocks and photodissociation regions the collisional excitation (mostly by neutrals) of molecules like H_2, H_2O, CO, SiO etc., and of isotopically

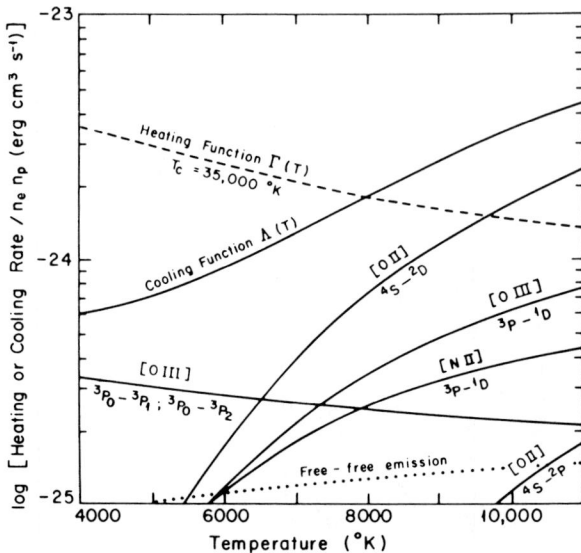

Fig. 8.6. Schematic heating and cooling functions for an H II region. $\Gamma/n_e n_p$ and $\Lambda/n_e n_p$ are plotted as a function of temperature for an H II region with a density of 100 cm^{-3} (no collisional de-excitation). The contributions to $\Lambda/n_e n_p$ from the main transitions of interest are indicated as well as that from free–free cooling. The heating function is an average over the whole H II region which is assumed to be ionization-bounded and ionized by a star with an effective temperature of 3.5×10^4 K (defined in the far-UV). From Spitzer [490], with the permission of John Wiley & Sons, Inc.

substituted species, is very important. This is treated in detail by Hollenbach & McKee [242].

In molecular clouds the most important cooling mechanism is the excitation of the rotational lines of CO and of its isotopically substituted variants[6]. The contribution of C I, whose first excited fine-structure level is 23.4 K above the fundamental, is far from being negligible (Goldsmith & Langer [204]): we will see in Chap. 9 that molecular clouds contain large quantities of C I. On a galactic scale, the cooling by C I lines is as important as the cooling by CO lines (Gérin & Phillips [194]). The corresponding cooling rate can easily be calculated once the abundance of C I and temperature are given, as we did for C II.

Several authors, such as Goldsmith & Langer [204] and de Jong et al. [116], [117] that we will follow here, have treated in an approximate way the CO cooling

---

[6] The cooling of molecular clouds by H_2O has long been considered important. However recent observations of the fundamental rotation line 1_{10}–1_{01} of ortho-H_2O with the SWAS and ODIN satellites show that this line is weak in molecular clouds so that the cooling by H_2O is inefficient. Probably most of the water is condensed as ice on the grains. Similarly, the cooling by molecular oxygen O_2, considered by Goldsmith & Langer [204], is negligible because observations show that this molecule is not abundant: see further Sect. 9.4.

using the escape probability formalism (Sect. 3.2). Then the energy lost per unit volume by molecules in level J, with density n_J, through the transition $J \to J-1$ is, using (4.41):

$$\frac{dE_{J,J-1}}{dt} = (2hB_0 J)n_J A_{J,J-1}\beta_{J,J-1}\frac{S_{J,J-1} - B(\nu_{J,J-1}, T_{BB})}{S_{J,J-1}}, \tag{8.51}$$

where $\beta_{J,J-1}$ is the escape probability for this transition (see Sect. 3.2) and $S_{J,J-1} = (2h\nu_{J,J-1}^3/c^2)[(g_J n_{J-1})/(g_{J-1}n_J) - 1]^{-1}$ is the source function (cf. Sect. 3.1). The term $(S_\nu - B_\nu)/S_\nu$ indicates that the energy loss is measured above the blackbody radiation of the Universe with temperature T_{BB} and brightness $B(\nu_{J,J-1}, T_{BB})$ at the frequency of the transition. If the density is high we may consider, to a first approximation, that the levels are thermalized up to some level J_m and that the total energy loss is approximately the sum of the losses of all the thermalized levels, since the levels above J_m are not significantly populated. All the thermalized lines of CO are optically thick so that $S_{J,J-1} = B(\nu_{J,J-1}, T)$. For these lines we thus have, in the case where the temperature T of the gas is much larger than T_{BB},

$$\Lambda_{CO} \simeq 4\pi \sum_1^{J_m} B(\nu_{J,J-1}, T)\beta_{J,J-1}, \tag{8.52}$$

where the escape probability β is given by (3.66), (3.68), (3.69) or (3.70) depending on the geometry, as a function of the optical thickness τ_0 at the line centre. Note that the value of τ_0 must be evaluated. Let us consider, for example, the case of a cloud with a uniform velocity gradient $dv/ds = const. = V/R$. In this case the line width at a given point is $\Delta\nu = \Delta v/c = (dv/ds)\Delta s/c$, Δs being the distance travelled by a photon before escaping. We can then write the optical depth using (3.32), as

$$\begin{aligned}\tau_0(\nu) &= \frac{c^2 n_{J-1} g_J}{8\pi\nu^2 g_{J-1}} A_{J,J-1}\left[1 - \exp\left(\frac{-h\nu}{kT}\right)\right]\frac{\Delta s}{\Delta\nu} \\ &= \frac{c^3 n_{J-1} g_J}{8\pi\nu^3 g_{J-1}} A_{J,J-1}\left[1 - \exp\left(\frac{-h\nu}{kT}\right)\right]\left(\frac{dv}{ds}\right)^{-1},\end{aligned} \tag{8.53}$$

where we have written ν for $\nu_{J,J-1}$ in order to simplify the expression. This equation allows us to obtain τ_0 then β for each transition, the n_J being calculated at LTE by (3.72) (this is correct as long as the density is higher than the critical density corresponding to the level J_m). Note that the cooling rate is independent of the collisional excitation parameters. However, for the lines with $J > J_m$, which have a smaller optical depth and are not at LTE, these parameters intervene explicitly, but these lines give only a small contribution to the cooling. Figure 8.7 shows the cooling rate Λ_{CO} by CO for two kinetic temperatures. It is interesting to compare it to Fig. 4.12. Falgarone & Puget [170] (Appendix A.1.1) give a useful analytical approximation to the results of Goldsmith & Langer [204], which do not differ significantly from those of Castets et al. [80] displayed in Fig. 4.12.

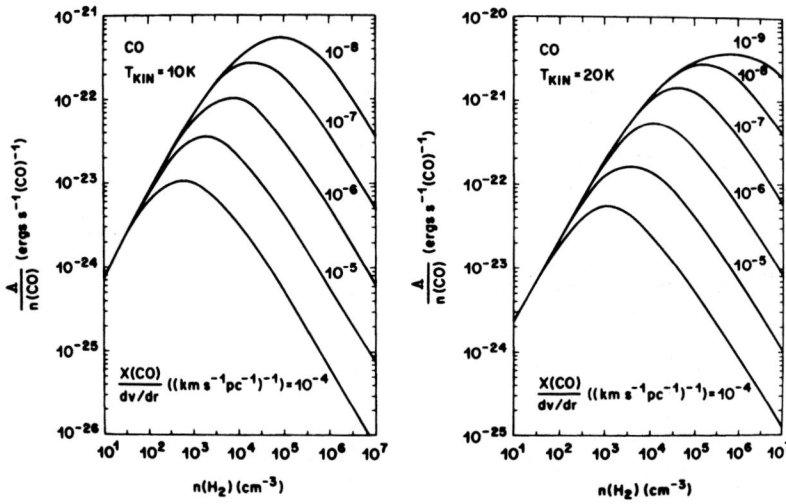

Fig. 8.7. The cooling rate of a molecular gas by CO line emission, for two temperatures. The gas density is the abscissa, and the cooling rate per CO molecule is the ordinate. To obtain the rate per cm^3, multiply by $n_{CO} = x(CO)n_{H_2}$, $x(CO)$ being the abundance of CO which is of the order of 3×10^{-5}. The calculations include the cooling by the isotopically substituted molecules ^{13}CO ($x = 10^{-6}$) and C^{18}O ($x = 10^{-7}$) which contribute signficantly at high densities. The different curves correspond to different values of $x(CO)/(dv/ds)$. Reproduced from Goldsmith & Langer [204], with the permission of the AAS.

For a very dense molecular gas the maser effect, discussed Sect. 3.3, can be a very efficient source of cooling, collisional or not according to the pumping mechanism.

The most important cooling mechanism for the warm and hot ionized media is the collisional excitation by electrons of permitted and forbidden lines of various ions. For this, free–free emission and recombination, which will be discussed just below, also contribute. This cooling has been treated by Raymond et al. [425], Fig. 8.5 is based on their calculations. These processes are included in models of the hot gas (cf. Böhringer [50]), but it is probably necessary to take into account the fact that this gas is not generally in ionization equilibrium (cf. Sect. 5.3 and further Sect. 8.3).

8.2.3 Cooling by Electron–Ion Recombination

Although recombination lines and continua are intense in H II regions their cooling role is negligible because the emitted energy essentially comes from the binding energy of the atom: the effect of recombination is only to make the free electrons disappear without affecting the temperature of the electron gas. However, in high-temperature plasmas, the dielectronic recombination of He and of heavier elements can bring an important contribution. This mechanism is included in models for the hot gas, as noted earlier.

8.2.4 Cooling by Dust

The cooling by the recombination of electrons on charged dust grains has already been treated in Sect. 8.1 and is displayed in Fig. 8.3. The cooling rate increases rapidly with increasing temperature.

The cooling by thermal exchange between gas atoms and dust was mentioned Sect. 8.1. It is generally negligible in the diffuse cold medium.

8.3 Thermal Equilibrium and Stability

We will now examine the thermal equilibrium of the different components of the interstellar medium. The cases of photodissociation regions and of shocks will be discussed in Chapters 10 and 11, respectively.

8.3.1 The Atomic Medium

For this medium the dominant heating mechanisms are the photoelectric heating by grains (Γ_{pe}, (8.32) and Fig. 8.2), the heating by X-rays (Γ_X, Fig. 8.4) at low column densities, and also the hydrodynamic and MHD heating that we will ignore for the moment. The main cooling mechanism is the emission by forbidden lines, essentially the [C II]λ158 µm line for the cold medium ($\Lambda_{e,\mathrm{CII}} + \Lambda_{\mathrm{H,CII}}$, (8.48) and (8.49). If we equate the gains (8.32) and the losses per hydrogen atom, forgetting for the moment the X-ray heating, we obtain the relation

$$10^3 \epsilon \chi = n_H d_C e^{-91.2/T}[1.23 d_C (T/100\,\mathrm{K})^{-1/2} + 7.9]. \tag{8.54}$$

This is a non-linear equation of state. It is approximately verified for the "standard" cold atomic gas[7] ($n = 25$ cm^{-3}, $T = 100$ K, $n_e = n_{\mathrm{CII}}$), with $\Lambda/n_H = \Gamma/n_H = 3 \times 10^{-26}$ erg s^{-1} (H atom)$^{-1}$. For the warm atomic gas, (8.47) has to be used for $\Lambda_{e,\mathrm{CII}}$ because $n_e \gg n_{\mathrm{CII}}$ and we must remember that several mechanisms contribute to the cooling. Complete calculations have been performed by Wolfire et al. [555], who take into account all microscopic mechanisms for heating and cooling. Figure 8.8 gives their results for thermal equilibrium, as a function of density, for the local radiation field. They include X-ray heating at a column density of 10^{19} atom cm^{-2}. In Fig. 8.8a, the density dependence of the pressure $P = n_{tot}kT$, or more exactly $P/k = n_{tot}T$, has been plotted. n_{tot} is the total density of particles $n_H + n_{He} + 2n_e$, the factor 2 in front of n_e coming from the fact that there are as many ions as electrons. Figure 8.8b shows the contribution of the various heating and cooling mechanisms. We can check the value $\Lambda = \Gamma = 3 \times 10^{-26}$ erg s^{-1} (H atom)$^{-1}$ just determined for the standard cold medium. It is of interest to mention that the heating by hydrodynamic waves, e.g. Ferrière et al. [179], would only increase the pressure by at most 5% at the maximum of the curve of Fig. 8.8a and is thus not very important for the diffuse medium.

---

[7] Note that for the cold, relatively dense atomic gas, the contribution of the photoelectric effect on grains to the electron density n_e is smaller than that of carbon ionization. However that of X-ray ionization can be non negligible: cf. Fig. 8.8.

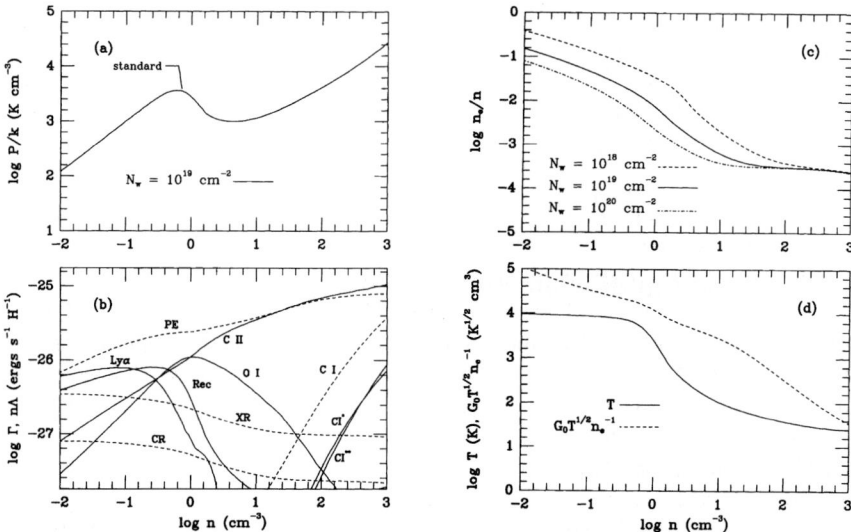

Fig. 8.8. (a) Thermal P/k at equilibrium as a function of the density n in the interstellar matter in the solar neighbourhood (standard model), the X-ray heating being evaluated for a column density of 10^{19} atom cm^{-2}. In the isobaric case the representative point for a medium out of equilibrium because its pressure is too high (above the equilibrium curve) would tend to move to the right. That for a medium with too low a pressure (less than the equilibrium pressure) would tend to move to the left. (b) Contributions of the different heating and cooling mechanisms as a function of density, under the same conditions. **Heating** $\varGamma$ per H atom ($\varGamma/n$ in our notation), is shown as a dashed line: PE = photoelectric effect on grains; XR = X-ray heating; CR = cosmic-ray heating; C I = heating by carbon photoionisation. **Cooling** $\varLambda$ per H atom ($\varLambda/n$ in our notation), is shown as a full line (remember that the cooling rates are proportional to n^2): C II = cooling by [C II]λ158 μm; O I = cooling by [O I]λ63 μm; CI* et CI** = cooling by [C I]λ609 μm and 370 μm respectively; Lyα = cooling by excitation of Lyman α and of other transitions; Rec = cooling by recombination on the grains. (c) Ionization fraction n_e/n as a function of the density n, under the same conditions, for three different values of the hydrogen column density, showing the role of X-rays. (d) Temperature (full line) and ionization parameter for the grains $\chi T^{1/2}/n_e$ as a function of density, under the same conditions. Reproduced from Wolfire et al. [555], with the permission of the AAS.

The [C II] cooling rate is approximately known through direct observations of the intensity of the 158 μm line. Wolfire et al. [555] discuss relatively old observations, that have recently been revised. The best one is probably that obtained with the FIRAS instrument aboard the COBE satellite (Bennett et al. [31]): the intensity of the 158 μm line is well correlated at high galactic latitudes with that of the 21-cm line of H I. The slope of the correlation gives a cooling rate of $(2.65 \pm 0.15) \times 10^{-26}$ erg s^{-1} per H atom, in good agreemeent with the prediction for the standard cold gas. Observations with the ISO satellite, when they are fully analysed, should allow us to refine the measurement of the cooling rate in different conditions.

It is interesting to compare the contributions of the different phases of the interstellar medium to the production of the C II line in the solar neighbourhood. Reynolds [429] estimates the C II line intensity at high galactic latitudes b emitted by the ionized diffuse medium as about 3.7×10^{-7} cosec$|b|$ erg cm^{-3} s^{-1} ster^{-1}. This is 4 times less than the value measured by Bennett et al. [31], $(1.43 \pm 0.12) \times 10^{-6}$ cosec$|b|$ erg cm^{-2} s^{-1} ster^{-1}, justifying their use of the [C II] - H I correlation to obtain the cooling rate per H atom. On the other hand Wolfire et al. [555] estimate that the C II line emissivity per H atom is about 8 times smaller in the warm atomic medium than in the cold atomic medium. As the masses of these two neutral components are similar (Table 1.1), we see that the cold atomic component dominates the emission of the C II line in the solar neighbourhood. The situation might be different in other regions of the Galaxy. In the central regions the ionized diffuse medium might dominate (Heiles [225]). Heiles et al. [226] suggest that this medium has a more inhomogeneous structure than in the solar neighbourhood.

The articles by Wolfire et al. [555], [556] contain an interesting discussion of the effect of the different physical parameters on the equation of state of the diffuse medium. It is of interest to realize that the $P - n$ relation keeps, almost everywhere, the same shape as that shown in Fig. 8.8a. There is almost invariably a part with negative slope for densities around 1 atom cm^{-3}. In this part of the equation of state, the pressure decreases as density increases, and it therefore corresponds to an instability.

Field [180] has given a general discussion of thermal instability in low-density media like the interstellar medium or the solar corona, ignoring the effect of self-gravity. Let us define

$$\mathcal{L}(\rho, T) = \Lambda - \Gamma, \qquad (8.55)$$

the generalized cooling function, which depends on the density ρ and the temperature T of the medium. For a uniform and static medium with density ρ_0 and temperature T_0, $\mathcal{L}(\rho, T) = 0$. Let us introduce a perturbation of density and/or temperature while keeping constant a thermodynamic variable A, for example either the pressure or the density. The entropy S of the medium being defined by

$$\frac{1}{T} = \frac{\partial S}{\partial E}, \qquad (8.56)$$

E being the internal energy of the medium, the perturbation yields a change δS in entropy and $\delta \mathcal{L}$ in the cooling function. The variation dE of E for the duration dt of the perturbation is

$$dE = -\delta \mathcal{L} dt = T d(\delta S). \qquad (8.57)$$

The condition for thermal instability is thus

$$\left(\frac{\partial \mathcal{L}}{\partial S}\right)_A > 0. \qquad (8.58)$$

If the perturbation is at constant volume (*isochoric case*), $TdS = C_V dT$, where C_V is the specific heat at constant volume. If it is at constant pressure (*isobaric case*),

$TdS = C_P dT$, where C_P is the specific heat at constant pressure. The respective conditions for thermal instability can be written as

$$\left(\frac{\partial \mathcal{L}}{\partial T}\right)_\rho < 0 \text{ (isochoric), and} \tag{8.59}$$

$$\left(\frac{\partial \mathcal{L}}{\partial S}\right)_P = \left(\frac{\partial \mathcal{L}}{\partial S}\right)_\rho - \frac{\rho_0}{T_0}\left(\frac{\partial \mathcal{L}}{\partial \rho}\right)_T < 0 \text{ (isobaric).} \tag{8.60}$$

The perfect gas law is assumed for the latter relation. The isobaric case corresponds to a horizontal line in Fig. 8.8a. Wolfire et al. [555] checked that the instability criterion is true everywhere outside the equilibrium curve. In the region above the equilibrium curve the pressure is higher than the equilibrium pressure and the medium tends to contract, its representative point on Fig. 8.8a moving to the right until equilibrium is reached. Conversely, the pressure is smaller than the equilibrium pressure in the region under the equilibrium curve, and the medium tends to expand, its representative point moving to the left towards equilibrium. As a consequence, there is a single stable equilibrium configuration with a low density at low pressures ($P/k < 990$ K cm^{-3} in the standard model). At high pressures ($P/k > 3600$ K cm^{-3} in the standard model) there is also a single equilibrium configuration with a high density. At intermediate pressures there are three equilibrium configurations: an unstable one which has little chance to exist in nature, and two stable ones, with respectively low and high densities. These stable configurations, that are supposed to be in pressure equilibrium with each other, have been identified by Field, Goldsmith and Habing [181] as the cold atomic "clouds" and as the warm "intercloud" medium respectively (cf. Sect. 4.1). This is the *two-component model* of the interstellar medium. Several questions arise concerning this model:

i) is the observed pressure compatible with the existence of two phases in mutual equilibrium? We saw in Sect. 4.1 that the analysis of the populations of the fine-structure levels of C I implies $10^3 < P/k < 10^4$ cm^{-3} K. This is compatible with the model but there are regions with larger or smaller pressures.

ii) can the pressure equilibrium be reached? As a matter of fact, the interstellar medium is strongly perturbed, principally by supernova explosions that yield pressure fluctuations with characteristic intervals of 4×10^5 years at a given location (McKee & Ostriker [359]). These fluctuations propagate with the velocity of sound

$$\boxed{c_s = \left(\frac{\gamma P}{\rho}\right)^{1/2} \simeq 10^4 T^{1/2} \text{ cm s}^{-1},} \tag{8.61}$$

in the adiabatic case, with $\gamma \equiv C_P/C_V = 5/3$. They fully affect the clouds if the cloud sizes are sufficiently small so that they are crossed by the sound wave in times smaller than the characteristic interval of the perturbations. This occurs for sizes smaller than 0.4 pc for the cold medium and smaller than 4 pc for the warm medium. The fluctuations are averaged, and hence attenuated, in larger structures. Thus the relatively large dense structures can be in pressure equilibrium with the external, lower density medium, even if this is not the case at small scales.

iii) is there enough time to achieve thermal equilibrium? The cooling time t_{th} (which is equal to the heating time at equilibrium) is in the isobaric case

$$t_{th} = \frac{(3/2)(1.1 + x_e)n_H kT}{\Gamma} \approx 10^5 \left(\frac{T}{100\,\mathrm{K}}\right)\left(\frac{10^{-26}\,\mathrm{erg\,s^{-1}cm^{-3}}}{\Gamma}\right)\,\mathrm{yr}, \quad (8.62)$$

if $x_e \ll 1$, $t_{th} \leq 10^5$ yr for the cold medium and thermal equilibrium can be achieved. However, t_{th} is much larger for the warm medium and this medium can be well out of equilibrium. This complicates the analysis, nevertheless since the warm medium tends necessarily to evolve towards equilibrium, the description of the interstellar medium as a two-component medium remains qualitatively valid provided that we consider the equilibrium conditions as an average of the real conditions.

8.3.2 The Hot Ionized Gas

The hot gas is initially produced in the shock between the rapidly expanding envelope ejected by the explosion of a supernova and the surrounding intestellar medium (see later Sect. 12.1). It then interacts with the denser, colder medium that is compressed by the hot gas and is heated by conduction in the interaction zone. What remains of the hot gas at the end of the life of the supernova spreads out in the disk and halo of the Galaxy. The hot gas in the halo either escapes from the Galaxy or cools and falls back onto the disk: this is the principle of the *galactic fountain*. We will examine these processes in more detail in Chapters 12 and 15. They have been extensively discussed by McKee & Ostriker [359] and by Spitzer [491], but they are still poorly understood.

Chevalier & Oegerle [92] have suggested a dynamical *in situ* heating process for the hot gas, but its efficiency is not well established. In any case this gas either cools by conduction with the colder gas or via radiation. The radiative cooling rate is given in Fig. 8.5. With typical values of the parameters for the hot medium, $n_e = 10^{-3}$ cm^{-3}, $T = 3 \times 10^5$ K, the characteristic time $(3/2)n_e kT/\Lambda$ for cooling is of the order of 10^4 yr and is thus very short (however the cooling slows down when temperature falls below 10^4 K, see Fig. 8.5). Thus the hot medium tends to rapidly condense into colder clouds. It is probably during this cooling phase that recombination produces ions like N v, C iv and Si iv which are observed through absorption lines in the interstellar medium. Because the recombination of electrons with ions, with a rate varying as $T^{-1/2}$, is slower than cooling at these temperatures, this medium is probably not in thermal equilibrium, as we already mentioned Sect. 5.3.

8.3.3 H ii Regions

The principle of the determination of the equilibrium temperature in H ii regions is relatively simple because only a few processes are of importance. They are schematically summarized in Fig. 8.6. The heating is dominated by ionization and the cooling by the emission of forbidden lines and by free–free emission. The actual calculation is however rather complex as we must take into account the transfer of ultraviolet radiation and solve the equations of ionization equilibrium for a large number of atomic and ionic species. The results are to be found in specialized papers devoted to models of H II regions. We will only say here that the temperature, which is always of the order of 10^4 K, depends to some extent upon the spectral type of the ionizing star, but mainly upon the abundance of those ions that are responsible for the cooling. If these ions are underabundant the temperature is higher since the cooling is less efficient. This affects in turn the degrees of ionization of the various elements. For example, if the abundance of oxygen, the main cooling agent, is smaller the temperature is higher as is the degree of ionization O^{++}/O^+ because recombinations are less frequent, their rate being proportional to $T^{-1/2}$. T can reach 15 000 K if the abundances are low (in some extragalactic H II regions). As a consequence the intensity of the [O III]$\lambda 4\,959 + 5\,007$ lines increases with respect to that of the recombination lines such as Hβ at 4 861 Å. Another reason for this increase is that the upper levels of these fine-structure transitions are more populated due to the higher temperature. The [O III]/Hβ intensity ratio reaches a maximum with decreasing abundance of oxygen but eventually decreases when this abundance becomes extremely low. On the other hand the intensities of the [O II]$\lambda 3\,727 + 3\,729$ lines decrease continuously with the decreasing abundance of oxygen.

8.3.4 Molecular Clouds

The heating and cooling processes in molecular clouds are relatively simple if we neglect their outer regions which are actually photodissociation interfaces. These interfaces will be discussed in Chap. 10. The heating in the deep regions of these clouds is due to cosmic rays, with a contribution of gas–grain collisions if the density is high enough. The cooling is due to the rotational lines of CO and of its isotopic substitutions, with probably also an important contribution from the forbidden lines of C I. Goldsmith & Langer [204] and, subsequently, other authors have discussed the case where the heating is due only to cosmic rays. They find gas temperatures of the order of 10 K. Falgarone & Puget [170] add gas–grain collisions to cosmic-ray heating, assuming that the molecular cloud is immersed in a rather intense far-infrared radiation field. They find gas temperatures lower by about 5 K than the dust temperature (15 to 20 K in their example) for a density of 10^4 mol. cm^{-3}. If the density is 10^5 mol. cm^{-3}, the gas and grain temperatures are nearly equal. We should keep in mind that their model is rather specific but the order of magnitude values that they give are certainly representative.

9 Interstellar Chemistry

The physical chemistry that determines the formation and destruction of interstellar molecules is of great interest for understanding the interstellar medium and elementary chemical processes. This interest leads to the formation of an "astrochemistry" community, which gathers together astronomers, chemists (mainly theoreticians) and molecular physicists. This community is very active. Amongst relatively recent reviews of the subject we have used those of Lequeux & Roueff [316] and of van Dishoeck [530]. A good introduction to the field is the book by Emma Bakes [14]. A recent compendium of useful gas-phase reactions is to be found in Millar et al. [371]. This group maintains the UMIST (University of Manchester Institute of Science and Technology) data base. Gas-phase chemical reactions have been more deeply studied than reactions on the surface of interstellar dust grains. The latter are much less understood, although they are of comparable importance. We will examine in turn these two kinds of reactions.

9.1 Gas-Phase Chemistry

The physical conditions in the interstellar medium have profound implications for possible chemical reactions. At the low temperatures of molecular clouds and the cold atomic medium only exothermal reactions are possible. Moreover some reactions, even if exothermal, have a *potential barrier* (also called an *activation barrier*) that must be overcome by the particles before reaction. These reactions are thus not possible if the temperature is too low for this barrier to be overcome. The situation is different in photodissociation regions and shocks where the higher temperatures make possible some endothermal reactions and reactions with an activation barrier. We will see in Chapters 11 and 13 that even in the diffuse interstellar medium shocks and turbulence generate warm zones where reactions that are otherwise not possible can occur. Ion–molecule reactions most often have no activation barrier and are thus favored in the neutral medium, forming molecular ions of increasing complexity. Neutral molecules are formed by dissociative recombination of these ions with free electrons. If there are ultraviolet photons, molecules can be photodissociated into smaller fragments or photoionized. We will study the main processes we have just mentioned. This study is incomplete because some processes will only be mentioned in passing, e.g. collisional dissociation and reactions with negative ions which do not appear to be as important as the other processes.

9.1.1 Ion–Molecule Reactions

These reactions are of the type $A^+ + B \xrightarrow{k} C^+ + D$, where k is the reaction rate, defined in such a way that the number of reactions per cm^3 and per second is $k\, n_{A^+} n_B$, n_{A^+} and n_B being the respective densities of the reaction partners. k is thus expressed in units of cm^3 s^{-1}.

An ion approaching a neutral molecule induces an electric dipole that in turn creates an attractive force on the ion. The ion can thus be captured by the molecule. Ion–molecule reactions are possible at low temperatures, provided that they are exothermal and without an activation barrier which would hinder the approach of the ion. If the molecule has no permanent dipole moment (H$_2$ is an example), the long-distance interaction potential is

$$V(R) = -\frac{\alpha q^2}{2R^4}. \tag{9.1}$$

where α is the polarizability of the molecule, such that $\langle \mathbf{p} \rangle = \alpha \mathbf{E}$, $\mathbf{E}$ being the electric field and $\langle \mathbf{p} \rangle$ the average of the induced dipole moment over all possible orientations of the molecule (molecules like H$_2$, N$_2$ etc. can in fact be considered as quasi-spherical). q is the charge of the ion and R the ion–molecule distance. A classical trajectory calculation gives the ion-neutral cross-section. Assuming that every encounter yields a reaction gives the reaction rate; this is called in this case the *Langevin rate*,

$$\boxed{k_L = 2\pi (\alpha q^2 / m_r)^{1/2},} \tag{9.2}$$

where m_r is the reduced mass of the collision partners (8.9). The Langevin rate does not depend upon temperature. If one of the partners is H$_2$, with a polarizability $\alpha = 4.5\, a_0^3$, a_0 being the Bohr radius (0.528 Å), k_L is close to 2×10^{-9} cm^3 s^{-1}. Laboratory experiments confirm this order of magnitude for ion–molecule reactions at room temperature. In the absence of measurements it is recommended to use the Langevin rate at relatively high interstellar temperatures (several tens of degrees). However, at low temperatures the rate becomes quite sensitive to the details of the potential surfaces between the neutral and the ion, and the reaction rate can be temperature-dependent. It is thus very important to measure the reaction rates in order to obtain results useful for low-temperature interstellar chemistry. Fortunately, this is presently possible for many reactions of interest.

If the neutral target molecule has a permanent dipole moment the long-distance interaction potentiel is anisotropic and takes the form

$$V(R, \theta) = -\frac{\alpha q^2}{2R^4} - \frac{qM_D \cos\theta}{R^2}, \tag{9.3}$$

where M_D is the modulus of the permanent dipole moment of the molecule and θ the angle between this moment and the ion–molecule direction. The trajectory calculation is complex because it is necessary to average over the possible orientations of the molecule. A quantum calculation is required for low temperatures because the partition function for the rotation levels of the molecule is affected by the interaction. Theory predicts an increase in the reaction rate at low temperatures. A useful approximation for the rate is

$$k(T) = k_1[1 - \exp(-k_0/k_1)], \text{ with}$$
$$k_0 = k_L[1 + M_D^2/(3\alpha B_0)]^{1/2}, \qquad k_1 \simeq k_L + 0,4qM_D\sqrt{8\pi/m_r kT}, \qquad (9.4)$$

B_0 being the rotation constant of the molecule (cf. Sect. 4.2).

9.1.2 Radiative Association

Radiative association is the direct combination of two particles, neutral or ionized, with de-excitation of the formed molecule by emission of a photon: $A + B \xrightarrow{k} AB + h\nu$. For this process to be efficient it is necessary that the molecule be formed in a state linked, by permitted transitions, to the fundamental level, so that it can get rid of its excess energy by emission of a photon. This is not the case for H_2, which therefore cannot form by direct association of two H atoms (cf. Fig. 4.8). It is only possible to form H_2 if a third H atom takes away the excess energy through a collision when the two first H atoms are close to each other; this is a 3-body reaction which requires very high densities ($n_H > 10^{11}$ cm^{-3}) which are not encountered in the interstellar medium. Actually H_2 is mainly formed on the surface of interstellar grains (see Sect. 9.2).

It is possible to obtain an estimate of the reaction rate in the following way, supposing that the two partners A and B are neutral atoms. A semi-classical expression for the radiative association cross-section is, assuming straight-line trajectories and integrating over the impact parameter b,

$$\sigma_{RA} = \int \left[\sum_i g_i A_i(R)\delta t\right] 2\pi b\, db, \qquad (9.5)$$

where g_i is the probability for A and B to approach each other on a particular potential energy curve i, $A_i(R)$ is the probability of spontaneous emission per second from this curve to the fundamental level of the molecule, and δt is the duration of the interaction. For an order of magnitude estimate, we can write

$$\sigma_{RA} \sim gA(R_0)t\pi b_c^2. \qquad (9.6)$$

$t \sim 10^{-14}$ s is the duration of the collision, $0 \leq g \leq 1$, and $A(R_0) \simeq 10^6$ s^{-1} for an electron dipole transition. πb_c^2 is a geometrical cross-section for which b_c is a critical impact parameter that can be obtained through classical considerations. A reasonable order of magnitude is $\pi b_c^2 \simeq 300 a_0^2$. Then, in the most favorable case for which $g = 1$, $\sigma_{RA} \simeq 3 \times 10^{-6} a_0^2$, a very small value. The corresponding reaction rate is $k = \langle v \rangle \sigma_{RA} \simeq 10^{-17}$ cm^3 s^{-1}. Radiative association is therefore a slow process. It can however be efficient if one of the partners is abundant (H or H_2), and if no other exothermal channel exists for the formation of the molecule.

A low temperature a long-lifetime complex can form if one of the partners is a polyatomic molecule. We can obtain an order of magnitude estimate for this lifetime as follows.

The first step of the process is the formation of the complex:

$$A + B \underset{k_d}{\overset{k_f}{\rightleftarrows}} AB^*. \tag{9.7}$$

The second step is the emission of a photon:

$$AB^* \overset{k_r}{\to} AB + h\nu. \tag{9.8}$$

At equilibrium the formation rate k_f of the complex is equal to its destruction rate either by separation of the components (k_d), or by formation of the stable molecule (k_r). The formation rate for the molecule being

$$d[AB]/dt = k_{RA}[A][B], \tag{9.9}$$

we then have

$$k_{RA} = \frac{k_f k_r}{k_d + k_r}, \tag{9.10}$$

and, as often $k_d \gg k_r$,

$$k_{RA} \simeq (k_f/k_d)k_r. \tag{9.11}$$

Thermal equilibrium implies that

$$k_f/k_d = \frac{Q'(AB^*)}{Q'(A)Q'(B)}, \tag{9.12}$$

where Q' is the partition function *per unit volume* (different from that of (3.71) which gives only the internal, non translational part of the partition function for a molecule). We find in Herbst [233] and in Bates & Herbst [24] how to calculate this expression.

Direct experimental determinations are very difficult for this process because three-body reactions, which imply collision with another particle during the lifetime of the AB^* complex, are unavoidable in the laboratory and are in general much faster than radiative de-excitation. However, as explained by Herbst [233], measurements of three-body reaction rates do allow us to indirectly obtain information on radiative de-excitation rates. Much remains to be done in this field.

9.1.3 Dissociative Recombination

Neutral molecules are often produced via *dissociative recombination* of a molecular ion with an electron: $AB^+ + e \overset{k}{\to} A + B$. Figure 9.1 illustrates this process. Its theoretical study is complex as can be seen on this figure, and the laboratory measurement is difficult. Some dissociative recombination rates of major importance, such as that of H_3^+ which as we will see plays a fundamental role in interstellar chemistry, are still controversial. The many measurements of the reaction give incompatible results. Another problem is that of the *branching ratio* between the various possible products A and B of dissociative recombination. Fortunately it is now possible to measure this ratio in the laboratory for a number of molecules (cf. for example Herbst & Lee [234]).

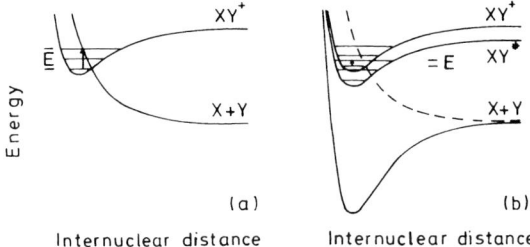

Fig. 9.1. Potential energy curves illustrating dissociative recombination. (a) In the direct process, an electron with energy E excites a transition of the stable ion XY^+ to a repulsive state of the neutral molecule XY which crosses the energy curve of the ion. This repulsive state most often leads to dissociation into X + Y. (b) In the indirect process, the electron excites an energy level of an excited electronic state XY* of the neutral molecule, which crosses a repulsive state of this molecule that can lead to dissociation.

9.1.4 Neutral–Neutral Reactions

Reactions between neutral atoms or molecules can play a role in low-temperature interstellar chemistry. They are fundamental at the higher temperatures reached in shocks, photodissociation regions and regions of intermittency in interstellar turbulence. The main difficulty in their study is the determination of the long-range potentials. In particular, the weak van der Waals attraction forces can be cancelled or overcome by small variations in the slope of the potential surface between the approaching particles. This produces an activation barrier E_a. The reaction rates then take the form

$$k = k_N e^{-E_a/kT}. \qquad (9.13)$$

When it is possible to measure the reaction rate at different temperatures the activation barrier can be determined experimentally.

There is generally no activation barrier between *free radicals* (molecules in which valence bonds are not all satisfied and which possess one or several single, unpaired electrons) or between a free radical and an unsaturated molecule. If one of these free radicals possesses a permanent dipole or multipole moment, long-range electrostatic interactions occur. One of the possible interaction potentials is generally attractive and the reaction can take place. In such cases the reaction rate generally increases at lower temperature, because the radicals can form a complex with low binding energy during the encounter. It is crucial for interstellar chemistry to directly measure these reaction rates at low temperatures because their predictions using quantum chemistry methods are often uncertain.

9.1.5 Photodissociation and Photoionization

These are the main mechanisms leading to the destruction of molecules in the diffuse interstellar medium and in the outer parts of molecular clouds where ultraviolet radiation can penetrate. There are still appreciable effects in clouds at a total visual extinction A_V of 2 to 10 magnitudes. These are the *translucent clouds*, that have been much studied theoretically (see e.g. van Dishoeck & Black [529]). Even in the deep regions of molecular clouds, ultraviolet photons are created by radiative de-excitations following collisional excitation of H_2 and He by cosmic rays or by secondary electrons from cosmic-ray ionization (see Sect. 8.1; Gredel et al. [208]).

Photodissociation is represented by a parameter β, expressed in s^{-1}, which defines the disappearance of molecule AB via the reaction $AB + h\nu = A + B$

$$dn(AB)/dt = \beta n(AB). \tag{9.14}$$

β depends upon the intensity of the UV radiation and upon the photodissociation cross-section σ_ν:

$$\beta = \int \sigma_\nu I_\nu \, d\nu. \tag{9.15}$$

The definitions and notation for photoionization are similar. Van Dishoeck & Black [528] gives a comprehensive review of these mechanisms. Many experimental determinations of cross-sections have been performed, in particular with UV synchrotron sources. In general photodissociation occurs via the absorption of UV continuum photons, with the very important exceptions of H_2 and CO (see Fig. 4.7 for a portion of the UV absorption spectra of these molecules). There are extensive studies of H_2 photodissociation due to its astrophysical and fundamental interest: see in particular van Dischoeck & Black [528] and Warin et al. [541] for astrophysical applications. Since photodissociation then occurs via absorption in strong lines, the UV radiation is considerably deprived of photodissociating photons after these lines become optically thick and the medium is thus self-shielded against photodissociation. This is the case for H_2 and for CO and isotopically substituted molecules. Self-shielding is less important for HD because this molecule is not very abundant. Conversely, H_2 is efficiently self-shielded, even in the diffuse interstellar medium, by saturation of its UV lines and also by UV extinction by dust as soon as the column density reaches a few 10^{20} H atoms cm^{-2}. This can be seen in Fig. 4.10. In photodissociation regions the column density above which H_2 is self-shielded depends upon the density and upon the UV radiation field, as discussed in Chap. 10. CO is photodissociated at greater depth, and its isotopomers still deeper because their abundances are smaller. We will examine the consequences of this in Chap. 10.

This is complicated by the line overlap between all these molecules (see Fig. 4.7) so that their photodissociations are not independent of one another.

The photodissociation rate of isolated H_2 molecules in the local interstellar radiation field is approximately $R_0 = 4.7 \times 10^{-11}$ s^{-1} (Abgrall et al. [3]). Inside a cloud or a photodissociation region the UV field is attenuated in the photodissociation lines. As a first approximation this attenuation is proportional to the square root of the column density $N_{H_2}(s)$ of H_2 molecules until depth s. This is because most of the lines are on the damping part of the curve of growth ((4.34) and Fig. 4.4). The destruction rate at depth s is then, the external parts of the cloud being illuminated by a UV field equal to χ times the local interstellar field,

$$D(s) = \frac{\chi R_0 \beta}{N_{H_2}(s)^{1/2}} n_{H2}(s), \qquad (9.16)$$

where $\beta = 4.5 \times 10^5$ cm^{-1} is the *self-shielding parameter* (Jura [274]) and $n_{H_2}(s)$ is the density of molecular hydrogen.

9.2 Chemistry on Dust Grains

It is clear that chemistry at the surface of dust grains plays a very important role in the formation and destruction of interstellar molecules. It has been known for a long time that H_2 can only form on grains because its formation in the gas phase is generally impossible as discussed Sect. 9.1.[1] There was a suspicion that other molecules would form in the same way but there was also a tendency to neglect such processes, mainly because they were poorly known. Such a neglect is unacceptable today, since there is much direct or indirect evidence for grain chemistry. The most direct evidence is the presence of CO_2 in ice mantles covering dust grains in the deeper regions of molecular clouds. This molecule is not abundant in the gas phase. It has also been claimed that the large abundance of deuterated molecules (HDO, HDCO, CH_3OD, etc.) and even of doubly deuterated ones like D_2CO and ND_2O (Turner [520], Loinard et al. [334],[335]) is evidence for grain surface chemistry. It is often considered that these molecules are difficult to form in the gas phase since deuterium enrichment occurs via isotope exchange reactions with HD, the main reservoir of deuterium: but deuterated molecules are observed in warm molecular clouds (*hot cores*), such as in some parts of the Orion molecular cloud, or in protostellar environments. The idea is that interstellar grains were previously at lower temperatures where the deuterium enrichment of molecules should be easy. However, it has recently been shown that deuterated and even doubly deuterated molecules can form in the gas

---

[1] It is however possible to form H_2 via the H^- ion: $H + e \rightarrow H^- + h\nu$, followed by $H^- + H \rightarrow H_2 + e$, or via the H^+ ion: $H^+ + H \rightarrow H_2^+ + h\nu$, followed by $H_2^+ + H \rightarrow H_2 + H^+$. However these mechanisms are negligible except in circumstances where the degree of ionization is exceptionally high: see Jenkins & Peimbert [262]. Petrie & Herbst [404] give some examples where negative molecular ions can play a role in interstellar chemistry.

phase in a medium where CO is very deficient (Tiné et al. [515]). Moreover, at least some molecules in grain ice mantles do not seem to be much enriched in deuterium (d'Hendecourt, private communication). The question is therefore still open.

We will now examine the formation of H_2 on grains and then other chemical reactions.

9.2.1 H_2 Formation on Grains

The formation of H_2 on grains was studied as early as 1971 by Hollenbach & Salpeter [241], then by Jura [275], Duley & Williams [145], etc. The principle is very simple. Two H atoms stick onto a dust grain, encounter one another and form an H_2 molecule, the excess energy being transferred non radiatively to phonons in the grain and then in part to excitation of the H_2 molecule. We saw in Sect. 9.1 that the H_2 molecule is unable to radiate away its excitation energy so that it cannot form in the gas phase at interstellar densities. The details of the formation of H_2 on grains are rather complex and still poorly known. Here we will only describe some of the basic principles, first in the unrealistic case of a perfect surface, following Hollenbach & Salpeter [241] where a deeper study can be found. Then we will discuss the processes that occurs on more realistic surfaces.

An H atom striking a dust grain has a probablity S to stick on it (*physisorption*). In principle, S depends upon the gas temperature, the grain temperature and the binding energy D for adsorption, and also upon the nature of the grain. D being probably much higher than the kinetic energy of H atoms in the cold interstellar medium, the gas temperature intervenes only through the collision rate. The average time for the sticking of H atoms onto the surface of a dust grain with geometrical cross-section $\sigma_d = \pi a^2$ is

$$t_s = (Sn_H \langle v_H \rangle \sigma_d)^{-1}, \qquad (9.17)$$

where $\langle v_H \rangle = (8kT/\pi m_H)^{1/2}$ is the mean velocity of the H atoms. The adsorbed atom can evaporate within a characteristic time given by

$$t_{ev} \simeq v_0^{-1} \exp(D/kT_d), \qquad (9.18)$$

where $v_0 \sim 10^{13}$ Hz is the characteristic vibration frequency of the lattice and T_d the grain temperature. However, the H atom can also hop across the surface of the grain until it encounters another H atom and forms a molecule. Hopping is a quantum process and depends only very weakly upon temperature. This is a fast process because hydrogen atoms are very mobile. However, for H_2 formation to be efficient, the atoms must have time to move appreciably before evaporating, implying that the grain temperature is lower than some critical temperature T_{crit}.

H_2 molecules in the surrounding gas can also stick on the grain and evaporate. Their evaporation is function of a binding energy D_2 slightly larger than that for H atoms. The evaporation time for molecules is then $t_{ev2} \simeq v_0^{-1} \exp(D_2/kT)$. Below a critical grain temperature T'_{crit}, most of the N sites where molecules can stick are occupied by them. T'_{crit} is estimated by considering a steady state where the sticking time $t_{s2} \simeq (S_2 n_{H_2} \langle v_{H_2} \rangle \sigma_d)^{-1}$ of molecules is equal to $N^{-1} t_{ev2}$:

$$kT'_{crit} = \frac{D_2}{\ln(Nt_{s2}v_0)}. \tag{9.19}$$

To be able to form a molecule, H atoms must find the grain surface sites that are not occupied by H_2 molecules, for this we must have $T_d > T'_{crit}$. The reason for this is that the binding energy on a H_2 monolayer is considerably smaller than on any other solid surface, so that atoms can stick on it only if the grain temperature is smaller than 6.5 K, which is never the case (cf. Sect. 7.2). On the other hand, the atom must have time to move to find a partner before evaporating, requiring $T_d < T_{crit}$. In this simple model, H_2 formation is only possible within a small range of grain temperatures, from 11 K to 13 K in the numerical example given by Hollenbach & Salpeter [241].

More realistic surfaces possess privileged sites where atoms preferentially stick by *chemisorption*. What occurs there is more complex. The main result is that T_{crit} is larger because the binding energy for these sites is larger than the average D. It perhaps reaches 25 K. Then the formation temperatures become compatible with the observed grain temperatures in a moderate radiation field. It is, however, difficult to obtain really quantitative results because of our poor knowledge of the exact nature of grains (silicates, graphite, PAHs, ice or organic grain mantles...) and of the state of their surfaces (amorphous, crystalline, fluffy...). We cannot acknowledge here the many studies on the subject made during the last 30 years. We will only say that no major difficulty arises to account for the observed H_2/H I ratios in the interstellar medium. A particularly clear study of the formation/photodissociation equilibrium of H_2 is that of Jura [274]. For a recent experimental study, see Katz et al. [280].

In the favorable temperature range, we may assume that every H atom adsorbed on a grain will form an H_2 molecule with a partner H atom, and that the formation rate of H_2 can simply be written as

$$k = 0.5 \, n_H \langle n_d(a)\pi a^2\rangle \langle v_H\rangle S \text{ cm}^3 \text{ s}^{-1}, \tag{9.20}$$

where $n_d(a)$ is the density of grains with radius a. Numerically, we obtain by integration over the MRN grain size distribution, with a grain/gas ratio similar to that in the solar neighbourhood,

$$k \simeq 8 \times 10^{-17} S n_H (n_H + 2n_{H_2})(T/100\text{ K})^{1/2} \text{ cm}^3 \text{ s}^{-1}. \tag{9.21}$$

The main unknown here is the sticking probability S, for which values between 0.3 and 1 could be adopted. Uncertainties are so large that it is provisionally preferable to use the formation rate determined empirically by Jura [274] from the column densities of H and H_2 observed in different directions. This is given by

$$\boxed{k = 3 \times 10^{-17} n_H (n_H + 2n_{H_2}) \text{ cm}^3 \text{ s}^{-1}.} \tag{9.22}$$

This expression will be used later in Chap. 10.

An important question is the distribution of the energy released by the formation of H_2 (4,48 eV) between the different possible forms. This question was shortly discussed in Sect. 8.1. In particular, it would be important to know in which excited states the molecule is formed. A specific prediction has been made by Duley & Williams [145] for the formation of H_2 on amorphous silicates. They conclude that the molecule is, to a large extent, in vibrationally excited states but in the fundamental rotation state. They mention however that the situation could well be different on other types of grain. Le Bourlot et al. [304] calculate the infrared spectrum resulting from the de-excitation of H_2 in a molecular cloud, under different hypotheses about its excitation. Confrontation with observation might yield progress although other excitation mechanisms (UV fluorescence and collisional excitation in shocks) might render the conclusions ambiguous. Laboratory experiments are badly needed.

9.2.2 Formation of Other Molecules on Grains

Clearly many molecules other than H_2 can also form on grains. Some can form inside mantles, stay imprisoned and eventually escape to the gas by mantle evaporation due to heating or after total or partial grain disruption by grain–grain collisions (see further Chap. 13 and 15). Others are formed on the grain surface and have some mobility allowing them to take part in chemical reactions. They might be evaporated into the surrounding gas or photo-desorbed after absorption of a UV photon by the grain. A good introduction to these processes is given by d'Hendecourt et al. [122]. The possible reactions are essentially between neutral species, and are limited by the possible presence of activation barriers. These are above all reactions with free radicals, but some such reactions like $H + HCO \rightarrow H_2CO$ appear to have an activation barrier. Examples of permitted reactions are:

$H + C \rightarrow CH$
$H + N \rightarrow NH$
$H + O \rightarrow OH$
$H + CH \rightarrow CH_2$, and so on until CH_4
$H + NH \rightarrow NH_2$, and so on until NH_3
$H + CO \rightarrow HCO$
$H + OH \rightarrow H_2O$
$O + C \rightarrow CO$
$O + CO \rightarrow CO_2$ [2]
$OH + OH \rightarrow H_2O + H$
$HCO + OH \rightarrow CO + H_2O$
$HCO + HCO \rightarrow H_2 + 2CO$

Note however that an activation barrier can be crossed by the tunnelling effect. The characteristic time for barrier crossing is

$$t_{tun} = v_0^{-1} \exp\left[\frac{2a}{\hbar}(2mE_a)^{1/2}\right], \qquad (9.23)$$

---

[2] This reaction has often considered as having an activation barrier. However recent observations show that this is not the case: see Sect. 7.4 and Fig. 4.11.

where $\nu_0 \sim 10^{13}$ Hz is the vibration frequency of the lattice. a, of the same order as the distance between atoms in the lattice, i.e. $\sim 10^{-8}$ cm, and E_a are respectively the width and the height of the barrier. m is the mass of the particle. The crossing time is of the order of 2×10^{-10} s for hydrogen if the barrier corresponds to 350 K. Reactions with a small activation barrier are therefore possible, but they are much slower than reactions with free radicals.

In order for molecules to return to the gas phase without total or partial destruction of the grain, or without external heating, a desorption process is necessary. For H_2 a part of the formation energy is used to eject the molecule from the grain. This is also possible for molecules formed on the surface of very small grains, because their formation energy transferred to the grain might heat it sufficiently for the thermal ejection of the molecule. The arrival of a UV photon can also produce desorption, but this phenomenon is poorly understood. It is probably important in the external regions of molecular clouds where it can prevent the formation of ice mantles on grains. Inside clouds this process is inefficient but the transient heating of the grain, when hit by a cosmic ray, might produce limited evaporation. Finally the accumulated energy of several chemical reactions in the grain mantle might lead to an explosion and then to the ejection of its constituents: see Tielens & Hagen [512] and d'Hendecourt et al. [122] for a study of these processes.

9.3 Equilibrium Chemistry and Chemical Kinetics

After a network of chemical reactions as complete as possible has been set for a given astrophysical situation, and all the corresponding reaction rates have been compiled or estimated (they might be very uncertain in some cases), it is possible to calculate the equilibrium abundances of the various chemical species. Alternatively, the time evolution of these abundances can be calculated starting from some initial composition. The latter method, which is almost exclusively used today, is undoubtly preferable because some crucial reactions, for example some radiative association reactions, can be so slow that the equilibrium is never reached in times comparable to the typical lifetime of the medium (say 10^6 years for an interstellar cloud).

The equilibrium treatment is relatively simple. In an irreversible reaction, steady state simply implies that the formation and the destruction rates are equal. Suppose that a molecule X is formed by the reaction $A + B \xrightarrow{k_f} X + \ldots$, and is destroyed by the reaction $X + Z \xrightarrow{k_d}$ products. The variation with time of the concentration of X is

$$d[X]/dt = k_f[A][B] - k_d[X][Z], \qquad (9.24)$$

hence in steady state

$$[X]_S = \frac{k_f[A][B]}{k_d[Z]}. \qquad (9.25)$$

Similarly, if the destruction of X is due to photodissociation with rate β, the destruction rate is $d[X]/dt = -\beta[X]$ (9.14), so that in steady state

$$[X]_S = \frac{k_f[A][B]}{\beta}. \tag{9.26}$$

For a reversible reaction (a rarer case in the cold interstellar medium) of type $A + B \rightleftarrows C + D + \Delta E$, chemical equilibrium implies if there are no other reactions:

$$\frac{[C][D]}{[A][B]} = \exp-(\Delta E/kT). \tag{9.27}$$

However, we have most often to deal with a complicated network of reactions. An interesting example is the formation and destruction of some deuterated molecular ions of type AD^+ in molecular clouds (Guélin et al. [212]). This is based on isotopic exchange with HD, the main reservoir of deuterium:

$$AH^+ + HD \underset{\leftarrow}{\overset{k_a}{\rightarrow}} AD^+ + H_2 + \Delta E. \tag{9.28}$$

Destruction of AD^+ occurs i) through dissociative recombinaison with electrons, with a rate k_b; ii) through the inverse reaction of formation, with a rate $k_a \exp(-\Delta E/kT)$; iii) through other reactions with species i of density n_i, with rates k_i. Setting the formation and the destruction rates equal, we obtain

$$R = \frac{n(AD^+)}{n(AH^+)} = \frac{n(HD)k_a}{n_e k_b + n(H_2)k_a} \exp(-\Delta E/kT) + \sum_i n_i k_i. \tag{9.29}$$

Knowing the reaction rates, the abundance of HD with relative to H_2 (about 3×10^{-5}, twice the cosmic abundance of deuterium), the n_i and ΔE, we can derive from the measurement of the AD^+/AH^+ abundance ratio, the electron density n_e in the cloud.

In steady state, it suffices to solve a set of algebraic equations similar to (9.28), which raises no difficulty. In the non-steady state, we must solve a set of differential equations like (9.23), a more complex problem. If we also take into account grain surface reactions and grain–gas exchanges, the treatement is even more complicated, although perfectly feasible on a modest work station. Grain chemistry can be handled either using kinetic equations similar to those used for the gas, or by Monte-Carlo methods in which the arrival of particles on the grain is simulated, leading to their displacement on the surface, molecule formation and ejection. The Monte-Carlo methods are more exact but they are more difficult to couple with the kinetic chemical equations for the gas: see the dicussion in Shalabiea et al. [467]. There exist many compilations of reactions, of their rates and of other useful parameters, for example that of Hasegawa & Herbst [222] for reactions on solid surfaces, or those of Le Bourlot et al. [302], Bettens et al. [38], Warin et al. [541] and Millar et al. [371] (the UMIST data base) for gas-phase reactions. A compendium of results of recent chemical models in dense clouds is given by Lee et al. [309].

9.4 Some Results

It is out of question to give here even a superficial review of the works made in the domain of interstellar chemistry. We will only describe some principles that apply to the diffuse interstellar medium and to molecular clouds. The chemistry in photodissociation regions, in shocks and in intermittency regions of the turbulent medium is left to the corresponding chapters.

9.4.1 Chemistry in the Diffuse Interstellar Medium

We consider the diffuse medium as characterized by a density of a few tens of particles per cm^3, a temperature of the order of 100 K (we exclude here the warm diffuse medium where chemistry is negligible due to the low density), and the presence of a UV radiation field. Carbon is almost entirely ionized and supplies most of the free electrons. Molecular hydrogen is present as soon as the column density reaches a few 10^{20} H atoms cm^{-3}. Besides the formation of H_2, grain chemistry does not appear to play an important role in this medium. The chemistry is thus essentially a gas-phase chemistry starting from C^+ and H_2; for a detailed study see van Dishoeck & Black [527]. We do not expect *a priori* to find complex molecules in the diffuse medium: the lifetime of molecules is not long enough, because they are rapidly photodissociated by UV radiation. As a matter of fact, in this medium we find diatomic molecules like H_2, CO, CN, CH, CH^+, OH, C_2, CS and SiO. But it was a surprise to also find more complex molecules like HCO^+, N_2H^+, HCN, HNC, C_2H, C_3H_2 and H_2CO. The abundance of these molecules is generally well determined because they are often observed through their absorption lines, either in the UV and visible wavelength range or at millimetre to decimetre wavelengths. An exception is H_2CO for which complicated excitation effects prevent accurate abundance determinations. For a discussion, see the papers by Liszt and Lucas, in particular [328], [329], [339], [340], and references in these papers. The most striking results are the excellent correlations between the abundances of HCO^+, OH and C_2H, molecules that appear as soon as H_2 is present at an extinction of $A_V \sim 0.25$ mag., and also the very high abundance of several multiatomic species. In translucent clouds where A_V is of the order of 1 magnitude or more, even more complex molecules are found (Turner et al. [521]). For recent results on this topic, see the series of papers by Liszt and Lucas (Liszt & Lucas [330] and references herein).

The two initial reactions for carbon chemistry are radiative association reactions, they are very slow:

$C^+ + H \rightarrow CH^+ + h\nu$ ($k = 10^{-17}$ cm^3 s^{-1}).
$C^+ + H_2 \rightarrow CH_2^+ + h\nu$.

The rate of the second reaction has been determined by many laboratory experiments and theoretical calculations, with discordant results. In spite of these discrepancies it is clear that none of the rates for these two reactions is large enough to account for the observed abundance of CH^+ in the diffuse medium. Another reaction that can be thought of, $C^+ + H_2 \rightarrow CH^+ + H$, is endothermic ($\Delta E = -0,4$ eV). One possibility to get out of this deadlock is that CH^+ is formed at high temperature by this fast reaction, either in shocks or in turbulence intermittent regions (Chapters 11 and 13). Once CH^+ is formed, other carbonaceous molecules can be formed without difficulty. However, it is not certain that there will be enough of its progenitor to form C_3H_2. This progenitor is most probably $C_3H_3^+$, the formation reaction being $C_3H_3^+ + e \rightarrow C_3H_2 + H$.

The formation of oxygen-bearing molecules like OH, CO and HCO^+ starts in principle from H^+ and H_2^+, which are produced by cosmic-ray or X-ray ionization. OH is the first product to be formed and is the progenitor of CO and HCO^+. The corresponding reactions are (van Dishoeck & Black [527], Viala et al. [535]):

$H^+ + O \rightarrow O^+ + H$ (charge-exchange reaction), followed by
$O^+ + H_2 \rightarrow OH^+ + H$.
Also,
$H_2^+ + H_2 \rightarrow H_3^+ + H$, then
$H_3^+ + O \rightarrow OH^+ + H_2$.
Then
$OH^+ + H_2 \rightarrow H_2O^+ + H$, and finally
$H_2O^+ + e \rightarrow OH + H$.

OH is rapidly destroyed by the neutral–neutral reaction without activation barrier:
$O + OH \rightarrow O_2 + H$, as well as H_2O by various reactions, which are fast even at low temperatures because of its large permanent dipole moment.

CO and HCO^+ are formed from OH:
$C^+ + OH \rightarrow CO + H^+$, and
$C^+ + OH \rightarrow CO^+ + H$, followed by
$CO^+ + H \rightarrow CO + H^+$, and by
$CO^+ + H_2 \rightarrow HCO^+ + H$.
HCO^+ can also be formed by
$C^+ + H_2O \rightarrow HCO^+ + H$.
Finally HCO^+ is destroyed by
$HCO^+ + e \rightarrow CO + H$.

This set of reactions cannot account for the absolute abundances and for the abundance ratios observed for OH, HCO^+, CO and H_2CO (cf. Lucas & Liszt [338]). However the last reaction is sufficient to account for the abundance of CO, the problem being in the formation of HCO^+ (Liszt & Lucas [329]). Therefore, there are major difficulties with understanding the chemistry of the diffuse interstellar medium. The solution is perhaps, like for CH^+, in shock chemistry or in chemistry in intermittent regions, that will be examined later. However, Liszt & Lucas [329] observe no anomalies in the profiles of the molecular lines of the diffuse interstellar medium, such as splitting, asymmetries or wings, that would indicate such dynamical processes. However, they see an ubiquitous weak, broad absorption component that might correspond to the superimposition of such processes along the line of sight. It may be that these processes are so fast and so localized that the chance to observe them individually is small, while some of their products can be seen due to their longer lifetimes. In any case, the problem cannot be considered as solved yet.

9.4.2 Chemistry in Dense Molecular Clouds

The situation in dense molecular clouds is very different. Here hydrogen is essentially molecular, and there is little UV radiation so that direct ionization of carbon is negligible. The source for ions and free electrons is cosmic-ray ionization of molecular hydrogen and of what remains of atomic hydrogen and of helium, producing respectively H_2^+, H^+ and He^+. We have already mentioned the following fundamental reaction in the preceding section:

$$H_2^+ + H_2 \rightarrow H_3^+ + H.$$

Carbon chemistry is then initiated by

$$C + H_3^+ \rightarrow CH^+ + H_2.$$

A series of exothermal reactions follow:

$$CH^+ + H_2 \rightarrow CH_2^+ + H,$$
$$CH_2^+ + H_2 \rightarrow CH_3^+ + H.$$

But the following such reaction starting from CH_3^+ is endothermic and only radiative association allows this to proceed further:

$$CH_3^+ + H_2 \rightarrow CH_5^+ + h\nu.$$

This reaction is known in the laboratory but exhibits a considerable complexity. Methane is then obtained by dissociative recombination:

$$CH_5^+ + e \rightarrow CH_4 + H.$$

The ion chemistry of oxygen is similar to that described in the previous section for the diffuse medium, but now there are no great discrepancies between the predictions of this chemistry and observations. We can also produce CO via other, rather more efficient channels like $C + H_3O^+ \rightarrow HCO^+ + H_2$ or $CH_3^+ + O \rightarrow HCO^+ + H_2$, followed by dissociative recombination of HCO^+. As CO is a very resilient molecule and the destruction processes inefficient, we expect that most of the gaseous carbon is in CO.

Although the preceding description gives a qualitatively correct idea of the gas-phase chemistry, it is only schematic. Figure 9.2 and 9.3 show diagramatically more complex, but still simplified versions of the carbon and of the oxygen chemistries, which are in fact very much coupled. Similar diagrams can be drawn for the chemistry of nitrogen, sulphur, etc.

However the predictions of the gas-phase chemistry models, even time-dependent, are not in good quantitative agreement with observations. For example:

- observations of the lines of C I in the direction of molecular clouds suggest that there is a large quantity of neutral carbon in these clouds, while models predict that almost all the carbon should exist as CO;
- the abundances of O_2 and of H_2O are considerably smaller than predicted by the models (actually O_2 has never been convincingly detected). If the small abundance of H_2O could be explained by condensation on grains, this explanation cannot hold for O_2 which is not found in the solid phase either [525].

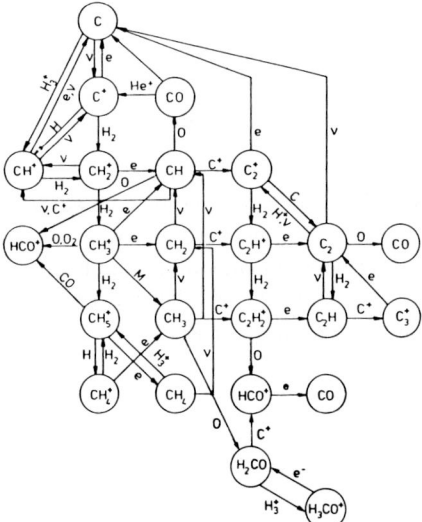

Fig. 9.2. A somewhat simplified version of the gas-phase chemistry of carbon in dense molecular clouds, and of its coupling with oxygen chemistry. Exothermic reactions are symbolized by arrows joining one of the parent products to the resulting one, the other parent partner being indicated near the arrow. From Prasad et al. [415], with the kind permission of Kluwer Academic Publishers.

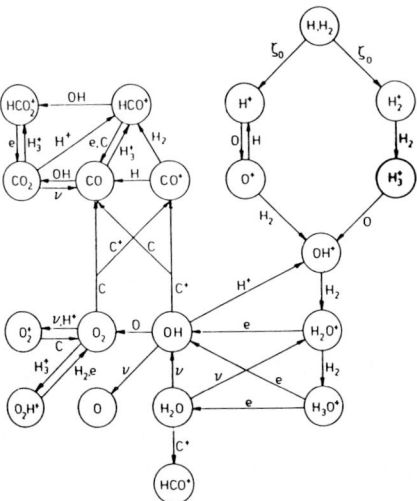

Fig. 9.3. A somewhat simplified version of the gas-phase chemistry of oxygen in dense molecular clouds, and of its coupling with carbon chemistry. Exothermic reactions are symbolized by arrows joining one of the parent products to the resulting one, the other parent partner being indicated near the arrow. From Prasad et al. [415], with the kind permission of Kluwer Academic Publishers.

Several explanations can be given for these discrepancies. One is of course that grain–surface reactions have been neglected. Actually those models which take them into account give better results (see for example the models of Hasegawa & Herbst [222]), however without solving the abundance problems we just mentioned. Bergin et al. [35] (see also Bergin & Langer [36]) also obtain interesting results while taking only into account gas-phase reactions and depletion/desorption of molecules on grains. Another possibility is that molecular clouds are very fragmented so that UV photons can penetrate deep inside the clouds where they photodissociate CO into large amounts of C I. However the observations mentioned Sect. 6.3, which favor a relatively uniform distribution of dust in some dark clouds, contrary to that of CO, throw doubt on this idea. An interesting suggestion is that of a dynamical (turbulent) mixing between the superficial regions of clouds which are exposed to UV radiation and the deep parts (Chièze & Pineau des Forêts [97], Xie et al. [559]). Then the overall chemical composition of the cloud resembles that predicted by time-dependent models at relatively early epochs, of the order of 10^5 years. In particular we find a large abundance of C I and a small abundance of O_2 and H_2O in agreement with observations. We will come back to this suggestion in more detail in Sect. 13.4, where the results are plotted in Figs. 13.3 and 13.4. Pineau des Forêts et al. [410] have modified this model by taking into account the deposition of CO and of other molecules on grains.

To end this chapter, we mention the possibility that the solution of chemical equations is not necessarily unique, but that two solutions can exist in a rather large range of physical conditions. This is studied in the papers by Le Bourlot et al. [303] and Lee et al. [311], who give numerical examples. This bistability is sometimes observed in laboratory chemistry, and seems to exist also in nature. In models of dense clouds where ionization is due to cosmic rays, one solution corresponds to a relatively high ionization fraction x, of the order of a few 10^{-7}, and the other one to a fraction of ionization smaller by one order of magnitude. This is illustrated in Fig. 9.4. This phenomenon is very sensitive to the gas-phase abundance of the heavy elements and appears only for very low abundances, i.e. for a large depletion of these elements on grains. The abundance of sulphur, which unfortunately is poorly known in molecular clouds, appears to play a key role. The bistability depends also of the reaction network used in the model. However, for a given network and a given set of abundances, it depends only upon the ζ/n ratio, where ζ is the ionization rate by cosmic rays and n the total density of hydrogen nuclei. In the high-ionization branch, H^+ is much more abundant than H_3^+ and charge-transfer reactions with H^+ dominate. Remembering the reaction of formation of molecular oxygen

$O + OH \rightarrow O_2 + H$, we now have

$O_2 + H^+ \rightarrow O_2^+ + H$, followed by dissociative recombination

$O_2^+ + e \rightarrow O + O$, leading to the destruction of O_2. This can reduce the abundance of O_2 by a factor 30 with respect to what occurs in the other "normal" solution with a low degree of ionization, which was implicitly the situation discussed until now. H_2O and OH are also underabundant and C is overabundant by a factor 20 because its main destruction mechanism, $C + H_3^+ \rightarrow CH^+ + H_2$, is inefficient due

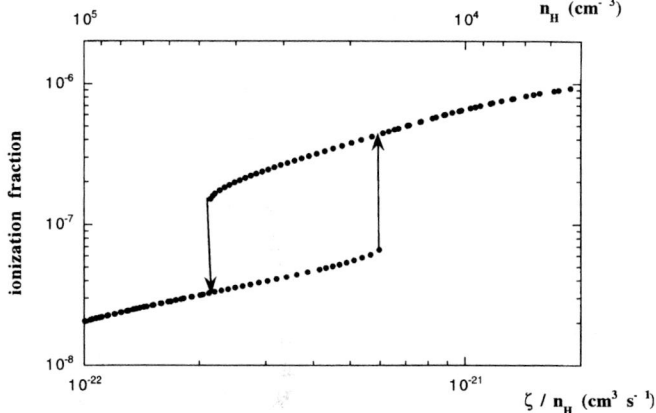

Fig. 9.4. Bistability of chemical equilibrium in a molecular cloud. The ionization fraction n_e/n_{total} is plotted as a function of the ratio between the cosmic-ray ionization rate ζ and the density. The upper scale in the abscissae corresponds to $\zeta = 10^{-17}\,\mathrm{s}^{-1}$. Two solutions appear in the region between the arrows. From Lee et al. [311], with the permission of ESO.

to the underabundance of H_3^+. Therefore the C/CO ratio can reach 0.1, while it is much smaller in the low-ionization branch. We conclude that this hypothesis leads to a reduced abundance of O_2 and H_2O, and to an enhanced abundance of C I, as observed in molecular clouds. The relative abundances of deuterated molecules is also much affected, as it can be foreseen given their high sensitivity to the ionization fraction (cf. Sect. 9.3; Gérin et al. [193]). It also appears that different molecular clouds, and even different parts of the same cloud, can exhibit very different molecular and atomic abundances. Small variations in the initial conditions can lead, after some time, to one branch or the other because we are dealing with a chaotic phenomenon. Time variations of the chemical composition can be relatively fast, so that it is necessary to consider non-stationary solutions. Oscillations might even occur between the two equilibrium solutions, and "chemical waves" with very different chemical compositions might propagate in a cloud. Le Bourlot et al. [305] find that these properties are still valid when taking into account the influence of the grains.

10 Photodissociation Regions

10.1 General Presentation

Photodissociation regions (abbreviation: PDR) are those parts of the interstellar medium where the ultraviolet radiation field is strong enough to photodissociate molecules. They are also called *photon-dominated regions*, with the same abbreviation PDR. In the most general sense, this definition encompasses most of the interstellar medium, with the exception of the inner parts of molecular clouds and of H II regions which are, by convention, treated separately. Historically however, photodissociations regions have been defined as the interfaces between regions and molecular clouds, where both the density and the UV radiation field are large. These are often called "dense photodissociation regions". They are observationally characterized by strong [C II]λ158 μm and [O I]λ63 μm lines, by intense rotation and vibration lines of H_2 and, finally, by strong aromatic band emission in the mid-infrared. We will only discuss these "standard" regions in the present chapter. However, let us point out the fact that there is no fundamental difference in the physics and chemistry between these dense photodissociation regions and those in the neutral medium in general. Good reviews of dense photodissociation regions are given by Hollenbach & Tielens [244], [245]. Plates 9, 16, 17, 20, 21, 22, 24 and 31 show some examples of photodissociation interfaces.

In PDR interfaces, the ultraviolet radiation field decreases continuously with distance into the molecular cloud. The depth into a PDR from the region is often quantified by the visual extinction A_V[1]. Schematically, there is a stratified structure (Fig. 10.1) in which hydrogen is first ionized as H^+ in the H II region, then recombines into atomic hydrogen H and finally forms H_2 molecules. The H→H_2 transition occurs at a UV optical depth $\tau_{UV} \approx 0.6$ corresponding to a visual extinction $A_V \approx 0.2$, for a density of about 10^3 cm^{-3} and a moderate UV radiation field ($\chi \approx 100$). We will see later how this depth depends upon the parameters of the photodissociation region: see Fig. 10.5. Similarly, carbon recombines with electrons from C^{++} in the H II region into C^+ in the outer parts of the photodissociation region. At a depth $A_V \approx 1$, C^+ recombines with electrons to give C; then slightly deeper it forms CO molecules. Actually, as we will see in Fig. 10.4, none of these transitions is sharp

---
[1] The column density of matter N_H from the surface is related to A_V by (7.5), assuming the convention $A_V/E(B-V) = 3.1$, $A_V = 1$ then corresponds to $N_H \simeq 1.87 \times 10^{21}$ H atom cm^{-2}.

and there is a coexistence of the various forms of hydrogen and carbon over a range of depths. Oxygen is everywhere atomic outside the H II region and its abundance decreases somewhat where CO forms. Other forms of oxygen, essentially O_2 and H_2O, are not abundant as described in the preceding chapter.

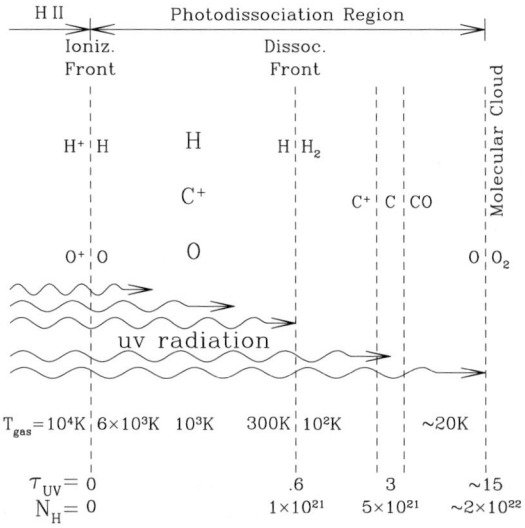

Fig. 10.1. Schematic structure of a photodissociation interface. UV radiation from a H II region penetrates into a neutral cloud from the left. We first meet the ionization front. To the right of this front, H and O become neutral while C remains ionized. Then we encounter the progressive transition H→H_2, schematized here by a vertical line, and the similarly progressive transition C^+→C→CO. Deeper, the cloud is fully molecular. The gas temperature, the gas column density and the optical depth in the far UV are indicated on the lower part of the figure. The conditions correspond to a dense cloud submitted to an intense UV radiation field. From Draine & Bertoldi [140], with the kind permission of the authors.

The lines of H_2, C II, O I and CO are very intense in photodissociation regions because both the density and the temperature are high. All these lines are in the infrared to submillimetre range. The corresponding energy comes from the ultraviolet radiation of stars through the photoelectric effect on the grains and the excitation of H_2 by far-UV photons. These lines cool the medium very efficiently: its temperature goes from about 10^4 K in the H II region to a few tens of K at $A_V \approx 1$. CO lines are very intense at this depth and take a large part in the cooling. The absorption of UV and visible radiation by dust also heats the grains which then emit an infrared continuum and aromatic bands. Most of the energy emitted by young stars in star-forming regions is converted into these forms. The emission from photodissociation regions dominates the respective line and contiuum emission of a galaxy, except perhaps for the low-excitation rotation lines of CO and the emission by big dust grains in the far infrared. However, a large fraction of the emission in the [C II]$\lambda 158$ μm line can also come from the diffuse interstellar medium (cf. Sect. 8.3). Papers by Crawford et al. [106], Madden et al. [344] and Pierini et al. [407] discuss the origin of this line.

Many models for photodissociation regions have been constructed since the first detections of the infrared cooling lines in the 70's. The reviews by Hollenbach & Tielens [244] [245] give an updated list for the literature up to 1999. Most models assume a steady state, although the photodissociation interface more or less rapidly penetrates into the molecular cloud. However the most important chemical reactions are rather fast, justifying this approximation, with some exceptions. Some models adopt a plane-parallel geometry, which has the advantage of making clear the impact of the different parameters, and we will here discuss only these models. Others adopt a spherical geometry, treat the case of inhomogeneous media, etc. Some models are entirely self-consistent, since they include a calculation of temperature. Before examining the models in Sect. 10.3, we will treat the basic processes in Sect. 10.2. Sect. 10.4 briefly discusses the non-equilibrium models.

10.2 Physico-Chemistry

10.2.1 The Penetration of Far-UV Radiation and Photodissociation

The penetration of far-UV radiation is the key to the physico-chemistry of photodissociation regions. The transfer of this radiation is determined by dust absorption and to a lesser extent by dust scattering, and by absorption by molecules.

The effect of dust is relatively simple for a homogeneous medium (for a rather complete discussion and calculations of the photodissociation of various molecules, see Roberge et al. [437]). As a first approximation, we can neglect scattering because it is strongly forward-directed in the UV (see Chap. 7). Extinction at any wavelength can be obtained from the column density of matter, the visual extinction and the extinction law (Table 7.1, Fig. 2.5 and Fig. 7.1). Extinction at a wavelength λ is $A_\lambda = k_\lambda A_V$. There is however a problem with the value of k_λ. Standard values correspond to the diffuse interstellar medium but there are large variations from direction to direction in the UV (cf. Fig. 7.3). If interstellar clouds are fragmented the penetration of UV is more complex than in a uniform cloud (Boissé [51])[2]. Then the UV radiation can penetrate much more deeply and the radiation density can have large fluctuations from place to place. This is the case for the photodissociation region of M 17 (Plate 24). In plane-parallel models and for a relatively thin slab we must also account for photons that penetrate from the other side of the slab. In general a two-exponential approximate solution for the transfer equation suffices in this case (Roberge et al. [437]). The absorbed UV radiation heats the dust, whose temperature then decreases with depth into the photodissociation region.

---

[2] We can treat the problem in an approximate way by considering a mixture of media with different effective opacities: cf. Meixner & Tielens [364].

The decrease in the UV radiation due to absorption by molecules is a more complicated phenomenon. However, only H_2 and CO are abundant enough to play a role. Absorption by these molecules occurs in lines which correspond to numerous electronic transitions (Fig. 4.7). This absorption often leads to photodissociation, for H_2 this occurs in approximately 1/10 of the cases. The optical thickness of these lines produces a self-shielding against photodissociation for the molecules which are located deep inside the region. This self-shielding is extremely efficient for H_2 because of its large abundance (Sect. 9.1): it is already appreciable at column densities as small as 10^{14} H_2 molecules cm^{-2} and shielding is complete at about 10^{20} H_2 molecules cm^{-2}. The numerical treatment of self-shielding is somewhat involved, but useful approximations are given for H_2 by Jura [274], Federman et al. [175], van Dishoeck & Black [528], Draine & Bertoldi [138], Lee et al. [310], etc. Tables and curves concerning CO can be found in van Dishoeck & Black [528] and in Lee et al. [310]. In spite of the importance of self-shielding, the absorption of the UV radiation by dust cannot be neglected in studies of the photodissociation of H_2 and CO.

10.2.2 Chemistry

The chemistry in photodissociation regions differs from that in the cold diffuse interstellar medium and in molecular clouds because of the high temperatures in dense photodissociation regions. Many endothermal reactions and reactions with activation barriers can occur as well as reactions with H_2 in excited states. A very complete study was undertaken by Sternberg & Dalgarno [495]. Figure 10.2 shows a scheme for the most important reactions in the oxygen and carbon chemistries. Note the similarities, but also differences, with the schemes of Figs. 9.2 and 9.3 relative to molecular clouds. For example, reactions with excited H_2 (noted H_2^*) enhance the formation of OH thanks to the reaction $O + H_2^* \rightarrow OH + H$. Similarly, the reaction $C^+ + H_2^* \rightarrow CH^+ + H$ produces more CH^+, although still not enough to account for carbon chemistry (see Sect. 9.4). In the presence of shocks or turbulence, the endothermal reaction $C^+ + H_2 \rightarrow CH^+ + H - 0{,}4$ eV can be another important source of CH^+ because the kinetic energy of H_2 can be large enough to overcome endothermicity (Spaans et al. [485]; cf. also Falgarone et al. [173]). Some recent chemical models take into account polycyclic aromatic hydrocarbons or PAHs (Bakes & Tielens [15], see also Sect. 9.4). The main effect of introducing PAHs is to enhance the abundance of neutral atoms like C and S due to the neutralization of the corresponding positive ions on PAH$^-$s.

10.2.3 Heating Processes

As in the diffuse interstellar medium, the photoelectric heating by grains plays a very important role in photodissociation regions. We discussed this in Sect. 8.1, and the physics is just the same for photodissociation regions. H_2 formation on grains (Sect. 8.1) is also an important source of heating. We should also add to this

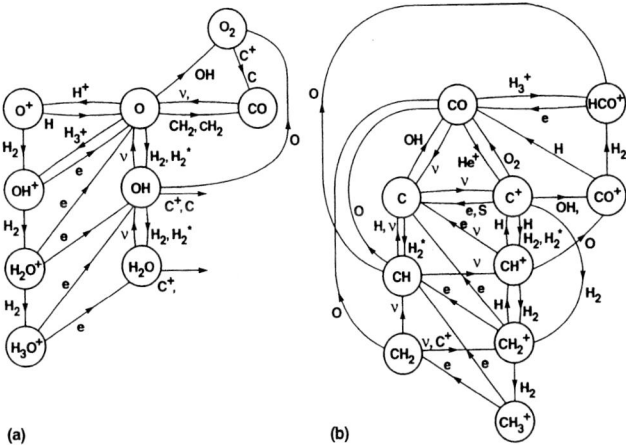

Fig. 10.2. The most important chemical reactions in photodissociation regions: (a) carbon chemistry; (b) oxygen chemistry. Compare to Figs. 9.2 and 9.3. Reproduced from Sternberg & Dalgarno [495], with the permission of the AAS.

the chemical heating by the electrons released by the dissociative recombination of several ions (H_3^+, HCO^+, etc.) and by various exothermal reactions with He^+, H_2^+ and H_3^+ (see Fig. 10.3).

The collisional de-excitation of the excited levels of H_2 is another heating mechanism that we only briefly mentioned in Sect. 8.1 because it is not important in the diffuse interstellar medium. The absorption of far-UV photons by H_2 promotes the molecule to excited electronic states. These excited states rapidly decay either into the vibrational continuum of the ground electronic state with a probability of approximately 10%, leading to dissociation[3], or into a vibrationally excited level of the ground electronic state, with a probability of about 90%. At low densities these excited vibrational levels de-excite radiatively by emission of a characteristic ro-vibrational spectrum in the near infrared, that we will discuss later, and heating is negligible. However, at densities larger than 10^4 H atoms cm^{-3} collisions with H atoms or H_2 molecules can de-excite the vibrational levels, the excess energy being transferred to the colliding atom or molecule as kinetic energy. This produces heating after thermalization. Complete studies of this can be found in Martin et al. [348] and Le Bourlot et al. [306]. This heating by H_2, and the photoelectric heating by grains, are both important to a depth corresponding to $A_V \sim 0.6$ mag. At greater depths the heating by H_2 decreases considerably because the photodissociation (formation) rate for H_2 becomes small due to the of UV photons at the required wavelengths.

Figure 10.3 shows an example of the contributions of the various processes to the heating of a photodissociation regions with a density of 10^3 cm^{-3} submitted to a UV

---

[3] The H atoms resulting from this photodissociation are suprathermal and contribute to the heating, but this process is generally much less important than the heating by collisional de-excitation.

radiation field 1 000 times larger than in the solar neighbourhood ($\chi = 10^3$). In this case, the H→H_2 transition takes place at $A_V \approx 1$ mag. and the C^+→C transition at $A_V \approx 2.5$ mag.

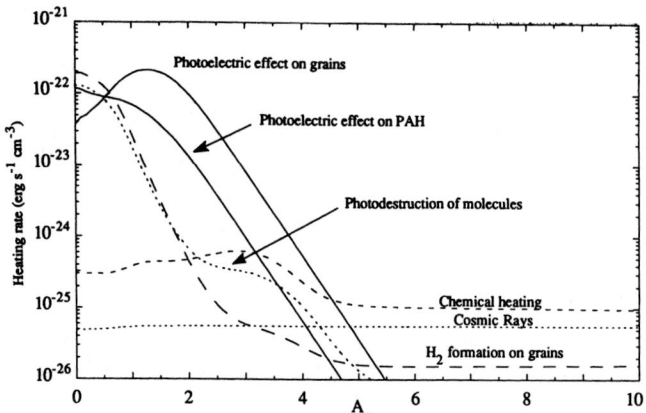

Fig. 10.3. The contribution of different processes to the heating of a photodissociation region as a function of the depth expressed in terms of the visual extinction A_V. The external UV radiation field is 1 000 times the local interstellar field ($\chi = 1\,000$) and the density is 1 000 cm^{-3}. The temperatures of the gas and of the grains are calculated self-consistently. C and O are assumed to be underabundant by a factor 10. The density is not large enough for collisional de-excitation of H_2 to be important. From le Bourlot et al. [302], with the permission of ESO.

10.2.4 Cooling Processes

Gas cooling in photodissociation regions is dominated by the emission of fine-structure lines and by emission from the rotational lines of CO, as well as by the collisional excitation of the rotational levels of H_2. The temperature can be high, allowing strong emission in the [O I]$\lambda 63$ and 146 μm lines, and even sometimes in the [OI]$\lambda 6\,300$ Å, [S II]$\lambda 6\,730$ Å and [Fe II]$\lambda 1.26$ and 1.64 μm lines, that dominate the cooling at temperatures greater than 4 000 K. Similarly, the emission from high-J CO lines can be strong. At high densities cooling by the collisions of atoms and molecules with the colder dust grains (cf. Sect. 8.1), and by the recombination of electrons with positively charged grains (Sect. 8.2), can be of some importance.

The emission of molecular hydrogen deserves a special mention. As we said, it corresponds to radiative cascades from high electronic states excited by UV photons. At low densities collisional de-excitation of the vibrational levels populated through cascades is negligible and the spectrum is pure fluorescence. Such a fluorescence spectrum has been observed in a variety of photodissociation regions, with a very good agreement between observation and theory. As remarked in Sect. 10.2, fluorescence emission does not correspond to a change in kinetic energy, hence to heating or cooling. At densities higher than $\approx 10^4$ cm^{-3} collisional excitation takes place and the spectrum is different. This corresponds to the heating that we mentioned earlier. Conversely, collisions can populate higher rotational levels at the expense of the kinetic energy of the colliding particles, compensating for this heating to some extent. This cooling is taken into account, together with the heating, in H_2 excitation models. The corresponding line emission has been observed with the ISO satellite: see e.g. Draine & Bertoldi [140]. The cooling rate is given by Martin et al. [348] or by Le Bourlot et al. [306] as a function of density and temperature.

Atoms also exhibit interesting fluorescence effects, however with no consequence for the thermal balance. For example, O I lines and the Balmer recombination lines of deuterium are observed in emission in the direction of the Orion nebula. They are excited by fluorescence, not in the H II region, but in the photodissociation interface behind it (Hébrard et al. [223]). The fluorescent Balmer lines of deuterium should be useful for a determination of the D/O ratio and hence, of the abundance of deuterium D/H.

10.3 Stationary Models

Building a model of a photodissociation region requires a number of choices and calculations, i.e.:

– the choice of the geometry (plane-parallel with illumination on one side or on both sides, spherical, etc.);
– the choice of density and temperature distributions. The simplest models have these quantities fixed, with an inhomogeneous density requiring, in general, the use of a Monte-Carlo method. Other models calculate iteratively the temperature at fixed density. The most sophisticated models fix the pressure, or consider a self-gravitating cloud in hydrostatic equilibrium, etc., and calculate all the other parameters;
– the transfer of UV radiation, which implies the choice of the abundance and the optical properties of dust. For H_2 and CO photodissociations iterations are necessary;
– the chemistry, which requires hypotheses about the initial abundances of the most important elements in the gas phase;
– the transfer of CO rotation lines and, possibly, of other optically thick lines; these calculations usually use the escape probability approximation (Chap. 3), and hence involve a choice of the velocity dispersion which is generally obtained from observations of the width of a CO line. This choice is actually not very critical.

Let us now discuss some results from steady-state photodissociation models. As an example, Fig. 10.4 shows the results of a plane-parallel model with a uniform density and one-sided illumination by a strong UV field, which corresponds roughly to the photodissociation region of the Orion bar (Plates 9 and 31).

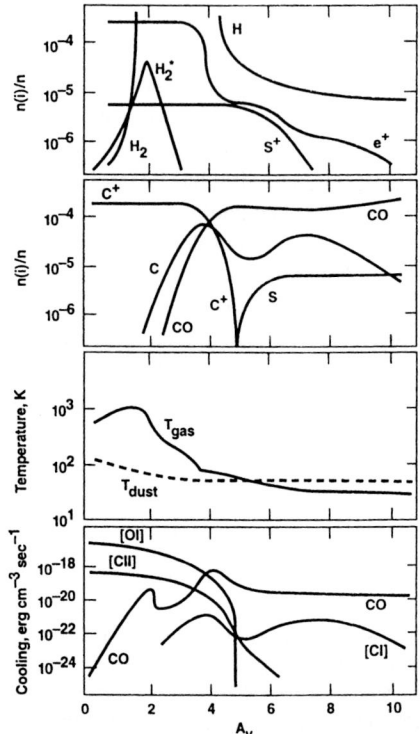

Fig. 10.4. Structure of a photodissociation region with a uniform density $n = 2.3 \times 10^5$ cm^{-3}, illuminated from one side by a UV field of 10^5 times the local one ($\chi = 10^5$). The conditions are approximately those in the Orion bar (Plates 9 and 31). The results are given as a function of the visual extinction A_V from the illuminated surface. Top: the relative abundances of atomic hydrogen, molecular hydrogen and molecular hydrogen H$_2^*$ excited by UV. The transition H$\leftrightarrow$H$_2$ occurs near $A_V = 3$ magnitudes. The abundances of free electrons (noted e$^+$ by mistake) and of ionized sulfur are also given. Middle, the relative abundances of C$^+$, C, CO and S; note that the transitions C$^+\leftrightarrow$C and C$\leftrightarrow$CO occur almost at the same depth. The temperatures of the gas and of the big dust grains are also plotted. Bottom, the contribution to the cooling of the fine-structure lines [C II]λ158 μm, [O I]λ63 and 126 μm, [C I]λ609 and 370 μm and of the CO lines. Taken from Hollenbach & Tielens [244].

Combining (9.15) and (9.20) or (9.21) we can obtain the fraction of hydrogen as H$_2$, as a function of depth into a cloud or in a photodissociation region. It is interesting to note that the key quantity for this problem, and for many other problems concerning photodissociation interfaces, is the ratio of density to the far-UV radiation field, n/χ. We can see that the extinction by dust has little effect on the abundance of H$_2$ as long as $n/\chi \geq 130$ H atom cm^{-3} (Pak et al. [391]). Figure 10.5 shows the column density at which half of the hydrogen is molecular as a function of the density and the UV radiation field, for uniform-density photodissociation regions. It also shows the

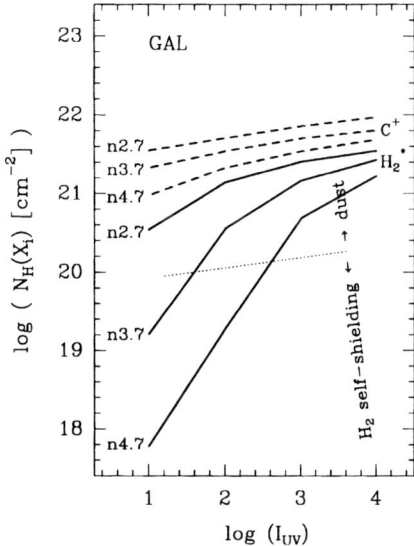

Fig. 10.5. Depths of the transitions H↔H$_2$ and C$^+$↔C↔CO for a uniform-density photodissociation region as a function of the UV radiation field expressed in units of the local interstellar field. The depth is expressed in terms of the total column density of the gas from the surface. A column density $N_H = 1.87 \times 10^{21}$ atom cm^{-2} corresponds to 1 magnitude of visual extinction. The depth is calculated for different densities. The notation n4.7, for example, means $n = 10^{4.7}$ cm^{-3}. The full lines correspond to $n_H = 2n_{H_2}$; much molecular hydrogen outside this region is excited hence the notation H$_2^*$. The dashed lines have the same meaning for the C$^+$/C/CO transition. The dotted line corresponds to the limit between the outer region where H$_2$ is mainly protected by self-shielding, and the inner region where the dust extinction dominates. Reproduced from Pak et al. [391], with the permission of the AAS.

zones where the photodissociation of H$_2$ is preferentially limited by self-shielding or by dust extinction, and the depth of the C$^+$/C/CO transition.

The most complete and most recent systematic study of photodissociation regions was made by Kaufman et al. [281]. It considers plane-parallel models with constant density ranging from 1 to 10^7 cm^{-3}, illuminated on one side by radiation fields of $\chi = 10^{-0.5}$ to $10^{6.5}$. These models include the PAH chemistry of Bakes & Tielens [15]. The effect of this chemistry is to produce a relatively uniform gas temperature from the surface of the cloud to a column density of 10^{21} atom cm^{-2} at which the far-UV optical depth is about unity (Fig. 10.6). The temperature depends on χ and decreases at larger depths. Figure 10.7 illustrates the different regimes for the emission of the cooling lines in a photodissociation region.

From (4.16), (4.18) and (4.19), we see that the intensity of the [C II] line emitted by a photodissociation region is approximately given by

$$I(\text{C II}) \propto N(\text{C}^+) \frac{exp(-92\,\text{K}/T)}{1 + n_{cr}/n}, \tag{10.1}$$

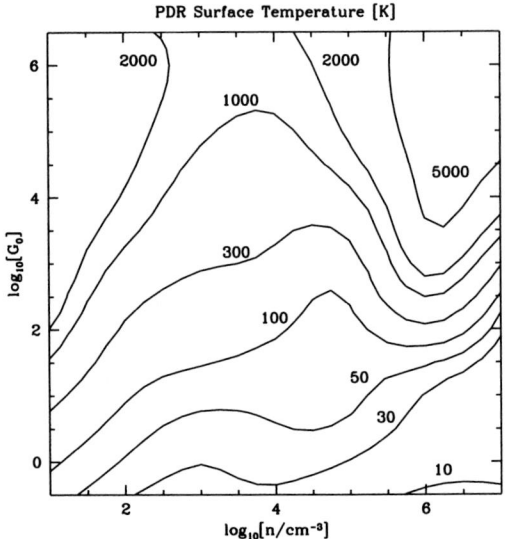

Fig. 10.6. Gas temperature at the surface of a photodissociation region as a function of the density and the incident UV radiation field, represented by $G_0 \equiv \chi$. The temperature is approximately uniform until the column density reaches 10^{21} atom cm^{-2}, corresponding to a far-UV optical depth close to 1. Reproduced from Kaufman et al. [281], with the permission of the AAS.

where $N(C^+)$ is the column density of C^+ in the photodissociation region and $n_{cr} \simeq 3\,000$ cm^{-3} the critical density for collisions with neutrals for this line.

At densities less than $n_{cr} \simeq 3\,000$ cm^{-3}, which is roughly the critical density for the [C II]λ158 μm, [C I]λ370 μm and CO(1-0) lines, the intensities of these lines per ion or atom are proportional to the total density, as long as they are optically thin. Conversely, they become independent of the total density for densities larger than n_{cr} because the energy levels are then in LTE (4.16) to (4.20). In both cases, of course, the intensities per unit volume are proportional to the density of the emitting particles. If their abundances are uniform in the medium the cooling rate per unit volume is therefore proportional to n^2 for densities $n < n_{cr}$, or to n if $n > n_{cr}$. The heating rate per unit volume always increases faster than n so that the temperature is higher at higher densities, at least as long as the above lines dominate the cooling: this can be seen in Fig. 10.6. In Fig. 10.7 the curve for which the temperature is 92 K is plotted. This is the excitation temperature $\Delta E/k$ for the [C II]λ158 μm line. At lower temperatures the energy emitted in this line depends upon temperature, and hence upon the incident radiation field, while at higher temperatures the level populations tend to LTE and the emitted energy does not depend upon the UV radiation field. The cooling is then dominated by the [O I]λ63 μm line.

Figure 10.8 gives the intensity of the [C II]λ158 μm emission perpendicular to the surface of the photodissociation region and Fig. 10.9 shows the intensity ratio between the [O I] λ63 μm line and the [C II]λ158 μm line. The [O I] line is generally

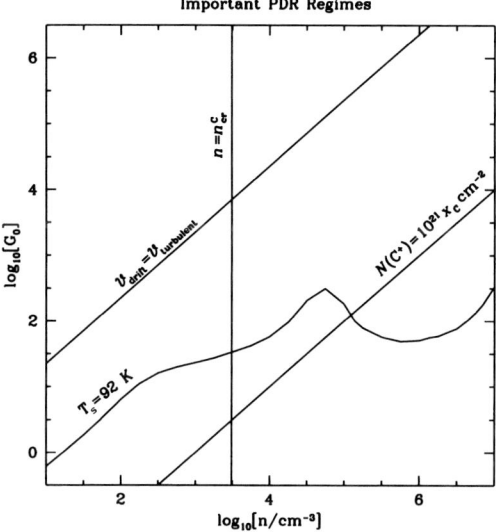

Fig. 10.7. The different regimes for the emission of the fine-structure lines in a photodissociation region. The abcissae and ordinates are as for Fig. 10.6. The vertical line $n = n_{cr}$ corresponds to the critical density for collisions with neutrals, which is approximately the same for the [C II]λ158 µm, [C I]λ370 µm and CO(1-0) lines. The line $v_{drift} = v_{turbulent}$ shows the conditions for which radiation pressure drives the dust grains with a velocity equal to the turbulent velocity of the gas. At higher radiation fields, a steady-state is not possible. The curve $T_s = 92$ K indicates the conditions such that the gas temperature at the surface of the photodissociation region is equal to $\Delta E/k$, $\Delta E = hc/158$ µm being the excitation energy for the C II line. Above this curve, the intensity of this line depends only weakly upon the intensity of the radiation field for a given density. Finally, the line $N(C^+) = 10^{21} x_C$, x_C being the abundance of gaseous carbon, indicates the conditions such that carbon is ionized up to a hydrogen column density of 10^{21} cm^{-2}. Below this line the absorption of UV radiation by dust is large and the column density of C^+ as well as the intensity of the 158 µm line depend upon the radiation field and upon density. Reproduced from Kaufman et al. [281], with the permission of the AAS.

optically thick and this introduces uncertainties in the results. The article by Kaufman et al. [281] contains several other useful diagrams which give, as a function of the density and radiation field, the energy emitted by the [O I]λ145 µm and the [C I]λ370 and 609 µm lines, as well as that emitted in the (1-0), (2-1), (3-2) and (6-5) (433 µm) lines of CO. Figure 10.10 shows the intensity ratio between the [C II]λ158 µm and the CO(1-0)λ2.6 mm lines, in energy units.

We note that, due to their large optical depths, the observed CO lines come only from a thin layer of the photodissociation region, while the lines of the isotopically substituted CO molecules, which are less abundant, come from deeper parts. This is illustrated in Fig. 10.11.

Figure 10.4 to 10.10 assume standard abundances and the usual depletions of the heavy elements in the local galactic interstellar medium. However, the gas-phase

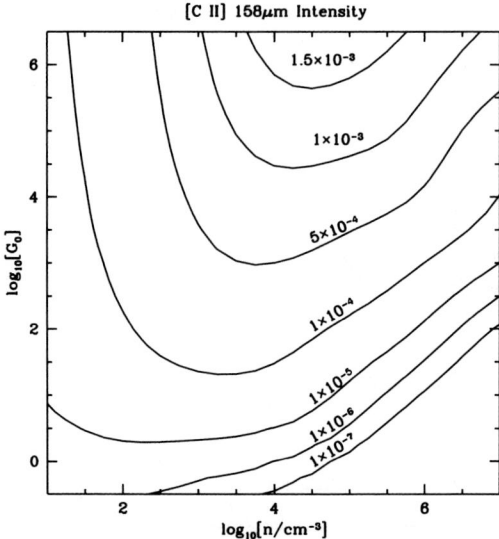

Fig. 10.8. Intensity of the [C II]λ158 µm line emitted perpendicularly to the surface by a photodissociation region, as a function of density and incident UV radiation field. The contours are labelled by the intensity in units of erg cm^{-2} s^{-1} ster^{-1}. Reproduced from Kaufman et al. [281], with the permission of the AAS.

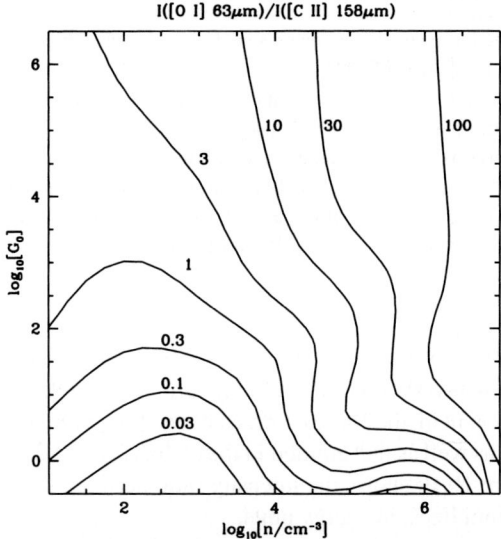

Fig. 10.9. Ratio between the energies of the [O I]λ63 µm and [C II]λ158 µm lines emitted by a photodissociation region, as a function of the density and the incident UV radiation field, the contours are labelled by this ratio. The [O I] line is optically thick over almost all of the parameter space displayed in the figure. Reproduced from Kaufman et al. [281], with the permission of the AAS.

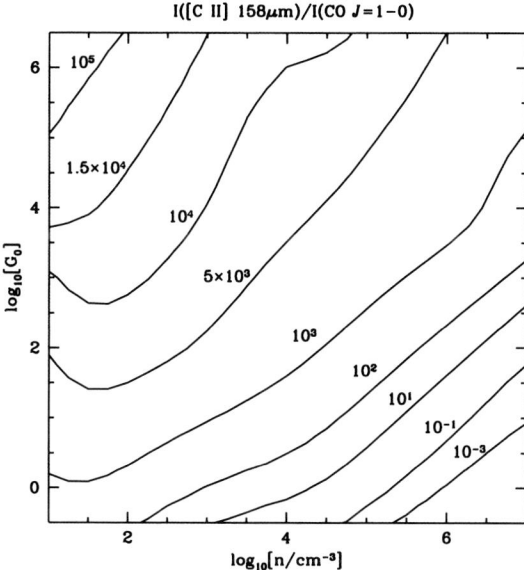

Fig. 10.10. Ratio between the energy emitted, perpendicularly to a photodissociation region, in the [C II]λ158 µm and the CO(1-0)λ2.63 mm lines as a function of the density and the incident UV radiation field, the contours are labelled by this ratio. Reproduced from Kaufman et al. [281], with the permission of the AAS.

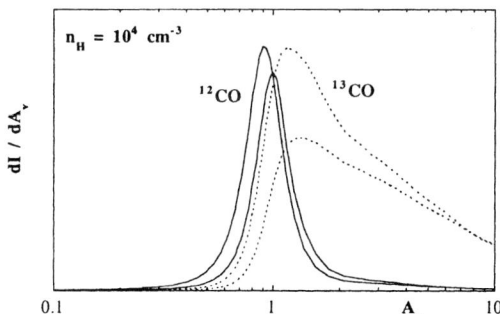

Fig. 10.11. The contribution of the different depths of a photodissociation region to the emission of the CO lines as observed perpendicular to the surface by an external observer. This contribution is given in arbitrary units, which are proportional to the flux, as a function of the visual extinction A_V, for a photodissociation region of density 10^4 cm^{-3}. The illuminating radiation field is 10 times that of the local interstellar radiation field ($\chi = 10$). The abundances of the heavy elements and of dust are those typical of the Small Magellanic Cloud, i.e., 10 times smaller than the galactic abundances. The full lines correspond to the ^{12}CO(2-1) and ^{12}CO(1-0) lines and the dotted lines to the ^{13}CO(2-1) and ^{13}CO(1-0) lines. The (2-1) lines reach higher levels than the (1-0) lines. From Lequeux et al. [317], with the permission of ESO.

abundances of heavy elements, in particular of carbon, have important consequences for the physics of photodissociation regions, as well as the abundance of dust which is generally assumed to be proportional to that of the heavy elements. Photodissociation models with different abundances, in particular for those appropriate to the Magellanic Clouds, can be found in Lequeux et al. [317], Allen et al. [8], Pak et al. [391], Bolatto et al. [52] and Kaufman et al. [281].

The structure of photodissociation regions also plays a major role in their physics. A fragmented structure, such as observed in the M 17 photodissociation region (Stutzki et al. [498], Stutzki & Güsten [499], see Plate 24), allows the UV radiation to penetrate deeper into the medium. This enhances the effective area of the photodissociation region and favors the emission of the [C I] lines. Models of inhomogeneous photodissociation regions using Monte-Carlo calculations of the radiation transfer can be found in Spaans & van Dishoeck [485].

10.4 Out of Equilibrium Models

We saw, in the previous section, a case where a steady state is not possible in a photodissociation region, i.e., where the radiation pressure on grains is large enough to decouple them from the gas. Another case of non-stationarity is that in which the radiation field rapidly changes during the characteristic formation time of H_2, $\tau_{H_2} \approx 10^9 (n/\mathrm{cm}^{-3})^{-1}$ yr: this time determines the chemistry. This occurs when a massive, hot O or B star switches on inside a molecular cloud (see references in Hollenbach & Tielens [244]; a detailed study has been made by Bertoldi & Draine [37]).

Another, related phenomenon is the progression of an ionization front into a molecular cloud. If this progression is fast with respect to τ_{H_2}, the thermal and chemical structures may reach a steady state, but they differ from those in a photodissociation region at equilibrium because the chemical reactions do not have time to reach equilibrium before the gas is photodissociated and ionized. This occurs, in particular, for small neutral clouds embedded within a H II regions, as observed for example in the Orion nebula (Störzer & Hollenbach [496]). This will be discussed later in Chapter12 (Sect. 12.3).

11 Shocks

In this chapter, and the three following, we discuss the dynamics of the interstellar medium. The present chapter recalls the general equations that control the motion of a compressible fluid and discusses shocks. Shocks intervene in many situations in the interstellar medium. Here we limit ourselves to supernova remnants and bubbles, that are treated in the following chapter. We will also discuss in Chap. 12 ionization fronts, the evolution of H II regions and the acceleration of cosmic rays.

For more details on shocks, we refer to the general monographs by Kaplan [278] (unfortunately not easy to find) and Spitzer [490], and the review by Draine & McKee [137]. We will follow the latter review rather closely.

11.1 The Equations of Gas Dynamics

If the various components of a gas (atoms, molecules, ions and free electrons) all have the same bulk velocity **v**, we can define it as a *single fluid*. On the other hand, it is possible that these components are partly decoupled from each other and possess different bulk velocities: in this case, we speak of a *multi-fluid medium*.

11.1.1 A Single-Fluid Medium

The motion of a fluid is controlled by the conservation of mass, momentum and energy, and also by the Maxwell equations if it is partly or totally ionized. We will assume here that the fluid is a perfect conductor of electricity and everywhere electrically neutral, which corresponds to the magnetohydrodynamic (MHD) approximation. This allows a considerable simplification of the dynamical equations. These equations are, in cartesian coordinates x_j ($j = 1, 2, 3$):

$$\frac{\partial \rho}{\partial t} + \frac{\partial(\rho v_k)}{\partial x_k} = 0, \tag{11.1}$$

$$\frac{\partial}{\partial t}(\rho v_j) + \frac{\partial}{\partial x_k}\left(\rho v_j v_k + P\delta_{jk} - \sigma_{jk} + \frac{B^2}{8\pi}\delta_{jk} - \frac{1}{4\pi}B_j B_k\right) = 0, \tag{11.2}$$

$$\frac{\partial}{\partial t}\left(\frac{1}{2}\rho v^2 + u + \frac{B^2}{8\pi}\right) +$$
$$\frac{\partial}{\partial x_k}\left[\left(\frac{1}{2}\rho v^2 + u + \frac{B^2}{8\pi}\right)v_k + Pv_k - \sigma_{jk}v_j - \frac{B_j B_k}{4\pi}v_j + Q_k + F_k\right]$$
$$= 0. \tag{11.3}$$

Here, we have written these equations in developed form, in projection on the j axis for (11.2). This implies summation over indices $k = 1, 2, 3$, and also over j for (11.3). δ is the usual Kronecker symbol ($\delta_{jk} \equiv 1$ for $j = k$, $\delta_{jk} \equiv 0$ for $j \neq k$). ρ, **v** and P designate respectively the density, the fluid velocity and the pressure. The pressure includes the gas pressure $P_g = \rho kT/\mu m_H$, where μ is the molecular mass in units of the hydrogen atom mass $m_H = 1.67 \times 10^{-24}$ g, and the pressure P_{cr} of high-energy particles if present. **B** is the magnetic field. u is the total energy density of the fluid: $u = (3/2)P + u_{int}$, where u_{int} is the density of internal energy in forms other than pressure. **Q** is the heat flux due to thermal conduction, **F** the radiative flux and σ_{jk} an element of the stress tensor due to viscosity. Let us comment on these equations in turn.

Equation (11.1) is the continuity equation, which describes the conservation of mass during the motion of the fluid. It can easily be demonstrated by considering a slab of the medium with thickness dx and unit area, perpendicular to the velocity **v**. During a time dt, a quantity of matter $\rho v\, dt$ enters the slab and a quantity of matter $[\rho + (\partial \rho/\partial x)\, dx][v + (\partial v/\partial x)\, dx]dt$ leaves the slab. The slab thus loses a the quantity of matter $(\partial \rho/\partial t)\, dx\, dt = [\rho + (\partial \rho/\partial x)\, dx][v + (\partial v/\partial x)\, dx]dt - \rho v\, dt = [\partial(\rho v)/\partial x]\, dx\, dt$. Dividing by $dx\, dt$, we obtain the 1-dimensional version of the continuity equation.

We can obtain in the same way the equation of conservation of momentum (11.2). A complication arises from the coexistence of scalar terms (pressure P and magnetic pressure $B^2/8\pi$) with tensor terms (viscosity σ_{jk} and Laplace force $-B_j B_k/4\pi$). The Laplace force $\mathbf{F}_B$ is $\mathbf{j} \times \mathbf{B}$, where $\mathbf{j}$ is the density of the electric current. Since we have assumed the medium to be a good conductor, $\mathbf{j} = \nabla \times \mathbf{B}/4\pi$, from the first Maxwell equation, in which we neglect the displacement current (the MHD approximation): therefore $\mathbf{F}_B = -\mathbf{B} \times (\nabla \times \mathbf{B})/4\pi$ whose component along x_j is $-(1/4\pi)\partial(B_j B_k)/\partial x_k$.

The energy conservation equation (11.3) can be derived in the same way. The energy per unit volume is the sum of the kinetic energy $(1/2)\rho v^2$, the gas total energy u and the magnetic energy $B^2/8\pi$. This magnetic energy can be in the form of plasma waves that have to be averaged over time. The successive terms in square brackets correspond, respectively, to the variation of the energy, the work of the pressure force, the viscous heating and the work of the Laplace force, the heat flux and the radiative flux. The divergence of the radiative flux is $\nabla \cdot \mathbf{F} = \Lambda$, where Λ is the net rate of energy loss per unit volume due to radiative cooling (or heating). The internal energy u_{int} can be eliminated from (11.3) by using an effective rate of energy loss

$$\Lambda_{eff} = \Lambda + \frac{\partial}{\partial x_k}(u_{int} v_k) + \frac{\partial u_{int}}{\partial t}. \tag{11.4}$$

Λ_{eff} includes the energy "losses" associated with, for example ionization or dissociation of molecules. If Λ_{eff} is used to evaluate **F** in (11.3) it is then possible to replace u by $(3/2)P$ in this equation.

11.1.2 A Multi-Fluid Medium

For a multi-fluid medium, the equations of conservation of mass, momentum and energy have to be written for each component, taking into account exchanges between the components. For a fluid with two components, the neutrals n and the ions i (the electrons follow the motion of the ions), the basic equations become:

- for the neutrals

$$\frac{\partial \rho^{(n)}}{\partial t} + \frac{\partial [\rho^{(n)} v_k^{(n)}]}{\partial x_k} = S^{(n)}, \tag{11.5}$$

$$\frac{\partial}{\partial t}[\rho^{(n)} v_j^{(n)}] + \frac{\partial}{\partial x_k}[\rho^{(n)} v_j^{(n)} v_k^{(n)} + P^{(n)} \delta_{jk} - \sigma_{jk}^{(n)}] = \mathcal{F}_j^{(ni)}. \tag{11.6}$$

- for the ions an equation similar to (11.6) is obtained by replacing the index n by i, $P^{(e)}$ being the pressure of the electrons:

$$\frac{\partial}{\partial t}[\rho^{(i)} v_j^{(i)}]$$
$$+ \frac{\partial}{\partial x_k}\left\{\rho^{(i)} v_j^{(i)} v_k^{(i)} + \left[P^{(i)} + P^{(e)}\right] \delta_{jk} - \sigma_{jk}^{(i)} + \frac{B^2}{8\pi} \delta_{jk} - \frac{1}{4\pi} B_j B_k\right\}$$
$$= -\mathcal{F}_j^{(ni)}, \tag{11.7}$$

- and the energy conservation equations, with $(\alpha) = (n)$ or (i):

$$\frac{\partial}{\partial t}\left\{\frac{1}{2}\rho^{(\alpha)} \left[v^{(\alpha)}\right]^2 + u^{(\alpha)} + \frac{B^2}{8\pi}\delta_{\alpha i}\right\}$$
$$+ \frac{\partial}{\partial x_k}\left[\left\{\frac{1}{2}\rho^{(\alpha)} \left[v^{(\alpha)}\right]^2 + u^{(\alpha)} + \frac{B^2}{8\pi}\delta_{\alpha i}\right\} v_k^{(\alpha)} \right.$$
$$\left. + P^{(\alpha)} v_k^{(\alpha)} - \sigma_{jk}^{(\alpha)} v_j^{(\alpha)} - \frac{B_j B_k}{4\pi} v_j^{(\alpha)} \delta_{\alpha i} + q^{(\alpha)} F_k^{(\alpha)}\right] = \sum_{\beta \neq \alpha} G^{(\alpha\beta)}. \tag{11.8}$$

In these equations, $S^{(n)} = -S^{(i)}$ is the net rate per unit volume of conversion of ions to neutrals (and vice-versa). $\mathcal{F}_j^{(ni)}$ is the component along axis j of the net rate of momentum transfer from the ions to neutrals, per unit volume. It includes the effects of recombination, elastic collisions, etc. $G^{(\alpha\beta)}$ is the net rate, per unit volume, of the variation of the energy u of one fluid due to the effects of the other fluid, for example the variation of kinetic energy arising from elastic collisions between the ions and neutrals.

11.2 Different Types of Shocks

Shocks occur in the interstellar medium as a result of a violent increase of pressure, which produces supersonic motions. The (adiabatic) sound velocity is

$$c_s = \left(\frac{\gamma kT}{\mu m_H}\right)^{1/2}, \tag{11.9}$$

where $\gamma = 5/3$ is the ratio between the specific heats at constant pressure and at constant volume[1]. μ is the molecular mass, about 1.5 for the atomic gas, 2.7 for the molecular gas and 0.7 for a fully ionized gas, taking helium and heavy elements into account. In the H I medium, $c_s \simeq 1.2$ km s^{-1} for $T = 100$ K and in H II regions $c_s \simeq 14$ km s^{-1} for $T = 10^4$ K. The observed velocities in the H I gas are generally highly supersonic. The velocities in H II regions are of the order of the sound velocity or somewhat higher. The gas in supernova remnants, bubbles and jets expelled by protostars or young stars reaches far higher velocities and is supersonic. In these conditions *discontinuities* form, i.e., surfaces where the velocity and the physical parameters of the gas exhibit jumps. The discontinuities through which there is a flow of matter are the *shock waves*, or more simply the *shocks*. The discontinuities without flow are called the *contact discontinuities*.

In order to describe a shock it is sufficient to chose to control surfaces, one on each side of the discontinuity, and to use the shock as the reference frame. The control surfaces are therefore stationary with respect to the shock. All the changes occur between these two surfaces. Of course, momentum and energy are conserved when crossing the shock. Let us indicate with index 1 the quantities before the shock and those after the shock with index 2, and define the "parallel" direction as the direction perpendicular to the shock front. For a single fluid, we can now obtain the limiting conditions for a stationary shock, i.e., a shock where all the physical parameters are time-independent in the shock reference frame, hence $\partial/\partial t = 0$.

- The equation which follows expresses the equality of the fluxes of matter through the two control surfaces. It is obvious, but we can also derive it from the continuity equation (11.1) with $\partial/\partial t = 0$, integrated along the parallel direction:

$$\rho_1 v_{\|,1} = \rho_2 v_{\|,2}. \tag{11.10}$$

- The next two equations express the conservation of magnetic flux ϕ_B and of electromotive force $-\partial \phi_B/\partial t + \mathbf{v} \times \mathbf{B}$ through the shock:

$$B_{\|,1} = B_{\|,2}, \tag{11.11}$$

---

[1] γ is equal to 5/3 in practice for an atomic gas *and* for a molecular gas because the energy contained in the rotation and vibration of the hydrogen molecule is generally much smaller than its translational energy: only a small fraction of molecules are in excited rotational levels. We may therefore consider for the purposes of thermodynamics that H_2, by far the most abundant molecule in the interstellar medium, has only three degrees of freedom, like an atom.

$$v_{\|,1} B_{\perp,1} - v_{\perp,1} B_{\|,1} = v_{\|,2} B_{\perp,2} - v_{\perp,2} B_{\|,2}. \tag{11.12}$$

- The following two equations express the conservation of the momentum flux. They are derived from (11.2) by integration over space, neglecting viscosity between the control surfaces:

$$\rho_1 v_{\|,1}^2 + P_1 + \frac{1}{8\pi} B_{\perp,1}^2 = \rho_2 v_{\|,2}^2 + P_2 + \frac{1}{8\pi} B_{\perp,2}^2, \tag{11.13}$$

$$\rho_1 v_{\|,1} v_{\perp,1} - \frac{1}{4\pi} B_{\|,1} B_{\perp,1} = \rho_2 v_{\|,2} v_{\perp,2} - \frac{1}{4\pi} B_{\|,2} B_{\perp,2}. \tag{11.14}$$

- The last equation expresses the conservation of the energy flux and is derived from (11.3) in the same way (note that there is no energy flux due to $v_\perp$):

$$v_{\|,1}\left(\tfrac{1}{2}\rho_1 v_1^2 + P_1 + u_1\right) + \tfrac{1}{4\pi}(B_{\perp,1}^2 v_{\|,1} - B_{\|,1} B_{\perp,1} v_{\perp,1}) + Q_{\|,1} + F_{\|,1}$$
$$= v_{\|,2}\left(\tfrac{1}{2}\rho_2 v_2^2 + P_2 + u_2\right) + \tfrac{1}{4\pi}(B_{\perp,2}^2 v_{\|,2} - B_{\|,2} B_{\perp,2} v_{\perp,2}) + Q_{\|,2} F_{\|,2}.$$
$$\tag{11.15}$$

11.2.1 Shocks with no Magnetic Field

The shocks for which there is a discontinuity (jump) in the properties of the gas are called the *J shocks*. This is the case for shocks with no magnetic field.

If the magnetic field can be neglected, and neglecting also the heat flux and the radiative losses (it is sufficient for this to locate the control surfaces very near the shock), the preceding equations can be simplified to:

$$\boxed{\mathcal{J} = \rho_1 v_1 = \rho_2 v_2,} \tag{11.16}$$

$$\boxed{\mathcal{P} = P_1 + \rho_1 v_1^2 = P_2 + \rho_2 v_2^2,} \tag{11.17}$$

$$\boxed{\mathcal{W} = w_1 + \frac{1}{2}v_1^2 = w_2 + \frac{1}{2}v_2^2.} \tag{11.18}$$

$w = (u + P)/\rho$ is the enthalpy. As a simplification we have written $v = v_\|$, because the transverse velocity $v_\perp$, if any, is conserved in an oblique shock with no magnetic field and can be neglected provided that the reference frame moves along the shock front with this velocity. Equation (11.18) comes from (11.15) with no magnetic field, after each term has been divided by $\rho_1 v_1 = \rho_2 v_2$. $\mathcal{J}$, $\mathcal{P}$ and $\mathcal{W}$ are, respectively, the flux of matter, momentum and energy per unit mass. For a perfect gas, we have

$$P = \frac{\rho k T}{\mu m_H}, \tag{11.19}$$

$$w = \frac{\gamma}{\gamma - 1}\frac{P}{\rho} = \frac{5}{2}\frac{P}{\rho}. \tag{11.20}$$

It is convenient to introduce the specific volume, i.e. the volume occupied by 1 gram of matter, $V = 1/\rho$. From (11.16) and (11.17), we then obtain the general formula

$$v_1 - v_2 = [(P_2 - P_1)(V_1 - V_2)]^{1/2}, \tag{11.21}$$

which governs the change in velocity, independently of any energy consideration.

When dealing with a shock problem, the parameters ρ_1, T_1 and P_1 of the gas before the shock are usually given, as well as the pressure P_2 after the shock or the shock velocity v_1. From (11.16), (11.18) and (11.21) we then obtain, after some simple algebra,

$$\mathscr{J}^2 = \frac{P_2 - P_1}{V_1 - V_2} = \frac{2\gamma}{\gamma - 1} \frac{P_2 V_2 - P_1 V_1}{V_1^2 - V_2^2}, \tag{11.22}$$

hence, taking the same value for $\gamma (= 5/3)$ on either side of the shock

$$\frac{\rho_2}{\rho_1} = \frac{V_1}{V_2} = \frac{(\gamma - 1)P_1 + (\gamma + 1)P_2}{(\gamma + 1)P_1 + (\gamma - 1)P_2} = \frac{P_1 + 4P_2}{4P_1 + P_2}. \tag{11.23}$$

The temperature jump is given by

$$\frac{T_2}{T_1} = \frac{P_2 V_2 \mu_2}{P_1 V_1 \mu_1} = \frac{P_2}{P_1} \frac{4P_1 + P_2}{P_1 + 4P_2} \frac{\mu_2}{\mu_1}, \tag{11.24}$$

and the gas velocities on either side of the shock are given by

$$v_1^2 = \frac{1}{3\rho_1}(P_1 + 4P_2), \quad v_2^2 = \frac{1}{3\rho_1} \frac{(4P_1 + P_2)^2}{P_1 + 4P_2}. \tag{11.25}$$

These are the *Rankine–Hugoniot relations*.

For a strong shock ($P_2/P_1 \gg (\gamma + 1)/(\gamma - 1) = 4$) we have

$$\boxed{\frac{\rho_2}{\rho_1} \to \frac{\gamma + 1}{\gamma - 1} = 4,} \tag{11.26}$$

$$\boxed{v_1 \to \left(\frac{4}{3}\frac{P_2}{\rho_1}\right)^{1/2},} \tag{11.27}$$

$$\boxed{\frac{v_2}{v_1} \to 1/4,} \tag{11.28}$$

$$\boxed{v_2 \to \left(\frac{kT_2}{3\mu_2 m_H}\right)^{1/2}.} \tag{11.29}$$

The temperature jump is given by

$$\boxed{\frac{T_2}{T_1} \to \frac{\gamma - 1}{\gamma + 1}\frac{\mu_2 P_2}{\mu_1 P_1} = \frac{\mu_2 P_2}{4\mu_1 P_1}.} \tag{11.30}$$

11.2 Different Types of Shocks

The conditions for a strong shock are very often met in the interstellar medium, for example in supernova remnants. The increase in temperature after the shock can produce different effects. It can cause collisional ionization of the gas, changing its mean molecular mass μ and absorbing energy behind the shock, which requires a change in some of the above equations. A simple analysis of this phenomenon can be found in Kaplan [278] p. 44–45. Also, the temperature increase is often such that the shocked medium emits radiation due to the collisional excitation of atoms, ions and molecules, to ion recombination, etc. The corresponding radiation losses progressively cool the gas behind the shock. On the other hand, since the radiation from this gas propagates with the velocity of light, it reaches the region not yet affected by the shock, forming a *radiative precursor*[2]. The structure of a strong shock is shown schematically by Fig. 11.1, which illustrates the phenomena we have just described.

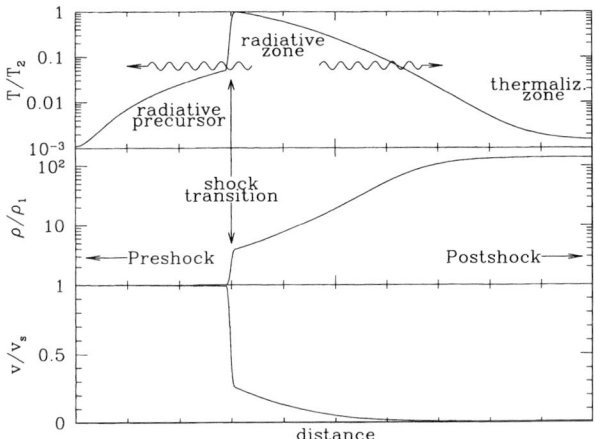

Fig. 11.1. Schematic structure of a single-fluid strong shock. The unshocked medium is to the left. The different panels show, respectively from bottom to top, the velocity v relative to that of the shock front v_s, the density ρ normalized to the pre-shock value ρ_1, and the temperature T normalized to the value T_2 immediately after the shock. We see that the velocity is one quarter and the density multiplied by a factor of 4 after crossing the shock. The radiative zone behind the shock sends radiation into the medium, creating a radiative precursor in the pre-shocked region that ionizes the gas, photodissociates it if it is molecular, and raises its temperature. On the other hand this radiation progressively cools the shocked medium until it is thermalized, thus affecting its velocity and density. From Draine & McKee [137], with permission from Annual Reviews, http://www.AnnualReviews.org.

---

[2] There is sometimes in the literature a confusion between radiative shocks, in which the radiation comes from the shocked medium, and shocks with radiative precursors, in which it is emitted by the upstream medium heated by for example some unobservable UV or X-ray emission from behind the shock. We will be careful in indicating the origin of the radiation when appropriate.

The numerical calculation of the structure of a radiative shock is complex, requiring iterations for a self-consistent solution. Some simple relations can however be obtained by taking two control surfaces, one on either side of the region where most of the radiation is emitted and absorbed. Now the control surfaces are far from the shock front. Equations (11.16) and (11.17), which give $\mathcal{J}$ and $\mathcal{P}$, are still valid for these surfaces, but not (11.18) which involves energy. If the degree of ionization and of dissociation of the molecules have not appreciably changed during the passage of the shock, the medium returns to the initial state and $T_1 = T_2$ for the two surfaces. Such a shock is said to be *isothermal*. If not, $T_1 \neq T_2$. If these temperatures are known from observation, as well as the densities, we can calculate the velocity v_1 of the radiative shock relative to the gas initially at rest. We have, for a perfect gas,

$$v_1^2 = \frac{\rho_2}{\rho_1} \frac{(k/m_H)(\rho_2 T_2/\mu_2 - \rho_1 T_1/\mu_1)}{\rho_2 - \rho_1}, \tag{11.31}$$

while the velocity of the gas behind the front (more precisely behind the radiative zone) is, relative to the unshocked gas at rest

$$v = v_1 - v_2 = \left[\frac{k}{m_H}\left(\frac{\rho_2 T_2}{\mu_2} - \frac{\rho_1 T_1}{\mu_1}\right)\frac{\rho_2 - \rho_1}{\rho_2 \rho_1}\right]^{1/2}. \tag{11.32}$$

For a radiative shock with a *Mach number* $M = v/c_{s,2} = v/\sqrt{\gamma k T_2/\mu_2 m_H} \gg 1$, we have simply

$$\boxed{\frac{\rho_2}{\rho_1} \simeq \frac{\mu_2 m_H v^2}{k T_2}.} \tag{11.33}$$

This equation shows that the gas can be compressed to very high densities, thanks to the loss of the compression energy by radiation. This might happen in the neutral interstellar gas. Taking for example $T_2 \approx 1\,000$ K, with $\mu_2 = 1.5$, we find $\rho_2/\rho_1 = (v/2.4 \text{ km s}^{-1})^2$; if T_2 is smaller, ρ_2 is still larger. We saw that the 1-dimension velocity dispersion between structures in the H I medium is of the order of 9 km s^{-1} (Sect. 4.1). Collisions between these structures must often take place with relative velocities of this order, producing strong radiative shocks which create localized regions with high densities. However, these shocks are not easily observable in spite of their radiation. On the other hand, it is not correct to neglect the magnetic field in the H I medium. The effects of the magnetic field in a weakly ionized medium like the H I gas will be examined later in Sect. 11.2.

An interesting point is that the structure of a radiative shock, with or without a radiative precursor, differs from that of a non-radiative shock (but do non-radiative shocks exist in the interstellar medium?). The latter shock is only a discontinuity surface whose thickness is of the order of the mean free path of the particles between consecutive collisions

$$\lambda = 1/(n\pi a_0^2) \simeq 10^{16} n^{-1} \text{ cm}, \tag{11.34}$$

for hydrogen, where $a_0 \simeq 0.5$ Å is the radius of the first Bohr orbit. In practice, it is impossible to measure such a small thickness in the interstellar medium. In a radiative shock, this thin layer also exists and it is there that the temperature of the neutrals and ions is strongly increased, although the electron temperature does not have time to be much enhanced[3].

11.2.2 Shocks in a Magnetized Medium

The presence of a magnetic field considerably complicates the properties of shocks. It is not only the magnetic pressure which adds to the thermal pressure of the gas, but we must take into account the variation of magnetic energy when crossing the shock. Even more important, the tangential velocity changes when crossing the shock if the magnetic field is not parallel to the flow. This can be seen from (11.12) to (11.14). The angle between the normal to the shock and the magnetic field plays a prominent role. Shocks with a transverse magnetic field are very peculiar (we will not discuss them in this book). Another complication arises from the existence of three different types of waves in a magnetic fluid, each with a different velocity. *Alfvén waves*, which correspond to transverse vibrations of the magnetic tubes, similar to those of vibrating strings, propagate along the magnetic field with a velocity

$$v_A = B/\sqrt{4\pi\rho}. \tag{11.35}$$

In a direction making an angle θ with the direction of the magnetic field there are three types of waves called, respectively, *slow*, *intermediate* and *fast magnetosonic waves*. Their respective phase velocities $u_S < u_I < u_F$ are given by

$$u_{F,S} = (1/\sqrt{2})[(v_A^2 + c_s^2) \pm (v_A^4 + c_s^4 - 2v_A^2 c_s^2 \cos 2\theta)^{1/2}]^{1/2}, \tag{11.36}$$

$$u_I = v_A \cos\theta, \tag{11.37}$$

(see, for example, courses in plasma physics, or Priest [417]).

We will assume for the moment that the medium is strongly ionized. What takes place in a weakly ionized medium will be described in the next section. The dynamics of a MHD shock is determined by the comparison between the flow velocity and the velocity of the plasma waves. For a shock to exist, the flow velocity must be faster than the velocity of one of these waves before the shock, and slower than this velocity after the shock. Therefore, shocks can be classed according to the value of the velocity of the flow through the shock compared to that of the various wave velocities (in passing, it now becomes clear why the transverse shocks, with $\theta = 90°$,

---

[3] The observed post-shock electron temperature in some supernova remnants is too high to result from thermalization with the ions heated in the shock; in such cases, there is probably a direct, non-collisional heating of electrons similar to that which occurs in collisionless shocks in solar-system plasmas: cf. Sect. 12.4, and Draine & McKee [137] Sect. 2.3. The thermalization of electrons takes place in a layer 10 to 15 times thicker than the shock. Finally we find the layer where radiation is emitted, with an even larger thickness. The book by Kaplan [278] p. 50–58, gives a simple discussion of the properties of this layer. They are dependent on the state of ionization of the medium before the shock.

are so peculiar). For fast shocks ($v_\| > u_{F,1}$) the magnetic field increases after the shock, while it decreases for slow shocks ($u_{I,1} > v_\| > v_{S,1}$). For intermediate shocks, the field changes direction and can either increase or decrease in strength. We cannot here describe the complexities of these shocks; the details can be found in Kaplan [278] and in Draine & McKee [137]. Our understanding of these shocks is in fact still incomplete. We will only remark that the properties of *strong* radiative shocks are not essentially changed by a magnetic field, so that (11.26) to (11.30) are still approximately valid for strong shocks.

11.2.3 Multi-Fluid Shocks in a Weakly Ionized Gas

In the "neutral" interstellar medium, even in molecular clouds, there always exists some weak ionization. Even if it is not abundant, the ionized component plays an important role in shocks if there is a magnetic field because it is the only one which is coupled with this field. This affects the dynamics of the shock. On the other hand, collisions between the neutral and the ionized components are rare if the ion density is low, and these two components can be considered as dynamically distinct fluids. If the cyclotron angular frequency (the gyrofrequency) of the ions in the magnetic field, $\omega_i = eB/(m_i c)$, is much higher than the collision frequency of the ions with the neutrals $n_n \langle \sigma v \rangle$, σ being the cross-section for the elastic scattering of the ions with the neutrals and v the velocity of the ions, we may consider that the magnetic field is frozen into the charged fluid. The three types of plasma waves we mentioned (cf. (11.36) and (11.37)) can propagate in the ion–electron plasma. In a molecular cloud the phase velocity of these waves is very high. The magnetic field being approximately $B \approx [n_n/(1~\text{cm}^{-3})]^{1/2}~\mu\text{G}$ (see Fig. 2.6), the Alfvén velocity in the plasma is $v_A^{(i)} = B/[4\pi\rho^{(i)}]^{1/2} \approx 100(x/10^{-4})^{-1/2}$ km s^{-1}; $x = n_i/n_H$ is the degree of ionization. The sound velocity c_s is about 1 km s^{-1} (there is no need here to distinguish between the neutral and the ionized media). Therefore, the ionized medium can propagate the fast and intermediate plasma waves which have almost the same velocity $v_A^{(i)} \cos\theta$ (cf. (11.36) and (11.37)). A discontinuity can affect the neutral medium if this velocity is supersonic, while there is not necessarily a discontinuity for the plasma provided it is sub-alfvénic, i.e., that the perturbation is not very strong[4]. There is necessarily a velocity difference between the neutral flow and the ionized flow close to the discontinuity. This systematic velocity difference between the ions and neutrals is called *ambipolar diffusion*. It will be studied Sect. 14.1. Because information on the physical parameters and in particular on the presence of a shock in the neutral medium is transmitted rapidly through the ionized component, at the velocity of the plasma waves, there is a *magnetic precursor* to the shock (Draine [133]), which affects the pre-shocked medium and begins to dynamically separate the ionized medium from the neutral one.

The shock structure can exhibit one of the three configurations displayed in Fig. 11.2. In this figure, is is assumed that the velocity of the neutrals is supersonic

---

[4] If the perturbation is very strong we are back to the case of the previous sections: there is a discontinuity for the whole medium but there is a separation in velocity of the ions and the neutrals in a relatively thin layer behind the shock.

before the shock while the velocity of the ions is sub-alfvénic. If the neutral fluid stays cold, because the shock is weak or because the ion-neutral collisions efficiently transfer energy, the velocity of the neutral fluid can everywhere stay supersonic, and there is no discontinuity for the neutral and ionized fluids, although their velocities are different. These shocks without a discontinuity are called *C shocks* (C for "continuous"). If the ion-neutral collisions are numerous enough, due to the velocity difference between the two fluids, and sufficiently raise the temperature of the gas and hence the sound velocity, then the flow can become subsonic. There are now two possibilities: either the supersonic→subsonic transition occurs at a discontinuity, and there is a J shock, or this transition is smooth, which might be possible under some conditions, and the shock is said to be a *C* shock*. In Sect. 11.3 we note the possible non-existence of such C* shocks.

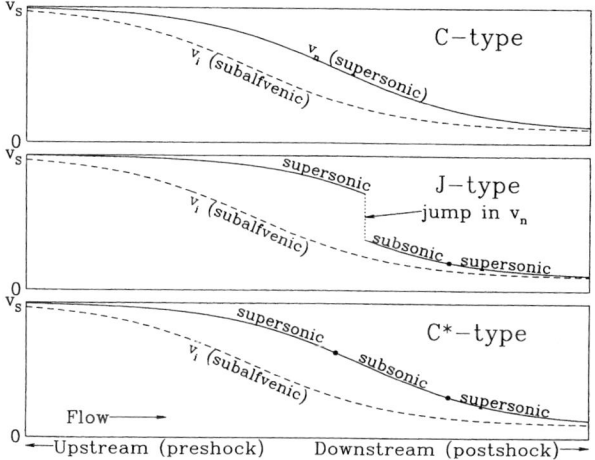

Fig. 11.2. Schematic structure of two-fluid MHD shocks. The pre-shock medium is to the left. The velocity of the neutrals (full line) and that of the ions (dashed line), relative to the velocity of the shock front, are plotted as a function of position. See text for an explanation of the three types of shocks. From Draine & McKee [137], with permission from Annual Reviews, http://www.AnnualReviews.org.

Note that at some distance behind the shock the velocity difference between the two fluids vanishes, due to the ion-neutral coupling which corresponds to a viscosity[5]. Viscous dissipation heats the fluid, but this heating stays small because the ions are rare, so that radiative cooling keeps the temperature low. We can then say that the shock "transition" and the radiative zone coexist. Since the kinetic temperature stays low, molecules can be accelerated to rather high velocities without dissociation. The radiation from such shocks comes from low-energy transitions: atomic and ionic

---

[5] The relationships between the gas conditions far upshock and far downshock are thus the same as for a single fluid, except of course if there are chemical changes in the gas.

fine-structure lines and the rotational or vibrational lines of molecules, in particular the rotational lines of H$_2$. These lines are characteristic of two-fluid shocks, and important progress in our understanding of these shocks presently results from their observation, in particular with the ISO satellite.

Ion-neutral coupling can arise in three different ways:

- via elastic scattering, the force per unit volume which affects species α under the influence of species β being

$$F^{\alpha\beta} = \frac{\rho^{(\alpha)}\rho^{(\beta)}}{m_\alpha + m_\beta} \langle \sigma v \rangle_{\alpha\beta} \Delta v_{\alpha\beta}. \quad (11.38)$$

$\Delta v_{\alpha\beta}$ is the relative bulk velocity between the flows of the two kinds of particules, whose respective masses are m_α and m_β, and $\langle \sigma v \rangle_{\alpha\beta} \sim 1.9 \times 10^{-9}$ cm^3 s^{-1} for $\Delta v_{\alpha\beta} \leq 15$ km s^{-1};
- through charge exchange between H$^+$ and H, an important phenomenon if H$^+$ makes up a significant fraction of the ions; the corresponding force is given by (11.38) but now with $\langle \sigma v \rangle_{\alpha\beta} = 1.6 \times 10^{-8} (T/10^4$ K$)^{0.4}$ cm^3 s^{-1};
- through elastic scattering by charged grains.

Both the ions and the neutrals are heated through these processes, as well as by the chemical reactions that take place in the shock (see Sect. 11.4). We must also take into account the energy exchanges between the ions and electrons. Exchanges between neutrals and electrons are generally negligible due to the small mass of the electron (cf. Sect. 8.1). Although electrons and ions have the same bulk velocity due to electrical neutrality, their temperatures can be very different. Although, in principle, we only have to consider two fluids as far as the dynamics is concerned, we have nevertheless to take three fluids into account (neutrals, ions and electrons) when treating chemical and energy processes, in particular recombination. These processes in turn affect the gas properties and thus the dynamics. The formulation of the heating rates and of the radiative cooling rate is somewhat complex, although of no great difficulty in principle; see the paper of Flower et al. [185] for a treatment of this. For the role of grains in the energetics, see Draine & McKee [137].

11.3 Non-Stationary Shocks

Up to now, we assumed that the shock was stationary, i.e., that its properties are independent of time because a state of dynamical equilibrium has been reached. It is legitimate, however, to have some doubts about this hypothesis, in particular due to the slowness of the ion-neutral coupling. Abandoning stationarity causes complications in the numerical treatment of the problem, which is probably the reason why non-stationarity has only been assumed sporadically in the past. It requires in particular solving simultaneously the equations of dynamics and those of chemistry, which are coupled. This has been performed in a few cases by Chièze et al. [98]. We will summarize very briefly their results.

11.3 Non-Stationary Shocks

For a J shock, the slowest process is the cooling of the gas after crossing the shock and it is this process which defines the condition for stationarity. Chièze et al. [98] find that for a shock propagating at 10 km s^{-1} in a medium with density 10^3 cm^{-3}, without a magnetic field, equilibrium is reached fast enough that the shock is stationary 2 000 years after the beginning of the perturbation. It is thus likely that the J shocks of supernova remnants can be considered as stationary.

This is no longer true if there is a magnetic field strong enough to produce a C shock. For the same conditions as above, but now with a magnetic field of 10 µG, the stationary state is reached only after 10^5 years (Fig. 11.3). In Fig. 11.3, the shock was supposed to have been created by a piston which penetrates the medium at 10 km s^{-1}. This figure shows the temperature variation of the neutrals and the ions in front of this piston. The initial shock structure is interesting. A temperature and velocity discontinuity arises in the neutral component, similar to that for a J shock. On the other hand, the ionized component is heated and accelerated by the magnetic precursor in a large upstream zone without exhibiting a discontinuity. The ion-neutral coupling occurs slowly and progressively quenchs the neutral discontinuity. We thus see a range of structures intermediate between a J shock and a C shock.

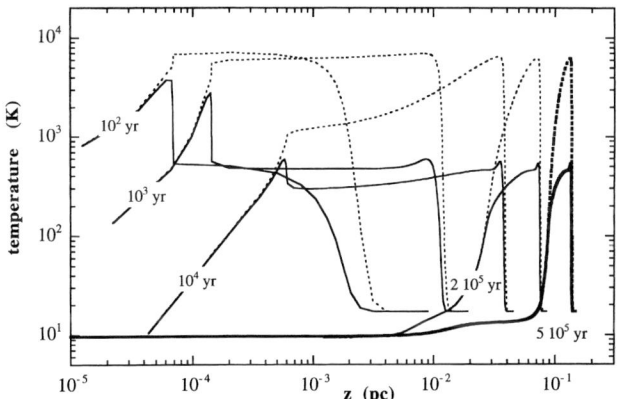

Fig. 11.3. The time evolution of a C shock propagating in a medium with density 10^3 cm^{-3}, in the presence of a transverse magnetic field of 10 µG. The abscissae shows the distance to a piston located to the left, which is supposed to push the medium at 10 km s^{-1}. The plot is in the reference frame of this piston and the shock velocity is slightly larger than 10 km s^{-1}. The temperature of the neutrals (full line) and of the ions (dashed line) is shown at different epochs after the motion of the piston started. Note the discontinuity in the temperature of the neutral fluid (there is also a discontinuity in its velocity, not plotted) which disappears with increasing time. A stationary structure is reached after 5×10^5 years (bold lines). From Chièze et al. [98], with the permission of Blackwell Science Ltd.

Similar calculations for conditions in the diffuse interstellar medium ($n_H = 25$ cm^{-3}, $v_S = 10$ km s^{-1}, $B = 5$µG) show that the stationary state is reached more rapidly thanks to efficient ion-neutral coupling, which is due to the higher degree of ionization. However, a discontinuity is still present in the neutral flow, which exhibits some J character. These calculations do not appear to confirm the existence of C* shocks for which the supersonic→subsonic transition in the neutral fluid would not show a discontinuity.

Shock non-stationarity can have major consequences for the chemistry and excitation of molecules such as H_2 (Flower & Pineau des Forêts [187]). This could explain the differences between the observations and the predictions from stationary shock models, differences that will be detailed in the next section. Unfortunately, we lack space to discuss this topic here, which is in any case still in infancy.

11.4 Physico-Chemistry in Shocks

Shocks in the neutral interstellar medium can have important consequences for the chemistry. The temperature rise after a shock permits many reactions that are impossible at low temperatures. If the shock is very fast (a J shock, see Fig. 11.1), H_2 is dissociated by collisions. This requires shock velocities of at least 45 km s^{-1}. However, H_2 reforms after the shock when the gas has cooled sufficiently, either by electronic attachment (cf. Sect. 9.2, note 1), or on the surface of dust grains. The article of Wilgenbus et al. [548] contains a detailed study of the chemistry and the emission of H_2 in C and J shocks propagating in a molecular cloud.

Shocks are often invoked to account for the abundance of CH^+ in the diffuse medium and in translucent clouds. We saw in Sect. 9.4 that it is not possible to explain the formation of CH^+ by low-temperature reactions. The endothermic reaction $C^+ + H_2 \rightarrow CH^+ + H$ ($\Delta E = -0,4$ eV) is possible in shocks. The idea that it is this reaction which produces CH^+ is supported by the observed association of this molecule with the warm component of H_2 seen in absorption in the far-UV (cf. Fig. 4.9). However, the velocity of the CH^+ absorption lines does not appear to reflect the velocity difference between the ions and neutrals that is expected to arise in a shock, so that the CH^+ abundance problem in the diffuse medium cannot yet be considered solved, in spite of the many theoretical and observational efforts. Conversely, it is clear that the above reaction is important in molecular clouds.

Other molecules can form in shocks by endothermic reactions, or reactions with activation barriers, for example

$O + H_2 \rightarrow OH + H$ ($\Delta E = -0.08$ eV, 0.25 eV barrier), followed by

$OH + H_2 \rightarrow H_2O + H$ (exothermic but with a 0.13 eV barrier). HCO^+ is formed from these molecules (cf. Sect. 9.4).

We are here again in the same situation as for CH^+. These reactions are certainly of importance behind the shocks which arise in molecular clouds and in molecular jets accelerated by protostars or young stellar objects where we find lines of OH and H_2O lines with intensities in good agreement with two-fluid shock models.

In any case, the formation of these molecules and the energy they dissipate in their rotational and vibrational lines have a considerable influence upon the shock thermodynamics. Any realistic shock model must take them into account.

Such models have been developed, for example by Flower et al. ([185] and other papers in this series), and by Hollenbach & McKee [243], the latter being a very detailed paper. Such models usually calculate the spectrum emitted by the shock. Figure 11.4 and 11.5 illustrate respectively the structure of the shocked medium and its chemical composition after the passage of a J shock through a molecular cloud. A recurrent problem, however, is that these models are not always in agreement with observations, as discussed for example by Draine & McKee [137] Sect. 7.3.

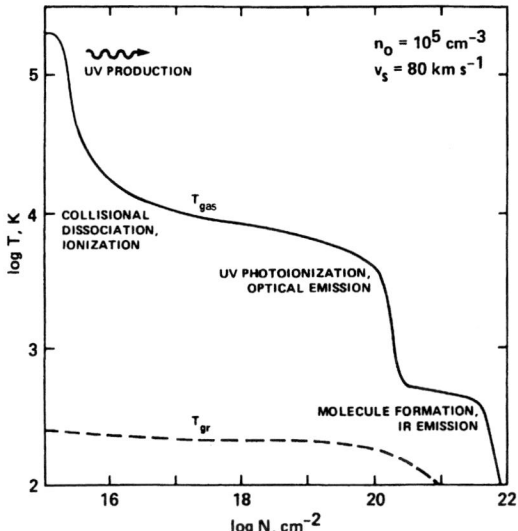

Fig. 11.4. The shock structure and the temperature of the gas and dust grains in a fast J shock propagating at 80 km s^{-1} in a molecular cloud with density 10^5 cm^{-3}. The abscissae gives the distance from the shock front, expressed as the column density of hydrogen nuclei $N(H + 2H_2)$, in logarithmic units. The gas temperature (full line) and the dust temperature (dashed line) are given in the ordinates. There are three different regions behind the shock: a narrow, hot region at about 10^5 K, where the gas is collisionally dissociated and ionized; a region where hydrogen recombines, at a temperature of about 10^4 K, where the optical lines are emitted; a region where molecules reform, at about 200 K, where the fine-structure lines in the mid- and far-IR are emitted. The energy liberated by the formation of H_2 maintains this temperature over an appreciable thickness in the medium, in spite of the radiative losses. Models show that the column density of the two first regions is not too sensitive to the initial density. At the chosen density the grains are only weakly coupled thermally to the gas (cf. Sect. 8.1), so that the dust temperature $T_d \ll T_{gas}$. Reproduced from Hollenbach & McKee [243], with the permission of the AAS.

Another effect is the partial destruction of grains in two-fluid MHD shocks if their velocity is larger than about 20 km s^{-1}. If the grains are not charged, they are driven by the neutral fluid and collide with the ions of the ionized fluid; He^+ is the most effective ion, due to its larger mass than that of H^+ (remember that the mean

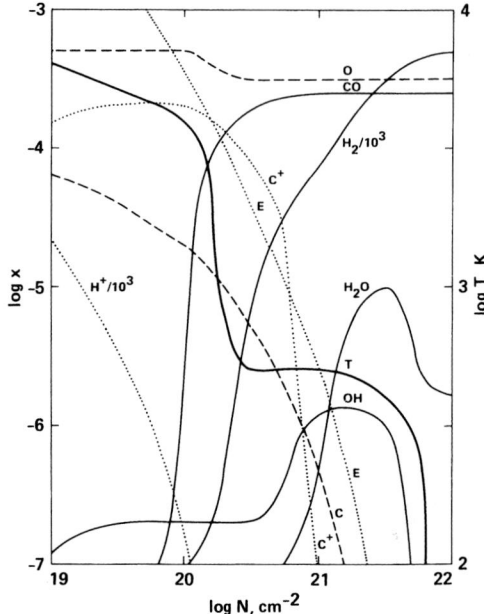

Fig. 11.5. The temperature and chemistry behind a fast J shock propagating at 80 km s^{-1} in a molecular cloud with density 10^5 cm^{-3} (same conditions as for Fig. 11.4). The abscissae gives the distance from the shock front, expressed as the column density of hydrogen nuclei $N(\mathrm{H} + 2\mathrm{H}_2)$, in logarithmic units. The gas temperature is given by the full line labelled "T" (scale to the right). The fractional abundances x (by number) of different atoms, ions and molecules are given with respect to total hydrogen; those of H and H_2 are plotted as $x/10^3$. The curve labelled "E" corresponds to the free electrons. Reproduced from Hollenbach & McKee [243], with the permission of the AAS.

velocity is the same for all ions in a two-fluid shock). If the grains are charged, they have a systematic velocity difference with the neutral atoms and the molecules which strike them. The impact of gas particles erodes the grains, starting with their ice mantle, if any, in the deep regions of molecular clouds (Sect. 7.4). The volatile molecules forming this mantle are then released to the gas, possibly in excited states. Flower et al. [186] show that this mechanism can account for the population of the excited metastable levels of the released NH_3 (Sect. 4.2 and Fig. 4.13). It also accounts for the presence of relatively abundant deuterated molecules in regions such as the Orion hot core, where there are strong shocks (see Sect. 9.2). If the kinetic energy of the gas particles which strike the grains is high enough to overcome the binding energy of silicates, necessitating a velocity greater than 40 km s^{-1}, the silicate grains can be eroded, releasing SiO molecules (or Si atoms which will form SiO in reactions with oxygen). SiO is indeed an abundant molecule in shocks. Carbon can also be released by the sputtering of carbonaceous grains in shocks with similar velocities, but the results are difficult to observe. The erosion of silicate grains has been studied in detail by May et al. [354], whose model is in good agreement with observations of SiO in protostellar molecular flows. Similarly, shocks appear to be able to release silicon atoms in the diffuse interstellar medium, which can then be observed through their UV absorption lines (cf. e.g. Gry et al. [211]). It is very likely that it is, to a large extent, the effect of shocks that causes the observed differences between the chemical composition of the gas in the warm diffuse medium and that in the cold neutral medium, the former exhibiting lesser depletions of the heavy elements (cf. Sect. 4.1). We will come back to these points in Sect. 15.4.

11.5 Radiation and the Diagnosis of Shocks

It is not really possible to directly observe shocks. However, the radiation of the medium heated by a shock is observable and can serve as a diagnostic for the existence and properties of a shock. For J shocks the radiation is easily detectable when the shock velocity is greater than about 50 km s^{-1}. For the strong, very fast J shocks in young supernova remnants (several thousands of km s^{-1}), the temperature as given by (11.30) can be greater than 10^6 K, X-rays are then emitted. At the lower temperatures met in later phases of the evolution of supernova remnants, the emission shifts to the far-UV and then to the visible range. A complication is that far-UV or soft X-ray radiation is rapidly absorbed by the gas and dust of the medium before and after the shock and is therefore not observable: this radiation then only heats the medium, producing the radiative precursor that was mentioned earlier. The emission from J shocks of moderate velocities (say 30–150 km s^{-1}) has been studied by several authors (references in Draine & McKee [137]). At velocities greater than ~ 100 km s^{-1} the shock dissociates the molecules and ionizes the atoms. The production of an unobservable UV radiation results, which creates the radiative precursor and heats the shocked gas. At lower velocities, only dissociation is of importance and there is less UV radiation. The calculation of the physical parameters of the shocked gas and of the line emission is somewhat involved. We reproduce in Figs. 11.6 and 11.7 some results from the very complete study of Hollenbach & McKee [243] concerning line emission in the mid- and far-IR.

In the visible range J shocks with velocities 50–300 km s^{-1} emit recombination lines and forbidden lines. Some of these lines can be used to distinguish shocks from H II regions. The line ratios [S II]$\lambda 6\,717,6\,731$/Hα and [O I]$\lambda 6\,300$/Hα are much larger in shocks than in H II regions, because sulphur and oxygen in H II regions are essentially in the form of [S III] and of [O II] or [O III], respectively. Conversely, in shocks we observe a large range of ionization states, for example the lines of [O I], [O II] and [O III] are observed from the same gas. The line ratio [O III]$\lambda 4\,363$/[O III]$\lambda 4\,949,5\,007$, which is an indicator of high electron temperatures, is larger in shocks than in H II regions, because the region that emits these lines in shocks is at a temperature of some 2×10^4 K, twice the typical temperature for H II regions.

The infrared lines can also supply interesting diagnostics. For example, the line ratio [Fe II]$\lambda 1.64$ µm/Br$\gamma\lambda 2.17$ µm is larger in shocks than in H II regions. The reason is that in H II regions iron is mainly in ionization states higher than Fe II.

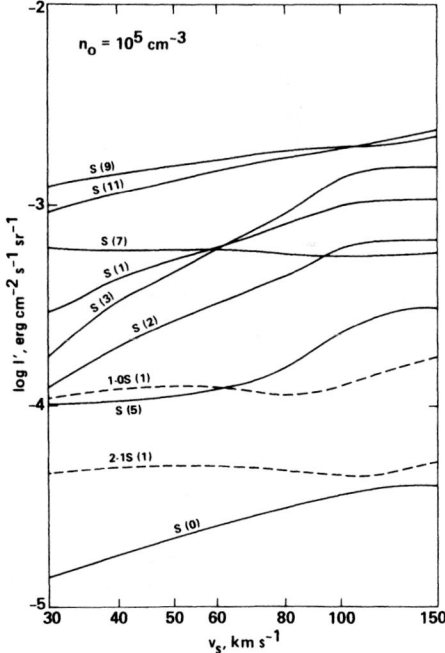

Fig. 11.6. The integrated intensities, perpendicular to a fast J shock, for the rotational (full lines) and ro-vibrational (dashed lines) lines of H_2. The shock propagates in a gas with initial density 10^5 cm^{-3} (same condition as for Figs. 11.4 and 11.5), the velocity is the abscissa. For the identification and wavelengths of the rotational and vibrational lines of H_2 in its ground electronic state, see Sect. 4.2 and Table 4.5. Reproduced from Hollenbach & McKee [243], with the permission of the AAS.

Also, iron is released from dust grains by sputtering in shocks. The [O I]λ63 μm line is very intense in radiative shocks and the [O I]λ63 μm/[C II]λ158 μm line ratio is much larger in shocks than in photodissociation regions, giving an excellent discriminator between shocks and PDRs.

Low-velocity shocks and C shocks do not emit visible lines because the gas does not reach high enough temperatures, however they do emit [O I]λ63 μm, [C II]λ158 μm and the rotational et ro-vibrational lines of H_2 (Table 4.5): see Wilgenbus et al. [548] for the H_2 line emission. However, the forbidden atomic or ionic lines of elements less abundant than O and C are only detectable if the density is relatively high, in which case the ionization degree is generally small and the multi-fluid effects become important. If the velocity is low, there is never, in practice, a J shock in the interstellar medium and only a C shock can result. The rotational lines of H_2 are then intense and characterize these C shocks if there is at the same time emission in the [O I] and [C II] far-IR lines but no emission of other fine-structure lines. Conversely, many fine-structure and other lines are emitted by J shocks from the far-IR to the visible.

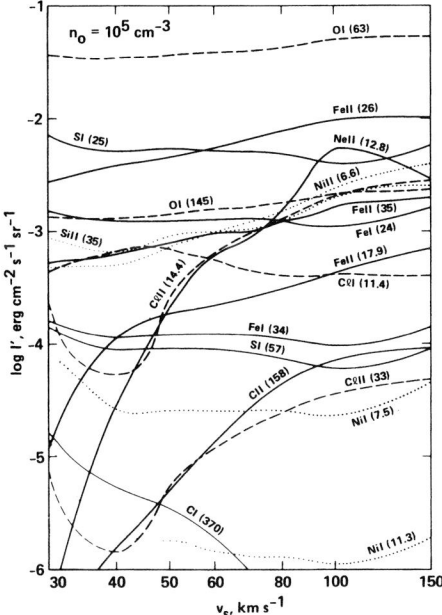

Fig. 11.7. The integrated intensities, perpendicular to a fast J shock, for the most important fine-structure lines. The shock propagates in a gas with initial density 10^5 cm^{-3} (same condition as for Figs. 11.4 to 11.6), the velocity is the abscissa. The intensities for the lines of S, Fe, Cl and Ni are lower limits because the network of chemical reactions used for these species is incomplete. Reproduced from Hollenbach & McKee [243], with the permission of the AAS.

As expected, the lines emitted by shocks exhibit "anomalous" velocities and are often broad. This helps in distinguishing them from the lines originating in photodissociation regions, which all have the same (low) velocity and have widths of only a few km s^{-1}.

The infrared continuum emitted by the dust heated in strong J shocks can be intense and has been observed in supernova remnants (cf. Sect. 8.1; Lagage et al. [296]; Arendt et al. [11]).

11.6 Instabilities in Shocks

There are various types of instabilities in shocks. A thermal instability can arise in strong shocks, producing temperature inhomogeneities which complicate the interpretation of the observed emission lines (cf. Draine & McKee [137], Sect. 5.1). Various "mechanical" instabilites can also arise (same reference, Sect. 5.2–5.4, and Chandrasekhar [89]).

Here we will only discuss the *Rayleigh–Taylor instability*. This instability arises in any dense fluid supported against gravity by a lighter fluid underneath, or pushed by such a fluid. It is obvious that, for example, a layer of water cannot be supported by a layer of oil which is lighter: an instability arises and water globules fall through the oil. The situation is often similar behind a shock. The shock is generally created by a "piston" which pushes the fluid at a supersonic velocity. This is the case for a supernova remnant, in which gas is ejected by the explosion of a massive star, forming an expanding and approximately spherical envelope which pushes the ambient interstellar medium at supersonic velocity. From outside to inside we find in succession the ambient gas, the shock, and the dense shocked gas separated by a contact discontinuity from the ejected piston gas which is less dense. It is at this discontinuity that instability arises. We will present a simplified analysis following Spitzer [490].

Let us assume that initially the contact discontinuity is a plane and take it as the origin for z perpendicular to this surface (which will be later deformed by the instability). The inertia of the shocked medium can be expressed in this reference frame as an acceleration g along z, similar to gravity, that we can describe by a potential $\Phi = gz$. We will assume, for simplicity, that both fluids above and below the surface are incompressible. Near the interface they have, respectively, a velocity and a pressure equal to $\mathbf{v}_a$ and P_a above and to $\mathbf{v}_b$ and P_b below. We will also assume that the densities ρ_a and ρ_b on each side are independent of time t and of z. The gas on each side of the discontinuity obeys the continuity equations for mass (11.1) and for momentum (11.2). In the present case, ρ is uniform and constant, the equations can be written as

$$\nabla \mathbf{v} = 0 \qquad (11.39)$$

$$\rho \frac{\partial \mathbf{v}}{\partial t} + \rho \nabla \mathbf{v} = -\nabla P - \rho \nabla \Phi, \qquad (11.40)$$

where the new term $\rho \nabla \Phi$ corresponds to the acceleration due to the inertia of the shocked medium.

Let us also assume for simplicity that the velocity derives from a potential Ψ_v (this introduces no restriction)

$$\mathbf{v} = -\nabla \Psi_v. \qquad (11.41)$$

Equation (11.39) then implies

$$\nabla^2 \Psi_v = 0. \qquad (11.42)$$

A linearized perturbation solution for this equation, satisfying the condition that Ψ_v is zero at large $|z|$, is

$$\begin{aligned}\Psi_v &= K_a\, e^{-i\omega t - kz} \sin kz \qquad \text{above,} \\ &= K_b\, e^{-i\omega t + kz} \sin kz \qquad \text{below,}\end{aligned} \qquad (11.43)$$

where K_a, K_b, ω and k are, for the moment, arbitrary quantities.

On the other hand, equation 11.40 yields, after integrating over z, with $\nabla \mathbf{v} = 0$

$$\rho \frac{\partial \Psi_v}{\partial t} = P + \rho g z, \qquad (11.44)$$

11.6 Instabilities in Shocks

a relation valid both above and below the interface. The integration constant is zero because there must be hydrostatic equilibrium at $z = \pm\infty$ where $\partial \Psi_v / \partial t = 0$.

The interface is now perturbed and moves to z_i. The value of z_i can be obtained by integration of the component v_z of the velocity along z over time, using (11.41) and (11.43):

$$z_i = -\frac{1}{i\omega}\frac{\partial \Psi_v}{\partial z} = \begin{array}{ll} \frac{k}{i\omega}\Psi_v & \text{above,} \\ -\frac{k}{i\omega}\Psi_v & \text{below.} \end{array} \quad (11.45)$$

The integration constant is zero since z_i and Ψ_v are both zero for $t = 0$ if ω is real and for $t = -\infty$ if ω is imaginary. The above equation is only exact to first order in Ψ_v because we have neglected, in the exponential of (11.43), the small variation of z_i with time.

The boundary conditions at the interface are that v_z and the pressure P are continuous because the interface is not a shock but a contact discontinuity.

The continuity condition for v_z is equivalent to the condition that z_i is the same above and below, which gives

$$k\Psi_v(a) = -k\Psi_v(b) \qquad \text{for } z = z_i. \quad (11.46)$$

If we are once again interested in the first-order terms in Ψ_v we can evaluate this equation in the plane $z = 0$. It is simply

$$K_a = -K_b. \quad (11.47)$$

In order to determine ω we use the continuity of pressure across the interface. Starting from (11.44) we obtain

$$P = i\omega\rho\Psi_v - \rho g z. \quad (11.48)$$

Equation (11.45) allows us to express z to first order, in terms of Ψ_v. The condition $P_a = P_b$, evaluated again in the plane $z = 0$, gives using (11.47)

$$\omega^2 = gk\frac{\rho_b - \rho_a}{\rho_b + \rho_a}. \quad (11.49)$$

Since $\rho_b < \rho_a$, ω is imaginary and there is a growing instability.

The Rayleigh–Taylor instability is considered to be important during the expansion of supernova remnants in the final isothermal stage (see further Sect. 12.1). It can create turbulence which will enhance the magnetic field. A difference in the tangential velocities on either side of an interface yields another instability called the *Helmholtz instability*. Other instabilities can arise when the velocity distribution function of some particles is not maxwellian, for example if there are high-energy particles. We have finally to mention the existence of instabilities in the magnetic field. The study of these instabilities is a complex affair, and it is all but too easy to forget one or several of them in a dynamical study.

12 Shock Applications

In this chapter, we discuss some astrophysical applications of discontinuities and shocks, including supernova remnants, bubbles, the dynamics of H II regions and the acceleration of cosmic rays. Regardless of their interest, we will not discuss molecular jets from forming stars, which are outside the topics covered by this book.

12.1 Supernova Remnants

The evolution of massive stars ($M > 8$ M$_\odot$ approximately) ends with an explosion which ejects at high velocity a substantial fraction of the stellar mass into the surrounding medium. The star would have previously ejected matter into the ISM, in a more or less continuous way, as lower-velocity winds, and the ejecta from the explosion first encounters this circumstellar matter and then the interstellar medium itself. A supernova remnant then forms, this is essentially an hollow shell. Those supernovae which result from the explosion of massive stars are called Type II supernovae or SN II. Observationally, they are defined by the presence of hydrogen lines in their optical spectrum (for the classification of the optical spectra of supernovae, see e.g. Weiler & Sramek [543]). SN Ib which do not exhibit hydrogen lines, but do have an absorption line of Si II at 6 355 Å blue-shifted toward about 6 150 Å, probably result from the explosion of massive stars in a late Wolf–Rayet stage where hydrogen has disappeared. In both cases the material which was not ejected ends up as a neutron star or a black hole. The neutron star, which is a pulsar, can emit high-energy particles after the explosion, supplying extra energy to the supernova remnant. In this case, the spherical shell of the remnant is filled with a relativistic gas, in which the electrons emit synchrotron radiation. Such remnants are called *plerions*.

A white dwarf belonging to a close binary system can explode after it has accreted material from the other component, giving rise to a SN Ia. The spectra of SN Ia do not exhibit hydrogen lines, nor the Si II line. There is no compact remnant after the explosion in this case because the white dwarf is completely destroyed. Such supernova remnants cannot become plerions. Since there is little or no circumstellar matter in such systems, SN Ia interact only with the interstellar medium.

The total kinetic energy in the ejected envelope is of the order of 4×10^{50} ergs for the different types of supernovae, although their optical luminosities can be quite different: the SN Ia are more luminous than the SN Ib and the SN II.

Various aspects of supernova remnants are illustrated in Plates 4, 11, 25, 26, 27 and 28.

Schematically, we can distinguish three successive phases in the evolution of supernova remnants (Woltjer [557]):

1.- a phase of free expansion, the density of the ejected matter being much larger than the density of the surrounding medium: this phase ends when the mass of the matter swept by the expanding envelope is of the same order as the initial mass of this envelope;

2.- a phase of adiabatic expansion; the temperature of the shocked gas is so high that its radiation is relatively weak (cf. Fig. 8.5), so that the only important energy losses are through adiabatic expansion of the gas; this phase terminates when the temperature drops below about 10^6 K because the radiative losses become important at this stage;

3.- a late isothermal phase, in which energy is lost through radiation; this phase ends with the dispersion of the envelope as its velocity falls to about 9 km s^{-1}, which is roughly the velocity dispersion between interstellar structures.

We now study these three phases in succession, mostly following Spitzer [490]. We will suppose for simplification that the gas ejected by the explosion forms a spherical shell expanding isotropically. This is a highly idealized situation but it allows us to understand the basic physics. We will see later in Sect. 12.1 what occurs if the surrounding medium is inhomogeneous. The analytical approximations that we use have limitations and a realistic study requires numerical simulations.

12.1.1 The Free Expansion Phase

The ejection velocity of the matter at the explosion is very large, from several thousands to several tens of thousands km s^{-1}. We might expect that a shock would form when the ejected matter has swept a radius equal to the mean free path of the ejected particles in the ambient medium. A simple calculation shows that this cannot be the case. With an ejection velocity of 20 000 km s^{-1}, the mean free path of protons in a gas with density 1 cm^{-3} is as large as 400 pc. A classical shock clearly cannot form. However, in the presence of a magnetic field of say 5 µG these protons have a gyration radius smaller than 10^{11} cm and cannot escape from the envelope. The shock that forms is a MHD one but its exact nature is poorly understood. This phase ends at a radius r_s of the shock front, when the swept-up mass is of the same order as the initially ejected mass M_{ej}:

$$\frac{4}{3}\pi r_s^3 \rho_1 \approx M_{ej}, \tag{12.1}$$

where ρ_1 is the density of the surrounding medium. Assuming $M_{ej} = 0.25$ M$_\odot$ (SN Ia) and $\rho_1 = 2 \times 10^{-24}$ g cm^{-3} ($n \approx 1$ cm^{-3}), we find $r_s = 1.3$ pc, a radius that will be reached 60 years after explosion if the ejection velocity is 20 000 km s^{-1}. The shock velocity then decreases due to conservation of momentum.

In the case of the SN II and SN Ib, the expansion takes place into the circumstellar medium formed by the matter ejected by the massive star before explosion. The corresponding theory was developed by Chevalier [93]; see also Rohlfs & Wilson [439]. We will not discuss it here because its consequences for the interstellar medium are minor, the expansion velocity being barely decreased by this first interaction. It is however interesting to mention that there is synchrotron emission in this phase, indicating the early acceleration of particles to relativistic energies and also the presence of a magnetic field, probably enhanced by Rayleigh–Taylor instabilities.

12.1.2 The Adiabatic Phase

This phase is characterized by a non-radiative shock because, as we have already said, the temperature behind the shock is so high that the radiation from the matter is weak, the radiative energy losses therefore being negligible. There is conservation of the energy E liberated by the explosion. This energy is the principal parameter of the problem. A detailed treatment can be found in the book of Sedov [463]. In particular, Sedov has shown that there is a self-similar solution, which means that the structure of the supernova remnant stays constant with time. In this solution a part $K_1 E$ of the total energy is thermal, the rest being kinetic energy, with $K_1 = \text{const.} = 0.72$. This solution also indicates that the ratio K_2 of the pressure P_2 immediately behind the shock to the average pressure $\langle P \rangle$ of the heated gas in the spherical volume inside the shock is constant, with $K_2 = P_2/\langle P \rangle = 2.13$. The gas being assumed to be a perfect gas, the pressure is equal to 2/3 of its specific internal energy u, $P = (2/3)u$. We then have

$$P_2 = K_2 \frac{2}{3} \frac{3 K_1 E}{4\pi r_s^3} = \frac{KE}{2\pi r_s^3}, \tag{12.2}$$

defining $K = K_1 K_2 = 1.53$. The shock velocity $v_s \simeq v_1$ is given by (11.27) and is

$$v_s = \left(\frac{4}{3}\frac{P_2}{\rho_1}\right)^{1/2} = \left(\frac{2}{3\pi}\frac{KE}{\rho_1 r_s^3}\right)^{1/2}. \tag{12.3}$$

Since $v_s = dr_s/dt$ this equation is easily integrated and yields

$$\boxed{\begin{aligned}r_s &= \left(\tfrac{5}{2}\right)^{2/5} \left(\tfrac{2KE}{3\pi\rho_1}\right)^{1/5} t^{2/5} \\ &= 0.26 \left(\tfrac{n_H}{\text{cm}^{-3}}\right)^{-1/5} \left(\tfrac{t}{\text{yr}}\right)^{2/5} \left(\tfrac{E}{4\times 10^{50}\,\text{ergs}}\right)^{1/5} \text{pc.}\end{aligned}} \tag{12.4}$$

The temperature behind the shock is given by (11.29) and is, taking (11.28) into account,

$$T_2 = \frac{3\mu m_H}{16k} v_s^2 = 1.5 \times 10^{11} \left(\frac{n_H}{\text{cm}^{-3}}\right)^{-2/5} \left(\frac{t}{\text{yr}}\right)^{-6/5} \left(\frac{E}{4 \times 10^{50}\,\text{ergs}}\right)^{1/5} \text{K.} \tag{12.5}$$

The temperature behind the shock front *increases* towards the centre because the gas was shocked there at an earlier epoch when v_s, and hence T_2, were larger, and because the cooling is very slow. Detailed calculations by Chevalier [90] which neglect thermal conduction in the hot gas show that $T(r)/T_2(r_s) \propto (r/r_s)^{-4.3}$, where r is the radius. On the other hand, the density decreases towards the interior even more rapidly than the temperature increases. Half of the gas inside the shock is in the outer 6.1% of the shock radius and 3/4 in the outer 12.6%. The density and the temperature in these layers are respectively $0.50\rho_2$ and $1.31T_2$, and $0.28\rho_2$ and $1.78T_2$. ρ_2 is related to the ambient density ρ_1 by (11.26). It is in this zone that most of the X-ray continuum and lines are emitted. All the interior of the remnant is thus filled with low-density hot gas, which cools almost exclusively through adiabatic expansion and will eventually spread out into the general interstellar medium at the end of the life of the remnant.

However, thermal conduction in the hot gas changes these results considerably, though without much affecting the shock dynamics. In a general way, the heat flux $\mathbf{Q}$ per unit area in the presence of a temperature gradient ∇T is given by

$$\mathbf{Q} = -\kappa \nabla T, \tag{12.6}$$

where κ is the coefficient of thermal conductivity.

If there is no magnetic field, κ is given by

$$\kappa = \frac{5}{3} \frac{kT\lambda}{v_{rms}} n \left(\frac{3k}{2m} \right), \tag{12.7}$$

where λ is the mean free path, $v_{rms} = (3kT/m)^{1/2}$ the r.m.s thermal velocity, n the density and m the mass of the gas particles. For a neutral medium, λ is given by (11.34). For a fully ionized medium, which is the case here, the mean free path is (Lang [299])

$$\lambda = \frac{m^2 v_{rms}^4}{Z^2 n_e e^4 \ln \Lambda} = \frac{3.2 \times 10^6 T^2}{Z^2 n_e \ln \Lambda}, \tag{12.8}$$

with

$$\Lambda = 1.3 \times 10^4 \frac{T^{3/2}}{n_e^{1/2}}. \tag{12.9}$$

Here, m is the mass of the particles of interest, Z their charge, v_{rms} their r.m.s. velocity, e the elementary charge and n_e the electron density[1]. Conductivity is mostly due to electrons and we have, to a first approximation, $\kappa \approx 1 \times 10^{-6} T^{5/2}$ erg s^{-1} deg^{-1} cm^{-1}. Using this numerical value, we find that the temperature is almost uniform inside the supernova remnant (Chevalier [91]). We can then easily obtain the temperature of the hot gas. We saw that in the self-similar Sedov solution the thermal energy E_{th} is 0.72 times the total energy E of the supernova, so that the average gas temperature $\langle T \rangle$ inside the remnant is (McKee & Ostriker [359])

---

[1] Note that λ is the same for electron–electron collisions and for proton–proton collisions for the same kinetic temperature, the mass of the particle being eliminated between m^2 and v_{rms}^4 in (12.8) for λ.

$$\langle T \rangle = \frac{2\mu m_H E_{th}}{3kM} \simeq 1.4 \times 10^{10} \left(\frac{n_0}{\text{cm}^{-3}}\right)^{-1} \left(\frac{r_s}{\text{pc}}\right)^{-3} \left(\frac{E}{10^{51} \text{ ergs}}\right) \text{ K}, \quad (12.10)$$

where μ is the mean atomic weight $\simeq 0.7$ and M is the mass of the remnant, which is expressed as a function of the ambient density n_0 and of the radius of the remnant r_s.

However, thermal conduction is very strongly reduced by a magnetic field, perpendicularly to the field lines, while it does not change much along the field lines. The preceding case is thus extreme and the reality should be between the two extreme cases we have described. Unfortunately we know very little of the properties of the magnetic field inside supernova remnants. This problem was discussed by Spitzer [491], Sect. 2.2.

We emphasize the fact that the description we have given is only approximate and that transitions between the different phases produce additional effects (cf. Spitzer [491]). The most important effect is the *reverse shock*, which propagates towards the centre in the ejected matter. It arises during the transition between the free expansion phase and the adiabatic expansion phase. The pressure behind the shock, which results from the interaction between the ejected gas and the interstellar medium, is high and tends to propagate downstream in the ejected matter. As this matter cools radiatively the sound velocity decreases and the pressure perturbation turns supersonic forming the reverse shock. The emission of visible lines in young supernova remnants is to a large extent produced in this shock. After the passage of the reverse shock the remnant enters the adiabatic phase.

12.1.3 The Isothermal, or Radiative, Expansion Phase

When the post-shock temperature T_2 falls to about 10^6 K the nuclei of abundant elements like C, N and O begin to recombine with electrons. The collisional excitation of the ions so formed and the lines they emit considerably increase the cooling rate of the electron gas, by some two orders of magnitude (cf. Fig. 8.5). The density is sufficiently large to allow a fast energy equipartition between the electrons and ions. The cooling of the gas as a whole is thus efficient and the shock becomes radiative. Its propagation is no longer maintained by the thermal energy, which is radiated, but by the momentum of the gas. This regime is often called the *snowplough regime*. The transition takes place at a radius of about 15 pc in an interstellar medium with a density of 1 cm^{-3}, with an expansion velocity of the order of 85 km s^{-1}. The age of the remnant is then about 4×10^4 yrs. The shell compressed by the shock is much thinner than during the adiabatic phase.

Due to the conservation of momentum, the product Mv_s of the total mass and the mean velocity of the shell is constant after entering the isothermal phase. Most of the mass of the shell comes from the interstellar medium, and increases as r_s^3. Then,

$$\frac{4}{3}\pi r_s^3 \rho_1 v_s = \text{const.,} \quad (12.11)$$

which gives by integration

$$\boxed{r_s = r_{rad}\left(\frac{8}{5}\frac{t}{t_{rad}} - \frac{3}{5}\right)^{1/4}},\qquad (12.12)$$

r_{rad} and t_{rad} being, respectively, the radius and the age of the remnant at the beginning of this phase, say when half of the initial energy of the supernova has been radiated away.

The hot medium which fills the remnant at the end of the adiabatic phase cools adiabatically during the isothermal phase. It is still rather hot (10^5 to 10^6 K) at the end of the life of the supernova remnant, when it eventually spreads out into the interstellar medium, contributing to the hot phase of this medium (see further Sect. 15.2). The internal hot medium has some effect on the dynamics of the isothermal phase: the above simple treatment is thus only approximate. We have also ignored the effect of the magnetic field, which in this phase can play an important dynamical role: see Spitzer [491], Sect. 2.2.

The supernova remnant vanishes when the final shock velocity v_f is of the same order as the velocity dispersion in the interstellar medium, about 9 km s^{-1}. The kinetic energy of the remnant, $(1/2)M_f v_f^2$, is then transferred to the general interstellar medium. The radius of the remnant is then of the order of 40 pc and its age of 10^6 years. It is interesting to give an order of magnitude for the efficiency η of conversion of the initial energy E of the supernova into kinetic energy of the interstellar medium. We have

$$\eta = \frac{M_f v_f^2}{2E} = \frac{M_f v_f^2}{M_{rad} v_{rad}^2}\frac{M_{rad} v_{rad}^2}{2E},\qquad (12.13)$$

where M_{rad} and v_{rad} are, respectively, the mass and the velocity of the shell at the beginning of the radiative (isothermal) phase. We saw that the kinetic energy during the adiabatic phase, thus at the beginning of the radiative phase, is 28 % of the total energy. Since we also have $M_f v_f = M_{rad} v_{rad}$,

$$\boxed{\eta = 0.28\frac{v_f}{v_{rad}}}.\qquad (12.14)$$

Suppose that the transition between the two phases occurs when the temperature T_2 is of the order of 10^5 K. Then $v_{rad} \approx 85$ km s^{-1} and

$$\eta \approx 0.030.\qquad (12.15)$$

Identified supernova remnants are generally smaller than the radius r_{rad} at which the snowplough model becomes applicable. But this phase must undoubtly exist.

12.1.4 The Evolution of Plerions

The evolution of plerions has been discussed by Reynolds & Chevalier [434]. In these supernova remnants the central pulsar supplies energy continuously, at a rate that decreases with time because the energy is borrowed from the rotational momentum of the neutron star. This energy takes the form of a more or less spherical and homogeneous bubble of high-energy particles and magnetic field, which is seen through the synchrotron emission of the relativistic electrons. The pressure is approximately uniform in this bubble because the sound velocity is very high in the relativistic fluid and rapidly damps pressure fluctuations. The bubble forms within the matter ejected by the explosion, as can be seen for example in the Crab nebula (Plate 25). The expansion of the bubble through the ejected matter can eventually engulf it entirely if the pulsar has a sufficient supply of energy. Then the acceleration continues until the energy production ceases. If the pulsar stops supplying energy before the shell has engulfed all the matter the expansion slows down. After the pulsar emission ceases the evolution of plerions does not differ from that of other supernova remnants.

Due to the rapid slowing down of the pulsars, whose luminosity evolves as

$$L(t) = \frac{L_0}{(1 + t/\tau)^p}, \qquad (12.16)$$

with $p \approx 2$ and $\tau \approx 300$–700 years, plerions are rather young objects with small radii. We can legitimately consider that the plerion phase is rather common in the early evolution of supernova remnants.

In general, we must acknowledge that our knowledge of supernova remnants is still insufficient at the time of writing. The theory we have described is rudimentary and many uncertainties remain. There are not enough observations, in particular in the crucial X-ray wavelength range where the best diagnosis of the evolution of supernova remnants can be found. Fortunately, the situation is rapidly improving thanks to observations with the X-ray satellites CHANDRA and XMM–NEWTON, as can be seen in Plates 11, and 25 to 28.

12.1.5 The Expansion of Supernova Remnants in an Inhomogeneous Medium

Inhomogeneities in the interstellar medium obviously produce distorsions in supernova remnants. The most interesting effects arise in dense inhomogeneities engulfed by the remnant.

The passage of a shock through an inhomogeneous medium compresses the clouds or the filaments. The pressure increase at the surface of such a cloud creates an internal shock which propagates from the surface into the dense medium. The medium is generally dense enough for radiation to occur and to lower the temperature, so that the shock is isothermal. The pressure being higher on the side directed towards the centre of the supernova remnant, the cloud is both accelerated and flattened. The pressure behind the external shock, whose velocity is $v_{s,e}$, is of the order of $\rho_{e1} v_{s,e}^2$ (11.27), ρ_{e1} being the density at the surface of the cloud. This pressure is approximately equal to the pressure behind the internal shock; this shock propagates inside the cloud where it meets a density ρ_{e2}, such that the velocity $v_{s,i}$ of the internal shock is approximately

$$v_{s,i} \approx \left(\frac{\rho_{e1}}{\rho_{e2}}\right)^{1/2} v_{s,e}. \quad (12.17)$$

These internal isothermal shocks emit visible and infrared lines.

The acceleration of small, pre-existing interstellar condensations through the effect of supernova remnant shocks is probably the reason why cloudlets with a velocity ~ 200 km s^{-1} are observed in the supernova remnant Cassiopeia A. These cloudlets have the typical chemical composition of the interstellar medium. The expansion velocity of this remnant, about 6 000 km s^{-1}, is observed in other cloudlets, whose very different chemical composition shows that they are made from supernova material.

Theory and numerical simulations show that, as expected, the interstellar clouds affected by the shock of a supernova remnant experience instabilities, in particular of Rayleigh–Taylor type, which eventually destroy them completely but for their densest parts.

Thermal conduction acts on those dense structures that survive the crossing of the shock of the supernova remnant. They are now immersed in the hot gas where they can either evaporate or accrete material. These phenomena have been studied by Cowie & McKee [105] and McKee & Cowie [358]. The evaporation rate of a spherical dense globule of radius a immersed in a hot gas with temperature T_h is, if there is no magnetic field (cf. McKee & Ostriker [359])

$$\frac{dM}{dt} = 10^{4.44} T_h^{5/2} \left(\frac{a}{\text{pc}}\right) \text{ g s}^{-1}. \quad (12.18)$$

However, if the cloud radius a is larger than $0.16(T_h/10^6\text{K})^2 n_h^{-1}$ pc, where n_h is the density of the hot medium, radiative losses exceed conductive heating and the cloud grows by the condensation of the hot medium. Such phenomena can deeply affect the supernova remnant dynamics and play a very important role in the physics of the three-component interstellar medium as described by McKee & Ostriker [359]. We will discuss this further in Sect. 15.2. However, as we have seen, the magnetic field would significantly decrease the thermal conduction in a poorly understood way and make these processes less efficient: cf. Spitzer [491], Sect. 3.

12.1.6 Non-Thermal Radiation of Supernova Remnants

Supernova remnants have long been known to be very intense radio sources. The radio emission is clearly due to the synchrotron radiation from relativistic electrons (cf. Sect. 2.2), as revealed by a spectrum decreasing towards higher frequencies, and above all by the linear polarization which can be very high in some regions, for example in some parts of the Crab nebula. Recently, it was shown that supernova remnants are non-thermal sources of X-rays and of gamma rays reaching energies as high as several TeV (see references in Bykov et al. [74]). The presence of these emissions shows that supernova remnants are extremely powerful and efficient particle accelerators.

For observational reasons, the radio emission of supernova remnants was the first studied and best understood non-thermal radiation. In SN II and SN Ib radio emission appears a few days after the explosion, reaches a maximum after about 10 days at high frequencies, later at lower frequencies, and then decreases (Weiler & Sramek [543]). The apparent radio diameter of some extragalactic supernova remnants has been measured with the Very Large Array soon after the explosion and exhibits an increase with time. The delay of the emission at lower frequencies is due to the synchrotron optical thickness, or to free–free absoption by the ionized gas in the remnant (5.26), which are both larger at lower frequencies. SN Ia have no radio emission at these early stages, showing that the synchrotron radiation is due to electrons and magnetic fields which are, respectively, accelerated and created by a shock in the pre-explosion circumstellar medium, and this exists only for SN II and SN Ib (cf. Chevalier [93]).

More evolved SN II or Ib remnants are also radio emitters, for example Cassiopeia A (Plates 26 and 27) with a present age of about 340 yrs[2]. The case of Cassiopeia A, the youngest supernova remnant known in our Galaxy, is of interest because its radio flux decreases slowly with time. Shklovsky [468] has suggested that this decrease is due to the fact that the magnetic field and the relativistic electrons were born soon after the explosion and fill up the cavity of the remnant. In his model the expansion causes a decrease in the magnetic field, and also adiabatic energy losses for the relativistic electrons, which both contribute to the decrease of the radio flux. The book of Rohlfs & Wilson [439] reproduces the calculation of Shklovsky, which presents a great pedagogical interest. However this model does not explain why the emission comes from a thin shell, rather than from all the volume, and suggests that the flux decrease would make old supernova remnants unobservable in the radio, which is contrary to observations. Another old model due to van der Laan [526] supposes that the emission comes from the compression of the interstellar magnetic field and of the cosmic-ray electrons by the shock wave of the remnant. It can account for the old supernova remnants but not for the very strong radio emission of the young remnants, because the density increases only by a factor of 4 after the passage of the strong J shock, and the magnetic field can barely be increased by a larger factor (except through secondary effects that will be mentioned later). Van der Laan thus assumed that other relativistic electrons were accelerated during the initial explosion itself, but his model then has the same deficiency than that of Shklovsky. It is thus certain that relativistic electrons are accelerated in the shock and that the magnetic field is strongly enhanced there (for a recent paper on this subject see Berezhko & Völk [33]). This problem was first treated quantitatively by Gull [215]. He assumed that the Rayleigh–Taylor instability in the contact discontinuity behind the shock (Sect. 11.5) produces a turbulent zone which confines and accelerates

---

[2] While the explosion of the Crab supernova and of other supernovae has been reported in historical annals, this is not the case for the explosion of Cassiopeia A, probably because of the large amount of interstellar extinction in front of this object. The age of Cassiopeia A has been calculated from the observation of the proper motion of its filaments (Kamper & van den Bergh [277]).

those relativistic electrons that survive from the initial phase or come from the surrounding interstellar medium. The magnetic field is also amplified because of the strong interaction between turbulence, the magnetic field and relativistic particles, which leads to an approximate equipartition of particle and magnetic energies with the energy of turbulence. The relativistic electrons and the magnetic field can also come from the central pulsar in the case of plerions, like the Crab nebula. This case was discussed by Reynolds & Chevalier [434].

Since the work of Gull considerable progress has been made in the theory of the acceleration of charged particles in shocks and we have realized that this acceleration is extremely efficient. We will examine it later in Sect. 12.4, when dealing with the acceleration of cosmic rays. Satisfactory models exist that can account for all the properties of the non-thermal radiation of supernova remnants with accelerated particles radiating either by the synchrotron mechanism or by Bremsstrahlung, inverse Compton effect and nuclear interactions (Sect. 6.2). We lack space here to describe these models: the interested reader should consult the papers by Baring et al. [21], by Ellison et al. [155] and by Berezhko & Völk [32].

12.2 Bubbles

Most of the contents of this section come from the review of Tenorio-Tagle & Bodenheimer [507].

The interstellar medium contains shells, or bubbles, of neutral or ionized hydrogen which are apparently large and empty (sizes from 100 pc to more than 1 kpc): cf. Heiles [224]. They are easily seen in H I maps and Hα images of external galaxies (Plates 10 and 29), but they also exist in our Galaxy, although here they are more difficult to see (Plates 3 and 32). The Local bubble is one example. The expansion velocity of these bubbles has been measured in some cases. Their total kinetic energy can be larger than 10^{53} ergs, showing that these bubbles cannot be supernova remnants since the total energy produced by a supernova is smaller than 10^{51} ergs. They therefore might result from the collective effect of many supernova explosions, to which the winds of massive stars add energy (Bruhweiler et al. [72]). It is therefore necessary that many such objects occur in a relatively short time interval. This does not raise particular difficulties in galaxies with active star formation like the Large Magellanic Cloud, where hundreds or even thousands of massive stars can be formed in a single cluster or association during a period of only a few million years. The situation is, however, less favorable in less active galaxies, like our Galaxy or the Andromeda galaxy M 31, although even these galaxies possess populous star clusters or associations like NGC 3603 in the Galaxy (Brandner et al. [66]) or NGC 206 in M 31. The number of bubbles observed in such external galaxies is irreconcilable or only marginally reconcilable with star formation rates (McClure-Griffiths et al. [355]). Alternative ways to create bubbles are the expansion of supergiant H II regions, this will be discussed in Sect. 12.3, or the collision between an infalling gas cloud and the disk of the Galaxy, this will not be discussed here. Alternatively, Wada & Norman [538] suggest that a thermally and gravitationally unstable disk can generate large shell-like structures in which dense clumps and filaments surround a hot gas interior, a suggestion apparently supported by the two-dimensional simulations carried out by Wada et al. [539].

Given the large dimensions of the bubbles, their relatively low expansion velocity (a few tens of km s^{-1}) and the fact that their shells do not radiate appreciably at X-ray wavelengths and are thus relatively cold, we may assume that they are in the isothermal phase, if indeed there are created by collective explosions of supernovae. Assuming that all the events that created the bubble occurred in a short time interval compared to the present age of the bubble, t, which is reasonable since $t > 10^7$ years for a large bubble, we can then evaluate the total energy E_E of these events in the same way as we evaluated the efficiency η for a single supernova (12.15). Taking a mass for the shell, given by the empirical relation $M \approx 8.5(r_S/\mathrm{pc})^2$ M$_\odot$, where r_S is the radius, and an expansion velocity of 20 km s^{-1}, a 1-kpc radius bubble has a kinetic energy of 3.4×10^{52} ergs. This implies, using (12.15), an initial energy $E_E = 5 \times 10^{53}$ ergs. This is equivalent to the energy of about 1 000 supernovae. However, this does not take into account stellar winds. The total kinetic energy of a stellar wind of 3×10^{-6} M$_\odot$/year emitted for 3×10^6 years with a velocity of 2 000 km s^{-1}, values typical of a massive star, is of the order of 4×10^{50} ergs, comparable to the energy of a supernova. We can then divide by a factor ~ 2 the necessary number of supernovae, the energy per star being about 10^{51} ergs.

A simple calculation, which is perhaps more realistic than the previous one, was presented by McCray & Kafatos [356]. They supposed that supernovae explode at a constant rate with an interval between explosions $\Delta\tau = (50/N_*)10^6$ years, where N_* is the total number of massive stars that will produce supernovae. The injected power in the bubble is then $E/\Delta\tau = 6.3 10^{35}(E/10^{51}\text{ ergs})$ ergs s^{-1}. With a uniform density n_0 for the interstellar medium the radius of the bubble is then found to be

$$r_s = 97 N_* \left(\frac{n_0}{\mathrm{cm}^{-3}}\right)^{-1} \left(\frac{E}{10^{51}\text{ ergs}}\right)^{1/5} \left(\frac{t}{10^7\text{ ans}}\right)^{3/5} \text{ pc.} \qquad (12.19)$$

However, the density decreases in going away from the galactic plane, and the pressure inside the bubble will also decrease because it tends to spread out into the halo. After the bubble radius reaches the scale height of the gas z_0 in the galactic disk we may assume the conservation of momentum in the Galactic plane, which leads to $r_s(t) = z_0(t/t_0)^{0.25}$, where t_0 is the age at which the radius reaches z_0. The bubble now has the shape of a hollow cylinder whose walls are almost perpendicular to the galactic plane, forming a chimney through which the hot gas which fills the inside of the bubble can escape. These results are confirmed by numerical simulations such as that presented in Fig. 12.1. Such chimneys are observed in the 21-cm line (Plate 32).

Another effect is the deformation of the bubbles due to the differential rotation of the disk (Tenorio-Tagle & Palous [509]). The bubbles become elliptical after some 10^7 years. These ellipses become progressively elongated and align themselves in the direction of rotation. Such deformations are effectively observed in external galaxies and can give interesting constraints on the ages of the bubbles.

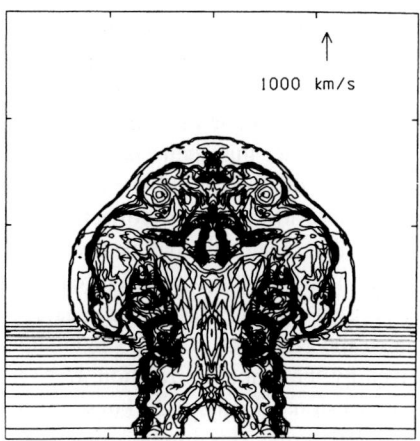

Fig. 12.1. Hydrodynamic simulation of a bubble formed by the sequential explosion of supernovae in the disk of a Galaxy where the density decreases with z according to a gaussian distribution. The density of the interstellar gas in the plane is $n = 1$ cm^{-3} and its scale height is $z_0 = 100$ pc. The interval between supernova explosions is 2×10^5 years and the first one was supposed to have occured 5.6×10^6 years ago. Each supernova supplies an energy of 10^{51} ergs. Tick marks in abscissae and ordinates are placed every 250 pc. The contours are isodensity in the plane of symmetry and the arrows indicate the velocity. The structures are due to the Rayleigh–Taylor instability. From Tenorio-Tagle et al. [508], with the permission of ESO.

Finally, the interaction between neighbouring bubbles produces a kind of dense wall of gas (Yoshioka & Ikeuchi [566]). Such a wall has been observed in the 21-cm line and through UV and X-ray absorption between the Local buble and another bubble (Egger & Aschenbach [152]).

12.3 The Dynamics of H II Regions

The interface between H II regions and the surrounding neutral gas was examined from the physico-chemical point of view in Chap. 10. We will now examine it from the dynamical point of view. A good review article is that of Yorke [564] but interesting information can also be found in the books by Kaplan [278] Sect. 3.11 and 3.15, and Spitzer [490] Sect. 12.1, and also in Tenorio-Tagle & Bodenheimer [507]. Here, we cannot discuss all the aspects of the problem and we will not, in particular, treat the formation of ultra-compact and compact H II regions around young stars in molecular clouds, for which the rotation and the wind of the star play an important role. The references just cited give some elements of their formation and evolution.

The ionized and the neutral gases around a H II region are separated by a discontinuity which moves into the neutral medium because the photons from the central hot star(s) continuously ionize new hydrogen atoms in a thin layer of the neutral medium. Furthermore, the pressure in the ionized gas is considerably higher than the pressure in the neutral gas, producing an expansion of the ionized medium. The discontinuity, called the *ionization front*, acts as a piston which moves at a velocity generally greater than the sound velocity in the neutral medium, and therefore produces a shock in this medium. This shock propagates in the neutral medium as a second discontinuity.

12.3.1 The Ionization Front

The equations that describe the motions of these two discontinuities are similar to the general equations for a shock. Let us first consider the motion of the ionization front. The flux of matter through this front depends only on the flux of ionizing photons. We will assume for simplicity that the neutral medium has no motion parallel to the front and we will neglect the magnetic field. These assumptions are well justified, except for the neglect of the magnetic field in a dense molecular cloud. The continuity equation for the mass flow (11.16) becomes

$$\boxed{\rho_1 v_1 = \rho_2 v_2 = \mathcal{J},} \tag{12.20}$$

where $\mathcal{J}$ is now the flux of hydrogen atoms through the front, which is (cf. Sect. 5.1)

$$\mathcal{J} = \frac{\mu_1 m_H S(0)}{4\pi r_S^2} e^{-\tau}, \tag{12.21}$$

where $\mu_1 m_H$ is the mean mass which crosses the front per ionization (about $1.5 m_H$ when taking into account helium and heavy elements) and $S(0)$ is the flux of photons able to ionize hydrogen, which is given by Table 5.1 as a function of the spectral type of the star. r_S is the distance to the ionizing star(s) (the Strömgren radius)[3]. The exp(-τ) factor represents the absorption of the Lyman continuum photons by dust inside the H II region. This is an uncertain quantity because the properties of the dust at very short wavelengths are poorly known, in particular inside H II regions. exp(-τ) is often assumed to be of the order of 0.5. A more exact expression is given by Bertoldi & Draine [37].

The equation of conservation of the momentum flux (11.17) can simply be written as

$$\boxed{\rho_1 \left(\frac{kT_1}{\mu_1 m_H} + v_1^2 \right) = \rho_2 \left(\frac{kT_2}{\mu_2 m_H} + v_2^2 \right),} \tag{12.22}$$

where μ_1 and μ_2 are the mean masses of the gas particles on either side of the ionization front (respectively about 1.5 et 0.8). We should in principle add to the right-hand side of the equation a term corresponding to the radiation pressure in the transition zone, which is $(h\nu_0/c)[S(0)/4\pi r^2]e^{-\tau}$, but this term is negligible with respect to the others.

The discussion of the different types of ionization fronts is rather involved. We present here, as an illustration, a simplified version following Kaplan [278], which

---

[3] Of course, if there are several ionizing stars their contributions must be summed.

uses a supplementary condition called the *Jouguet point condition*, used in the case of detonation waves which have properties close to those of ionization fronts: if the pressure in the H II region is not too high, the ionized gas behind the front expands freely in the H II region, at the velocity of sound c_2 with respect to the front[4]. We point out the fact that this condition is not always met, so that the cases we will describe are limiting cases. If it is met, we have

$$v_2 \approx c_2 = \left(\frac{\gamma k T_2}{\mu_2 m_H}\right)^{1/2}. \tag{12.23}$$

Substituting this equation in (12.20) and (12.21) we immediately obtain the values of the densities ρ_2 and ρ_1

$$\rho_2 = \mathcal{J}/c_2, \tag{12.24}$$

$$\rho_1 = \frac{\rho_2 v_2}{v_1} = \frac{\mathcal{J}}{v_1}. \tag{12.25}$$

The velocity v_1 of the front, with respect to the neutral medium, is given by

$$v_1 + \frac{1}{v_1}\frac{kT_1}{\mu_1 m_H} = (\gamma + 1)\left(\frac{kT_2}{\gamma \mu_2 m_H}\right)^{1/2}. \tag{12.26}$$

The solution of this equation for v_1 is

$$v_1 = \left[\frac{(\gamma+1)^2 kT_2}{4\gamma \mu_2 m_H}\right]^{1/2} \pm \left[\frac{(\gamma+1)^2 kT_2}{4\gamma \mu_2 m_H} - \frac{kT_1}{\mu_1 m_H}\right]^{1/2}, \tag{12.27}$$

or, since $T_1 \ll T_2$,

$$v_1 = \left[\frac{(\gamma+1)^2 kT_2}{4\gamma \mu_2 m_H}\right]^{1/2}\left\{1 \pm \left[1 - \frac{2\gamma}{(\gamma+1)^2}\frac{\mu_2 T_1}{\mu_1 T_2} + \ldots\right]\right\}. \tag{12.28}$$

The minus sign corresponds to a *rarefaction wave*, for which the density in the H II region is smaller than that in the neutral medium. The plus sign corresponds to a *compression wave*, for which the density in the H II region is larger than that in the neutral medium. Both cases are met in nature, as we will see later, but only the rarefaction waves are encountered in well-developed H II regions.

The parameters of the ionization front are fully determined by the above equations if the upstream temperature, T_1, and the downstream temperature, T_2, are known. T_1 has to be determined independently (we will see how later). T_2 plays a more important role. If the density behind the front is high enough, which is the case for a compression wave, T_2 is approximately equal to that in the H II region, and is of

---

[4] We here use the adiabatic sound velocity for simplification. However, the cooling being relatively fast in H II regions, the sound velocity might be closer to the isothermal velocity $c_{II,2} = (kT_2/\mu_2 m_H)^{1/2}$.

the order of 10 000 K. But if the density behind the front is low, we must calculate T_2 using the equation for the conservation of the energy flux,

$$\rho_1 v_1 \left[\frac{\gamma k T_1}{(\gamma - 1)\mu_1 m_H} + \frac{v_1^2}{2} \right] = \rho_2 v_2 \left[\frac{\gamma k T_2}{(\gamma - 1)\mu_2 m_H} + \frac{v_2^2}{2} \right] - \epsilon_0 \mathcal{J}. \quad (12.29)$$

The last term in this equation accounts for the heating of the gas by ionization, where ϵ_0 is the mean energy given to a photoelectron by an ionizing photon, which depends on the effective temperature of the ionizing star(s). Here, we neglect the cooling of the gas since it is supposed to have a low density. Using (12.20) and (12.22), we find

$$v_1^2 = \frac{\gamma + 1}{\gamma - 1} \frac{k T_2}{\mu_2 m_H} - \frac{2\gamma}{\gamma - 1} \frac{k T_1}{\mu_1 m_H} - \frac{\epsilon_0}{m_H}. \quad (12.30)$$

Equating this value for v_1 to that given by (12.28) we can determine T_2 given T_1. Finally, we have the values of all the important quantities as follows.
- Compression wave (*critical solution of type R*)[5]:

$$T_2 = \frac{\gamma(\gamma - 1)}{\gamma + 1} \frac{\mu_2 \epsilon_0}{k}, \quad v_1 = \left[\frac{(\gamma^2 - 1)\epsilon_0}{m_H} \right]^{1/2},$$

$$\frac{\rho_2}{\rho_1} = \frac{v_1}{v_2} = \frac{\gamma + 1}{\gamma}, \quad \rho_1 = \mathcal{J} \left[\frac{m_H^3}{(\gamma^2 - 1)\epsilon_0} \right]^{1/2}. \quad (12.31)$$

- Rarefaction wave (*critical solution of type D*)[6]:

$$T_2 = \frac{\gamma - 1}{\gamma(\gamma + 1)} \frac{\mu_2 \epsilon_0}{k}, \quad v_1 = \frac{\mu_2 T_1}{\mu_1 T_2} \left[\frac{(\gamma - 1)\epsilon_0}{(\gamma + 1)^3 m_H} \right]^{1/2},$$

$$\frac{\rho_2}{\rho_1} = \frac{v_1}{v_2} = \frac{1}{\gamma + 1} \frac{\mu_2 T_1}{\mu_1 T_2}, \quad \rho_1 = \mathcal{J} \left[\frac{(\gamma + 1)^3}{\gamma - 1} \frac{m_H^3}{\epsilon_0} \right]^{1/2}. \quad (12.32)$$

With $\gamma = 5/3$, the density jump ρ_2/ρ_1 is 8/3 for a compression front. It can be a small as 1/60 for a rarefaction front. The velocity of a R front is about 26 km s^{-1} (but see later a more exact value), while that of a D front is much smaller, of the order of 0.2 km s^{-1}. R fronts are therefore supersonic and D fronts are subsonic.

Given the usual ranges for S_0, and for ϵ_0 (4 to 7 × 10^{-12} erg), and taking $T_1 = 1\,000$ K, we find values of the temperature T_2 from 9 000 to 15 000 K behind a compression front and from 3 000 et 6 000 K behind a rarefaction front. The gas then heats or cools accordingly until it reaches its equilibrium value in the H II region, which is of the order of 10 000 K.

---
[5] The notation R is for "rarefied gas" since the front meets in this case a gas of relatively low density.
[6] D is for "dense gas" since the front meets a denser gas than that of the H II region.

12.3.2 The Shock

Given (12.31) or (12.32), the density ρ_1 upstream of an ionization front is fixed by the ionizing photon flux and by the mean energy of the photoelectrons. It is generally different from the mean density ρ_0 in the neutral medium through which the ionization front progresses. We therefore expect that the ionization front is preceded by a density perturbation, then equivalently a pressure perturbation, which can be either a compression wave, if $\rho_1 > \rho_0$, or a rarefaction wave, if $\rho_1 < \rho_0$. These waves propagate with the velocity of sound in the neutral medium. However, in the case of a compression ionization front (R front), with a supersonic velocity, such a perturbation can only propagate if it is of rarefaction type and adiabatic, in which case it might possibly reach a velocity greater than that of the front.

Conversely, if the ionization front is a D-type (rarefaction) front, it is subsonic and a perturbation can always precede it. This perturbation is in general a compression wave because the upstream compression by the ionization front is such that $\rho_1 > \rho_0$ in most cases. It can become supersonic, forming a shock, because the sound velocity decreases in the neutral medium due to the fact that its temperature is much lower than T_1. This brings to mind the reverse shock that we encountered in supernova remnants (Sect. 12.1). The analytical treatment of this configuration is difficult but there exists a self-similar solution for a planar flow, when the photon flux in the Lyman continuum is constant in time: see Kaplan [278] Sect. 15. Another approximate solution, that deals with spherical symmetry, is due to Spitzer [490] Sect. 12.1.c. Its results can be summarized in a qualitative way (Fig. 12.2), for a $H\,\textsc{ii}$–$H\,\textsc{i}$ transition. The D ionization front is preceded by a shock which compresses the gas in the $H\,\textsc{i}$ region and heats it without producing collisional ionization: the temperature, which is the temperature T_1 upstream of the ionization front, is limited by the emission of the fine-structure lines, in particular of those of $O\,\textsc{i}$, and is probably of the order of 1 000 K or slightly higher. This temperature determines in turn, as we have seen, the compression ratio for the ionization front. The ratio between the thickness of the compressed region and that of the Strömgren radius is of the order of $T_1/4T_2$, i.e., a few hundredths. All of the compressed region is expanding. In the self-similar model of Kaplan its velocity, with respect to the ambient $H\,\textsc{i}$ medium at rest, is constant and of the order of

$$v_s = 2 \left(\frac{\rho_2}{\rho_0} \frac{kT_2}{\mu_2 m_H} \right)^{1/2} \simeq 20 \text{ km s}^{-1}. \tag{12.33}$$

Spitzer gives the following expression for the spherical case, which is probably more realistic:

$$\frac{r}{r_0} = \left(1 + \frac{7}{4} \frac{c_{II} t}{r_0} \right)^{4/7}, \tag{12.34}$$

which implies that the velocity of the medium cannot be larger than the isothermal sound velocity $c_{II} = (kT_2/\mu_2 m_H)^{1/2}$ in the ionized gas, which is of the order of 10 km s^{-1}.

Note that the velocity v_s of the compressed medium is not the velocity v_1, which is the relative velocity of the compressed gas with respect to the ionization front

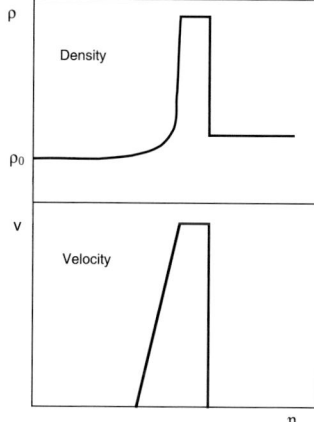

Fig. 12.2. Schematic distribution of the gas density ρ and of its velocity v with respect to the surrounding medium, for a system consisting of a compression shock preceding a D-type (rarefaction) ionization front. These quantities are given as a function of the non-dimensional parameter $\eta = r/[t\sqrt{(kT_2)/(\mu_2 m_H)}]$, where t is the time, and the other quantities are defined in the text. Taken from Kaplan [278].

and is a very low velocity. Just upstream of the ionization front there is a flow of freshly ionized gas whose velocity decreases downstream of the front. The average density in the H II region is in the end not very different from the density ρ_0 in the surrounding medium: this is an expected consequence of mass conservation. It is possible to generalize these results to the case of a non-negligible magnetic field. They are not qualitatively different.

12.3.3 Neutral Globules in a H II Region

The structure of the compressed gas that we just described contains the photodissociation region that was already treated in Chap. 10. A dynamical study, including the magnetic field and accounting for ionization and photodissociation at the surface of a molecular cloud submitted to the UV radiation field of hot stars, is given by Bertoldi & Draine [37].

A similar structure exists near the surface of neutral globules inside an H II region. A full non-stationary treatment is, in principle, necessary in this case, as well as in the case of dense globules immersed in a more diffuse medium at the surface of an inhomogeneous molecular cloud. There, the chemical composition of the photodissociation region is significantly modified because slow chemical reactions do not have time to reach completion. Conversely, in the case of a well-developed classical H II region, the ionization front propagates slowly into the neutral medium (the inner part of which is also the photodissociation region) because the Lyman continuum flux is not very large, and the stationary models of photodissociation regions of Chap. 10 are valid.

A simple calculation, following Hollenbach & Tielens [244], gives the conditions for which a non-stationary study becomes necessary. Assuming for simplification pressure equilibrium and temperatures of 10^4 K in the H II region and 1 000 K in the photodissociation region (PDR), the density in this region is $n_{\rm PDR} = 2n_{\rm HII} T_{\rm HII}/T_{\rm PDR} \simeq 20 n_{\rm HII}$ (remember that there are twice as many particles in the H II region due to ionization, hence the factor 2). Assuming that evaporation of the newly ionized gas occurs at the isothermal sound velocity in the H II region, $c_{II} \simeq 10$ km s^{-1}, the velocity of the flow through the photodissociation region is $v_{\rm PDR} \simeq (n_{\rm HII}/n_{\rm PDR}) c_{II} \approx 0.5$ km s^{-1}. The time taken by this flow to cross the photodissociation region is $\tau_{\rm PDR} \simeq (N_{\rm PDR}/n_{\rm PDR})/v_{\rm PDR}$, where $N_{\rm PDR}$ is the column density in the active part of the photodissociation region, which can be obtained from photodissociation models. Equating $\tau_{\rm PDR}$ to τ_{H_2}, the characteristic destruction time of H_2 by photodissociation, we obtain a critical velocity for the flow of ≈ 0.7 km s^{-1}, above which the abundance of H_2 becomes significantly different from its "stationary" abundance (with no flow of matter). In fact, the abundance of H_2 is larger because the flow continuously brings new molecules.

The fact that the critical velocity is so close to the estimated flow velocity through the photodissociation region shows that the effect of this flow on chemistry is important. This has been studied in detail by Bertoldi & Draine [37], and later by Störzer & Hollenbach [496]. In reality, our hypothesis of uniform pressure is not exact. Upstream of the ionization front there is a zone compressed by the pressure of the flow of gas just ionized. This is the zone in which the shock, mentioned in the previous section, propagates. In particular, the pressure can increase strongly and suddenly when a neutral region is exposed to strong UV radiation, for example when a star switches on or when a dense fragment is exposed to radiation due to the dispersion of a more tenuous surrounding medium. This pressure increase creates a shock which propagates into the neutral medium.

Figure 12.3 describes the structure of the surface of a neutral globule immersed in a tenuous medium and submitted to the UV radiation field of a hot star. It is limited by a D-type ionization front, from which freshly ionized gas evaporates with a velocity close to the velocity of sound in the H II region. The photodissociation zone is located inside the compressed region upstream, and we can define a photodissociation front, much thicker than the shock front, where the abundances of H I and H_2 are roughly the same. In Fig. 12.3 this front is located between the ionization front and the shock but, in an earlier phase, it might have been upstream of the shock. The dissociation front propagates into the molecular cloud with the velocity

$$v_D = \frac{2\eta f_d F_{\rm UV}}{n_1} = 18.2 \left(\frac{\eta}{0.5}\right) \left(\frac{f_d}{0.15}\right) \frac{\chi}{n_1} \text{ km s}^{-1}, \quad (12.35)$$

in which $\eta \simeq 0.5$ is the fraction of the UV photons that are absorbed by H_2 (the rest being absorbed by dust) and $f_d \simeq 0.15$ is the fraction of these photons that lead to photodissociation. $F_{\rm UV}$ is the photon flux between 912 and 1 110 Å, which is χ times the flux in the solar neighbourhood. The dissociation front slows down with time due to a piling up of the matter upstream. The velocity of the ionization front is given by (12.33) (see also (12.34)) and in general differs from that of the dissociation front. If it is higher, the dissociation front cannot separate from the ionization front and a photodissociation region cannot form. This requires very hot exciting stars so that the flux ratio between the Lyman continuum and the wavelengths between 912 and 1 110 Å is large. Such merged ionization/dissociation fronts might well exist at the surface of the *cometary globules* in the Gum nebula. These are small, neutral, evaporating globules localized inside an extended, low-density H II region and which are illuminated by the very hot stars ζ Puppis and γ^2 Velorum. The Orion nebula also contains cometary globules (Plate 30), many of which are evaporating protostellar disks which might never form stars. Very little atomic hydrogen is expected around these globules and the chemistry is different from the usual chemistry in photodissociation regions, with formation of H_2^+ and also of H_3^+ produced by $H_2^+ + H_2 \rightarrow H_3^+ + H$.

The photodissociation front and the shock can also merge in some cases. Even outside these extreme cases, the physico-chemistry of the photodissociation region can be affected by non-stationary phenomena and requires a time-dependent analysis: see e.g. Bertoldi & Draine [37] for the details.

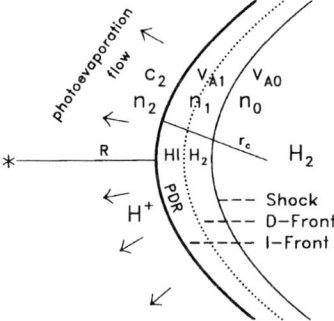

Fig. 12.3. Schematic structure of the surface of a dense molecular globule exposed to the ionizing and dissociating radiation of a hot star. A dense, neutral gas layer between the ionization front and the shock propagates toward the inside of the globule, pushed by the pressure of the ionized gas which escapes through the ionization front and forms a photo-evaporation flow. A dissociation front is also present and this separates the zones dominated, respectively, by atomic and molecular hydrogen. In this example the dissociation front is located between the ionization front and the shock but it might merge with either of these. See the text for the meaning of the parameters. v_{A1} and v_{A2} are the Alfvén velocities in the compressed zone and in the upstream medium, respectively. Reproduced from Bertoldi & Draine [37], with the permission of the AAS.

The non-stationary structures we have described can be identified with the curved bright rims that are often observed at the extremes of H II regions, some of these look like fingers or elephant trunks (see Plates 20 and 22). The brightest part of these rims corresponds to the relatively dense layer which, according to spectroscopic observations, has a velocity of the order of $c_{II} \simeq 10$ km s^{-1} with respect to the ionized gas. A simple calculation by Spitzer [490] gives the orders of magnitude for the parameters of these objects. Let J_0 be the ionizing flux outside the dense ionized layer (12.21) and J the remaining ionizing flux at the ionization front, the rest being absorbed by recombinations in the dense layer. Suppose that the ionization front has a radius R_i and that the ionized gas is expanding spherically. The density in the ionized medium at some radius $R > R_i$ is, given the conservation of matter,

$$n_i(R) = \frac{J}{c_{II}} \left(\frac{R_i}{R}\right)^2. \tag{12.36}$$

Neglecting the absorption by dust, the difference between J_0 and J is simply the number of recombinations in a 1 cm² column through the ionized layer. We thus have

$$J_0 - J = \int_{R_i}^{\infty} \alpha [n_i(R)]^2 dR \approx \frac{\alpha J^2 R_i}{3 c_{II}^2}, \tag{12.37}$$

where (12.36) was used in the integration. The recombination coefficient α is intermediate between the quantities a and $\alpha^{(2)}$ given by (5.8) and (5.15), respectively, because a part of the Lyman continuum photons resulting from recombination escape from the layer. Furthermore, (12.37) is only strictly valid in the direction of the ionizing star. We will assume that it applies to all of the internal hemisphere. The solution of (12.37) for J_0/J is

$$\frac{2 J_0}{J} = 1 + \left(1 + \frac{4}{3} \frac{\alpha J_0 R_i}{c_{II}^2}\right)^{1/2}. \tag{12.38}$$

Let us take $R_i = 0.2$ pc, $\alpha = 4 \times 10^{-13}$ cm³ s⁻¹ and $c_{II} = 10$ km s⁻¹. At a distance of 9 pc from a O7V star, for which Table 5.1 gives $S(0)$, we find $J_0/J \approx 8$ and $n_i \approx 100$ cm⁻³ immediately behind the ionization front. From (12.24) the density of neutral atoms just upstream of the front is of the order of 3×10^4 cm⁻³. This region has probably been compressed by a shock. The thickness of the dense ionized gas is of the order of 15 to 20% of R_i, which is in reasonable agreement with observations.

The rocket effect of the ionized gas on the side of the cloud which faces the star accelerates this cloud, the momentum exchange per second being of the order of $c_{II} dM/dt$, where M is the mass of the cloud. dM/dt is simply the ionization rate of the cloud and is

$$\frac{dM}{dt} = -\pi R_i^2 J \mu_1 m_H, \tag{12.39}$$

while the velocity v of the globule increases such that

$$M = M_0 e^{-v/c_{II}}, \tag{12.40}$$

where M_0 is the initial mass of the globule.

It is clear that, if the H II region forms in a fragmented medium, the above phenomena will yield density irregularities and complex velocity fields inside the H II region, that will persist even after the gas is entirely ionized. This is probably the cause of the large density fluctuations that are observed in many H II regions, although turbulence can also play a role here (cf. Chap. 13).

12.3.4 The Evolution of H II Regions

The evolution of a classical, spherical H II region involves the following steps.

Formation Phase

One or several stars switch on and begin to emit ionizing photons inside a neutral medium, that we will assume for simplification to be uniform and infinite, although this is certainly not the case in reality. A R-type ionization front forms and moves through the gas with a supersonic velocity

$$\frac{dr}{dt} = \frac{S(r)}{4\pi r^2 n_H} = \frac{1}{4\pi r^2 n_H}\left[S(0) - \frac{4}{3}\pi r^3 n_H^2 \alpha^{(2)}\right], \quad (12.41)$$

$S(r)$ being the total number of ionizing photons that reach the radius r of the front per second. This number is related to the total flux $S(0)$ of photons emitted by the star by an equation that can easily be derived from equations of Sect. 5.1. Its integration gives

$$r^3 = r_S^3[1 - \exp(-n_H \alpha^{(2)} t)], \quad (12.42)$$

where r_S is the Strömgren radius given by (5.16). When the front velocity falls below a critical value of the order of $2c_{II}$, where c_{II} is the sound velocity in the H II region, the ionization front changes to a D-type front and is preceded by a shock wave. This occus at a radius r such that

$$\left(\frac{r_S}{r}\right)^3 = 1 + \frac{6c_{II}}{n_H \alpha^{(2)} r}. \quad (12.43)$$

We can see that this critical value for the radius is only a few per cent smaller than the Strömgren radius. This transition marks the end of the formation phase. Detailed numerical studies of the initial phase show that the temperature is higher in the front compared to the equilibrium temperature in the rest of the H II region. This gives a diagnostic for this phase.

Expansion Phase

This phase corresponds to the configuration D-front - compression shock that we already described. The velocity of the ensemble is initially of the order of 10 km s^{-1} and reduces by half when $r = 2r_S$. Expansion ends when pressure equilibrium sets in between the neutral and the ionized regions, for $r \simeq 5r_S$, or when the stars end their main-sequence evolution and cease to emit ionizing photons. This marks the beginning of the recombination phase.

Recombination Phase

The evolution of "fossil" H II regions during this phase depends on the way the ionizing flux decreases with time. From the results of numerical simulations of by various authors, Yorke [564] makes the following general remarks. The decrease of the ionizing flux causes a retrogradation of the ionization front, which might be supersonic if the flux decreases fast. Due to its inertia, the neutral gas which results from recombination keeps expanding, preceded by the shock which is now maintained by inertia, that propagates in the neutral surrounding medium. The H II region keeps the appearance of an ionized sphere surrounded by a thick shell of neutral gas with density comparable to that of the ionized gas, which expands at a few km s^{-1}.

The two last phases correspond to the formation of bubbles of neutral gas which add to those we described Sect. 12.2. They were studied in detail by Tenorio-Tagle & Bodenheimer [507].

Expansion in an Inhomogeneous Medium: the Champagne Phase

In this section, we have assumed that the expansion of the H II region occurs in a homogeneous medium. This is obviously a very idealized situation. What occurs in a clumpy medium at small scale was examined in Sect. 12.3. Expansion in a medium inhomogeneous at large scales will obviously produce spectacular effects. Because massive stars often tend to form near the surface of molecular clouds, the initially spherical H II region around them will puncture the surface and expand through the hole in the low-density surrounding medium. This is the *champagne phase*, so called by analogy with what occurs when a champagne bottle is opened. The expansion of the H II region is essentially free and takes place at the sound velocity of this gas. It forms a blister of ionized gas at the surface of the neutral cloud. The H II region is now bounded by ionization inside the cloud, and by density (lack of matter) outside. Figure 12.4 shows an example of numerical simulation of such a flow. Champagne flows are very important in practice because they can efficiently disperse molecular clouds after star formation.

12.4 The Acceleration of Cosmic Rays

With the development of radioastronomy it became obvious that shock waves in supernova remnants are able to accelerate electrons very easily, and probably also ions. They carry an energy about ten times larger than that contained in cosmic rays, which makes the idea attractive from the energetic point of view. Although it is known that particles are also accelerated in solar and stellar bursts, these bursts do not carry enough energy to produce the galactic cosmic rays.

We will first remind the reader of some basic notions on the propagation of charged particles in a moderately inhomogeneous magnetic field, in particular in the interstellar medium which can be considered as perfectly conducting due to its ionization and low density. The electric field is everywhere zero, and the magnetic field is assumed to be frozen in the gas.

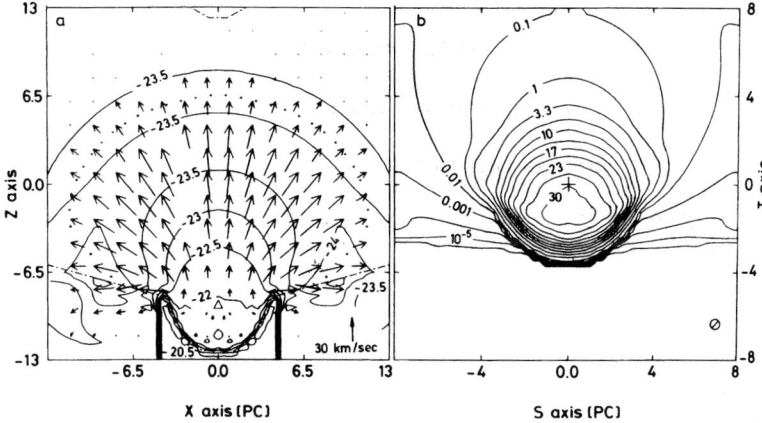

Fig. 12.4. Numerical simulation of the velocity field and the density (left) in a champagne flow with cylindrical symmetry, $t = 6.6 \times 10^5$ years after switch-on of an O star producing 7.6×10^{48} Lyman continuum photons per second. The star is at about 1 Strömgren radius of the cloud surface (triangle on the left of the figure). The contours in this figure give the density in g cm^{-3} in logarithmic scale. The arrows indicate the velocity (scale at the bottom right). The figure to the right shows the radio continuum brightness of this structure at 11 cm wavelength, as seen by a distant observer located in the equatorial plane. The contours give the brightness in units of 10^{-18} erg s^{-1} cm^{-2} Hz^{-1}. The cross indicates the direction of the exciting star. Note that the origins and the coordinate scales are different in the two figures. From Yorke et al. [563], with the permission of ESO.

12.4.1 Propagation of Charged Particles in a Magnetic Field

Homogeneous Magnetic Field

A particle of charge e, rest mass m and momentum **p**, moving with velocity **v** in a magnetic field **B** directed along **x** is submitted to the Lorentz force (Jackson [258])

$$\frac{d\mathbf{p}}{dt} = \gamma m \frac{d\mathbf{v}}{dt} = \frac{e}{c} \mathbf{v} \times \mathbf{B}. \tag{12.44}$$

This yields for the components of **v**, respectively perpendicular, $\mathbf{v}_\perp$, and parallel, $\mathbf{v}_\parallel$, to **B**:

$$\frac{d\mathbf{v}_\perp}{dt} = \left(\mathbf{v}_\perp \times \omega \frac{\mathbf{B}}{|B|} \right); \quad \frac{dv_\parallel}{dt} = \text{constant}, \tag{12.45}$$

in which we have introduced the Larmor angular frequency[7]

$$\omega \frac{\mathbf{B}}{|B|} = \frac{e\mathbf{B}}{\gamma m c} = \frac{ec\mathbf{B}}{E}. \tag{12.46}$$

---

[7] In the non-relativistic case, $\gamma \to 1$, $\omega \to (eB/mc)$ and depends only upon B. The Larmor frequency is then 2.83 MHz/gauss for electrons and 1.61 kHz/gauss for protons.

Integrating twice over time, we obtain successively:

$$\mathbf{v}(t) = v_\parallel \mathbf{u}_3 + \omega r(\mathbf{u}_1 - i\mathbf{u}_2)e^{-i\omega t}, \text{ and} \tag{12.47}$$

$$\mathbf{x}(t) = \mathbf{X}_0 + v_\parallel \mathbf{u}_3 + ir(\mathbf{u}_1 - i\mathbf{u}_2)e^{-i\omega t} \tag{12.48}$$

in which $\mathbf{u}_3, \mathbf{u}_1, \mathbf{u}_2$ are the unit vectors respectively parallel (index 3) and perpendicular (indices 1 and 2) to $\mathbf{B}$. The trajectory of the particle is a helix with radius

$$\boxed{r = \frac{cp_\perp}{eB}.} \tag{12.49}$$

The gyration centre moves parallel to the axis $\mathbf{x}$. This is the *guiding axis* or *guiding centre* of the particle whose origin is $\mathbf{X}_0$. The pitch angle of the helix is $\varphi = \arctan(v_\parallel/\omega r)$.

More generally, for partly ionized ions with atomic mass $\mathcal{A}$ and charge q, we introduce the momentum per nucleon, $p_n = p/\mathcal{A}$, and the gyration radius is written $r = cp/qB = cp_n \mathcal{A}/qB$. The quantity $cp_n \mathcal{A}/q$ is called the *rigidity*.

At high energies, $cp \to E$ and the gyration radius of protons, for $\mathbf{p}$ perpendicular to $\mathbf{B}$, is

$$\boxed{r = \frac{E}{eB} = 3.33 \times 10^{12} \frac{(E/\text{GeV})}{(B/\mu\text{G})} \text{ cm}.} \tag{12.50}$$

Magnetic Field with a Transverse Gradient

Let us now consider a magnetic field with a small transverse gradient $(1/B_0)(\partial B/\partial \xi)$, where ξ is the distance along $\mathbf{n}$, the unit vector perpendicular to $\mathbf{B}$. By small, we here mean that the relative variation of B is not large with respect to the gyration radius r, allowing a first-order analysis. The Larmor frequency is:

$$\omega(\mathbf{x})\frac{\mathbf{B}}{|B|} = \frac{e}{\gamma mc}\mathbf{B}(\mathbf{x}) = \omega_0 \frac{\mathbf{B}}{|B|}\left[1 + \left(\frac{1}{B_0}\frac{\partial B}{\partial \xi}\right)(\mathbf{n}\cdot\mathbf{x})\right]. \tag{12.51}$$

As we know that the guiding centre moves parallel to $\mathbf{B}$ during the motion of the particle, we will only consider the component of the velocity perpendicular to $\mathbf{B}$, $\mathbf{v}_\perp = \mathbf{v}_0 + \mathbf{v}_1$, where we assume $\mathbf{v}_0$ to be constant and $\mathbf{v}_1$ to be small with respect to $\mathbf{v}_0$. Introducing (12.51) in (12.45), and neglecting the higher-order term in $\mathbf{v}_1(1/B_0)(\partial E/\partial \xi)$, we obtain:

$$\frac{d\mathbf{v}_\perp}{dt} = \left[\mathbf{v}_1 + \mathbf{v}_0\left(\mathbf{n}\cdot\mathbf{x}_0\right)\left(\frac{1}{B_0}\frac{\partial B}{\partial \xi}\right)\right] \times \frac{\omega_0 \mathbf{B}}{|B|}. \tag{12.52}$$

From 12.47 and 12.48, we then get:

$$\mathbf{v}_0 = -\frac{\omega_0 \mathbf{B}}{|B|} \times (\mathbf{x}_0 - \mathbf{X}), \tag{12.53}$$

where $\mathbf{X}$ is the gyration centre in a uniform field corresponding to $\mathbf{B}_0$. Let us take $\mathbf{X} = 0$. Substituting (12.53) in (12.52) and neglecting the second term of higher order in $\mathbf{v}_1$, we obtain

$$\frac{d\mathbf{v}_1}{dt} = \left[\mathbf{v}_1 - \left(\frac{1}{B_0}\frac{\partial B}{\partial \xi}\right)\frac{\omega_0 \mathbf{B}}{|B|} \times \mathbf{x}_0(\mathbf{n} \cdot \mathbf{x}_0)\right] \times \frac{\omega_0 \mathbf{B}}{|B|}. \tag{12.54}$$

We expect the motion to be approximately circular, so that the time average of $d\mathbf{v}_1/dt$ is zero, hence

$$\mathbf{v}_1 = \left(\frac{1}{B_0}\frac{\partial B}{\partial \xi}\right)\frac{\omega_0 \mathbf{B}}{|B|} \times \mathbf{x}_0(\mathbf{n} \cdot \mathbf{x}_0). \tag{12.55}$$

We know that $\mathbf{x}_0$ moves on a circle with radius r. Only the component of $\mathbf{x}_0$ perpendicular to $\mathbf{n}$ intervenes, given the scalar product $(\mathbf{n} \cdot \mathbf{x}_0)$. We then have for the average of $\mathbf{v}_1$

$$\langle \mathbf{v}_1 \rangle \equiv \mathbf{v}_D = \left(\frac{1}{B_0}\frac{\partial B}{\partial \xi}\right)\frac{\omega_0 \mathbf{B}}{|B|} \times \langle (\mathbf{x}_0)_\perp (\mathbf{n} \cdot \mathbf{x}_0)\rangle, \tag{12.56}$$

with $\langle (\mathbf{x}_0)_\perp (\mathbf{n} \cdot \mathbf{x}_0)\rangle = \langle \mathbf{n} r_0^2 \sin^2 \omega t \rangle = (1/2)r_0^2 \mathbf{n}$, hence finally we have

$$\mathbf{v}_D = \frac{r_0^2}{2}\left(\frac{1}{B_0}\frac{\partial B}{\partial \xi}\right)\left(\frac{\omega_0 \mathbf{B}}{|B|} \times \mathbf{n}\right), \tag{12.57}$$

or, independent of the coordinates,

$$\frac{\mathbf{v}_D}{\omega r} = \frac{r}{2B^2}\left(\mathbf{B} \times \mathbf{n}\frac{\partial B}{\partial \xi}\right). \tag{12.58}$$

$\mathbf{v}_D$ is the *drift velocity* of the gyration centre. If $r(1/B_0)(\partial B/\partial \xi) \ll 1$, it is clear that the drift velocity $\mathbf{v}_D$ is small with respect to the orbital velocity ωr. It is perpendicular to the direction $\mathbf{n}$ of the gradient and to the direction of the magnetic field $\mathbf{B}$. After one gyration of the particle, the guiding centre has moved by

$$2\pi \frac{\mathbf{v}_D}{\omega} = \pi \frac{r^2}{B^2}\left(\mathbf{B} \times \mathbf{n}\frac{\partial B}{\partial \xi}\right). \tag{12.59}$$

The trajectory of the particle is displayed in Fig. 12.5.

Magnetic Field with a Longitudinal Gradient

In a magnetic field $B(z)$ with a longitudinal gradient the lines of force are compressed in the direction of increase of the field. The angular momentum J of the particle around its guiding centre is

$$J = rp_\perp = \frac{cp_\perp}{eB(z)}p_\perp = \frac{c}{e}\frac{p^2 \sin^2 \theta}{B(z)}. \tag{12.60}$$

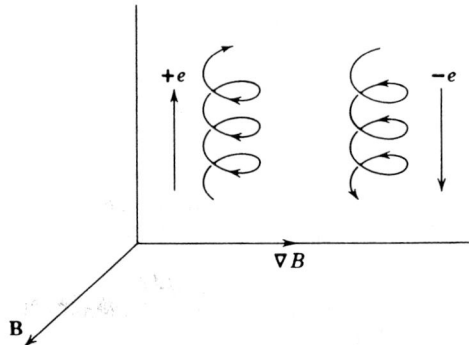

Fig. 12.5. Trajectory of a positively or negatively charged particle in a magnetic field with a transverse gradient $(1/B_0)(\partial B/\partial \xi)$ in the direction **n** perpendicular to the field. The component of the velocity parallel to the field is assumed to be zero. The motion of the particle is therefore in the plane of the figure. The vectors **v**, **n** and **B** form a trihedron with positive orientation. From Jackson [258].

If there is no change of energy the angular momentum J and the modulus of the particle momentum p remain constant. As a consequence, the quantity $\sin^2\theta/B(z)$ remains constant. It is called the *adiabatic invariant*.

Replacing $p_\perp$ by $eB(z)r/c$ in (12.60) we also find $r^2 B =$ constant. This means that the magnetic flux inside the trajectory of the particle is also invariant. Coming back to the general case with arbitrary v, we have $v^2 = v_\parallel^2 + v_\perp^2$. At the point $z = 0$, let us take $B(z) = B_0$ and $v_\perp{}^2 = v_{\perp 0}^2$. We can immediately derive from (12.60) that

$$\frac{v_\perp^2}{B} = \frac{v_{\perp 0}^2}{B_0}, \qquad (12.61)$$

while the parallel component is everywhere

$$v_\parallel^2 = v_0^2 - v_{\perp 0}^2 \frac{B(z)}{B_0}. \qquad (12.62)$$

It follows that a particle which moves towards the direction of increasing $B(z)$ can only reach a point where the longitudinal component of its velocity $v_\parallel$ becomes zero. The longitudinal component reverses and the particle is repelled. This configuration is a *magnetic mirror* (Fig. 12.6). Two similar but opposite configurations form a *magnetic bottle*.

12.4.2 Diffusion of Charged Particles in a Disordered Medium

The structure of the interstellar magnetic field is far more complex than the elementary cases we have just considered. In the general case of a disordered magnetic field the treatment becomes quite heavy and has been discussed by many authors

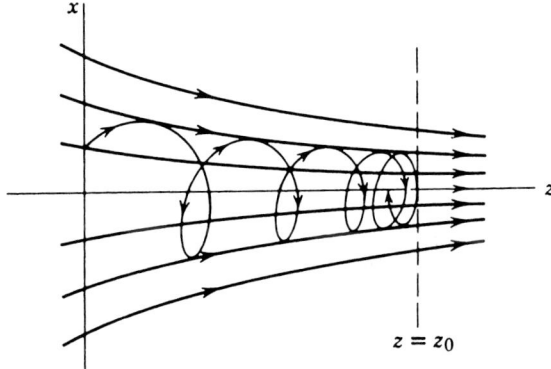

Fig. 12.6. Trajectory of a particle in a magnetic field with a longitudinal gradient. When the pitch angle $\theta = \pi/2$, the particle is repelled by a magnetic mirror. Two similar, opposite configurations form a magnetic bottle where the particle is trapped. From Jackson [258].

in the past. In principle, we should calculate the exact trajectories after the statistical properties of the field have been defined (see e.g. Axford [12]). However, it is possible to produce an interesting, although non quantitative, description of particle diffusion using a simple phenomenological model with two diffusion coefficients, respectively, parallel and perpendicular to the field (Drury [143]).

The fluctuations of the transverse component of the mean magnetic field B_0 are given by

$$(\Delta\varphi)^2 B_0^2 = kE(k), \tag{12.63}$$

φ representing the direction of the magnetic field and $E(k)$ being the spectral density of the fluctuations with a spatial frequency k. It is often assumed, although this is not necessarily true, that $E(k) \propto k^{-5/3}$, which means that the spectral density obeys the Kolmogorov distribution characteristic of turbulence (see later Chap. 13).

While the particle follows its spiral trajectory, its guiding axis changes direction progressively. After N turns its direction has changed by an angle $\varphi \approx N^{1/2}\Delta\varphi$. When φ reaches approximately 1, we may consider that the particle has no memory of its initial direction. In other words, the correlation function between the initial and present directions is zero. We can then say that the particle has experienced a diffusive shock. Therefore, we have

$$N \approx \frac{1}{(\Delta\varphi)^2} \approx \frac{B_0^2}{kE(k)}. \tag{12.64}$$

Le us define as $\lambda_\parallel \approx Nr$ the mean free path of the particle between two such diffusive shocks[8]. The particle velocity being v, the frequency of diffusing shocks

---

[8] More correctly, we should have written $\langle\lambda_\parallel\rangle = 2\pi rN\langle\cos\theta\rangle$, where θ is the pitch angle, with $\langle\cos\theta\rangle = 2/\pi$. In our order-of-magnitude approach we have thus neglected a factor ~ 4.

is $v_\| = v/\lambda_\|$. Let us introduce the *diffusion coefficient* D for a quantity with density n, defined in a general way such that

$$\phi_z = -D\frac{\partial n}{\partial z}, \qquad (12.65)$$

where ϕ_z is the flux of the quantity n in the direction z along which there is a gradient $\partial n/\partial z$. Using the equation for conservation of matter in an infinitely thin layer perpendicular to the z axis, $\partial n/\partial t = -\partial \phi_z/\partial z$, we obtain by differentiating (12.65) with respect to z

$$\frac{\partial n}{\partial t} = D\frac{\partial^2 n}{\partial z^2}. \qquad (12.66)$$

The diffusion coefficient being related to the mean free path by the relation $D = (1/3)\lambda v$, we can write the longitudinal diffusion coefficient (parallel to the field) as

$$D_\| = (1/3)\lambda_\| v = (1/3)\lambda_\|^2 \nu_\|, \qquad (12.67)$$

and with $\lambda_\| \approx Nr$ and (12.64), we have

$$D_\| = \frac{1}{3}r\frac{vB_0^2}{kE(k)}. \qquad (12.68)$$

The elementary transverse displacement at each diffusion is $\Delta\varphi\lambda_\perp \approx \Delta\varphi r$. The transverse mean free path which, as above, corresponds to a zero correlation function is given by $\lambda_\perp \approx N^{1/2}\Delta\varphi r$. Since $N^{1/2}\Delta\varphi \approx 1$ (12.64), it follows that $\lambda_\perp \approx r$. After N revolutions of the particle, its guiding centre has moved by $Nl_\perp$ in a time $\tau \sim rN/v$. Putting $\nu_\perp = 1/\tau$, for the transverse diffusion coefficient we obtain, with $\lambda_\perp \approx r$ and using (12.64),

$$D_\perp = (1/3)\lambda_\perp^2 \nu_\perp = \frac{1}{3}\frac{rvkE(k)}{B_0^2}. \qquad (12.69)$$

The geometric mean of the two coefficients $(D_\| D_\perp)^{1/2} = (1/3)rv$ is called the *Bohm diffusion coefficient*. We also note that $D_\perp/D_\| = 1/N^2$. The transverse diffusion is therefore *a priori* smaller than the longitudinal diffusion, as expected intuitively from the fact that the particles tend to stay coupled to a magnetic field line or to a field tube. This is used as a starting assumption by many authors. However, it was soon realized that this does not account for the observed isotropy of the distribution of cosmic rays in galaxies (cf. Section 6.1). In order to overcome this difficulty, the notion of *diffusion* or *random walk of the field lines* was introduced at the end of the 60's in order to allow for a supplementary, transverse transport of particles.

This term is unfortunately ambiguous. In fact it designates a different mechanism that we will describe without demonstration. Let us consider in a plane perpendicular to the mean field $\langle B \rangle$ the section of the cylinder circumscribed by the helical trajectory of a particle. Putting the origin of a coordinate s along the lines of force, at $s = 0$

the magnetic field forms a bundle of field lines (a *braided magnetic field*) with an approximately circular section corresponding to the trajectory of a test particle. When following s, the shape of the section deforms due to the irregular topography of the field, although its area remains constant due to the conservation of magnetic flux. If the field lines diverge in some direction perpendicular to $\langle B \rangle$, they must converge in the other direction perpendicular to $\langle B \rangle$. The force tube then flattens to a width $l_\parallel \exp(s/l_L)$ and a thickness $l_\parallel \exp(-s/l_L)$, with

$$l_L = \frac{l_\perp^2}{l_\parallel^2} \frac{(\Delta B)^2}{B^2}. \tag{12.70}$$

This is a way to quantify the scale of the fluctuations ΔB of the magnetic field (Duffy et al. [144]). When the thickness becomes smaller than the gyration radius r, the helical trajectory of a particle moving along s necessarily overlaps with the neighbouring force tubes, and the guiding axis is no longer in the initial force tube. This causes an efficient diffusion. If later longitudinal diffusion brings the particle back to the vicinity of $s = 0$, it enters another field tube and thus experiences the equivalent of a transverse diffusion. We should add that the interstellar medium is crossed by Alfvén waves which correspond to a real motion of the field lines. All these effects can be expressed globally by a diffusion tensor: see for example Bieber & Matthaeus [39] who give an elaborated theory of the phenomenon.

Numerical methods have given a new impulse and much more realism to diffusion calculations. Much work continues on the subject and recent studies quantitatively solved several problems, for example, to obtain *ab initio* values for the diffusion coefficients on the sole basis of the statistical properties of the magnetic field. As an illustration, we cite the work of Giacalone & Jokipii [195], who showed that $D_\parallel / D_\perp \approx 0.02$ to 0.04 for energies smaller than or equal to approximately 1 GeV. A realization of the field is constructed, given an adopted fluctuation law, and the trajectories can be calculated exactly. Illustrative examples are presented in graphical form by these authors.

The interest in these studies is not purely academic. They are useful for a correct interpretation of satellite observations in the Solar system, where the magnetic field is highly inclined with respect to the radial directions, and for which a good knowledge of the transverse diffusion is critical in order to infer the elemental abundances of interstellar cosmic rays.

12.4.3 Energy Losses

A high-energy particle circulating in the interstellar medium ionizes the gas through electromagnetic interactions. The corresponding energy loss is given by the Bethe–Bloch formula (8.21). This formula indicates that the energy loss dE/dx per unit length decreases as $1/v^2$ at non-relativistic energies, goes through an *ionization minimum* for a value of kinetic energy $\approx 2.7 Mc^2$, where M is the rest mass of the particle, and slowly increases at higher energies. It is usual to give the energy losses as a function of grammage L, i.e., the mass in a column of matter of unit surface followed by the particle, expressed in g/cm^2. We have $dL = n\mathcal{A}m_H dx$, where n is the number of atoms per cm^3 in the traversed medium and $\mathcal{A}$ their atomic number. Let us introduce the Lorentz factor for the primary particle $\gamma = 1/\sqrt{1-v^2/c^2} = 1/\sqrt{1-\beta^2}$ and express the energy loss in MeV/(g cm^{-2}). We then define the characteristic function

$$f(\gamma) = 0.307 \frac{\gamma^2}{\gamma^2-1}\left[\ln\frac{7.53 \times 10^5}{(\langle \Delta_{eff}\rangle/\text{eV})}(\gamma^2-1) - \frac{\gamma^2}{\gamma^2-1}\right], \quad (12.71)$$

where $\langle \Delta_{eff}\rangle$ is the mean ionization potential of the traversed medium, which is in general taken to be ≈ 15 eV (cf. (8.21)). The Bethe–Bloch equation can then be written as

$$\frac{dE}{dL} = Z^2 \frac{z}{\mathcal{A}} n\, f(\gamma), \quad (12.72)$$

where z and $\mathcal{A}$ are, respectively, the charge and atomic number of the particles in the medium. Their ratio is 1 for hydrogen and $\sim 1/2$ for the other elements. Z is the charge of the primary particle. The function $f(\gamma)$ (12.71) is plotted in Fig. 12.7, for which $\langle \Delta E_{eff}\rangle$ is taken to be 13.6 eV, the ionization potential of hydrogen.

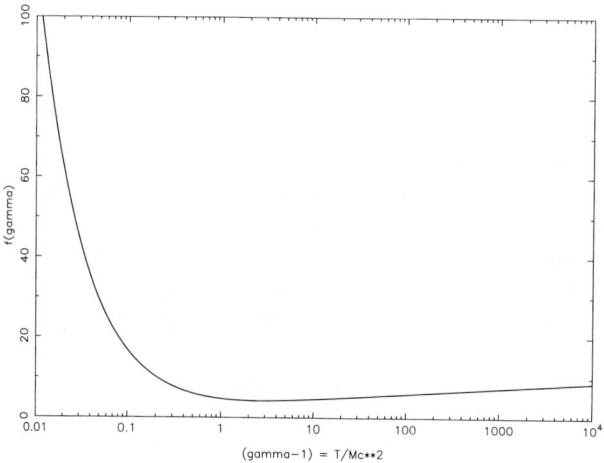

Fig. 12.7. The characteristic function for the ionization energy losses by a primary particle of mass M as a function of its kinetic energy T represented as $T/Mc^2 = (\gamma - 1)$ (12.71). The energy loss is given by (12.72), in MeV per g cm^{-2} for a particle of charge Z, circulating in a medium of atomic number $\mathcal{A}$ and charge per atom z.

12.4.4 The Acceleration of Charged Particles

The Fermi Mechanism

Fermi [177] proposed a statistical acceleration mechanism in the interstellar medium, that we mention for historical reasons. Consider a particle with initial energy E and velocity v ($\beta = v/c$) circulating in a low-density medium with a turbulent magnetic field. Let us also consider a magnetic perturbation of this medium (e.g. a cloud) moving along the axis x with velocity U. When encountering this perturbation the particle is repelled by a magnetic mirror (Sect. 12.4) and comes back with the same pitch angle θ, the direction of motion of its guiding centre being inverted by this interaction.

If the collision is head-on (or direct) the particle gains energy. Conversely, it looses energy if the collision is trailing. Applying the classical relativistic transformations for a fixed reference with respect to the reference frame of the perturbation (noted with a ′ sign), the energy and momentum of the particle are respectively $E' = \gamma_U(E + Up_x)$ and $p'_x = \gamma_U(p_x + UE/c^2)$, with $\gamma_U = \sqrt{1 - U^2/c^2}$. If the collision is head-on, U is negative. Using the same relations to come back to the fixed reference frame (noted with a ″ sign although this frame is superimposed on the initial reference frame), after changing p'_x into $-p'_x$ to account for the inversion of the direction of propagation of the particle, we find

$$E'' = \gamma_U^2(E \pm 2Up_x + U^2)E/c^2). \tag{12.73}$$

Then, noting that $p_x/E = p\cos\theta/E = v\cos\theta/c^2$,

$$\frac{E''}{E} = \frac{(1 \pm 2(U/c)(v/c)\cos\theta + (U/c)^2}{1 - (U/c)^2}. \tag{12.74}$$

The + or − signs correspond to the sign of U hence, respectively, to the approaching or receding motion of the magnetic mirror, and hence to a gain or a loss of energy. Since the direct collisions (approaching) are statistically more numerous that the inverse ones (receding), due to the relative velocities, there is a net energy gain for the particle, whose energy increases continuously due to the accumulated collisions. In the interstellar case, the mechanism proposed by Fermi is not very efficient but the fact that particles can gain energy in collisions with the irregularities of the magnetic field is at the basis of all acceleration theories.

Basic Theory of Acceleration in Shock Waves

"Fortunately the real world is not as simple as first-year physics" (Drury [143] p. 976).

It can be shown that a shock wave propagating in a uniform medium has little effect on charged particles (Drury [143]). However, the associated magnetic irregularities diffuse the particles on either side of the shock and are extremely efficient. We will assume here that the magnetic field is longitudinal (parallel to the x axis), and that the shock is moving along in the direction of x with velocity v_s. This shock is a transition zone, assumed to be thin compared to the gyration radius of the particles, which separates a gas at rest (the unperturbed interstellar medium) from a denser gas with velocity v_2 (Sect 11.2). We consider that the magnetic field has no effect on the shock, which is supposed to be a strong J shock (Sect. 11.2), but acts only on the charged particles.

Let us consider test particles with density $f(x, p, t)$ in the phase space, injected isotropically into the shock with a velocity v much higher than that of the shock. Since they are test particles, we will assume that their density is sufficiently low and that there is only a negligible reaction on the shock or the gas.

An upstream test particle which meets the shock crosses it without any problem but is then trapped in the magnetic field in the compressed medium moving with a velocity $\sim v_2$, itself close to v_s with respect to the gas at rest. The particle then gains energy with respect to the pre-shock reference frame 1. Since its velocity is much larger than v_2, it has a non-zero probability of crossing the shock and coming back into the upstream region. It has then gained an energy given by (12.74), in which $U = v_2 - v_1$. In this medium diffusion reduces its velocity parallel to the velocity of the shock and it can again cross the shock front. After another period of diffusion behind the shock it can again cross the shock in the upstream direction with further energy gain, and so on.

In order to treat this mechanism analytically it is convenient to chose as in Chap. 11 a reference frame comprising the plane of the shock with its perpendicular axis x along the velocity of the shock. The gas flows coming from the upstream direction ($x < 0$) with a velocity $v_1 \approx v_s$, crosses the plane and flows downstream with velocity $v_2 < v_s$ (11.28). The velocities are now relative to that of the shock, as in Chap. 11. We assume that both the upshock and the downshock gas contain magnetic irregularities able to scatter the charged particles.

The problem is therefore that of diffusion in a fluid moving with different velocities on either side of the shock. Mathematically, this means solving the *diffusion-advection* equation near a discontinuity. The evolution of the density $f(x, p, t)$ of particles in the phase space is given by:

$$\frac{\partial f}{\partial t} + v\frac{\partial f}{\partial x} = \frac{\partial}{\partial x}\left[D(x, p)\frac{\partial f}{\partial x}\right] + \frac{1}{3}\frac{\partial v}{\partial x} p \frac{\partial f}{\partial p}. \tag{12.75}$$

The second term of the left-hand part includes the general motion of the diffusion centres. The second term of the right-hand part expresses the conservation of energy (the Liouville theorem). The term

$$D(x, p) = \frac{\lambda(r)v}{3} \tag{12.76}$$

is the diffusion coefficient in the x direction for particles with velocity v, and is equivalent to the $D_\parallel$ coefficient given by (12.67). If the magnetic field is perpendicular to the shock front, the problem is one-dimensional and the transverse diffusion coefficient does not intervene. Oblique shocks for which B is not parallel to x will not be examined here: see for example Drury [143].

The particular velocities of the magnetic diffusion centres with respect to the general flow, which are of the order of the Alfvén velocity (10 to 50 km/s), will be neglected with respect to the velocities v_1 and v_2, and *a fortiori* with respect to the velocities of the accelerated particles.

12.4 The Acceleration of Cosmic Rays

The gyration radius r of particles with atomic mass $\mathcal{A}$ and charge q is

$$r = \frac{p_n}{eB} \frac{\mathcal{A}}{q}, \qquad (12.77)$$

where p_n is the momentum per nucleon. The mean free path of these particles is $\lambda(r) \sim r \mathcal{I}(r)$, where $\mathcal{I}(r)$ is the spectral power of the magnetic irregularities. Let us assume for simplicity that $\mathcal{I}(r)$ does not depend on r and is close to 1 (a white spectrum). Then $\lambda(r) \approx r$. The mean free path does not depend upon the nature of the particle but only upon its gyration radius (or its rigidity). Our assumption that $\lambda(r) \approx r$ is rather natural because the particle ignores the irregularities that are small with respect to r, and follows without diffusion the irregularities that are large with respect to their gyration radius.

Equation (12.75) can be solved exactly (see e.g. Drury [143] and references herein). However, we will here follow the more suggestive presentation of Bell [28] in which the relevant microscopic physics is made clear.

In the chosen reference frame, it is natural to search for a stationary solution. Since $\partial f/\partial t$ and $\partial v/\partial t$ are zero outside the shock, (12.75) downstream reduces to

$$v_2 \frac{\partial f}{\partial t} = \frac{\partial}{\partial x}\left[D(x,p) \frac{\partial f}{\partial x}\right] = 0, \qquad (12.78)$$

whose general solution, obtained by double integration over x, is

$$f(x,p) = a + b \exp\left(\int^x \frac{v_2}{D(x',p)} dx'\right), \qquad (12.79)$$

where a and b are the integration constants. $D(x,p)$ being finite but x extending to infinity, the second term diverges. A physical solution therefore requires that b is zero. The flow of test particles downstream, far from the shock, is $v_2 f(x,p) - D(\partial f/\partial x)$, which is equal to $v_2 f(0,p)$. This is the flow of particles that escape from the shock at large downstream distance. Assuming an isotropic diffusion, the number of particles that are diffused backwards, cross the shock and cross it again downstream is $(1/4)v f(0,p)$ per unit time. Of these particles, $v_2 f(0,p)$ escape downstream while the rest diffuses upstream again. The (small) escape probability is thus

$$\eta = 4 \frac{v_2}{v}. \qquad (12.80)$$

When a particle with energy E_k goes from region 1 to region 2 its energy is (12.73) and (12.74)

$$E_2 = \gamma_U [E_k + (v_1 - v_2) p_{k1} \cos \theta_{k1}] = \gamma_U E_k \left[1 + (v_1 - v_2) \frac{v_{k1} \cos \theta_{k1}}{c^2}\right], \qquad (12.81)$$

then, when it comes back in region 1, its energy is such that

$$\frac{E_{k+1}}{E_k} = \gamma_U^2 \left[1 + (v_2 - v_1) \frac{v_{k2} \cos \theta_{k2}}{c^2}\right]\left[1 + (v_2 - v_1) \frac{v_{k1} \cos \theta_{k1}}{c^2}\right]. \qquad (12.82)$$

In these equations v_k is the particle velocity and the indices 1 and 2 indicate that the particles goes from 1 to 2, or from 2 to 1 respectively[9].

Particles injected with energy E_0 have an energy after l cycles

$$\frac{E_l}{E_0} = \prod_{k=0}^{l-1} \frac{E_{k+1}}{E_k}. \tag{12.83}$$

Introducing the values of E_k and E_{k+1} and taking the logarithm in order to express this in terms of a sum, we have

$$\ln\left(\frac{E_l}{E_0}\right) = \ln\left\{\frac{1}{1 - [(v_1 - v_2)/c]^2}\right\}^l$$
$$+ \sum_{k=1}^{l-1} \ln\left(1 + \frac{v_1 - v_2}{c}\cos\theta_{k1}\right) + \sum_{k=1}^{l-1} \ln\left(1 + \frac{v_2 - v_1}{c}\cos\theta_{k2}\right). \tag{12.84}$$

For having an appreciable energy gain, l must be at least of the order of $c/(v_1 - v_2)$, a condition that we can write as

$$l = O\left(\frac{c}{v_1 - v_2}\right) = \frac{c}{v_1 - v_2}\left[1 + 1/O\left(\frac{c}{v_1 - v_2}\right)\right]$$
$$= \frac{c}{v_1 - v_2}\left[1 + O\left(\frac{v_1 - v_2}{c}\right)\right], \tag{12.85}$$

where $O()$ means *of the order of*. Assuming that the velocity of the particle $v_k \approx c$ simplifies the writing and is in any case valid at very high energies. Because the successive terms in l are not very different from each other we may replace the sums by the products of l with the mean values of the terms with index 1 and 2, so that

$$\ln\left(\frac{E_l}{E_0}\right) = l\ln\left\{\frac{1}{1 - [(v_1 - v_2)/c]^2}\right\}$$
$$+ l\left\langle\ln\left(1 + \frac{v_1 - v_2}{c}\cos\theta_{k1}\right)\right\rangle + l\left\langle\ln\left(1 + \frac{v_1 - v_2}{c}\cos\theta_{k2}\right)\right\rangle. \tag{12.86}$$

The number of particles that cross the shock between the angles θ and $\theta + d\theta$ is proportional to $2\pi \sin\theta \cos\theta d\theta$. The first term can be neglected. Averaging over angles from 0 to $\pi/2$ for index 1 and from π to $\pi/2$ for index 2, expanding the logarithm and retaining only the first-order term in $(v_1 - v_2)/c$, we find

$$\ln\left(\frac{E_l}{E_0}\right) = \frac{4}{3}l\frac{v_1 - v_2}{c}\left[1 + O\left(\frac{v_1 - v_2}{c}\right)\right], \tag{12.87}$$

---

[9] Bell gives the ratio (his formula 4, p. 149) and not the product of the quantities between brackets in (12.82). This is apparently a misprint, with no consequences for what follows.

because the average of $\cos\theta$ weighted by $2\pi \sin\theta \cos\theta$ is $(4/3)(\cos^3\theta)/(\sin 2\theta)$ within the above limits.

The probability P_l that a particle has experienced l cycles and therefore has reached an energy E_l is

$$P_l = \ln(1-\eta) = \ln\left(1 - \frac{4v_2}{c}\right)$$
$$= \frac{3v_2}{v_1 - v_2} \ln\left(\frac{E_l}{E_0}\right)\left[1 + O\left(\frac{v_1 - v_2}{c}\right)\right]. \quad (12.88)$$

When $l \to \infty$, the normalized spectrum is given by

$$\int_{E_0}^{\infty} P(E)dE = C E_0 \int_{E_0}^{\infty} \left(\frac{E}{E_0}\right)^{-\mu} \frac{dE}{E_0} = 1, \quad (12.89)$$

which determines the constant C, and finally

$$N(E) \propto \left(\frac{\mu - 1}{E_0}\right)\left(\frac{E}{E_0}\right)^{-\mu}, \quad (12.90)$$

with

$$\mu = \frac{2v_2 + v_1}{v_1 - v_2}\left[1 + O\left(\frac{v_1 - v_2}{c}\right)\right]. \quad (12.91)$$

The term $1 + O[(v_1 - v_2)/c]$ is small and can be neglected in what follows.

The important result we have obtained is that the spectrum of the accelerated particles follows a power law. Note that we did not specify the distribution of the momentum $f(x, p)$ (or of the velocity v) of the test particles. There is no effect of the initial momentum distribution upon the final result. The memory of the initial spectrum of the particles has thus been totally erased. This is characteristic for a stochastic acceleration mechanism derived from the Fermi mechanism. For a strong J shock whose velocity v_s is large with respect to the Alfvén velocity v_A, $v_1/v_2 \approx 4$ (11.28), hence a value of $\mu = 2$ for the exponent, which is close to the observed exponent (actually slightly smaller).

Despite of these very encouraging conclusions, we should not forget the hypotheses and the important simplifications we have made.

1. We assumed that the test particles have an initial velocity larger than that of the shock, so that they can cross into the upstream direction. We might of course imagine that they correspond to a suprathermal energy tail of the distribution of ions in stellar coronae. However, these ions should be able to diffuse until they reach the shocks but the energy losses along the path are too large. It thus seems necessary that the shocks find particles pre-accelerated in the regions they cross (Eichler [153]). This condition is met if we assume that interstellar ions, when entering the shock, have a relative velocity at least equal to that of the shock (which corresponds to $\sim 10^6$ K). This is the basic assumption of the acceleration models of Blandford & Ostriker [42] and followers.

2. The calculation of the spectrum that we have presented is, in principle, valid for ultra-relativistic particles only. It is possible to lift this restriction but in any case the initial velocity of the particles must be larger than that of the shock. The correct spectrum is found to be (Bell [28])

$$N(T) \propto (\mu - 1)(T_0^2 + 2E_R T_0)^{(\mu-1)/2} (T_0^2 + 2E_R T_0)^{(\mu+1)/2}. \qquad (12.92)$$

$E_R = mc^2$ is the rest energy of the particle and T_0 its kinetic energy. It follows that the slope of the spectrum is flatter by a value of 1 below ~ 1 GeV for protons, and ~ 0.5 MeV for electrons.

3. We have not taken into account the reaction of the accelerated particles on the nature and evolution of the shock. However, this reaction cannot be negligible in the real life because the energy contained in cosmic rays is approximately $1/10$ of the initial energy in supernova remnants. Those particles which escape upstream generate Alfvén waves which exert some pressure on the gas upstream and produce a magnetic precursor. These waves interact between themselves, become rapidly isotropic and produce the magnetic irregularities that are required for the diffusion of charged particles. Downstream, it is indeed observed that the medium is very hot and turbulent. It is interesting to note that the accelerated particles generate, by themselves, the magnetic irregularities that are necessary for their acceleration.

Acceleration by Modified Shocks

The restrictions we mentioned can be, to a large extent, lifted by numerical methods which, in particular, allow to account for the diffusion of particles from their lowest energy, at the very beginning of the acceleration, and to take into account the reaction of the accelerated particles on the shock. When considered in this way, the accelerating shocks are sometimes called *modified shocks*. Based on measurements made with the ULYSSES space probe on the interplanetary medium, where the magnetic field and the particle energy spectrum are both measured *in situ*, Baring et al. ([20] and references herein) have shown that the shock is able to accelerate the interstellar ions themselves without preliminary injection. Additionally, the acceleration mechanism involving the turbulence associated with the shock can explain the observations in a satisfactory fashion.

As an illustration we will present in more detail the study made by Ellison et al. [154] for the origin of galactic cosmic rays. It assumes a stationary interaction between the shock and the accelerated particles. Particle diffusion is treated by a Monte-Carlo method, starting with the thermal energy: no injection is postulated. Moreover, while the essence is the same, the modelling of the acceleration mechanism is considerably more complete than the one we described at the beginning of this Section. Particles are no longer test particles but their pressure is now taken into account. It is found that this pressure progressively repels and compresses the gas upstream and that the shock is no longer a sharp discontinuity but a zone where the velocity changes progressively. The velocity profile is represented in Fig. 12.8. The abscissae x in this figure are given in units of $\lambda_0 = \eta r_0$, where r_0 is the gyration radius of a proton which would have the velocity v_s of the shock. $\eta = 1$ is assumed,

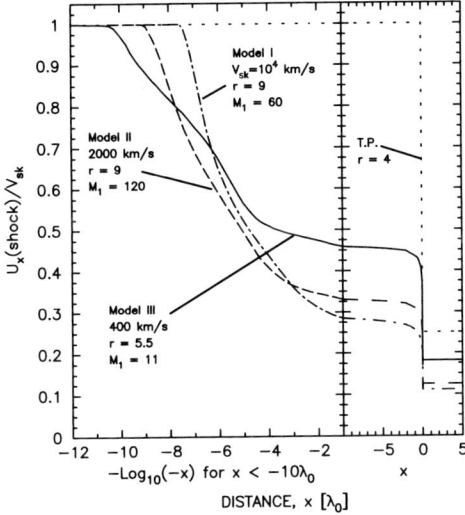

Fig. 12.8. Distribution of the gas velocity U in a shock, in the shock reference frame, as a function of the distance x from the shock front. The gas comes from the left (negative x). The scale, in units of $\lambda_0 = \eta r_0$ (see text), is linear from $x = 0$ to $-10\lambda_0$, then logarithmic to $-\log x = -\log 10^{12}$. The vertical scale gives the gas velocity normalized to that of the shock (here v_{sk}) The three curves correspond to three values of the shock velocity, as indicated. Reproduced from Ellison et al. [154], with the permission of the AAS.

which would mean $N = 1$ in (12.64), and corresponds to very strong turbulence. λ_0 is thus the mean free path of a proton having the same velocity as the shock. In a typical case with $v_s = 400$ km s^{-1} (full line), the particles initially penetrate the shock at $x \approx -10^{10} \lambda_0$. The initial gyration radius, and consequently the mean free path of a proton, is

$$\lambda_0 \approx 10^7 \left(\frac{v}{\text{km/s}}\right)\left(\frac{B}{\mu\text{G}}\right) \text{ cm.} \qquad (12.93)$$

For particles with atomic mass $\mathcal{A}$ and charge q, a shock velocity of 400 km s^{-1} and a magnetic field $B = 5\,\mu$G, the initial mean free path is

$$\lambda(\mathcal{A}, q) \approx r_0 \frac{\mathcal{A}}{q} \approx 0.8 \times 10^9 \frac{\mathcal{A}}{q} \text{ cm.} \qquad (12.94)$$

The thickness of the acceleration zone ($\sim 10^{10}\lambda_0 \approx 8 \times 10^{18}$ cm) can reach 1 pc or more. The velocity gradient indicates a progressive compression of the gas. Let $v(x)$ be the gas velocity at a point x upstream of the shock. This matter moves with the velocity $v_s - v(x)$ in the shock reference frame; on each side, the velocities $v(x \pm \Delta x)$, and thus the matter, converge towards x. This is an ideal condition for an efficient Fermi mechanism where particles are confined between two converging diffusion zones. Such a contraction zone crosses the shock in a few hundred years.

An examination of Fig. 12.8 allows us to intuitively understand the consequences of this approach.

1. Given (12.77), partly ionized atoms heavier than hydrogen have an initial gyration radius (at the shock velocity v_s, but with $A/q > 1$) larger than that of protons. As a consequence, they penetrate deeper than protons into the shock, find in the shock reference frame a higher velocity difference and are thus accelerated more efficiently. As their energy grows, they loose more and more electrons and their gyration radius, at some 10^6 GeV/n, is $\approx 1.7 \times 10^9 \lambda_0$ (Eq. 12.77 with $A/q \approx 2$). They can still be confined to the shock and thus be accelerated up to this energy. Conversely, thermal electrons only see a very small velocity gradient at the scale of their gyration radius. This could explain their very small abundance, i.e., only $1/100$ of that of the protons, but they can be accelerated to energies of about 10^6 GeV, which are ultra-relativistic for them.

2. Similarly, we can assume that dust grains, even slightly electrically charged and despite their enormous masses compared to those of ions, can be accelerated. This question is examined in detail by Ellison et al. [154]. We will here only show its feasibility.

The electric potential of a grain with radius a and charge qe is $\Psi = qe/4\pi\epsilon_0 a = 1.4q/(a/\text{nm})$ volts. The high-energy tail of the electrons in a gas with temperature T reaches $2.5kT$, thus $(E_e/\text{eV}) \approx 2.2 \times 10^{-4} T$. Equating E_e to Ψ, we have $q = 1.5\,10^{-4} Ta$. For grains with $a = 100$ nm, $\Psi \approx 2.2$ volts. However, the grain charge can be very different depending on the ambient radiation field (Bakes & Tielens [13]). Assuming a grain density of 1 g cm^{-3}, we find:

$$\frac{\mathcal{N}}{q} = 1.7 \times 10^4 \left(\frac{a}{\text{nm}}\right)^2 \left(\frac{T}{10^4 \text{K}}\right), \qquad (12.95)$$

where $\mathcal{N}$ represents the total number of nucleons in the grain, all elements included, and T the gas temperature.

A 100 nm grain in a plasma at 10^4 K has an initial gyration radius of about 10^{16} cm (12.94), small with respect to the thickness of the shock. It can then be accelerated like a heavy ion. However, the grains accelerated in this way experience sputtering by the ions of the gas and progressively release their constituent atoms as ions which can then be accelerated.

The possibility of accelerating dust grains, and from them refractory element ions, solves an old problem. It explains why elements like silicon (or even carbon), and many other elements that are not *a priori* expected to be accelerated from interstellar ions are in fact accelerated as easily as volatile ions. Recent cosmic ray abundance studies even show a relative overabundance of some of these refractory elements with respect to their cosmic abundances (Ellison et al. [154] Sect. 6.1 and Fig. 6.5), which can be explained by the composition of grains.

We can also conclude from this study that a shock wave is far from looking like the idealized image of a sharp discontinuity but that it must be considered as an extended turbulent medium with a large velocity gradient. Turbulence is probably generated by Alfvén waves which are created by the accelerated particles themselves (Kulsrud & Pierce [293]; Cesarsky [82]). It is thus necessary to self-consistently treat the shock waves and the particles that they accelerate, something that we have not attempted in this chapter.

13 Interstellar Turbulence

In this chapter we will see that the structure of the interstellar medium is fragmented and self-similar over more than four orders of magnitudes in size, from the galactic to the protostellar scales. This indicates the existence of a strong dynamical coupling between all these scales. Turbulence, which is the subject of this chapter, is at the origin of a large part of this coupling. Open systems, like the interstellar medium, which are out of equilibrium, extended and dissipative, tend to evolve towards self-organized states comparable to critical states. However, this requires that the large-scale processes which keep these systems out of equilibrium are quasi-stationary with respect to the processes that occur at the (smaller) dissipative scales. Such systems do not usually exhibit characteristic scales larger than the dissipative ones and are remarkably resilient: if they are perturbed, they come back to a self-similar structure after some transition period. We find all of these properties in the interstellar medium.

13.1 Velocity Structure and Fragmentation

As soon as the spectral resolution of optical observations allowed us to measure the velocity for the interstellar gas for a very large number of directions ($\sim 10\,000$) with an accuracy of the order of 1 km s^{-1} (cf. Wilson et al. [550]), it was found that the velocities are not purely thermal and that they obey scaling laws. In H II regions, the quantity $\langle [v(x) - v(x+r)]^2 \rangle$ averaged over an extended surface, where $v(x)$ and $v(x+r)$ are the radial velocities (velocities projected on the line of sight) measured at points x separated by r, increases with r as r^β, with β between 0.8 and 1 (Miville-Deschênes et al. [372] and references therein).

Following on from these initial optical studies, the first observations of the radio rotation lines of interstellar molecules showed that the linewidths are also not purely thermal. Observations of the ^{12}CO, ^{13}CO, and OH lines have allowed an approximate determination of the gas temperature, as explained Sect. 4.2. The temperatures determined in this way are often smaller than 10 K. At 10 K, the mean thermal velocity of a ^{12}CO molecule, with molecular mass $\mu = 28$, is $\langle v_{th} \rangle = \sqrt{2kT/\mu m_H} = 0.07$ km s^{-1}, smaller by about two orders of magnitudes than the observed ^{12}CO line widths, which are 1 to 10 km s^{-1}.

These observations soon raised considerable interest and several physical processes have been suggested in order to explain these non-thermal motions. The gravitational collapse of molecular clouds was eliminated because it would produce a star formation rate much larger than that observed in the Galaxy. Another, still plausible scenario, is that of matter divided into very small structures moving in gravitational equilibrium in the potential wells of stars and gas. Finally, there is the possibility that the suprathermal motions are the signature of turbulence in the interstellar velocity fields.

More recently, observations of molecular clouds, in particular those made with the 30 m radiotelescope of IRAM (Falgarone et al. [174]) have shown remarkable scaling laws between the non-thermal part of the internal velocity dispersion and the size of the observed structures, and also between their mass and their size. Two examples of these scaling laws are shown in Figs. 13.1 and 13.2, these come from a compilation of data obtained by different groups on galactic molecular clouds. We will come back to the possible interpretations of these laws but let us say from the start that, although they suggest the presence of turbulence in the interstellar medium, they do not give proof of it. The non-thermal part of the internal velocity dispersion of the molecular structures varies approximately as

$$\sigma_{v,NT} \propto L^{0.5(\pm 0.1)}, \tag{13.1}$$

where L is the size of the structure. In this expression, and the following three, the quantity between brackets in the exponent should not be considered as an error but rather as an indication of real variations in the exponent in different media and in different regions. We also should realize the difficulties that arise when trying to extract structures of a given size within a wide range of encapsulated structures. In any case, the scaling law (13.1) is remarkably similar to the first estimate made decades ago by Wilson et al. [550] in the Orion nebula, and to estimates made on molecular structures as early as 1981 by Larson [300]. The mass of molecular structures varies approximately as (Elmegreen & Falgarone [161], Heithausen et al. [227])

$$M_{H_2} \propto L^\kappa \approx L^{2.5(\pm 0.3)}. \tag{13.2}$$

The corresponding diagram can be seen in Fig. 13.2.

Although once again the relationship of these scaling laws to turbulence, and more generally to the structure of the interstellar medium, is not established, it is interesting to consider this further. Combining (13.1) and (13.2) we get

$$\sigma_v(NT) \propto M^{0.2(\pm 0.05)}. \tag{13.3}$$

The size distribution of the fragments is observed to obey a power law:

$$n(L)dL \propto L^{-3.3(\pm 0.3)}dL. \tag{13.4}$$

The scaling laws and the very appearance of the interstellar medium (Plate 7) suggest that this medium might have a fractal structure (cf. e.g. Bazell & Désert [25], Falgarone et al. [171], or Pfenniger & Combes [409]). In a fractal medium, i.e., in a medium where the structures are geometrically similar to each other at different size scales, the number $N(> L)$ of structures larger than the size L is such that

13.1 Velocity Structure and Fragmentation 303

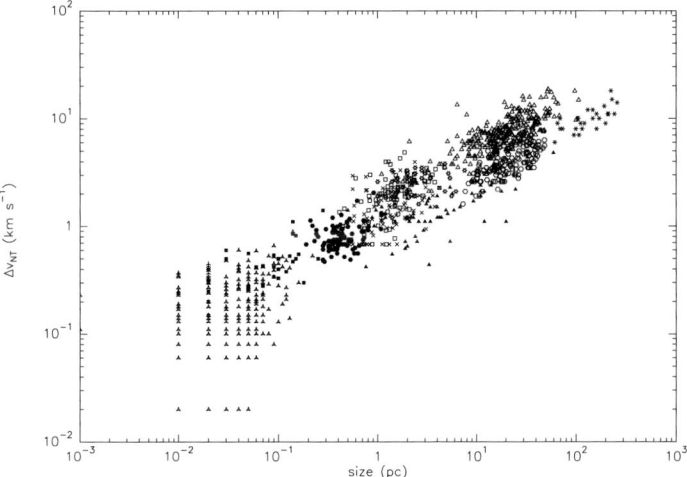

Fig. 13.1. The dispersion of the non-thermal component of interstellar velocities as a function of the size of molecular structures identified in the central region of the Galaxy (stars, empty triangles), in the third quadrant (hexagons), in the Rosette nebula (crosses), in the Maddalena cloud (small empty squares) and in other nearby clouds (other symbols). The small and large filled squares correspond to dense molecular cores in the solar neighbourhood.

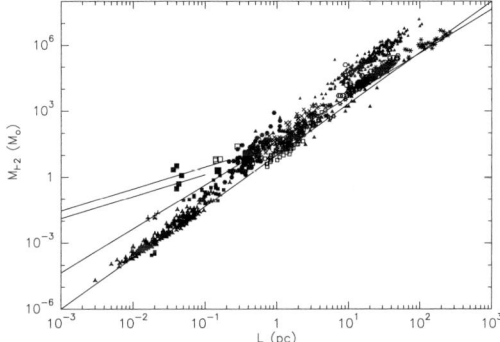

Fig. 13.2. The relationship between the mass of molecular structures and their size. The masses are generally obtained from the intensities of the ^{13}CO and ^{12}CO lines as explained Sect. 4.2. The symbols have the same meaning as in Fig. 13.1. The two long lines have slopes of 2 and 2.3. The two short lines correspond to self-gravitating clouds with temperatures of 10 and 40 K, which could correspond to dense cores.

$$n(>L) \propto L^{-D}, \tag{13.5}$$

which determines the *fractal dimension* D. Differenciating this expression with respect to L, we have

$$n(L)dL \propto L^{-(1+D)}dL. \tag{13.6}$$

Comparing this relation with (13.4) we find $D = 2.3(\pm 0.3)$. D is a three-dimension fractal dimension. It is remarkable that numerical simulations of a fractal medium with a D of about 2.3 look, in projection, very similar to maps of the interstellar medium, for example a CO map in the Orion–Monoceros region: see Plate 7.

Combining (13.2) and (13.6) we find the mass distribution

$$n(M)dM = n(L)\frac{dL}{dM}dM \propto M^{-(1+D/\kappa)}dM. \tag{13.7}$$

Using the above values of D and κ we obtain $n(M) \propto M^{-1.9(\pm 0.3)}$. Direct observations of this relationship in molecular clouds yield $n(M) \propto M^{-1.8}$, on average (see e.g. Heithausen et al. [227]), with rather large uncertainties. This exponent is consistent with the results of some fractal theories which predict $M \propto L^D$, hence $\kappa = D$ and $n(M) \propto M^{-2}$ whatever the fractal dimension D.

After this digression let us return to our subject. It is more than an academic problem to determine if the random supersonic motions of the interstellar gas and, more generally, its structure at different scale sizes are of turbulent origin. Turbulence in fact has very specific properties which have a considerable impact on the physics of the gas, as we will see later.

13.2 Incompressible Turbulence

What follows is only a short introduction to the very broad domain of turbulence. The interested reader will find more details in the books of Landau & Lifshitz [297] and of Guyon et al. [216], for example.

13.2.1 The Birth of Turbulence

Let us consider a fluid cell with density ρ, pressure P and velocity $\mathbf{v}$, subjected to an external force $\mathbf{F}$, for example gravity. In Eulerian coordinates, the equation for the continuity of momentum, that we will call for brevity the equation of motion *(the Navier–Stokes equation)*, that we have already seen in a slightly different form as (11.2), can be written for the component v_j of velocity along the j axis

$$\rho\frac{Dv_j}{Dt} = \rho\left(\frac{\partial v_j}{\partial t} + v_k\frac{\partial v_j}{\partial x_k}\right) = -\frac{\partial P}{\partial x_j} + F_j + \rho\nu\frac{\partial^2 v_j}{\partial x_k \partial x_k}, \tag{13.8}$$

which implies summation over the indices $k = 1, 2, 3$. Two similar equations have to be written for the two other components. In this equation the fluid is assumed to be incompressible. $\nu = (1/3)\lambda \langle v_{th}\rangle$, where λ is the mean free path (see footnote 2 in this chapter) and $\langle v_{th}\rangle$, the mean thermal velocity of the gas particles, is the *kinematic viscosity*[1]. The viscosity is assumed to be isotropic and constant in the medium. The last term is the stress originating from viscosity in the relative motion of the fluid cells. In a symbolic vector form, this equation can be written as

$$\rho\left(\frac{\partial \mathbf{v}}{\partial t} + \mathbf{v}\cdot\nabla\mathbf{v}\right) = -\nabla P + \mathbf{F} + \rho\nu\nabla^2\mathbf{v}, \qquad (13.9)$$

The last term (sometimes noted as $\rho\nu\Delta\mathbf{v}$) symbolizes the stress tensor explained earlier. Two terms express the transport of momentum $\rho\mathbf{v}$: a non-linear advection term $\rho\mathbf{v}\cdot\nabla\mathbf{v}$ and a linear diffusion term $\rho\nu\nabla^2\mathbf{v}$. In a flow where the ratio of advection/diffusion for these two terms becomes larger than about 100, either because the velocity of the flow increases or because the viscosity decreases, e.g. due to a decrease in temperature through the dependence of λ on $\langle v_{th}\rangle$, experiments show that the flow, which was initially laminar, becomes chaotic with large velocity fluctuations on all scales. Moreover, the flow becomes very sensitive to the boundary conditions. This is a transition towards turbulence where eddies (also called vortices) appear. The number which characterizes this transition to turbulence is the *Reynolds number Re*.

Although the theory of turbulence is fraught with uncertainties we can obtain simple results using a dimensional analysis, as follows. An order of magnitude for the gradient of the characteristic fluid velocity v_l at scale l is v_l/l, as can be intuitively understood by assimilating the scale l to an eddy with size l. An order of magnitude for the advection term at this scale l is $\rho v_l^2/l$ while that for the diffusion term is $\rho\nu v_l/l^2$. The Reynolds number for this scale is thus $Re = lv_l/\nu$. If this number reaches values much larger than 100 the turbulence is said to be developed. This transition to turbulence, while experimentally observed, was never inferred from the Navier–Stokes equation, although this equation is valid for describing the motions of a fluid.

13.2.2 The Developed Kolmogorov Turbulence

A phenomenological description of incompressible turbulence is due to Richardson (1922), and later to Kolmogorov [285]. It describes turbulence as an energy cascade produced by successive eddy instabilities, each eddy forming more numerous ones of smaller size. The energy therefore cascades from an eddy to smaller eddies, without dissipation, until the eddies become so small that diffusion dominates the other transport terms. These intermediate scales are called the *inertial scales*, which means that the dynamics is entirely dominated by the advection term $\rho\mathbf{v}\cdot\nabla\mathbf{v}$, dissipation (the diffusion term) being negligible. The cascade therefore occurs between the large scales called the *integral scales*, at which energy is injected, and the so-called *dissipative scales*, at which the diffusion term becomes comparable to the advection term ($Re \sim 1$). This description lead Kolmogorov to suppose (i) scale

---
[1] $\eta = \rho\nu$ is the *dynamic viscosity*, given in poise, symbol $P = \text{g cm}^{-1}\,\text{s}^{-1}$.

invariance, meaning that in the inertial domain the statistical properties of the flow are independent of the scale (self-similarity), and (ii) that the interactions are local in the Fourier space, meaning that the dynamics of a scale l is dominated by those of scales of adjacent sizes. From the first hypothesis we derive that the rate of transfer of specific energy ϵ (energy per unit mass) from scale to scale is independent of the scale in the inertial regime:

$$\epsilon \sim v_l^2/\tau_l, \tag{13.10}$$

the (kinetic) energy being of the order of v_l^2 per unit mass. τ_l is a transfer time which is of the order of the lifetime of a velocity fluctuation, which itself is the characteristic time for the inversion of the direction of the characteristic velocity in an eddy with scale l, $\tau_l \sim l/v_l$. Therefore $\epsilon \sim v_l^3/l$, hence

$$v_l \sim \epsilon^{1/3} l^{1/3}, \tag{13.11}$$

where ϵ is invariant in the cascade. In this description the final energy dissipation rate ϵ_d is thus equal to the transfer time from scale to scale, which is on average uniform in space and constant with time. The scale l_d at which viscous dissipation takes place corresponds, as we said, to a Reynolds number of order unity, so that

$$l_d \sim \left(v^3/\epsilon\right)^{1/4}. \tag{13.12}$$

Viscous dissipation transforms kinetic energy into heat and is more important if the shear in the velocity field is larger. The mean rate of viscous dissipation per unit mass is, for an incompressible fluid (see demonstration in Landau & Lifshitz [297] Sect. 16),

$$\langle \epsilon_d \rangle = \frac{1}{2} \nu \left(\frac{\partial v_j}{\partial x_k} + \frac{\partial v_k}{\partial x_j} \right)^2, \tag{13.13}$$

again summing over j and $k = 1, 2, 3$.

Let us introduce the *vorticity* $\boldsymbol{\omega} = \nabla \times \mathbf{v}$. It is possible to show that $\langle \epsilon_d \rangle$ is related to the vorticity by

$$\langle \epsilon_d \rangle = \nu \langle |\nabla \times \mathbf{v}|^2 \rangle = \nu \langle |\omega|^2 \rangle. \tag{13.14}$$

An essential property of turbulence is the correlation of the velocity fluctuations at all scales, this is described by the constancy of $\langle |\omega|^2 \rangle$ with scale.

Let us now introduce, instead of the scale l, the wavenumber $k = 2\pi/l$ and consider the power spectrum $E(k)$ of the kinetic energy, such that the average kinetic energy per unit mass between k and $k + dk$ is $E(k)dk$. The total specific energy at scale l is of the order of $\langle v_l^2 \rangle$ and is the integral of energy over scales smaller than l, or wavenumbers larger than k:

$$\langle v_l^2 \rangle = \int_k^\infty E(k')dk'. \tag{13.15}$$

This energy is of the order of $(\epsilon l)^{2/3} \sim (\epsilon/k)^{2/3}$ (13.11). Differentiating with respect to k, ϵ being constant, we find the *Kolmogorov law*

$$E(k) \sim \epsilon^{2/3} k^{-5/3}. \qquad (13.16)$$

As a consequence we see that most of the energy is at the largest scales.

It is also possible to demonstrate that the statistical law which relates the moments M_n of order n to the increments in the velocity field for a separation $\mathbf{r}$ (see the beginning of Sect. 13.1) is

$$M_n = \langle |\mathbf{v}(\mathbf{x}+\mathbf{r}) - \mathbf{v}(\mathbf{x})|^n \rangle \propto (\epsilon r)^{n/3}, \qquad (13.17)$$

the average being over all the positions $\mathbf{x}$ and directions $\mathbf{r}$. The Kolmogorov theory for incompressible turbulence predicts that the moment of order 2, also called the *structure function*, is $M_2 \propto r^{2/3}$. However the observations of H II regions give $M_2 \propto r^{0.8-1.0}$ (see the beginning of Sect. 13.1). One could think that the difference can arise from projection effects, since H II regions are three-dimensional structures. However, the corresponding correction for the Kolmogorov law gives $M_2 \propto r^{5/3}$, in disagreement with observations. This was already noted by Wilson et al. [550]; see also Miville–Deschênes et al. [372]. These authors suggest that compressible supersonic turbulence could give results in better agreement with observation. A similar problem arises for molecular clouds, for which we observe $M_2 \propto r^{0.86(\pm 0.3)}$ (Miesch & Bally [370]). However, the power spectrum of 21 cm emission agrees with the Kolmogorov prediction for incompressible turbulence (Miville-Deschênes et al. [373]). It is also possible for such a discussion to use the velocity dispersion/size relation ((13.1) and Fig. 13.1) for which the Kolmogorov theory predicts $\sigma_{v,NT} \propto L^{1/3}$, but its interpretation is much more uncertain (Scalo [454]). Finally, it is possible to use the power spectra of the density and of the velocity field, as recently done for H I in a high-latitude cirrus cloud by Miville-Deschênes et al. [373]: they find that these power spectra are similar and are in good agreement with the predictions of the Kolmogorov theory for incompressible turbulence. Further observational studies are badly needed for a better understanding of the statistical properties of the interstellar medium in its different phases.

13.2.3 Turbulent Viscosity and Pressure

A fluid with a turbulent flow can to some extent be described as a fluid possessing a *turbulent viscosity* v_t, differing from the usual viscosity which is also called the *molecular viscosity* to remind that it originates in collisions between the atoms or molecules of the gas. The turbulent viscosity does not directly produce energy dissipation. It depends on the scale l or on the wave number k. It is also possible to define a turbulent pressure P_t which is also scale-dependent and is obviously related to the characteristic velocity at this scale.

A powerful method has been developed to treat the dynamics of a turbulent medium at large scales. It consists of calculating renormalized transport coefficients which allow us to introduce the effect of the smaller scales $l < L$ (or $k > K$) in the equation of motion for the gas at scale $L \sim 1/K$. Using this method it is possible to define the turbulent viscosity at a scale with a wavenumber K as

$$v_t(K) = \left[\frac{4}{3}\int_K^\infty \frac{E(k)}{k^2}dk\right]^{1/2}, \tag{13.18}$$

and the turbulent pressure at this scale

$$P_t(K) = \frac{1}{3}\rho\langle v^2(K)\rangle, \tag{13.19}$$

$E(k)$, the power spectrum of the turbulence, being related to the mean velocity at the wavenumber K by

$$\langle v^2(K)\rangle = \int_K^\infty E(k)dk. \tag{13.20}$$

The turbulent pressure and the turbulent viscosity, contrary to the thermal pressure and viscosity, depend on the scale at which they are calculated. The equation of motion of the gas at wavenumber K then becomes

$$\rho(\partial \mathbf{v}/\partial t + \mathbf{v}.\nabla \mathbf{v}) = -\nabla P_t(K) + \mathbf{F} + \rho v_t(K)\Delta \mathbf{v}. \tag{13.21}$$

The turbulent viscosity is often larger by orders of magnitudes than the kinematic molecular viscosity v, in particular in the interstellar medium as we will see later. It can be the cause of the diffusive mixing of different parts of the medium. As an example, if a quantity exhibits an important fluctuation at some scale l, turbulent viscosity causes the disappearance of this fluctuation in a time $\tau = l^2/v_t(l)$.

The turbulent pressure helps in stabilizing scales which are gravitationally unstable (Bonazzola et al. [54]). This can be understood intuitively by the fact that pressure gradients, whatever their nature, counteract the collapse of a cloud. The expression for the turbulent pressure (13.19) is similar to that for the thermal pressure $P_{th} = nkT_K = (1/3)\rho\langle v_{th}^2\rangle$. Thus, for a cloud which can be considered as isolated, we can, as a first approximation, add the turbulent pressure *at the scale corresponding to its size* to the thermal pressure in order to determine its global gravitational stability using the virial theorem, as we will do in Sect. 14.1. This is approximately correct because, as we have seen, most of the energy occurs at large scales in a turbulent cascade. In practice, we simply add quadratically the r.m.s. velocity of the macroscopic motions due to turbulence (or to other bulk motions) to the r.m.s. thermal velocity.

However the effect is more complex if the component parts of a cloud are considered. Because the turbulent pressure is smaller at smaller scales, the unstable perturbations in a cloud are not necessarily those with the largest wave number. We will come back to this important point in Sect. 14.1.

13.2.4 Intermittency

Another property of turbulence, postulated as early as 1959 by Landau and Lifshitz, is *intermittency*. Initially, the idea of intermittency arose from doubts about the uniformity in space and time of the energy transfer from one scale to another. Later, it was realized that intermittency corresponds to the ephemeral existence in space and time of regions with large vorticity or velocity gradients. Gas dynamics experiments have shown that the moments of the velocity increments do not actually vary as $r^{n/3}$ (as in (13.17)), but diverge progressively from this statistical law when n increases. This is a consequence of intermittency at small scales: the distributions of the velocity gradients, of velocity increments and of their moments, and therefore of vorticity and dissipation do not obey gaussian statistics. Intermittency means a non-gaussian probability distribution of all the quantities where velocity gradients intervene, in particular the velocity differences (increments) between two points. An essential property of intermittency, also observed in the laboratory and in measurements of atmospheric turbulence, is that the departures from a gaussian distribution of the velocity increments are larger for smaller separations between the measurement points.

Intermittency corresponds to rare phenomena with extreme amplitudes, whose probability is much larger than that predicted by gaussian statistics. Many statistical models have been proposed for intermittency, with scaling laws of the type $M_n \sim \epsilon^{n/3} l^{\zeta_n}$, with $\zeta_n \neq n/3$, which describe the departure from the Kolmogorov law. Amongst them, Kolmogorov himself [286] assumed a log-normal distribution for the average energy dissipation at scale l and predicted $\zeta_n = n/3 - (\mu/18)n(n-3)$, where μ is the intermittency parameter. In the model of Frisch et al. [189], turbulence is assumed to be active only on a subset of the three-dimensional space, of filling factor smaller than the scales under consideration. It leads to $\zeta_n = n/3 - (\mu/3)(n-3)$. In this model, the turbulent activity is concentrated in a fractal structure with dimension D (in three dimensions), with $\mu = 3 - D$.

The general idea for the formation of intermittency is that of a non-linear amplification of the vorticity at small scales by the shear exerted by the large scales, which produces a stretch of the vorticity line, hence an increase of vorticity. The equation for the evolution of the vorticity in an incompressible fluid (the *Helmholtz equation*) is obtained by taking the curl of the Navier–Stokes equation:

$$\partial \omega/\partial t + (\mathbf{v} \cdot \nabla)\omega = (\omega \cdot \nabla)\mathbf{v} + \nu \nabla^2 \omega. \tag{13.22}$$

An equilibrium solution for vorticity therefore corresponds to the equality between the stretch $(\mathbf{v} \cdot \nabla)\omega$ and the vorticity diffusion $\nu \nabla^2 \omega$ produced by the velocity field.

13.3 Turbulence in the Interstellar Medium

Interstellar turbulence is, for several reasons, much more complex than the incompressible turbulence we presented in the preceding section. It is a compressible turbulence in a medium permeated by magnetic fields, that is supersonic and perhaps super-Alfvénic in some regions. This means that the gas motions are faster than both the sound velocity and the Alfvén velocity (see Sect. 11.2). Moreover, the magnetohydrodynamic approximation is not valid in many cases because the ambipolar diffusion between neutral and ionized species is sufficiently fast that the neutral and ionized fluids are decoupled. Finally – but this has nothing to do with turbulence – the self-gravity of the gas structures temporarily compressed in shocks, for example, can trigger gravitational instabilities and enhance the structures already present within the medium. We will come back to these aspects in the next section.

Another source of difficulty is that the size scales of viscous dissipation can in some media be quite close to the mean free path of the atoms or molecules. Then the fluid approximation used in models begins to break down. Order-of-magnitude estimates for the characteristic dimensions are given in Table 13.1. We see that in the diffuse neutral medium the viscous dissipation scale l_d is barely larger than the mean free path λ of the hydrogen atoms.

An additional problem, which is specific to the ISM, is the multiplicity of the energy sources, hence a difficulty in defining the integral scale at which energy is injected. The main energy sources are (i) the differential rotation of the Galaxy that produces shears at a large scale, of the order of a kpc; (ii) supernova explosions, which inject a large amount of kinetic energy at scales of the order of hundreds of pc; (iii) the expansion of H II regions, also a source of energy at a scale of several tens of pc, and (iv) bipolar flows and jets associated with star formation, which inject energy at much smaller scales (about one tenth of a pc). It is curious that

Table 13.1. Characteristic quantities for turbulence in the three components of the cold interstellar medium: the diffuse neutral medium, molecular clouds without star formation, and dense molecular cores. Explanations are given in the text.

Quantity	Unit	cold atomic medium	molecular clouds	dense molecular cores
$\langle n \rangle$	cm^{-3}	30	200	10^4
T_K	K	100	40	10
B	µG	10	20	100
l	pc	10	3	0.1
σ_l	km s^{-1}	≈ 3.5	1	0.1
$\lambda = 1/\langle n \rangle \sigma$	AU	2	0.03	6×10^{-4}
$c_S = \sqrt{kT_K/\mu m_H}$	km s^{-1}	0.8	0.5	0.2
$v_A = B/\sqrt{4\pi\rho}$	km s^{-1}	3.4	2.0	1.4
$\nu = \frac{1}{3}\lambda v_{th}$	cm^2 s^{-1}	2.8×10^{17}	1.8×10^{17}	9×10^{16}
$P_{th} = \langle n \rangle kT_K$	erg cm^{-3}	4×10^{-13}	10^{-12}	10^{-11}
$Re = lv_l/\nu$		5.7×10^7	8.1×10^6	5.4×10^4
$\frac{1}{2}\rho \langle v_l \rangle^3 / l$	erg cm^{-3} s^{-1}	2×10^{-25}	1.7×10^{-25}	2.5×10^{-25}
Λ	erg cm^{-3} s^{-1}	5×10^{-24}	4×10^{-24}	3.5×10^{-24}
$\epsilon = \frac{1}{2}\langle v_l \rangle^3 / l$	$L_\odot/M_\odot$	1.5×10^{-3}	1.1×10^{-4}	3.2×10^{-6}
l_d	AU	2.9	4.0	5.7
$\lambda_T = l_d^{1/3} L^{2/3}$	pc	0.34	0.38	0.42
$\nu_t = l \langle v_l \rangle$	cm^2 s^{-1}	2×10^{25}	5×10^{23}	5×10^{21}
$P_t = \frac{1}{3}\langle \rho v_l^2 \rangle$	erg cm^{-3}	3×10^{-11}	2×10^{-11}	10^{-11}

these four processes produce roughly similar mean energy injection rates in the Galaxy. However, it is very difficult to calculate their exact respective importance for gas turbulence because the coupling of these sources to the interstellar medium is complex and depends upon many factors, such as the topology and intensity of the magnetic field and the porosity of the medium, which are poorly known. Numerical simulations attempt to model these various processes.

Table 13.1 gathers the quantities which allow us to characterize the energy contained in turbulence for different media. The three first lines give the average density, the kinetic temperature and the magnetic field intensity that characterize the three considered media. These quantities are derived from observations. The fourth and fifth lines give, respectively, the scale l and the velocity dispersion σ_l that we adopt to characterize turbulence. We have been guided in their choice by the observed size scales and associated velocity dispersions.

The following lines give derived quantities based upon the preceding values. They are, in order:

- the mean free path λ of the gas particles[2];
- the isothermal sound velocity c_S;
- the Alfvén velocity v_A;
- the kinematic viscosity ν;
- the thermal pressure P_{th};
- the Reynolds number Re (note that the Reynolds numbers Re are very large in all three phases, even in the dense cores where v_l and c_S are close in value, suggesting that turbulence exists everywhere in the interstellar medium);
- the energy transfer rate of turbulence $\rho\epsilon$ per unit volume;
- for comparison, the dominant rate Λ of radiative energy loss, assumed to be due to the emission from the C^+ line with a fractional ionization $x = 10^{-3}$ in the cold neutral medium, and to the emission from the CO lines, including lines of isotopomers, in the molecular clouds and cores;
- the rate of energy transfer ϵ per unit mass in the turbulent cascade, to be compared with the power supplied to the interstellar medium by the stellar UV radiation, which is of the order of 1 $L_\odot/M_\odot$ in the solar neighbourhood;
- the Kolmogorov scale l_d for the dissipation of turbulence;
- the so-called Taylor scale λ_T, a mean scale obtained from l_d and the integral scale of turbulence, whose value adopted here is 100 pc;
- the turbulent viscosity ν_t;
- the turbulent pressure P_t.

A comparison of the different energy transfer terms calls for a few remarks. Firstly, we notice that the energy transfer rate per unit volume in the turbulent cascade is roughly the same in all three components of the cold neutral medium. This strongly suggests that the turbulent cascade is independent of the physical differences (density and temperature) between these components, and that energy propagates

---

[2] $\lambda = 1/\langle n\rangle\sigma$, σ being the elastic collision cross-section between hydrogen atoms ($\sigma_{H-H} \sim 10^{-15}$ cm^2) or between hydrogen molecules ($\sigma_{H_2-H_2} \sim 10^{-14}$ cm^2), for which we have neglected the temperature dependence.

between scales independently of these parameters. This rate, which is also the rate of the dissipation of turbulent energy in the Kolmogorov model, is approximately 1/20 of the radiative cooling rate. Therefore the heating due to turbulent dissipation can locally compensate for the cooling provided that it is dissipated in 1/20 of the volume. It is thus a potentially important source of local heating, while it is negligible globally. This heating is also negligible globally with respect to the UV and visible radiation coming from stars, which is of the order of 1 $L_\odot/M_\odot$ in the solar neighbourhood. We also note that the turbulent specific energy is considerably smaller in dense cores than in the diffuse neutral medium. This suggests that the condensation of the diffuse gas into a dense core requires a strong dissipation of the initial turbulent energy.

13.4 Some Effects of Interstellar Turbulence

We will see in the next chapter that turbulence can stabilize gravitationally unstable scales and may modify the criterion for gravitational instability. Here we examine some other effects of interstellar turbulence.

13.4.1 Turbulent Transport and Interstellar Chemistry

Turbulence leads to significant diffusion in the interstellar medium, which may deeply affect the chemistry. A discussion of this is given by Xie et al. [559], from whom we borrow what follows.

For a turbulent molecular cloud let us consider the chemical species i, of density n_i, whose (small) abundance with respect to molecular hydrogen is $f_i = n_i/n_{H_2}$, the density of hydrogen molecules being n_{H_2}. The diffusion of species i is different from that of hydrogen because this species undergoes chemical reactions which depend on time and location so that, in general, it has a density gradient different from that of H_2. As a consequence, there is, generally, a non-zero chemical composition gradient df_i/dz in the direction z of the density gradient. The net flux ϕ_i for the transport of species i in this direction is thus

$$\phi_i = -v_t n_{H_2} \frac{df_i}{dz} = -v_t n_i \left[\frac{1}{n_i} \frac{dn_i}{dz} - \frac{1}{n_{H_2}} \frac{dn_{H_2}}{dz} \right], \tag{13.23}$$

where v_t is a turbulent diffusion coefficient that can be defined, phenomenologically, as the mean value of the product of the random velocity of the gas and a mixing length, $v_t = \langle v_t L \rangle$. Note that both terms inside the square brackets may be considered as the inverse of the scale lengths of the z-distribution of species i and of H_2, respectively, $h_i = -n_i/(dn_i/dz)$ and $h_{H_2} = -n_{H_2}/(dn_{H_2}/dz)$. Putting $1/h = (1/h_i) - (1/h_{H_2})$, (13.23) reads

$$\phi_i = v_t n_i \frac{1}{h}. \tag{13.24}$$

The diffusion velocity for each species is thus $v_d \sim v_t/h$, and the characteristic diffusion time is $\tau_d \sim h/v_d \sim h^2/v_t$. h depends on the chemical species because it is a function of the density gradient specific to this species.

The evolution of the density, n_i, of species i is given by the continuity equation

$$\frac{\partial n_i}{\partial t} + \frac{\partial \phi_i}{\partial z} = G_i - L_i, \qquad (13.25)$$

in which G_i and L_i are, respectively, the creation rate and the destruction rate of species i due to chemical reactions at the current position.

Figures 13.3 and 13.4 show examples of some results obtained from a chemical model of a spherical, turbulent cloud. Initially, the considered elements are assumed to be entirely atomic, except hydrogen which is entirely molecular. Figure 13.3 shows the time evolution of the abundances of the oxygenated molecules at a given depth into the cloud, and Fig. 13.4 gives the variation of these abundances with radius as steady state is reached, about 3×10^6 years after the beginning of the simulation. We see that turbulent diffusion causes major changes in the chemical composition with respect to the non-turbulent case. In particular, chemical models without turbulence predict high abundances of molecular oxygen deep within the cloud but, on the other hand, predict little ionized carbon. This is contrary to observations, which show abundant C^+ but undetectable O_2, with very significant upper limits. Turbulent chemical models as illustrated by Figs. 13.3 and 13.4 give results in much better agreement with observation for values of the turbulent diffusion corresponding to the conditions of Table 13.1. Such models can be applied to the regions of a cloud where chemical abundance fluctuations exist, for example due to variations in the UV illumination. There are, however, other possible explanations for these abundance "anomalies", as discussed Sect. 9.4.

13.4.2 Intermittency of Turbulence Dissipation as a Gas Heating Source

The Characteristics of Interstellar Intermittency

Small-scale intermittency of turbulence has been observed in the laboratory and is also found in numerical simulations. It probably also exists in the interstellar medium, including H II regions (for molecular clouds see Pety & Falgarone [405], [406], and references herein). Turbulent intermittency in the cold interstellar medium seems to present the same statistical properties as intermittency in incompressible turbulence, in particular non-gaussian distributions of the quantities related to the velocity increments. It is unfortunately impossible to directly measure the velocity distribution in the interstellar medium. The only available tool to probe the velocity field is the observation, at high spectral resolution, of the profile of atomic or molecular lines. The observed quantities (the line profiles) are quantities integrated along the line of sight. Lis et al. [326] estimate that this is not a fundamental handicap for analysing the properties of interstellar turbulence. Based on simulations of compressible turbulence (Porter et al. [411]), they show that the probability distribution of the increments of the average velocity between two separated lines of sight, which are measurable quantities, have characteristics similar to those of the velocity increments between two points of the cloud. These distributions are non-gaussian and the departures from a gaussian distribution are larger when the two lines of sight are closer to each other.

Another important point of the work on numerical simulations is that the regions where the departures from a gaussian distribution are particularly large are also those where the modulus *of the measurable part* of the vorticity along the line of sight is

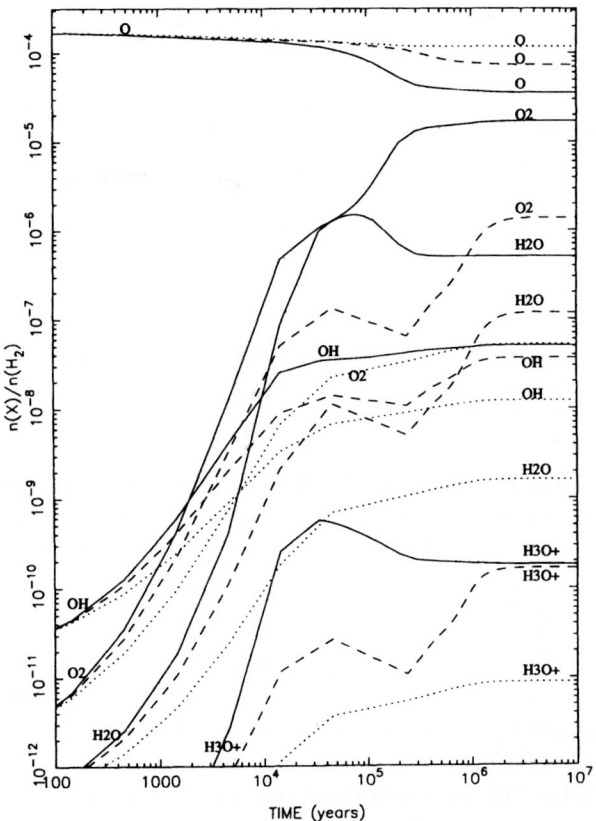

Fig. 13.3. The time evolution of oxygenated chemical species in a turbulent isothermal molecular cloud at a depth corresponding to an extinction of $A_V = 9$ mag. with a density $n_{H_2} = 3 \times 10^4$ cm^{-3}, submitted to the standard UV radiation field. The solid lines correspond to the case without turbulence, the dashed lines correspond to the case for a turbulent diffusion coefficient $\nu_t = 10^{23}$ cm^2 s^{-1}, and the dotted lines to $\nu_t = 10^{24}$ cm^2 s^{-1}. The latter value is close to that adopted in Table 13.1 for quiescent molecular clouds. Note in particular the strong decrease in the abundances of O$_2$ and of H$_2$O due to turbulent mixing. Reproduced from Xie et al. [559], with the permission of the AAS.

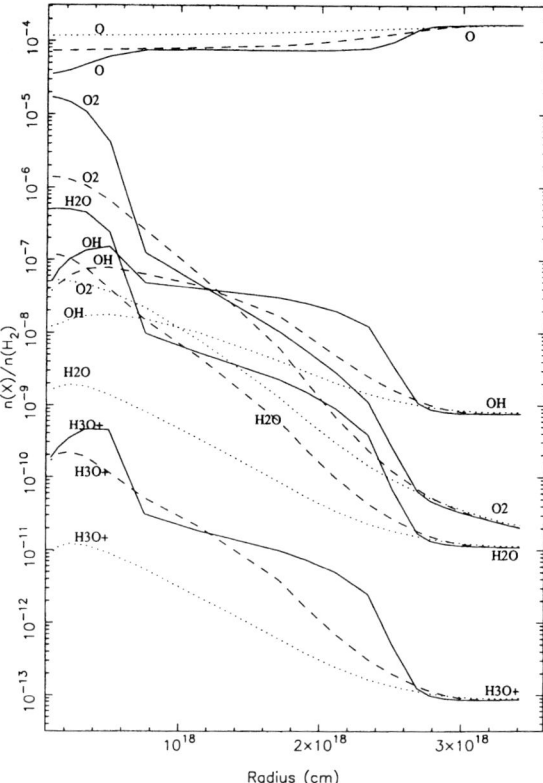

Fig. 13.4. The evolution with depth of the abundances of oxygenated chemical species in a turbulent isothermal molecular cloud. This cloud is subjected to the standard UV radiation field. The results are for a time of 3×10^6 years after the beginning of the simulation, a steady state having been approximately reached (cf. Fig. 13.3). The solid lines correspond to the case without turbulence, the dashed lines correspond to the case for a turbulent diffusion coefficient $\nu_t = 10^{23}$ cm^2 s^{-1}, and the dotted lines to $\nu_t = 10^{24}$ cm^2 s^{-1}. The latter value is close to that adopted in Table 13.1 for quiescent molecular clouds. At the cloud surface (radius 3.4×10^{18} cm), the abundances do not depend upon turbulence. Note the large decrease in the abundances of O_2 and H_2O deep inside the cloud. Reproduced from Xie et al. [559], with the permission of the AAS.

the largest. The measurable components of the vorticity are necessarily those which contain $\partial v_x/\partial y$ and $\partial v_x/\partial z$, where x is the coordinate along the line of sight, and y and z the perpendicular coordinates, these are therefore the components $(\nabla \times \mathbf{v})_y$ and $(\nabla \times \mathbf{v})_z$ of the vorticity.

The probability distribution functions for velocity increments have been constructed using very large samples of molecular line data, corresponding to maps containing almost 10 000 spectra of nearby molecular clouds. Figure 13.5 shows the probability distribution of the velocity increments for various separations between

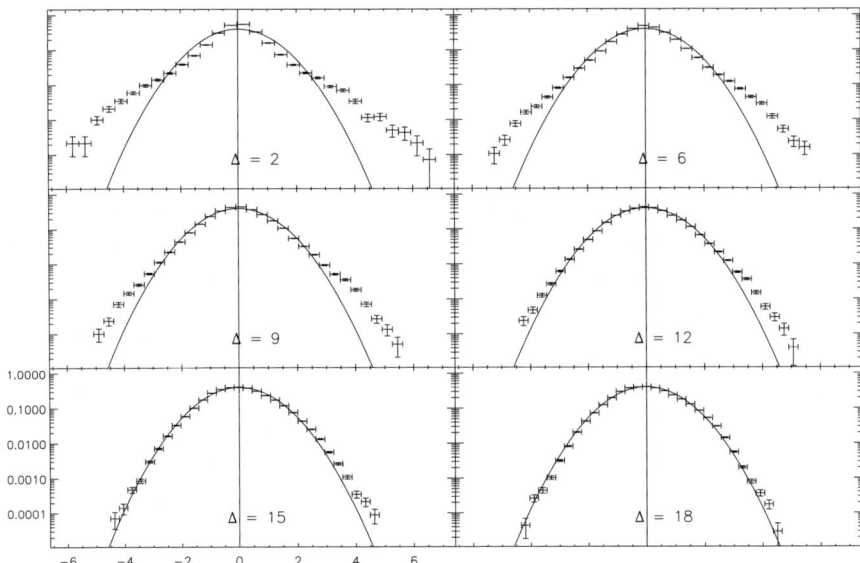

Fig. 13.5. The normalized distributions for the velocity increments (differences in velocity) observed between points separated by Δ, in a ^{12}CO(2-1) line map of the molecular cloud L 1512 S. Δ is in units of 16 arc seconds, half the angular resolution of the radiotelescope used for these observations. This corresponds to 0.012 pc at the distance of the cloud (150 pc). The velocities are defined as the centroid $\langle v \rangle = \int v' N(v') dv' / \int N(v') dv'$ measured in the line profile in each direction. Note the departures with respect to a gaussian distribution, for increments at small values of the distance Δ between the measurement points. The fact that these departures are more pronounced at small scales than at large scales provides support for the existence of intermittency in interstellar turbulence. From Pety & Falgarone [406], with the permission of ESO.

the two observed directions for a dense molecular core observed in the ^{12}CO(J=2-1) line.

The analysis of the positions where the departures from a gaussian distribution are the largest shows that they are not uniformly distributed but grouped in small regions. This suggests that the vorticity distribution in interstellar turbulence is very heterogeneous, as is the distribution of the viscous dissipation which is associated with regions with high vorticity.

The Local Dissipation of Energy through Intermittency

We have seen that, when averaged over space and time, the dissipation of the turbulent kinetic energy, even if supersonic, is a negligible source of heating for the interstellar gas (see Table 13.1). The *average* rate of transfer of specific energy ϵ

given in this table can be expressed as a luminosity per unit mass, which is much smaller than 1 $L_\odot/M_\odot$, the rate supplied by the average UV radiation of stars in the nearby interstellar medium. However, if concentrated in a fraction of the order of 1% of space or time, this dissipation becomes a locally dominant heating source, with important effects for the physics and chemistry of the gas.

The Intermittency of Viscous Dissipation

Small-scale intermittency of turbulence may be at the origin of the non-gaussian distribution of vorticity and of the viscous stress tensor, hence of viscous dissipation.

An analytical solution of the Helmholtz equation for vorticity (13.25), called the *Burgers vortex*, allows a quantitative description of the effect of viscous dissipation near coherent vorticity structures in the cold atomic medium. The interest in this solution is that the Burgers vortex is completely defined by only two parameters: the stretching rate α of vorticity by the large scales of turbulence, and the circulation C of vorticity, $C = \int_0^\infty \omega(r)\, 2\pi r\, dr$, where r is the distance from the rotation axis. The vortex is assumed to have a gaussian distribution of vorticity, which results from the equilibrium between the stretch of the vortex and the diffusion due to viscosity:

$$\omega(r) = \omega_0 e^{-(r^2/r_0^2)}. \tag{13.26}$$

The equilibrium radius r_0 and the vorticity circulation peak ω_0 can be expressed as a function of α and of C as

$$r_0^2 = \frac{4\nu}{\alpha}, \tag{13.27}$$

and

$$\omega_0 = \frac{\alpha C}{4\pi\nu}. \tag{13.28}$$

The stretching rate is imposed by turbulence at larger scales. This is suggested by experimental results in laboratory flows and numerical simulations which show that the maximum rotation velocity in vorticity filaments is of the same order as the velocity dispersion of the ambient turbulence. The only free parameter is thus, in fact, the circulation of vorticity.

Although it is very unlikely that such a simple description of intermittent regions is realistic for the interstellar medium, it is remarkable that, if the viscous dissipation is concentrated in high-shear regions at the edge of intermittent eddies, the temperature in the diffuse neutral gas can reach some 1 000 K in regions of a few 10^{14} cm in size (Joulain et al. [273]). Endothermic chemical reactions or reactions with high activation barriers then become possible. This model allows us to solve a number of long-standing problems encountered in the chemistry of the diffuse medium, such as the large abundances of CH^+, HCO^+ and OH. This abundance problem was mentioned at the end of Sect. 9.4. However, shocks give a similar result, as indicated in this section.

Decoupling of the Gas and Grains

Interstellar dust grains are in general very well coupled to the gas motions, which renders collisions between grains unlikely except in regions where the medium collides with itself, i.e., shocks. Turbulence, in particular in small-scale intermittency regions, plays a similar role in decoupling the gas and grains.

Let us consider a grain of radius a, of geometric cross-section $\sigma_g = \pi a^2$ and mass m_g, immersed in a medium with density n_H and a mean mass per particle μm_H. The rate of change of its velocity v_g is given, from (8.10), by

$$m_g \frac{dv_g}{dt} = -n_H \mu m_H v_{th} \alpha \pi a^2 (v_g - v), \qquad (13.29)$$

which accounts for the accomodation coefficient $\alpha \sim 0.35$ of atoms or molecules on the grain. v_{th} is the mean thermal velocity of atoms and v the bulk velocity of the gas. The solution to this equation is

$$v_g(t) = v(1 - e^{-t/\tau_f}). \qquad (13.30)$$

$\tau_f = 4a\rho_g/(3\alpha n_H \mu m_H v_{th})$, where ρ_g is the density of grain material, is the stopping time of the grain, i.e., it is the time required for a grain to meet a mass equal to its own mass. For times t, long with respect to τ_g, there is a good coupling between the gas and grains, while for times that are short with respect τ_g, $v_g \sim vt/\tau_f$.

An order of magnitude for τ_f, for a grain with $a = 0.1$ μm and density 3 g cm^{-3}, in a gas of density $n_H = 100$ cm^{-3} and temperature 100 K, is 10^5 years, a short time with respect to the usual dynamic scales for the gas.

However, turbulence in which the kinetic energy cascades to quite small spatial scales, and hence very short time scales, is able to produce differential motions between the gas and grains. Moreover, the gas velocity v is neither constant nor uniform and we can consider that the grain is randomly transported by a succession of eddies characterized by velocities v_k and lifetimes τ_k. Völk et al. [536] have shown that a grain experiences a number of random impulses $N = \tau_f/\tau_k$ from eddies with lifetime τ_k. If N is large enough, the grain can be stopped: its velocity in a rest reference system, which results from all these random collisions, is from the previous equations

$$v_g = \left(\frac{\tau_f}{\tau_k}\right)^{1/2} v_k \frac{\tau_k}{\tau_f} = v_k \left(\frac{\tau_k}{\tau_f}\right)^{1/2} \simeq \frac{v_k}{N^{1/2}}. \qquad (13.31)$$

The relative velocity of a grain with respect to an eddy with wavenumber k is thus of the order of v_k. In this case the gas and the grains are effectively decoupled. Between the large eddies, whose lifetime τ_k is large and which do not cause decoupling, and the very small eddies, which contribute to the stopping of the grain but cannot transfer much momentum to it, there exists a preferential turbulence scale k^* for which $\tau_{k^*} = \tau_f$, which gives a dominant contribution to gas–grain decoupling.

Intermittency generates time scales even smaller than those characteristic of a Kolmogorov spectrum. These very fast events effectively decouple grains from gas, even the very small grains, in spite of their small gas friction times τ_f. Falgarone & Puget [172] have shown, using the statistics of such events obtained in a wind tunnel (not well adapted to interstellar conditions but probably usable thanks to the universal properties of turbulence), that very small grains can efficiently decouple from the gas and that the relative velocities between these grains can be increased by almost one order of magnitude with respect to the situation within a turbulent flow in which intermittency is ignored.

These processes are of great interest for the evolution of the grain populations, mainly for the growth of grains in the interstellar medium, because they can alter the usual assumptions. In particular, we know that above some relative velocity threshold, grain collisions can lead to shattering while below this threshold they can lead to coagulation (see Chap. 15).

14 Equilibrium, Collapse and Star Formation

In this chapter we deal with the most complex problems raised by the interstellar medium: the gravitational collapse of clouds, their fragmentation and the final collapse of fragments to form stars. Despite a considerable amount of work, to which it is impossible to do justice here, these problems are not solved. We will only discuss some principles. We will stop the discussion at the end of collapse, leaving for others the burden of dealing with protostars, their disks and jets, and their evolution to the main sequence.

The interstellar medium evolves in the gravitational potential of the Galaxy. It forms stars through local gravitational instabilities. One of the most critical questions concerning this evolution is thus: what are the conditions that trigger or inhibit gravitational collapse? This question underlies all of this chapter and will be discussed again in the next and final chapter of this book. We will begin by a study of the equilibrium conditions for an interstellar structure (a "cloud"), we will then discuss collapse, fragmentation and star formation.

14.1 Stability and Instability: the Virial Theorem

14.1.1 A Simple Form of the Virial Theorem with No Magnetic Field nor External Pressure

The virial theorem is the fundamental theorem for the equilibrium of self-gravitating structures in the Universe. Its simplest formulation is due to Clausius (1870) and to Poincaré (1911).

Let us consider a particle with mass m located at position $\mathbf{r}$, relative to some arbitrary origin, and submitted to a force $\mathbf{F}$. Its equation of motion is

$$m\frac{d^2\mathbf{r}}{dt^2} = \mathbf{F}. \tag{14.1}$$

Let us take the scalar product of this expression with $\mathbf{r}$, obtaining in this way quantities with dimension of energy

$$m\frac{d^2\mathbf{r}}{dt^2} \cdot \mathbf{r} = \mathbf{F} \cdot \mathbf{r}, \tag{14.2}$$

which can be written

$$m\frac{d}{dt}\left(\mathbf{r}\cdot\frac{d\mathbf{r}}{dt}\right) = m\left(\frac{d\mathbf{r}}{dt}\right)^2 + \mathbf{F}\cdot\mathbf{r}. \tag{14.3}$$

The quantity $\mathbf{F}\cdot\mathbf{r}$ has been called the *virial* by Clausius. Consider now a system of particles all of the same mass m. The moment of inertia of this system is

$$I = \sum_j m\mathbf{r}_j^2, \tag{14.4}$$

where the summation is over all the particles. Its second derivative with respect to time is

$$\ddot{I} = 2\sum_j m\mathbf{r}_j \frac{d^2\mathbf{r}_j}{dt^2} + 2\sum_j m\left(\frac{d\mathbf{r}_j}{dt}\right)^2, \tag{14.5}$$

so that, combining with (14.3),

$$\frac{1}{2}\ddot{I} = \sum_j m\left(\frac{d\mathbf{r}_j}{dt}\right)^2 + \sum_j \mathbf{F}_j \cdot \mathbf{r}_j. \tag{14.6}$$

The kinetic energy of the system is

$$\mathcal{T} = \frac{1}{2}\sum_j m\left(\frac{d\mathbf{r}_j}{dt}\right)^2. \tag{14.7}$$

If the system is a gas at rest $\mathcal{T}$ is the thermal energy $\mathcal{T}_{therm}$. If there are also macroscopic motions of the gas, for example due to turbulence, $\mathcal{T} = \mathcal{T}_{therm} + \mathcal{T}_{macro}$, with $\mathcal{T}_{macro} = (1/2)\sum m v_{rms}^2$, where v_{rms} is the r.m.s. value of the macroscopic velocities. This assumes that v_{rms} can be defined, which might raise some problems in the case of turbulence. We will for the moment neglect global rotation because this rotation is slow for interstellar clouds and has no appreciable role in the first stages of their evolution. Conversely, we will take rotation into account for later stages.

The last term of (14.6), $\sum_j \mathbf{F}_j \cdot \mathbf{r}_j$, is the gravitational potential energy Ω of the system. If the only force is gravity, neglecting other interactions between particles and the rotational and vibrational energies of the molecules of the medium which are anyway internal energies, we can write:

$$\Omega = -\sum_j m_j \mathbf{r}_j \sum_k \frac{Gm_k(\mathbf{r}_j - \mathbf{r}_k)}{|\mathbf{r}_j - \mathbf{r}_k|^3}, \tag{14.8}$$

the j sum being taken over all the particles of the system, and the k sum over all masses interior or exterior to it. If the latter masses can be ignored, either because of the spherical or ellipsoidal symmetry of the system or because the system is isolated, each interaction will be counted twice in the double sum, and we obtain

$$\boxed{\Omega = -\sum_{j<k} \frac{Gm_j m_k}{|\mathbf{r}_j - \mathbf{r}_k|}.} \tag{14.9}$$

We can then write

$$\frac{1}{2}\ddot{I} = 2\mathcal{T} + \Omega, \qquad (14.10)$$

an equation that defines the dynamical behaviour of the system. The equilibrium condition is of course $\ddot{I} = 0$. It can therefore be written

$$\boxed{2\mathcal{T} + \Omega = 0,} \qquad (14.11)$$

the simple form of the virial theorem. This equation expresses the balance between the gravitational attraction between the different parts of the system, which tends to produce collapse, and the agitation of its particles (the pressure for a gas at rest, which is proportional to $\mathcal{T}$).

For a cloud of perfect gas at temperature T the thermal energy is

$$\mathcal{T}_{therm} = \frac{3}{2}\frac{MkT}{\mu m_H}, \qquad (14.12)$$

where M is the total mass of the cloud. If the cloud is spherical and homogeneous, its potential energy is

$$\Omega = -\frac{3}{5}\frac{GM^2}{R}. \qquad (14.13)$$

Note, however, that a cloud in equilibrium cannot be at the same time isothermal and homogeneous, and this expression may need to be modified.

14.1.2 The Jeans Length and Jeans Mass

Equating $2\mathcal{T}$ with $-\Omega$ (cf. (14.12) and (14.13)) and expressing R as a function of M and density n, we obtain an instability criterion. This criterion predicts that an hypothetical spherical, homogeneous and isolated cloud with no macroscopic motions, density n and temperature T is unstable with respect to contraction ($\dot{I} < 0$, $\ddot{I} < 0$) if its mass is larger than a critical mass $M_{crit,th}$

$$\boxed{M > M_{crit,th} \simeq \left(\frac{1}{\mu m_H}\right)^2 \left(\frac{5}{2}\frac{kT}{G}\right)^{3/2} \left(\frac{4}{3}\pi n\right)^{-1/2} \simeq 4.4\, T^{3/2} n^{-1/2}\, M_\odot,} \qquad (14.14)$$

assuming for the molecular gas of the cloud a mean molecular mass $\mu \approx 2.7$, accounting for helium and the heavy elements.

We have, numerically,

$$M_{crit,th} \simeq 1.4 \left(\frac{T}{10\,\mathrm{K}}\right)^{3/2} \left(\frac{n}{10^4\,\mathrm{cm}^{-3}}\right)^{-1/2}\, M_\odot. \qquad (14.15)$$

Note again that a cloud in equilibrium cannot be at the same time isothermal and homogeneous, so that the numerical factor is very approximate.

Another way to treat the instability problem is to perform a linear analysis of gravitational instability in a uniform, infinite medium. Although such a medium has no physical reality, there are enough disturbing observational facts to justify the presentation of this analysis. It was made by Jeans as early as 1902, whose instructive reasoning we will now reproduce, even though it is partly incorrect. Let us consider an infinite, isothermal medium, without magnetic field or macroscopic motions, with density ρ. The three equations which describe the dynamics of this medium are the continuity equation (11.1), the equation of motion (11.2) and the Poisson equation. Let us give them again in vector form

$$\frac{\partial \rho}{\partial t} + \nabla \cdot \rho \mathbf{v} = 0, \tag{14.16}$$

$$\rho \left(\frac{\partial \mathbf{v}}{\partial t} + \mathbf{v} \cdot \nabla \mathbf{v} \right) = -\nabla P - \rho \nabla \Phi, \tag{14.17}$$

$$\nabla^2 \Phi = 4\pi G \rho, \tag{14.18}$$

where Φ is the gravitation potential. The last equation is incorrect for an infinite medium because a uniform density would lead to a diverging potential at infinity. However, it allows a considerable simplification and the error is not too severe because we will only consider a perturbation. Using the index 0 for quantities at equilibrium, and the index 1 for the perturbed ones, we have $v = v_1$, $\rho = \rho_0 + \rho_1$ and $\Phi = \Phi_0 + \Phi_1$. Linearizing the three equations above, and with the further hypothesis of the constancy of the isothermal sound velocity c_S, such that $P/\rho = c_S^2 = k_B T_k / \mu m_{rmH}$, we obtain

$$\frac{\partial v_1}{\partial t} = -\nabla \Phi_1 - \frac{c_S^2}{\rho_0} \nabla \rho_1, \tag{14.19}$$

$$\frac{\partial \rho_1}{\partial t} = -\rho_0 (\nabla \cdot v_1), \tag{14.20}$$

$$\nabla^2 \Phi_1 = 4\pi G \rho_1. \tag{14.21}$$

Taking the divergence of (14.19) in order to eliminate $\nabla \cdot v_1$ and $\nabla^2 \Phi_1$ with (14.20) and (14.21), we find

$$\frac{\partial^2 \rho_1}{\partial t^2} = \rho_0 \nabla^2 \Phi_1 + c_S^2 \nabla^2 \rho_1. \tag{14.22}$$

For a periodic perturbation $\rho_1 = K e^{i(kx+\omega t)}$, from this equation it is easy to derive the dispersion relation between the angular frequency ω and the wave number $k = 2\pi/\lambda$ of the perturbation:

$$\omega^2 = k^2 c_S^2 - 4\pi G \rho_0. \tag{14.23}$$

The unstable modes ($\omega^2 < 0$, or ω imaginary) are those for which the wavenumber is such that

$$k < k_J = \left(\frac{4\pi G \rho_0}{c_S^2} \right)^{1/2}. \tag{14.24}$$

The *Jeans length* is $\lambda_J = 2\pi/k_J$. The mass included within a cube of side λ is the *Jeans mass* $M_J = \rho(2\pi/k_J)^3$. It is the largest gravitationally stable mass for a medium with given density and temperature. We can see that the Jeans mass is similar to the critical mass given by (14.14) and (14.15) within a numerical factor (2.4 instead of 1.4 in (14.15)).

It is worth noting that the velocity ω/k at which the perturbation propagates vanishes for $k = k_J$.

At a constant temperature, the Jeans mass decreases at increasing density. An isothermal gas in gravitational collapse stays unstable as long as its cooling is sufficient to maintain isothermality. With another, non-isothermal equation of state of the form $P \propto \rho^\gamma$, the density dependence of the Jeans mass is $M_J \propto \rho^{(3\gamma/2)-2}$. Thus, when $\gamma > 4/3$, the Jeans mass increases with density and an initially unstable mass can become stable at higher densities.

The Role of Turbulence

In a turbulent gas the dispersion equation (14.23) must be deeply modified (Sect. 13.2). According to Chandrasekhar [88], [89], the sound velocity c_S is replaced in this equation by an effective sound velocity depending on the wavenumber, $c_S^2(k) = c_S^2 + (1/3)\langle v(k) \rangle^2$, where $\langle v(k) \rangle$ is the characteristic velocity of turbulence for a wavenumber k. Bonazzola et al. [53] showed, writing the dispersion relation

$$\omega^2 - [c_S^2 + (1/3)\langle v(k) \rangle^2]k^2 + 4\pi G \rho_0 = 0, \tag{14.25}$$

that the unstable perturbations in a uniform, isothermal, turbulent medium with an average density ρ_0 and temperature T_K are those with scales λ so that the wavenumber $k = 2\pi/\lambda$ obeys the relation

$$\boxed{4\pi G \rho_0 k^{-2} > c_S^2 + \frac{A}{3(\alpha-1)} k^{1-\alpha},} \tag{14.26}$$

where $c_S = \sqrt{kT_K/\mu m_H}$ is the isothermal sound velocity and $E(k) = A k^{-\alpha}$ is the power spectrum of the turbulence (13.15). As a consequence, there is a critical slope $\alpha = 3$ for the turbulent spectrum. If this spectrum is steeper ($\alpha > 3$), unstable scales exist only in a limited range $[k_1, k_2]$ which depends upon ρ_0. In this case the large scales $k < k_1$ are stabilized. This effect was demonstrated analytically by Bonazzola et al. [53], [54]. It is confirmed by numerical simulations of compressible, supersonic turbulence in a self-gravitating gas, a more realistic situation for the interstellar gas (Klessen et al. [283]). These authors also show that a magnetic field would not fundamentally affect this property. It allows protostellar condensations to form in an apparently globally stable cloud. In spite of this, we will ignore turbulence in the rest of this chapter, while giving appropriate warnings.

We finally remark that the virial theorem implies that the system is isolated. It can, however, be bounded by an external pressure and even immersed in an external gravitational field which can be taken into account if necessary (see (14.8)).

These guidelines are an introduction to the key questions raised by star formation, such as those concerning the stability or instability of a gas in gravitational collapse, or the possibility of having increasingly small unstable fragments in a globally collapsing cloud. We will now elaborate on these questions.

14.1.3 The General Form of the Virial Theorem

The preceding discussion has ignored the external pressure and magnetic fields. In order to obtain a more general form for the virial theorem it is convenient to start from the general equation of motion (11.2), which we will write in vector form

$$\rho \frac{D\mathbf{v}}{Dt} = \rho \left(\frac{\partial \mathbf{v}}{\partial t} + \mathbf{v} \cdot \nabla \mathbf{v} \right) = -\nabla P - \frac{\nabla B^2}{8\pi} + \frac{1}{4\pi} \mathbf{B} \cdot \nabla \mathbf{B} - \rho \nabla \Phi. \quad (14.27)$$

$\mathbf{v} = D\mathbf{r}/Dt$, ρ, P, $\mathbf{B}$ and Φ are respectively, the bulk velocity, density, pressure, magnetic field and gravitational potential. Here, $\mathbf{v}$ is the *macroscopic* velocity (e.g. the turbulent velocity) of the gas. The *microscopic* (thermal) velocity of the atoms or molecules is taken into account through the pressure P. If there are cosmic rays, we should add to P their pressure $P_{CR} = (1/3)u_{CR}$, where u_{CR} is their energy density.

As before, we take the scalar product of this equation with $\mathbf{r}$ and integrate the result over the volume V of the cloud.

The successive terms of the integral obtained in this way are such that:

1. The left-hand term becomes $\int_V \rho \mathbf{r} \cdot (D\mathbf{v}/Dt) dV = \frac{1}{2}(D^2 I/Dt^2) - 2T_{macro}$. For this, we have used the second time derivative of $I = \int_V \rho r^2 dV$, which is $D^2 I/Dt^2 = 2\int_V \rho \mathbf{r} \cdot (D\mathbf{v}/Dt) dV + 2\int_V \rho v^2 dV$. $T_{macro} = (1/2) \int_V \rho v^2 dV$ is the kinetic energy of the macroscopic motions per unit volume.
2. The thermal pressure term becomes $-\int_V \mathbf{r} \cdot \nabla P dV = 3\int_V P dV - \int_S P\mathbf{r} \cdot d\mathbf{S} = 3V(\overline{P} - P_{ext})$. $\int_S$ is a surface integral and $\mathbf{S}$ is the vector normal to the surface of the cloud. We have assumed that the system is in pressure equilibrium with an external pressure P_{ext}. $\overline{P}$ is the average of the pressure inside the cloud. Given the vector relation $\nabla \cdot (X\mathbf{r}) = \mathbf{r} \cdot \nabla X + X\nabla \cdot \mathbf{r}$, this pressure term is indeed $-\int_V \nabla \cdot (\mathbf{r} P) dV + \int_V P\nabla \cdot \mathbf{r} dV = -\int_S P\mathbf{r} \cdot d\mathbf{S} + 3\int_V P dV$. Here we have used the relations $\int_S \mathbf{r} \cdot d\mathbf{S} = 4\pi V$ and $\nabla \cdot \mathbf{r} = 3$.
3. The term involving the magnetic pressure is
   $-(1/8\pi) \int_V \mathbf{r} \cdot \nabla B^2 dV = -(1/8\pi) \int_S B^2 \mathbf{r} \cdot d\mathbf{S} + (3/8\pi) \int_V B^2 dV$
   The derivation is similar to that for the thermal pressure.
4. The term arising from the magnetic tension is $(1/4\pi) \int_V \mathbf{r} \cdot (\mathbf{B} \cdot \nabla) \mathbf{B} dV = -(1/4\pi) \int_V B^2 dV + (1/4\pi) \int_S (\mathbf{B} \cdot \mathbf{r}) \mathbf{B} \cdot d\mathbf{S}$.
   This makes use of the vector relations $(\mathbf{B} \cdot \nabla)(\mathbf{B} \cdot \mathbf{r}) = \mathbf{r} \cdot (\mathbf{B} \cdot \nabla) \mathbf{B} + B^2$, and $(\mathbf{B} \cdot \nabla)(\mathbf{B} \cdot \mathbf{r}) = \nabla \cdot (\mathbf{B} \cdot \mathbf{r}) \mathbf{B} - (\mathbf{B} \cdot \mathbf{r})(\nabla \cdot \mathbf{B}) = \nabla \cdot (\mathbf{B} \cdot \mathbf{r}) \mathbf{B}$ since $\nabla \cdot B = 0$.
5. The term $-\int_V \rho \mathbf{r} \cdot \nabla \Phi = \Omega$ is the potential gravitational energy.

The complete virial equation, grouping together the surface terms and the volume terms, can then be written as

$$\boxed{\begin{aligned}\frac{1}{2}\frac{D^2 I}{Dt^2} &= 2T_{macro} + \Omega - \int_S (P + P_{mag})\mathbf{r}\cdot d\mathbf{S} \\ &\quad + \frac{1}{4\pi}\int_S (\mathbf{B}\cdot\mathbf{r})\mathbf{B}\cdot d\mathbf{S} + 3\int_V (P + \frac{P_{mag}}{3})dV,\end{aligned}}\qquad(14.28)$$

with $P_{mag} = B^2/8\pi$. $P = nkT_K$ is the internal thermal pressure. T_K is the kinetic temperature.

The equilibrium condition is $\ddot{I} = 0$ which gives, *without magnetic field*, the condition:

$$\boxed{2T_{macro} + \Omega + 3V(\overline{P} - P_{ext}) = 0.}\qquad(14.29)$$

The internal energy (without magnetic field) can be defined as $u = T_{macro} + (3/2)Mk\langle T_K\rangle$, where M is the total mass. If there is no external pressure the virial theorem can be written as $2u + \Omega = 0$, another form of (14.11), so that $u = -\Omega/2$. The total energy of the system $E = u + \Omega = \Omega/2$ is negative, because Ω is negative.

Let us insist again on the fact that the virial theorem is only really applicable to an isolated system, unless its interactions with the outer space can be expressed as pressure terms.

14.1.4 The Stability of the Virial Equilibrium

Let us first study this stability in the simple case of a spherical cloud with no external pressure and no magnetic field. At equilibrium we have $2u_0 + \Omega_0 = 0$. A perturbation δR of the cloud radius produces a change of $\ddot{I}$ such that

$$\frac{1}{2}\delta\ddot{I} = 2\delta u + \delta\Omega = \left(2\frac{\partial u}{\partial R} + \frac{\partial\Omega}{\partial R}\right)\delta R.\qquad(14.30)$$

The stability criterion around the equilibrium condition $\ddot{I} = 0$ writes $2\partial u/\partial R + \partial\Omega/\partial R < 0$. It is easy to check that a small increase of radius ($\delta R > 0$, $\dot{I} > 0$) has to generate $\ddot{I} < 0$ in order to allow the system to return to its equilibrium.

Let us illustrate this condition with a system where the internal energy is purely thermal, $u = (3/2)nkT_K = 3PV$. Let us assume that the equation of state for this system is polytropic:

$$P \propto \rho^\gamma,\qquad(14.31)$$

a relation that can also be written, for constant mass, as $PV^\gamma = $ constant. The quantity n such that $\gamma = 1 + 1/n$ is called the *polytropic index*. We then have $dP/P = -3\gamma dR/R$ and $du/u = dP/P + 3dR/R = -3(\gamma - 1)dR/R$. The above stability condition then becomes

$$[3(\gamma - 1) - 1]\frac{\Omega}{R} > 0,\qquad(14.32)$$

because $\partial\Omega/\partial R = -\Omega/R$ from (14.13).

This condition requires $\gamma > 4/3$. In particular, virialized isothermal systems ($\gamma = 1$) are unstable if isolated and without a magnetic field, while adiabatic systems

($\gamma = 5/3$) are stable. We can express this condition by saying that a system in virial equilibrium is only stable with respect to collapse (or expansion) if the gas is able to recover, as internal energy, a part of the gravitational energy it has gained (or lost). An external pressure P_{ext} changes this condition because $\gamma > 4/3 - 4\pi P_{ext} R^3 / \Omega$. Since $\Omega < 0$ the critical exponent is larger than $4/3$.

This illustrates, though only in a schematic way, the importance of the equation of state of the gas for virial stability. The problem is in fact complex because the polytropic index can vary enormously during the dynamical evolution and because the equation of state can even be non-polytropic.

Let us now introduce an external pressure. In the (hypothetical) case of an isothermal, uniform sphere with no macroscopic motions and no magnetic field, the virial theorem now becomes, from (14.12), (14.13) and (14.29),

$$\boxed{\frac{3MkT}{\mu m_H} - \frac{3}{5}\frac{GM^2}{R} - 4\pi R^3 P_{ext} = 0.} \qquad (14.33)$$

This equation can be considered in different ways. For a given mass M and temperature T, the radius R is a function of P_{ext}. If R is very large (a low-density cloud) the gravitational term can be neglected and the internal and external pressures are equal. If the external pressure increases, the radius shrinks and the gravitational term becomes appreciable, which causes a further decrease in the radius. If the external pressure is larger than some limit no equilibrium is possible and the cloud collapses (cf. Spitzer [490] Sect. 11.3a).

If we now fix P_{ext} and T we obtain the M, R relation shown in Fig. 14.1. The largest possible mass at virial equilibrium can be obtained by differentiating (14.33) with respect to R and setting $dM/dR = 0$. We then obtain $M_{max} = (20\pi P_{ext}/G)^{1/2} R^2$. However, as expected, the numerical coefficient is not correct because an isothermal sphere in equilibrium cannot be homogeneous, its density being larger in the central region as we will discuss in detail in the next section. Chièze [95] gives the correct relation:

$$\frac{M_{max}}{1 M_\odot} = 12.6 x_*^{-2} \left(\frac{P_{ext}/k}{3\,800\,\mathrm{K\,cm}^{-3}}\right)^{1/2} \left(\frac{R}{1\,\mathrm{pc}}\right)^2, \qquad (14.34)$$

with $x_* = 0.4466$ for an isothermal sphere. This expression was used in constructing Fig. 14.1. Chièze [95] also shows that this relation remains valid for any polytropic index n, provided that $n < -1$ or $n > 3$, i.e. $0 < \gamma < 4/3$. We then have:

$$x_* = \left(\frac{1}{8\pi}\frac{n-3}{n+1}\right)^{1/4}. \qquad (14.35)$$

The point which corresponds to the largest possible mass separates a gravitational equilibrium branch, for which the gravitational term dominates over the pressure term and which is thus gravitationally unstable for $\gamma < 4/3$, from a hydrodynamic equilibrium branch, on which gravity is small and which tends to pressure

14.1 Stability and Instability: the Virial Theorem

equilibrium ($\overline{P} \approx P_{ext}$), with $M \propto R^3$. In the M, R plane the equilibrium curve, on which $\ddot{I} = 0$, separates an upper region (large masses, large radii) where $\ddot{I} < 0$ from a lower region where $\ddot{I} > 0$. The gravitational branch is thus stable with respect to evaporation or to accretion, which corresponds to a decrease or to an increase of mass, respectively. However, it is instable for perturbations of radius (compressions) at constant mass, as we have just seen, because in the present case the cloud is isothermal. Conversely, the hydrodynamic branch is stable for this type of perturbation but unstable with respect to evaporation or accretion.

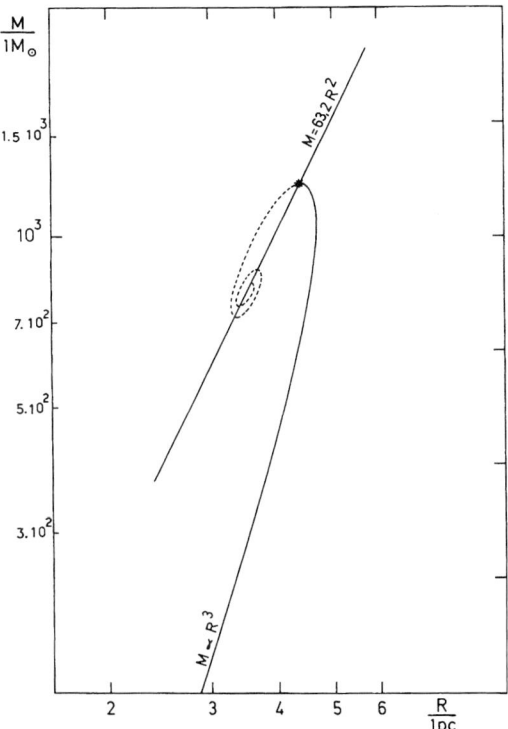

Fig. 14.1. The mass–radius relation for spherical, isothermal clouds with no magnetic field submitted to an external pressure $P_{ext}/k = 3\,800$ K cm^{-3}. The critical point (*) corresponds to the largest possible equilibrium mass. Its location as a function of temperature is a straight line of slope 2. This point separates gravitationally stable configurations (solid curve) from unstable configurations (dotted curve). From Chièze [95], with the permission of ESO.

Let us now introduce the thermal velocity dispersion $\sigma_{v,th}$, still assuming no macroscopic motions. The virial equilibrium equation for the critical (maximum) mass $M_{max} \equiv M_*$ becomes

$$M_* \sigma_{v,th}^2 - \alpha_* \frac{GM_*^2}{R_*} - 4\pi P_{ext} R_*^3 = 0, \tag{14.36}$$

in which the star symbol designates quantities taken at the limit of gravitational instability. $\alpha_* = 0.7323$ is a numerical parameter corresponding to a real, non-homogeneous, isothermal sphere. Eliminating R_* in combining (14.34) and (14.36), we obtain

$$\frac{\sigma_v}{1\,\text{km s}^{-1}} = 0.209 \left(\frac{P_{ext}/k}{3\,800\,\text{K cm}^{-3}}\right)^{1/8} \left(\frac{M_*}{1\,\text{pc}}\right)^{1/4}. \tag{14.37}$$

Similarly, we obtain by the elimination of M_*

$$\frac{\sigma_{v,th}}{1\,\text{km s}^{-1}} = 0.588 \left(\frac{P_{ext}/k}{3\,800\,\text{K cm}^{-3}}\right)^{1/4} \left(\frac{R_*}{1\,\text{pc}}\right)^{1/2}. \tag{14.38}$$

Assuming that P_{ext} is constant, these equations imply

$$\sigma_{v,th} = a_1 M_*^{1/4} = a_2 R_*^{1/2}. \tag{14.39}$$

These relations are also valid for a polytropic equation of state. The coefficients a_1 and a_2 are not very sensitive to the polytropic index and always stay within a factor 2 of the isothermal ones. a_1 is given by Table 1 of Chièze [95].

If there are now large macroscopic motions larger than the thermal agitation of atoms or molecules, which is the general case for interstellar clouds, we add a term $2\mathcal{T}_{macro} = M\sigma_{v,macro}^2$ in the virial equation (see (14.29)). This term has exactly the same form as the thermal term $3MkT_K/(\mu m_\text{H}) = M\sigma_{v,th}^2$. The previous study remains valid, including the numerical values, provided that the velocity dispersion is uniform in the cloud. However, some care should be exercised for turbulent clouds, as discussed previously.

The relationships in (14.39) are close to the observed relationships (cf. (13.1) and (13.3)). This led Chièze and other authors to assume that the observed interstellar structures are generally at the limit of gravitational instability.

We can imagine that if a cloud is gravitationally unstable it tends to fragment rather than collapsing as a whole. Indeed the isothermal Jeans mass decreases as $n^{-1/2}$ when density n increases (14.14). The decrease in the number of particles, i.e., of n, when the H I converts into H_2 if the density and the column density are large enough, reinforces the decrease of the Jeans mass and favors instability and fragmentation. However, fragmentation may prove impossible for non-isothermal equations of state. If the equation of state is $P \propto \rho^\gamma$ the Jeans mass varies with density as $M_J \propto \rho^{3\gamma/2-2}$. It becomes independent of density when $\gamma = 4/3$ and fragmentation is then impossible. In realistic cases it seems that fragmentation can intervene over a large range of interstellar conditions. Chièze [95] and others have suggested that fragmentation is at the origin of the hierarchical distribution of the interstellar structures and continues down to protostellar fragments (see also Chièze & Pineau des Forêts [96]). However, if clouds are turbulent, the situation is much more complex because turbulence can create instabilities and trigger fragmentation even in globally stable clouds, as we saw at the end of Sect. 14.1.2.

14.1.5 The Density Distribution in a Spherical Cloud at Equilibrium

A self-gravitating, isothermal cloud in hydrostatic equilibrium cannot have a uniform density since the gravitational force is a function of the radius r. The density dependence $\rho(r)$ is determined by two equations:
- the local equation of state, i.e. for a perfect gas

$$P = \rho \frac{kT}{\mu m_{\rm H}}, \qquad (14.40)$$

- the equation of hydrostatic equilibrium

$$-\frac{dP}{dr} = \frac{4\pi G \rho}{r^2} \int_0^r \rho y^2\, dy, \qquad (14.41)$$

or, differentiating with respect to r

$$-\frac{1}{r^2} \frac{d}{dr}\left(\frac{r^2}{\rho} \frac{dP}{dr}\right) = -4\pi G \rho. \qquad (14.42)$$

Combining these equations, we obtain

$$-\frac{1}{r^2} \frac{d}{dr}\left(\frac{r^2}{\rho} \frac{d\rho}{dr}\right) = -\frac{4\pi G \mu m_{\rm H} \rho}{kT}. \qquad (14.43)$$

In order to solve this equation analytically we introduce the change of variables $\rho = \lambda e^{\zeta}$, $r = \beta^{1/2} \lambda^{-1/2} \xi$, λ being an arbitrary constant and $\beta = kT/(4\pi G \mu m_{\rm H})$. Equation (14.43) then becomes a simple differential equation,

$$\frac{d^2\zeta}{d\xi^2} + \frac{2}{\xi} \frac{d\zeta}{d\xi} + e^{\zeta} = 0. \qquad (14.44)$$

The boundary conditions at the centre are $\rho = \rho_c$ and $d\rho/dr = 0$. Choosing $\lambda = \rho_c$, they become $\zeta = 0$ and $d\zeta/d\xi = 0$. The problem is now entirely determined once T, μ and the central density are chosen. We can find the solution in many astrophysical textbooks, for example in Chandrasekhar's *Introduction to the Study of Stellar Structures*. The density is found to be approximately proportional to r^{-2}, except in the central regions.

An isothermal sphere at equilibrium is expected to extend to infinity if there is no external pressure, because e^{ζ} hence P and ρ have no zero for positive finite values of ζ. If an external pressure is applied, the radius is finite. The expressions for the different parameters of an isothermal, pressure-bounded sphere are given by McCrea [357]. The corresponding numerical values were used in the preceding section in order to correct the results obtained in the simple, but unrealistic, case of an homogeneous sphere.

In reality, interstellar clouds are not isothermal because the heating and cooling processes detailed in Chap. 8 lead to temperature gradients. The equilibrium condition in more realistic clouds was studied by Falgarone & Puget [170] and by Chièze

& Pineau des Forêts [96]. Figure 14.2, taken from the latter authors, shows as an example the density structure of a molecular cloud whose surface is irradiated by the standard interstellar radiation field ($\chi = 1$) and which is subjected to the standard interstellar medium pressure. Its density varies approximately as $r^{-1.3}$ in this model, instead of r^{-2} as for an isothermal sphere. The equation of state for the gas is also reproduced in Fig. 14.3 for different conditions. An interesting result of this study is that a thermal instability (cf. Sect. 8.3) can occur in the low-density envelope of the clouds when the temperature is higher than about 92 K. It is probably this temperature which limits the extent of the clouds. But once again these somewhat academic results have to be revised for the case of a turbulent cloud.

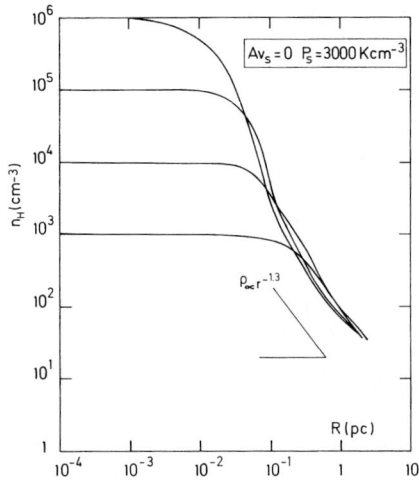

Fig. 14.2. The density profiles for molecular clouds subjected to the standard unattenuated interstellar radiation field ($A_{V,S} = 0$) and to an external pressure $P_S/k = 3\,000$ K cm^{-3}. From Chièze & Pineau des Forêts [96], with the permission of ESO.

14.1.6 Stability and Instabilities in the Presence of a Magnetic Field

The effects of magnetic fields on the stability, gravitational collapse and fragmentation of interstellar clouds are complex. We will here follow the pedagogical textbook of Spitzer [490]. We will only take into account the thermal pressure. Of course, if there are macroscopic motions, e.g. due to turbulence, we have to include their effect, for example by replacing the temperature T in the following equations by an effective, higher temperature. But here again turbulence can introduce complications that we will not examine.

If we assume for simplification that the magnetic field decreases very rapidly outside the cloud, the surface terms in (14.27) can be neglected, provided that the

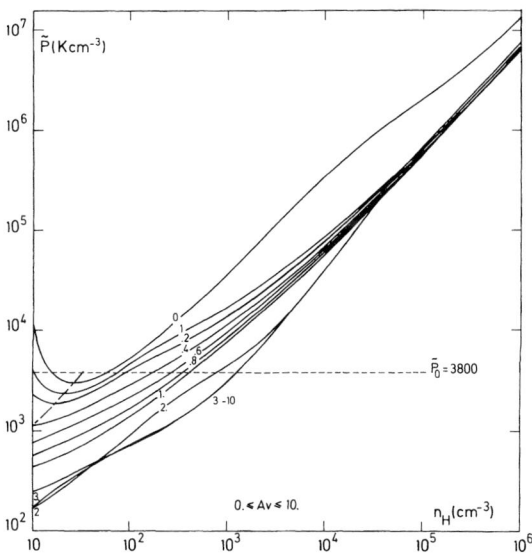

Fig. 14.3. Equations of state for various optical depths (expressed as A_V, as per the labels on the curves) in a molecular cloud subjected to the standard interstellar radiation field. The standard interstellar pressure $P/k = 3\,800$ K cm^{-3} is shown for reference. The pressure within the cloud is necessarily, and at least, equal to the external pressure. A region with thermal instability exists to the left of the dash-dotted line. Large variations in the slope of the equation of state can be seen, corresponding to variations in the polytropic index. From Chièze & Pineau des Forêts [96], with the permission of ESO.

reference surface S is at some distance from the actual surface of the cloud. More complete calculations (cf. Spitzer [490]) show that this approximation is reasonable. For a (always hypothetical) homogeneous, spherical cloud with a uniform magnetic field, the magnetic volume term is $(4\pi R^3/3)(B^2/8\pi)$, so that (14.27) takes a form similar to that of (14.33):

$$\frac{3MkT}{\mu m_H} - 4\pi R^3 P_{ext} - \frac{1}{R}\left(\frac{3}{5}GM^2 - \frac{1}{3}R^4 B^2\right) = 0. \qquad (14.45)$$

An important quantity is the critical mass M_c for which the magnetic energy is equal to the absolute value of the gravitational energy. It is

$$M_c = \frac{c_1}{\pi}\left(\frac{5}{9G}\right)^{1/2} \phi_B, \qquad (14.46)$$

where $\phi_B = \pi R^2 B$ is the total magnetic flux in the cloud and c_1 is a numerical factor that takes into account the real structure of the cloud. We then have numerically

$$\frac{M_c}{1 M_\odot} \approx 10^3 \left(\frac{B}{30\,\mu G}\right)\left(\frac{R}{2\,\text{pc}}\right)^2. \qquad (14.47)$$

Replacing ϕ_B by its value in (14.45), and using $R^2 = [3M_c/(4\pi\rho)]^{2/3}$, we find

$$M_c = \frac{c_1^3 \, 5^{3/2}}{48\pi^2} \frac{B^3}{G^{3/2}\rho^2}. \tag{14.48}$$

Collapse can occur only if $M > M_c$ *and* if the external pressure P_{ext} is larger than a critical value P_m. In order to obtain P_m, we fix the mass M in (14.45) and we examine the variation of the equilibrium external pressure P_{ext} with radius. This pressure reaches a maximum, P_m, for some value of the radius. P_m is thus obtained by differentiation of (14.45) with respect to P_{ext} and R, by putting $dP_{ext}/dR = 0$ and by expressing the corresponding radius as a function of M_c and of M. We then obtain

$$P_m = 3.15 c_2 \frac{c_S^8}{G^3 M^2 [1 - (M_c/M)^{2/3}]^3}, \tag{14.49}$$

where c_S is the isothermal sound velocity and in which we have inserted another numerical factor c_2.

Note that another possibility for collapse exists if there are macroscopic motions. A possible decrease in these motions, e.g. due to viscous dissipation, causes a decrease of the velocity dispersion or of the effective temperature which replaces T in (14.45) and (via c_S) in (14.49), and leads to a decrease of P_m, which might produce instability and collapse even if the external pressure is constant.

If the magnetic field is frozen into matter, the magnetic flux ϕ_B is conserved and the ratio between the gravitational energy and magnetic energy remains constant throughout the contraction of the cloud. Therefore, if the magnetic pressure does not hinder the initial contraction it will not be able to stop the ensuing collapse.

Observations show that magnetic flux is not conserved. Figure 2.6 suggests that $B \propto n^{0.45}$, hence $B \propto R^{-1.35}$ for a spherical cloud, so that $BR^2 \propto R^{0.7}$ for such a cloud. Then the magnetic flux decreases during contraction. The reason, as we will see later, is ambipolar diffusion, which is unavoidable if the medium is only weakly ionized, as it is the case for "neutral" atomic and molecular clouds. As a consequence, magnetic pressure is progressively less efficient at supporting a cloud during contraction.

However, the cloud might not remain spherical during contraction. If the magnetic field has a regular component, which seems to generally be the case, the matter slips along the field lines of this component and the cloud tends to flatten. This flattening increases the modulus of the gravitational energy and this even further decreases the effect of the magnetic field during contraction, with respect to the situation for a spherical cloud. This somewhat changes the initial equilibrium conditions, hence the numerical values of c_1 and c_2. Exact calculations of this case were done by Mouschovias & Spitzer [377], neglecting ambipolar diffusion. Their results are summarized by Spitzer [490]. They find $c_1 = 0.53$ and $c_2 = 0.60$. With this value of c_1 the critical mass is

$$M_c = 1.9 \times 10^4 \left(\frac{B_0}{\mu G}\right)^3 n_0^{-2} \, M_\odot. \tag{14.50}$$

For a standard diffuse cloud, if such a standard cloud exists ($B_0 = 5\mu\text{G}$, $n_0 = 20$ cm^{-3}), $M_c \approx 6 \times 10^3$ M$_\odot$. For the interstellar medium as a whole, $n_0 \approx 1$ cm^{-3} and $M_c \approx 2 \times 10^6$ M$_\odot$. This is indeed the order of magnitude of the masses of the large interstellar complexes (but see a discussion in Sect. 15.1).

To summarize, sub-critical magnetized clouds, i.e., clouds with $M < M_c$, cannot collapse if the external pressure increases, provided that the magnetic flux is conserved. Their contraction is only possible if the magnetic field is redistributed by ambipolar diffusion. Ambipolar diffusion is a slow phenomenon, so that contraction is quasi-static. If the ratio between magnetic pressure and thermal pressure is high (as in a cold molecular cloud) the Jeans mass can be much smaller than the mass of the cloud and fragmentation into dense cores can occur, these cores being perhaps able to form low-mas stars.

Conversely, super-critical clouds can collapse globally if the external pressure increases or if the internal pressure decreases. The magnetic field cannot oppose the collapse even if the magnetic flux is conserved. Super-critical clouds might form by coagulation of sub-critical clouds. A super-critical cloud probably collapses while flattening into a pancake along the regular component of the magnetic field. Fragments with a size comparable to the thickness of this pancake can themselves be super-critical and collapse to form a cluster of massive stars.

These ideas were developed by Lizano & Shu [332] in order to account for the existence of two modes of stellar formation: slow formation of low-mass stars and fast formation of stars of all masses. However, we will see later that their conclusions are considered to be somewhat controversial, and that the bimodal formation of stars might have a different origin.

To end this section, we wish to mention the models of molecular clouds built by McKee & Holliman [361], which include several superimposed polytropes representing the different types of pressure (thermal, turbulent and magnetic). This description is elegant but disputable because turbulence cannot be simply described by a polytropic equation of state.

14.1.7 The Coupling of the Gas and Magnetic Fields: Ambipolar Diffusion

In the preceding subsection we mentioned that the magnetic flux might not be conserved during the evolution of interstellar matter. This means that magnetic fields might not be well coupled with the gas, the MHD approximation no longer being valid. The ions and the magnetic fields are always well coupled. If the gas is only weakly ionized its dynamics is essentially that of the neutral particles. The coupling between magnetic fields and the gas is then due to collisions between the ions and neutrals.

Quasi-Static Case

A simple analysis of ambipolar diffusion is due to Spitzer [490] and Shu et al. [470]. It implicitly assumes equilibrium for the ions and a quasi-static magnetic field.

The Lorentz force applied by the magnetic field on the fluid of charged particles is

$$\frac{\mathbf{j}}{c} \times \mathbf{B} = \frac{1}{4\pi}(\nabla \times \mathbf{B}) \times \mathbf{B}, \tag{14.51}$$

where $\mathbf{j}$ is the density of electric current. This equation makes use of the Ampère law $\mathbf{j} = (c/4\pi)\nabla \times \mathbf{B}$. The Lorentz force causes a displacement of the ions with respect to the neutrals, at a mean velocity that can be obtained by equating the Lorentz force for the ions with the braking force due to the collisions between the ions and neutrals. Per unit volume, this force is

$$\mathbf{F}_d = \rho_i \rho_n \gamma (\mathbf{v}_i - \mathbf{v}_n), \tag{14.52}$$

where ρ_i and ρ_n are, respectively, the densities of the ions and of the neutrals, whose bulk velocities are $\mathbf{v}_i$ and $\mathbf{v}_n$. γ is a braking coefficient which is not too dependent upon the nature of the ions and neutrals, and is of the order of a few times 10^{13} cm^3 g^{-1} s^{-1}. The drift velocity $\mathbf{v}_d$ of the ions with respect to the neutrals is therefore

$$\mathbf{v}_d = \mathbf{v}_i - \mathbf{v}_n = \frac{1}{4\pi \gamma \rho_i \rho_n}(\nabla \times \mathbf{B}) \times \mathbf{B}. \tag{14.53}$$

The electrons play no role here because the momentum transfer between the electrons and neutrals is negligible with respect to that between the ions and neutrals. The electrons follow the ions in their bulk motion so that the medium remains electrically neutral. The ions are linked to the magnetic field. The time evolution of the magnetic field is then given by

$$\frac{\partial \mathbf{B}}{\partial t} + \nabla \times (\mathbf{B} \times \mathbf{v}_i) = 0. \tag{14.54}$$

Combining this equation with the previous one, we obtain

$$\frac{\partial \mathbf{B}}{\partial t} + \nabla \times (\mathbf{B} \times \mathbf{v}_n) = \nabla \times \left\{ \frac{\mathbf{B}}{4\pi \gamma \rho_i \rho_n} \times [\mathbf{B} \times (\nabla \times \mathbf{B})] \right\}. \tag{14.55}$$

If the right-hand part of this equation is zero, due to a high ionization degree, the magnetic field is well coupled to the motion of the neutrals. If not, it corresponds to a diffusion of the magnetic field with respect to the neutrals with an effective diffusion coefficient D equal to $(B^2/4\pi \rho_n)t_{ni}$ in order of magnitude, where $t_{ni} = (\rho_i \gamma)^{-1}$ is the average time between the collisions of a neutral particle with the ions. If the magnetic field is roughly uniform in the cloud, the characteristic time for ambipolar diffusion is $t_{AD} \approx R^2/D \approx R/v_d$.

The Non-Static Case

The following developments are inspired by a paper of Kulsrud & Pearce [293], Appendix C. We will consider that ions and neutrals have different masses, respectively $\mu_i m_H$ and $\mu_n m_H$. Let $\nu_{in} = n_n \langle \sigma v \rangle$ and $\nu_{ni} = n_i \langle \sigma v \rangle$ be, respectively, the collision frequencies of an ion with the neutrals, and of a neutral with the ions, where v is the relative velocity between an ion and a neutral. These frequencies are assumed to be averaged over a maxwellian velocity distribution. Note that if the bulk relative motion of the ions and the neutrals is small, $\langle \sigma v \rangle \simeq 2 \times 10^{-9}$ cm^3 s^{-1}, i.e., the Langevin rate (cf. Sect. 9.1).

The equations of motion for the ions and for the neutrals are, respectively, neglecting the effect of possible gradients of pressure and gravity since we only consider a small volume,

$$\rho_i \left(\frac{\partial \mathbf{v_i}}{\partial t} + \mathbf{v_i} \cdot \nabla \mathbf{v_i} \right) = \frac{1}{4\pi} (\nabla \times \mathbf{B}) \times \mathbf{B} - \rho_i v_{in} \frac{\mu_n}{\mu_n + \mu_i} (\mathbf{v_i} - \mathbf{v_n}), \quad (14.56)$$

$$\rho_n \left(\frac{\partial \mathbf{v_n}}{\partial t} + \mathbf{v_n} \cdot \nabla \mathbf{v_n} \right) = -\rho_n v_{ni} \frac{\mu_i}{\mu_n + \mu_i} (\mathbf{v_n} - \mathbf{v_i}), \quad (14.57)$$

with (14.54) describing the evolution of the magnetic field.

The total force applied by the ions on the neutrals per unit volume is equal to that applied by the neutrals on the ions:

$$|\mathbf{F}_d| = \rho_i \frac{\mu_n}{\mu_n + \mu_i} n_n \langle \sigma v \rangle = \rho_n \frac{\mu_i}{\mu_n + \mu_i} n_i \langle \sigma v \rangle. \quad (14.58)$$

Hence, comparing with (14.52), $\gamma = \langle \sigma v \rangle / [(\mu_n + \mu_i) m_H]$.

We recall that the ions are well coupled with the magnetic field, which implies that the time τ_{in} required for the neutrals to deviate an ion from its gyration around the magnetic field is much larger than the cyclotron (Larmor) period (12.46), so that

$$\tau_{in} = \frac{1}{v_{in}} \frac{\mu_i}{\mu_n} \gg \frac{2\pi \mu_i m_H c}{eB}. \quad (14.59)$$

Let us suppose that the medium is crossed by a (transverse) Alfvén wave characterized by its angular frequency ω and by its wavenumber k, propagating along the axis x, which is taken to be the direction of the stationary magnetic field $\mathbf{B_0}$. This wave perturbs the magnetic field. Using a linear perturbation analysis, we set $\mathbf{B} = \mathbf{B_0} + \delta \mathbf{B}$ with $\delta \mathbf{B}$ along the transverse direction y, so that $\delta B_y = B_{y0} \exp i(kx - \omega t)$, and $v_i = v_{iy}$. This induces a local bulk motion of the ions with a velocity $\mathbf{v_i}$ along y.

Linearizing (14.56) and (14.57), with $\mathbf{v_i} \cdot \nabla \mathbf{v_i} = 0$ because $\nabla \mathbf{v_i}$ is parallel to the x axis, thus perpendicular to $\mathbf{v_i}$, we obtain:

$$-i\omega \rho_i v_{iy} = i B_0 / 4\pi k B_y - \rho_i v_{in} \frac{\mu_n}{\mu_i} (v_{iy} - v_{ny}), \quad (14.60)$$

$$-i\rho_n \omega v_{ny} = -\rho_n v_{ni} (v_{ny} - v_{iy}). \quad (14.61)$$

The linearization of (14.54) gives

$$-i\omega B_y = ik B_0 v_{iy}, \quad (14.62)$$

hence

$$B_y = -\frac{kB_0 v_{iy}}{\omega}, \tag{14.63}$$

and

$$v_{ny} = \frac{\nu_{ni}}{\nu_{ni} - i\omega} v_{iy}. \tag{14.64}$$

Equation (14.61) allows us to obtain the dispersion equation we want. Let us introduce $\omega_k = k v_{Ai}$, the angular frequency of the Alfvén wave which would propagate with the velocity $v_{Ai} = B_0/(4\pi \rho_i)^{1/2}$ in the ions gas, supposed to be decoupled from the neutrals. After some algebra, we obtain the third-degree dispersion equation

$$\omega^3 + i\omega^2(\nu_{ni} + \nu_{in}\frac{\mu_n}{\mu_i}) - \omega_k^2 \omega - i\nu_{ni}\omega_k^2 = 0. \tag{14.65}$$

When writing the dispersion equation in this way we consider ω as a complex quantity, k being real. In this description, a wave exists everywhere and the dissipation causes a frequency cut-off. We could also assume that the wave is emitted by a source, for example by the collision between two clouds, and that it damps during propagation: in this case ω would be real and k would be complex.

In order to investigate this equation, let us first study the low-frequency solutions (ω small), then the high-frequency ones.

At low frequencies the dispersion relation simplifies into

$$\omega^2 + i\omega_k^2 \frac{\epsilon}{\nu_{ni}} \omega - \omega_k^2 \epsilon = 0, \tag{14.66}$$

where we define $\epsilon = \rho_i/\rho_n$, a quantity assumed to be $\ll 1$. The solution for this equation is

$$\omega = -i\omega_k^2 \frac{\epsilon}{2\nu_{ni}} \pm \sqrt{4\omega_k^2 \epsilon - \omega_k^4 \frac{\epsilon^2}{\nu_{ni}^2}}. \tag{14.67}$$

We see that propagation is impossible, ω being a pure imaginary quantity, when $\omega_k > 2\nu_{ni}/\sqrt{\epsilon}$, or in other words if the wavenumber is such that

$$k > \frac{2\nu_{ni}}{\sqrt{\epsilon} v_{Ai}}. \tag{14.68}$$

Conversely, at high frequencies the dispersion equation (14.65) reduces to

$$\omega^2 + i\omega \frac{\nu_{ni}}{\epsilon} - \omega_k^2 = 0, \tag{14.69}$$

with the solution

$$\omega = -i\frac{\nu_{ni}}{2\epsilon} \pm \sqrt{4\omega_k^2 - \frac{\nu_{ni}^2}{\epsilon^2}}. \tag{14.70}$$

Now, there is no wave propagation when $\omega_k < \nu_{ni}/2\epsilon$, or equivalently for wavenumbers such that

$$k < \frac{\nu_{ni}}{2\epsilon v_{Ai}}. \tag{14.71}$$

In summary, if the wavenumber $k = \omega_k/v_{Ai}$ of the Alfvén wave is such that

- $\omega_k > \nu_{ni}/2\epsilon$: the ions are unaffected by collisions with the neutrals because these collisions are rare with respect to the period of the wave, and in this case $\omega \sim \omega_k$ because the density which intervenes in the propagation of the waves is that of the ions;
- $\nu_{ni}/2\epsilon < \omega_k < 2\nu_{ni}/\sqrt{\epsilon}$: there is no wave propagation, all the energy being dissipated in the ion-neutral collisions;
- $\omega_k < 2\nu_{ni}/\sqrt{\epsilon}$: the neutrals couple increasingly to the ions at smaller values of k.

The drift velocity $v_d = v_{iy} - v_{ny}$ between the ions and the neutrals is derived from (14.61). At low frequencies, it reduces to

$$v_d = \frac{\omega v_{iy}}{\nu_{ni}}. \tag{14.72}$$

This drift velocity thus tends to zero when the wave frequency is much smaller than the collision frequency for a neutral with ions. Then the two kinds of particles are perfectly coupled. The more ionized, or the denser the medium, the larger is ν_{ni} and the better the neutrals are coupled with the waves that propagate along the magnetic fields.

The dissipation rate due to collisions between the ions and the neutrals is the product of the friction force $|\mathbf{F}_d|$ (14.58) with the modulus of the drift velocity $|\mathbf{v_i} - \mathbf{v_n}|$. It is, per unit volume,

$$\Gamma_{in} = \gamma \rho_i \rho_n |\mathbf{v_i} - \mathbf{v_n}|^2 = \frac{\rho_i \rho_n \langle \sigma v \rangle}{(\mu_i + \mu_n) m_H} |\mathbf{v}_d|^2. \tag{14.73}$$

In the diffuse medium, taking for example $B = 5\,\mu\text{G}$, $n_n = 30$ cm^{-3} and an ionization fraction $x = 10^{-4}$, the major ion is C$^+$ with $\mu_i = 12\mu_H$, we find $\nu_{in} = 6 \times 10^{-8}$ s^{-1}, $\nu_{ni} = 6 \times 10^{-12}$ s^{-1} and $v_{Ai} = 50$ km s^{-1}. The wavelengths between which no propagation is possible are 70 AU$< \lambda <$ 0.01 pc.

In the dense molecular medium, taking for example $B = 50\,\mu\text{G}$, $n_n = 10^4$ cm^{-3} and $x = 10^{-7}$, the major ion is HCO$^+$ with $\mu_i = 28\mu_H$, we have $\nu_{in} = 2 \times 10^{-5}$ s^{-1} and $v_{Ai} = 440$ km s^{-1}. The wavelengths which cannot propagate are such that 2 AU$< \lambda <$ 0.01 pc.

Strictly speaking, we should also consider that the small charged dust grains also take part in ambipolar diffusion. However, their cyclotron periods are far longer than those of the atomic or molecular ions: while they couple easily to the neutrals, they also decouple easily from the magnetic fields.

Ambipolar Diffusion and Magnetic Support

After this digression, let us come back to the problem of the stability of interstellar clouds. Consider a cloud supported by the magnetic field near the critical mass, thus obeying (14.45). If it is a molecular cloud, its ionization is due to cosmic rays, provided that the density is lower than 10^8 cm^{-3}. Otherwise, the column density is generally sufficiently high to shield the interior of the cloud from low-energy cosmic-ray particles and the ionization is due only to natural radioactivity. Since the cosmic-ray ionization rate is proportional to n_n and the recombination rate to $n_i n_e = n_i^2$, ionization equilibrium implies that $\rho_i = C\rho_n^{1/2}$ (cf. (4.33)), with C varying slowly with temperature and being proportional to the square root of the depletion of heavy elements in the gas. For a depletion of 0.1 and a temperature in the range 10–30 K, $C = 3 \times 10^{-16}$ cm$^{-3/2}$ g$^{1/2}$. Then the ratio between the time for ambipolar diffusion and the free-fall collapse time t_f (see its definition later in Sect. 14.2) is of the order of

$$\frac{t_{AD}}{t_f} \approx \frac{\gamma C}{2(2\pi G)^{1/2}}, \tag{14.74}$$

where γ is the braking coefficient defined earlier (14.52).

With the numerical value chosen for C, this ratio, which is independent of the mass and of the radius of the cloud, is of the order of 8. Therefore ambipolar diffusion is a slow phenomenon when the ionization is due to cosmic rays, and the simplified study we presented at the beginning of this section is valid.

It is interesting to consider the equation of motion for the neutrals where we no longer neglect pressure and gravity. It can easily be derived from (14.27) (see also (14.57)):

$$\rho_n \left[\frac{\partial \mathbf{v}_n}{\partial t} + \mathbf{v}_n \cdot \nabla \mathbf{v}_n \right] = -\nabla P - \rho_n \nabla \Phi + \mathbf{F}_d. \tag{14.75}$$

The neutrals being dominant, the support of the cloud by the magnetic field occurs through the intermediary of the braking force $\mathbf{F}_d$, which we also call the ion-neutral coupling force. Therefore in a weakly ionized medium such as a molecular cloud, the magnetic support is intimately linked to ambipolar diffusion.

14.2 Collapse and Fragmentation

Once the conditions for gravitational instability are met, the cloud collapses upon itself and, as we have seen, the magnetic fields cannot oppose this collapse. The Jeans mass decreases during the contraction and this leads to potential cloud fragmentation. We will now study these phenomena, with the proviso that turbulence could considerably change the conclusions.

14.2.1 The Free-Fall Time

We obtain as follows the characteristic time for collapse, which is called the *free-fall time*, for a spherical cloud which initially has a uniform density ρ_0, and with no magnetic field and no rotation (Spitzer [490]). We also neglect the pressure. This is obviously an academic case, but the result gives the correct order of magnitude.

The equation of motion for a shell with radius r whose initial radius was a can be written as

$$\frac{d^2r}{dt^2} = -\frac{GM(a)}{r^2} = -\frac{4\pi G \rho(0) a^3}{3 r^2}, \qquad (14.76)$$

where $M(a)$ is the mass inside the radius a. The mass inside r is invariant and is equal to $M(a)$. Multiplying the above equation by dr/dt we obtain

$$\frac{dr}{dt}\frac{d^2r}{dt^2} = -\frac{4\pi G \rho(0) a^3}{3}\frac{1}{r^2}\frac{dr}{dt}, \qquad (14.77)$$

which gives after integration over time

$$\frac{1}{2}\left(\frac{dr}{dt}\right)^2 = \frac{4\pi G \rho(0) a^3}{3}\left(\frac{1}{r} + C\right). \qquad (14.78)$$

The constant of integration C is defined by the boundary conditions; taking the square root we obtain,

$$\frac{1}{a}\frac{dr}{dt} = -\left[\frac{8\pi G \rho(0)}{3}\left(\frac{a}{r} - 1\right)\right]^{1/2}. \qquad (14.79)$$

Putting $r/a = \cos^2 \beta$ this equation becomes

$$\beta + \frac{1}{2}\sin 2\beta = t\left[\frac{8\pi G \rho(0)}{3}\right]^{1/2}, \qquad (14.80)$$

in which we take $t = 0$ at the beginning of collapse for which $dr/dt = 0$. β is the same for all shells at a given time t, and all the shells reach the centre at the same time t_f for which $\beta = \pi/2$. This is the free-fall time:

$$\boxed{t_f = \left[\frac{3\pi}{32 G \rho(0)}\right]^{1/2} = \frac{4,3\,10^7}{n_H^{1/2}(0)} \text{ years.}} \qquad (14.81)$$

These equations show that the density increases faster in the central regions than near the surface of the cloud (inhomogeneous collapse), provided of course that the initial density is uniform, or larger in the central parts, which is always true in practice. The free-fall time, which is often taken as $1/\sqrt{G\rho}$ given the uncertainties, is a useful order of magnitude estimate which remains usable to zeroth order in the presence of magnetic fields, for different cloud shapes or density distributions, etc.

14.2.2 Collapse Configurations

The preceding calculation, which neglects pressure, can only give an order of magnitude estimate. There are many more detailed studies, analytical or numerical, for isothermal clouds with or without a magnetic field.

With no magnetic field, the equations which govern the dynamical evolution of the system are, in Eulerian form:

- the equation of conservation of mass

$$\frac{\partial M}{\partial t} + v\frac{\partial M}{\partial r} = 0, \qquad \partial M/\partial r = 4\pi r^2 \rho, \tag{14.82}$$

where $M = M(r,t)$ is the mass inside radius r at time t and $v = v(r,t)$ is the gas bulk velocity at radius r and time t. These relations are equivalent to the usual continuity equation

$$\frac{\partial \rho}{\partial t} + \frac{1}{r^2}\frac{\partial}{\partial r}(r^2 \rho v) = 0, \tag{14.83}$$

- the equation of motion (the Euler equation)

$$\frac{\partial v}{\partial t} + v\frac{\partial v}{\partial r} = -\frac{1}{\rho}\frac{\partial P}{\partial r} - \frac{GM(r,t)}{r^2}, \tag{14.84}$$

- the equation of state, that we write in polytropic form

$$P = K\rho^\gamma. \tag{14.85}$$

It is possible to find self-similar analytical solutions to this system of differential equations: see e.g. Shu [469] or Blottiau et al. [44]. An exemple is shown in Fig. 14.4. Of course, the result depends to some extent upon the initial conditions. However, all realistic initial configurations, which are not very different from the solution labelled with a star (*) in Sect. 14.1, lead to a fast increase of density towards the centre, forming a kind of core with an approximately uniform density (the plateau in Fig. 14.2). The formation of this core can be accelerated by the formation of molecular hydrogen. The matter around this core falls onto it with essentially the free-fall velocity given by (14.81), the density varying approximately as $r^{-3/2}$ in this envelope if it is isothermal. A shock forms at the edge of the core, while the rapid infall of the matter around it generates a low-pressure region at some larger radius. This decrease in pressure causes a rarefaction wave which propagates outwards with the velocity of sound in the outer gas. Initially, the outermost zones are not reached by this rarefaction wave and they fall very slowly towards the centre, and more and more slowly at larger radii. They obey a self-similar solution as long as they are not affected by the rarefaction wave. Their infall is accelerated by the decrease of pressure caused by the passage of this wave. When the rarefaction wave arrives at the exterior of the cloud, the internal pressure becomes smaller than the ambient pressure. Then a compression wave forms which propagates towards the centre and soon forms a shock. This shock in turn helps the matter to fall onto the core.

We see that even in this elementary case the processes are complex. Moreover, we have neglected fragmentation, which is a possibility as mentioned earlier. Fragmentation is a natural phenomenon but we still do not really know the way it occurs. However, it is certain that it does occur and that it is efficient because stars are often observed to form in groups.

It is difficult to observe the collapse due to the low velocities. However, some rather convincing results were recently obtained (Williams et al. [549], Belloche et al. [29]).

The presence of rotation and of magnetic fields further complicates the problem. We will shortly summarize their effects.

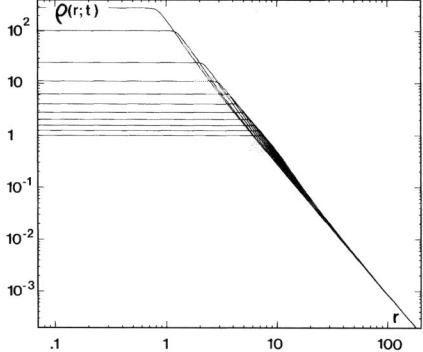

Fig. 14.4. An analytic self-similar solution for the collapse of a spherical cloud without a magnetic field, with an equation of state $P \propto \rho^{1.20}$. The density is plotted as a function of radius at different times. From Blottiau et al. [44], with the permission of ESO.

14.2.3 The Role of Rotation

In principle, rotation can considerably modify the collapse. In order to show this, let us assume that a cloud with an initial radius R_0 rotates with a uniform angular velocity $\omega(0)$. If the contraction was homologous, the conservation of the angular momentum **J** would imply

$$\omega(t)R^2(t) = \omega(0)R^2(0). \tag{14.86}$$

The centrifugal force at the equator $\omega^2 R$ would then vary as $1/R^3$, while the gravitational force varies as $1/R^2$. These forces would be equal at the equator for some value of the radius. Then the cloud would be unable to contract transversally due to **J** and would evolve towards a flattened disk. Fragmentation would be different from that in a spherical collapse, but would still occur.

While numerical simulations confirm these ideas, the contraction is not homologous and only the central core in fact becomes flattened, the envelope remaining approximately spherical. The core tends to form a "disk" (actually a ring) with fast rotation, which might fragment (Fig. 14.5): see for example Boss [58] and the review paper by Bodenheimer [46].

A major problem here is that the observed angular momentum of individual stars is considerably smaller than that expected from the contraction of an interstellar cloud, even if the cloud angular momentum was only due to the galactic differential rotation. Thus the angular momentum has to decrease during the contraction of the cloud. One possibility is a transfer to the orbital motion of double or multiple stars resulting from the fragmentation. This is seen in numerical simulations of cloud collapse. Figure 14.5 gives an example of such a simulation for clouds with different initial rotation velocities. It includes the magnetic fields, the role of which will be dicussed soon, but the results are not qualitatively very different from those without a magnetic field. The fission of a rotating protostar can allow further contraction of

Fig. 14.5. The fragmentation of a cloud with different degrees of rotation. Fig. a, b and c show the result of numerical simulations for clouds rotating with decreasing rotational velocities, after about 10 free-fall times. The contours give the density in the equatorial plane and increment by a factor 2. The magnetic fields and ambipolar diffusion have been taken into account. Cloud (a) has a ratio of rotational energy to gravitational energy of 0.012, and fragments into a binary protostar. Cloud (b) has a ratio of only 0.0080 and simply forms a bar. Cloud (c) stays almost axisymmetric with its ratio of 0.00012. Without a magnetic field, fragmentation occurs for smaller ratios. This shows the role of the magnetic fields in removing the angular momentum. Figure (d) shows the gas temperature for cloud (c), with contours at intervals of a factor 1.3. The central temperature is 25 K, substantially larger than the initial temperature of 10 K. Therefore, the collapse has become adiabatic, and heating does not favor further fragmentation. Reproduced from Boss [59], with the permission of the AAS.

the fragments. Another possibility, which is a necessity for forming isolated stars, with or without a protoplanetary disk, is the transfer of angular momentum to the external regions, either by turbulent coupling or by magnetic coupling. The latter seems to be more efficient. Let us now examine how this works with a simplified example, following Spitzer [490].

Let us consider a rigidly rotating spherical cloud with radius R surrounded by a less dense medium that is not rotating, the whole region being permeated by a magnetic field **B**. **B** is here assumed to be parallel to the rotation axis of the cloud (taken as vertical), whose angular velocity is ω. Rotation twists the field lines above and below the cloud. It can be shown that this deformation propagates along these field lines with the Alfvén velocity $v_A = B/\sqrt{4\pi\rho}$, where ρ is the density of the external medium. It accelerates the external medium to the angular velocity ω, in a cylinder with the same radius R as the cloud. The amount of matter accelerated per unit time in the fraction of this cylinder with radius r and thickness dr is $2\pi\rho r v_A\, dr$, the corresponding angular momentum is ωr^2. Assuming the cloud to be a rigid uniform sphere, whose moment of inertia is $2MR^2/5$, we obtain by integration over r

$$\frac{2MR^2}{5}\frac{d\omega}{dt} = -\pi\rho v_A R^4 \omega. \tag{14.87}$$

The term on the right has been multiplied by 2 to account for the propagation of the deformation on both sides of the cloud. We can define a braking time for the rotation

$$t_B = -\frac{\omega}{d\omega/dt} = \frac{4M}{5(\pi\rho)^{1/2} BR^2}, \tag{14.88}$$

in which v_A has been eliminated. The ratio $M/(BR^2)$ is close to $(M/M_c)^{1/2}G^{-1/2}$ (cf. (14.45) to (14.47)), where M_c is the critical magnetic mass, and does not vary during the contraction of the cloud, assuming that the magnetic flux is conserved. It even stays constant during fragmentation provided that the magnetic flux is divided equally between the fragments. Since M/M_c does not differ much from 1, t_B is close to the free-fall time t_f corresponding to the density ρ of the external medium (14.81). However, because the density is larger in the cloud, the cloud free-fall time is shorter and $t_B > t_f$(cloud). In order for the mechanism to be efficient, it has to act during the initial phases when the cloud density is smaller, and to last for several t_f. More realistic calculations give shorter braking times when the magnetic field is not parallel to the rotation axis, so that this difficulty is removed.

14.2.4 The Role of Magnetic Fields

We have just seen one of the effects of the magnetic fields. The analysis of their role is very complex, often with controversial results. We will first present the point of view of Shu and collaborators, but alternative points of view will be mentioned at the end of this section. As we saw, we should separate the discussion of the sub-critical and of the super-critical clouds.

Gravitational Sub-Critical Clouds

We saw in Sect. 14.1 that such clouds are gravitationally stable. From the point of view of Shu and collaborators, most molecular clouds should be sub-critical since only a small fraction of these clouds are forming massive stars. Their evolution is nevertheless possible thanks to the decrease of the magnetic flux through ambipolar diffusion, but it is slow and hence quasi-static. This was treated, in particular, by Lizano & Shu [332] who include turbulent support against gravity. They show the existence of a second critical mass that they call the *umbral mass* (from the spanish word umbrál, which means threshold). If the mass is super-umbral, ambipolar diffusion is sufficient to allow the formation of a dense core which might form a star. If the mass is smaller, ambipolar diffusion will create rectilinear magnetic field lines and no dense core will be formed. However, this scenario is controversial, as we will see later.

Gravitationally Super-Critical Clouds

These clouds can be unstable and can collapse upon themselves (Sect. 14.1). One possibility for forming such clouds is an agglomeration of sub-critical clouds to form a large complex. This can occur in spiral arms or in the central regions of some galaxies. This formation mode requires a rather large density of molecular material and could be at the origin of starbursts. A super-critical cloud collapses into a flattened system which can fragment into pieces with a size comparable to the thickness of this system (cf. Spitzer [490] Sect. 13.3.a). Another possibility, from the point of view of Shu and collaborators, is the formation of super-critical cores inside a contracting sub-critical cloud from fragments with super-umbral masses. These cores would slowly form low-mass stars in quiescent molecular clouds. The formation and evolution of these cores is relatively well understood as compared to the direct collapse of super-critical clouds. It was studied, in particular, by Galli & Shu [191] accounting for rotation and ambipolar diffusion. It is not possible here to summarize their study and we will only give a few results. They find that the general features of the collapse are not very different from those in the absence of a magnetic field: a central object (a protostar) forms onto which matter keeps falling at almost the same rate with or without a magnetic field. The main effect of the magnetic field is to generate a flattened structure around this object, resulting from the deflection of the accreted matter. This might be the disk that is often observed around protostars which is often, perhaps incorrectly, taken as a protoplanetary disk.

These phenomena have been investigated more recently in a series of papers by Boss (cf. [59]); see also Bodenheimer [46].

Alternative Points of View

There are alternative points of view to those of Shu and collaborators. For example, Nakano [381] considers that only a small fraction of the clouds are really sub-critical. He believes that most clouds are more or less super-critical and contract because their pressure becomes smaller than P_m (14.49) due to the decay of turbulence. The rate of decay of turbulence would then govern the contraction rate because it regulates the internal pressure of the cloud. However, recent numerical simulations (cf. for example Ostriker et al. [390] and references therein) suggest that turbulence decays faster than the free-fall time, so that the usual instability criterion described in Sect. 14.1 applies if turbulence is not maintained by some external mechanism. In any case (Myers [379]), the dense cores in low-activity clouds like the Taurus clouds show little turbulence and are associated only with a small number of low-mass, young stars. Conversely, the dense cores in active star-forming clouds are very turbulent and are associated with many young stars, often with high masses. It is important to note that the differences between the degree of turbulence cannot solely be due to the effect of the winds of the young associated stars. All this suggests that turbulence plays a major role in star formation. But once again these ideas are likely to evolve following future progress in our understanding of interstellar turbulence.

14.3 The End of Collapse: Star Formation

All that precedes is only valid if the thermal energy, into which the gravitational and turbulent energies are transformed, can be radiated away. After some time, this is no longer possible because, due to an increase in column density, the cloud envelope becomes opaque in the CO lines and in other molecular cooling lines, and later also in the dust emission continuum. The collapse is now adiabatic. Then the gas temperature rises. An increase in the thermal pressure results, and a decrease in the mean mass μ of the gas particles due to the dissociation of H_2, then to the ionization of H, can also contribute.

This phenomenon can stop the fragmentation of the cloud. This occurs because the energy radiated by a fragment with radius R cannot be larger than that of a blackbody of the same size at the fragment temperature (it would be equal if the fragment was optically thick at all infrared wavelengths). Most of the energy radiated by a fragment comes from the gravitational energy and the rate of decrease of this energy, in absolute value, is of the order of GM^2/Rt_f, where t_f is the free-fall time. Fragmentation then stops when

$$4\pi R^2 \sigma T^4 \approx \frac{GM^2}{Rt_f}, \tag{14.89}$$

where σ is the Stefan-Boltzmann constant. From (14.33), and with no external pressure, we obtain $GM/R = 5kT/\mu m_H$. Expressing t_f from (14.81) and R as a function of M and T in (14.33), we obtain the minimum mass of the fragments after some simple algebra:

$$M \approx \frac{0.03\, T^{1/4}}{\mu^{9/4}}\, M_\odot \approx 0.07\, M_\odot. \tag{14.90}$$

Although this is a very gross estimate, it seems that the mass of the fragments is unlikely to be less than 0.007 $M_\odot$. While brown dwarfs can be formed from such fragments, like ordinary stars, planets are necessarily formed in a different way, actually by the coalescence of small solid fragments and the accretion of gas.

More generally, the transition to the adiabatic regime during cloud collapse stops the contraction of the core and a star begins to form. However, the envelope material keeps falling onto the core and increases its mass as long as accretion occurs. On the other hand, the mass of the core is reduced by the emission of jets, a phenomenon that we will not discuss here.

We should remember that the adiabatic phase comes late in the collapse. This is because the energy that is radiated during the collapse is only a small fraction of the gravitational energy. For a perfect gas, the energy released in a variation of the volume ΔV at constant pressure is $-P\Delta V$, or $P\Delta\rho/\rho^2$ per unit mass. Integrating over the cloud, we have

$$\Delta E = \frac{kT}{\mu m_H} \int (\Delta \ln \rho) \, dM = \frac{MkT}{\mu m_H} \langle \Delta \ln \rho \rangle, \qquad (14.91)$$

the average being taken over the cloud mass M. We see that as long as the density remains small, ΔE is not much larger than the total thermal energy of the cloud $MkT/(\mu m_H)$, which is much smaller than the gravitational energy for a cold molecular cloud. During contraction, the gravitational energy is mostly converted into the kinetic energy of the macroscopic motions. It is only in the late stage that ΔE can become much larger than the thermal energy.

The contraction continues for some time in the adiabatic regime, the central temperature increases, and the star switches on when temperature and density are sufficient for thermonuclear reactions to take place. We will not discuss this stage in this book.

14.4 The Initial Mass Function and Its Origin

14.4.1 Determinations of the Initial Mass Function, and Related Problems

Stars, at least the relatively massive ones, do not appear to form in isolation but always in groups (see Plates 16 and 31). The *Initial Mass Function* (IMF) describes the relative fractions of stars of different mass at their birth. It can be considered for a star cluster or association, for field stars outside clusters, for a large region of a galaxy or even for a whole galaxy. In all cases the determination of the IMF is an operation full of pitfalls. Firstly, the masses of stars are not observed directly (except if they are members of well-studied binary systems), but only their luminosities and colours, and the use of models is required to calculate the mass from the observed quantities. For massive stars, the luminosity and colour are not sufficient to determine the mass, due to possible ambiguities, and their spectroscopic classification is necessary. Secondly, we have to be sure that the considered stars have not evolved much since their birth. If they have, we have to make corrections for this evolution, which is possible provided that their age is known. Thirdly and most importantly, we must work on complete

14.4 The Initial Mass Function and Its Origin

samples of stars with known distances. It is easiest to determine the initial mass function in young open clusters with known ages, but even in this case there can be a mass segregation between the inner and the outer parts of the cluster. Outside clusters, the determination of the IMF is very difficult, especially for our Galaxy, because of a poor knowledge of the distances and of the difficulties in ensuring complete sampling. This is easier in nearby galaxies, where at least all the stars are at approximately the same distance. A good recent review on this subject is given by Scalo [455]; see also Muench et al. [378].

On the average, the IMF is rather well approximated by a power law above about 1 or 2 $M_\odot$:

$$\Psi(M) = \frac{dn(M)}{d \ln M} = \frac{M dn(M)}{dM} \propto M^{-x}, \quad M_{min} < M < M_{max}, \qquad (14.92)$$

where $n(M)$ is the number of stars with masses between M and $M + dM$. Most recent observations have confirmed a slope of $x \approx 1.35$ given as early as 1955 by Salpeter [447], for masses larger than about 1 $M_\odot$ which can be taken as M_{min} in the above equation[1]. The upper mass limit is $M_{max} \approx 100$ $M_\odot$; a discussion of this limit is given later. The IMF is approximately flat for small masses, but its exact shape is still controversial (see e.g. Carpenter [240] and Luhman et al. [341] for recent determinations in Orion for relatively small masses). The minimum mass of luminous stars is slightly smaller than 0.1 $M_\odot$. At still lower masses we find the brown dwarfs with no thermonuclear reactions. Even in the very well studied case of Orion the poorly known multiplicity of a large fraction of the stars, which seems larger for more massive stars, causes a problem (Preibisch et al. [416]).

Observations at first glance seem to indicate variations in the initial mass function from cluster to cluster. However, in the best-studied clusters like Orion (Plate 31; Luhman et al. [341]) or the 30 Doradus cluster in the Large Magellanic Cloud (Plate 16; Massey & Hunter [350]), the initial mass function is compatible with the Salpeter function. Differences have been suggested for the initial mass functions in clusters and for the field stars, where the slope would be larger than for clusters (cf. references in Scalo [455] and in Elmegreen [164]). But the interpretation of these differences is not without problems, due to important systematic effects such as age effects, centre-to-edge segregation in clusters, and a differential drift in clusters between the massive, short-lived stars and the lower-mass, long-lived stars which might overpopulate the field after having escaped from clusters or associations. Finally, it is not certain that the observed differences between the luminosity functions of the different stellar populations correspond to real differences in the IMFs, as discussed in detail by Elmegreen [164].

The determination of the upper limit to the stellar masses that can form raises a different problem. Here, the difficulty is due to the paucity of stars with large masses, such that the probability of finding a massive star in a sample depends on

---

[1] Scalo [455] comes to a somewhat different average initial mass function, with a slope of 1.7 between 1 and 10 $M_\odot$; however, the difference with the Salpeter function that we use here is not large.

the size of the sample. The largest mass M_{max} to be expected, statistically, is such that, in this sample

$$\int_{M_{max}}^{\infty} n(M)\,dM = 1. \tag{14.93}$$

For the standard mass function with slope $-x$, M_{max} is related to the total mass M_{tot} of the sample by

$$M_{tot} = \int_{M_{min}}^{\infty} Mn(M)\,dM = \frac{xM_{max}^{x} M_{min}^{1-x}}{x-1} \approx 3\,10^{3} \left(\frac{M_{max}}{100\,M_{\odot}}\right)^{1.35} M_{\odot}. \tag{14.94}$$

For $M_{max} = 100\,M_{\odot}$, this already corresponds to the mass of a large star cluster.

For example, from such considerations we expect only one star with a mass larger than 30 $M_{\odot}$ in the Orion nebula cluster. In fact the four most massive stars in the cluster (the four Trapezium stars, see Plates 9 and 30) have masses of 45, 20, 17 and 7 $M_{\odot}$. But in a large cluster like 30 Doradus, or at the scale of a complex of more than $10^{6}\,M_{\odot}$ forming perhaps $3 \times 10^{5}\,M_{\odot}$ of stars, we expect to see stars much more massive than 100 $M_{\odot}$, perhaps as massive as 300 $M_{\odot}$ in the latter case. However, we know of no star more massive than about 100 $M_{\odot}$ (Heydari-Malayeri [239]), except perhaps in 30 Doradus where the most massive stars might reach 120 $M_{\odot}$ and perhaps even 150 $M_{\odot}$ if they are not multiple stars (Massey & Hunter [350]). This suggests the existence of a physical limit to the mass of stars. It has often been assumed that this limit comes from the effects of the radiation pressure of the protostar on dust grains, which would stop the accretion of matter. However, there is no relation between the maximum mass of stars and metallicity, which determines the dust/gas ratio. Also, there is no fundamental limit to the mass of a star at birth: stars with masses higher than a few $M_{\odot}$ are all near the Eddington stability limit at birth[2], which probably causes an important mass loss after their formation, but nothing appears able to stop the accumulation of gas when they form, especially if this gas forms an accretion disk or if there is coalescence between several fragments. Another mechanism therefore seems necessary to limit the mass of stars.

14.4.2 The Origin of the Initial Mass Function

The origin of the initial mass function has been discussed by many authors. It certainly has some relationship to the mass function of the fragments inside molecular clouds, the details of which are much discussed. Poorly answered or completely unanswered questions are: how to define a fragment? Which fraction of the mass of a fragment forms a star? What are the respective roles of fragmentation and of possible coalescence between protostellar condensation nuclei? Does the formation of the first stars inhibit or favor the formation of the next generation? What is the role of the characteristic times of the different physical phenomena? etc.

---

[2] No stability is possible for a star if the radiation pressure on the external layers cannot be balanced by gravity, i.e., if the luminosity of the star is larger than the Eddington luminosity $L_{Eddington} \simeq 3 \times 10^{4} (M/M_{\odot}) L_{\odot}$ (see Binney & Merrifield [41] p. 255).

It is not surprising that many different theories have been proposed to explain the initial mass function. We will only briefly mention here a theory due to Elmegreen ([164], and papers cited therein) which starts from a fractal description of the interstellar medium that we have mentioned at the beginning of Chap. 13. This theory has interesting pedagogical aspects because it illustrates several phenomena which probably take place during star formation.

The basic idea is that the initial mass function results from a random sampling, in a fractal medium, of the masses of the fragments which will form stars, the collapse of these fragments being triggered by some external effect, e.g., an increase in pressure corresponding to the crossing of a shock wave. This triggering is discussed in detail by Elmegreen [163]. It is not necessary to assume that the mass of each fragment is entirely converted into stars, but only that statistically each fragment produces stars according to the same mass distribution, with an upper mass limit proportional to the fragment mass. This does not affect the slope of the IMF if it is a power law, as observed. Evidence for the validity of this hypothesis is given by Motte et al. [376] who find, from observations of the dust continuum emission at 1.3 mm, that the distribution of fragments in the ρ Ophiuchi cloud is similar to the stellar IMF (however, their suggestion that these fragments are directly related to the protostellar fragments is controversial). The model assumes that a random fragmentation mechanism in the fractal medium occurs between a lower mass limit, which is related to the Jeans mass M_J in the molecular medium, and an upper mass limit.

The stellar initial mass distribution is, however, expected to be somewhat different from the mass distribution of the fragments because stars form more rapidly in smaller fragments, which appear to be denser (see (13.2)), the free-fall time being inversely proportional to the square root of density (14.81). There is also competition between the different masses in the fractal hierarchy: when a given mass has formed stars, the distribution of the remaining fragments masses is altered (Elmegreen [162]). For the details, see Elmegreen [164].

Elmegreen [164] attributes the presence of a maximum mass to a time effect. Since the low-mass stars form relatively rapidly, they use a fraction of the gas. It may be that the fractal structure of the remaining gas is redistributed, perhaps due to the turbulence created by the winds from the first stars, before the largest fragment has time to form a very massive star. This fragment would then be cut into smaller fragments and no extremely massive star could form.

All this is quite uncertain and we should not be surprised that different views are presented by other authors. For example, Bonnell et al. [55] estimate that the mass function directly resulting from star formation can be considerably modified in a cluster through the accretion of the residual gas onto the very young stars. The mass of stars in the centre of the cluster, where the gas density is larger, would then increase more rapidly than that of stars in the outer regions. This would preferentially enrich the initial mass function in massive stars in the centre of the cluster, in agreement with observations.

15 Changes of State and Transformations

In this chapter we briefly describe the global time-dependent evolution of the different components of interstellar matter and their mutual transformations. We only consider what happens on time scales of millions or tens of million years. On longer time scales, of the order of a billion years, the cycle *interstellar matter* → *star formation* → *stellar nucleosynthesis* → *stellar winds and supernovae* → *interstellar medium* leads to a progressive decrease of the ratio (mass of the interstellar medium)/(mass of stars) and to a progressive enrichment of the interstellar medium in heavy elements. We are here in the domain of the evolution of galaxies, which is outside the scope of this book, but for which a useful reference would be Pagel [394]. We will therefore assume that the *global* properties of the Galaxy, and in particular those of the interstellar medium, are constant for the time scales of interest to us.

We schematically distinguish four components of interstellar matter: an atomic component distributed into a dense, cold "cloud" medium and a tenuous, warm "intercloud" medium[1], a molecular component, a warm ionized component with a temperature close to 10 000 K (itself subdivided into dense H II regions and a diffuse ionized medium), and finally a hot ionized component with temperatures of the order of 10^6 K. This is only schematic because the observational distinction, and even the physical distinction, between these components is not always clear: this is the case between the neutral and ionized, warm, low-density components, as discussed Sect. 5.2. Also, where does circumstellar matter end and interstellar matter begin? Interstellar dust grains add to the gas and also exhibit variations in time and space in their composition and size distribution.

Our purpose is to summarize the interactions between these components and to study their balance. This is a very ambitious goal because many parameters which would allow a quantitative study are still poorly known, or are completely unknown. As a consequence, the conclusions are provisional and often uncertain. In this chapter, we will use either *formation or destruction rates*, expressed in yr^{-1}, or *lifetimes* which are the inverses of the rates, expressed in years. These quantities, in principle, imply exponential time variations and correspond to changes in the quantities by a factor e.

---

[1] The relative importance of the cold and warm neutral media varies throughout the Galaxy; in the internal regions, where the pressure is high, there is probably little warm neutral medium and the low-density gas is mostly ionized. Conversely, the neutral gas is probably mostly diffuse and warm in the external regions due to the low pressure.

15.1 Atomic, Molecular and Warm Ionized Gas

Exchanges between the atomic gas, the molecular gas and the warm ionized gas (the gas with a temperature of about 10^4 K, not the hot gas at $\sim 10^6$ K which will be discussed later) are dominated on the one hand by the dissociative and ionizing effects of the UV radiation from hot stars, which are rather well known, and on the other hand by recombination, condensation and molecule formation. The latter processes are much more difficult to observe and to model.

15.1.1 Ionized Gas and Exchanges with the Neutral Gas

The simplest rate that can be estimated in the interstellar medium is that of the creation of ionized gas from neutral gas. It is sufficient for this to make a census of hot stars in a given volume and to write that the number of Lyman continuum photons they emit per unit time ionizes as many atoms of hydrogen. This was done for the few kpc around the Sun by Torres-Peimbert et al. [517], Abbott [2] and others. During their census, Torres-Peimbert et al. found that only half of the ionizing O stars are located inside classical H II regions, the others are isolated in low-density regions[2]. Taking into account the fact that the isolated O stars are mostly single while many grouped O stars are in binary systems (Gies [197]), the real fraction of isolated stars should probably be reduced to 30%. The latter contribute to the ionization of the diffuse ionized phase described Sect. 5.2. It remains, however, to determine what fraction of the Lyman continuum photons is actually used to ionize the gas, the rest being absorbed by dust or escaping from the Galaxy.

In the Galaxy it is not possible to make a census of O stars beyond a few kpc, due to interstellar extinction. For this reason, Smith et al. [477] used the inverse procedure in which the distribution of ionized gas is determined quantitatively from its thermal emission, then the distribution of O stars and the formation rate for massive stars is inferred from the density of Lyman continuum photons. Once again, it is necessary to know the fraction of those photons which actually ionize the gas, a fraction that Smith et al. estimate to be of 50%. They obtain in this way a formation rate of 4 $M_\odot$/year for the Galaxy as a whole, neglecting the isolated O stars.

The Lyman continuum photons are only emitted by young, very hot stars with masses larger than about 10 $M_\odot$ (the contribution of evolved objects like white dwarfs or the nuclei of planetary nebulae is small). In order to derive from this the star formation rate, i.e., the amount of interstellar matter converted into stars per unit time, we need to know the IMF for newborn stars. Recent revisions of the star formation rate with the standard IMF described Sect. 14.4 give a value of about 3 $M_\odot$/yr for the Galaxy.

The ionization of classical H II regions poses no problem. Conversely, that of the diffuse ionized gas (Sect. 5.2) is not so obvious. Locally, the recombination rate for

---

[2] Many of these ionizing stars are *runaway* stars which were ejected from binary systems with high velocities following the explosion of the other component.

15.1 Atomic, Molecular and Warm Ionized Gas

this gas, for a 1 cm^{-2} section column perpendicular to the galactic plane (Reynolds [434]), is

$$r_G \approx 4 \times 10^6 I(b) \sin |b| \text{ recombinations s}^{-1} \text{ cm}^{-2}, \qquad (15.1)$$

where $I(b)$ is the average intensity of the Hα line at galactic latitude $|b|$, expressed in rayleighs (1 R = $10^6/4\pi$ photons cm^{-2} sr^{-1} s^{-1}), a relation that can be determined from the equations in Sect. 5.1. Observations lead to $r_G = 5 \times 10^6$ recombinations cm^{-2} s^{-1} (Domgörgen & Mathis [129]). Since each recombination corresponds to one ionization, the ionization rate is equal to this quantity. It corresponds roughly to 30 – 50% of the Lyman continuum photon flux emitted by O stars outside the classical H II regions. Some of these photons are absorbed by the edges of dense neutral clouds, some escape from the Galaxy and some are used for ionizing the low-density medium. This problem was treated by Domgörgen & Mathis in the framework of an inhomogeneous model of the interstellar medium. Their conclusions are in agreement with those of Hoopes & Walterbos [247] in their study of the galaxy M 33 (cf. Plate 10).

The recombination of the ionized gas is fast if the density is relatively large, as in classical H II regions. H II regions do not survive for very long without their ionizing stars, which end their life as supernovae after a rapid evolution through a red supergiant and/or Wolf–Rayet phase. The resulting supernova remnants and bubbles interfere with what remains of the H II region. The diffuse ionized medium survives for much longer once the ionization source disappears. With a density $n_e = 0.03$ cm^{-3}, a temperature $T_e = 8\,000$ K and a recombination coefficient $\alpha^{(2)} = 2.5 \times 10^{-13}$ cm^3 s^{-1} (5.15), the recombination time is

$$t_{rec} = \frac{n_e}{n_e^2 \alpha^{(2)}} = 4 \times 10^6 \text{ yr.} \qquad (15.2)$$

Nevertheless, this time is short with respect to most of the time scales of the problem. On the other hand the diffuse gas is continuously re-ionized by new hot stars, so that there is always a quantity of diffuse ionized gas in galactic disks.

The ionization of the gas produces a pressure increase with consequent dynamical effects on the interstellar medium, these we examined in Sect. 12.3. However, these effects are not very important compared to those of supernovae explosions and stellar winds, except for the champagne effect which contributes to the formation of the ionized phase and takes part in the destruction of molecular clouds.

15.1.2 Atomic Gas–Molecular Gas Exchanges

Molecular gas forms when the atomic gas is shielded from UV radiation, e.g. as a result of gravitational condensation which increases the column densities, while molecular clouds are destroyed by ionization, winds from massive stars and supernova explosions. We will first examine the gas condensation, in general, and the formation of H I complexes and clouds, followed by the formation of molecular clouds, and finally the destruction of the neutral structures.

Gravitational Condensation of the Gas and the Formation of H I Structures

The stability and gravitational condensation of spherical interstellar clouds was studied in Chap. 14.

The gravitational stability of an infinite plane or disk has been treated by Spitzer [490], his Sect. 13.3.a, for the case without a magnetic field. For an isothermal infinite gas layer with temperature T and density $\rho(0)$ in the plane of symmetry, instability occurs for perturbations with wavenumbers $\kappa = 2\pi/\lambda < \kappa_{crit}$, with κ_{crit} such that

$$\kappa_{crit}^2 = \frac{1}{h^2} = \frac{2\pi G \mu m_H \rho(0)}{kT}, \qquad (15.3)$$

where h is the scale height of the gas disk. Similarly, Spitzer showed that gravitational instability appears (taking differential rotation into account) in a disk rotating with the angular velocity ω when its density is such that

$$\rho > \rho_{crit} \simeq \frac{3}{2} \frac{\omega^2}{4\pi G}. \qquad (15.4)$$

This equation can be applied to the vicinity of the Sun, where $\rho_{crit} \simeq 5 \times 10^{-24}$ g cm^{-3}, or $n_H \simeq 2$ atoms cm^{-3} (taking helium into account). These instabilities occur on very large scales, comparable to the scale height of the disk as can be seen in (15.3). Their mass is of the order of 10^6 M$_\odot$. A more accurate criterion is the *Toomre criterion* (Toomre [516]; Binney & Tremaine [40]) which predicts that a gaseous disk with differential rotation is gravitationally stable if

$$\boxed{Q = \frac{v_{rms}\kappa_e}{\pi G \sigma} > 1,} \qquad (15.5)$$

where v_{rms} is the velocity dispersion of the random motions of the gas, κ_e is the *epicyclic frequency* and σ the surface density of the gas. The epicyclic frequency is such that

$$\boxed{\kappa_e^2 = \frac{2\omega(R)}{R} \frac{d}{dR}\left[R^2 \omega(R)\right],} \qquad (15.6)$$

where $\omega(R)$ is the rotational angular velocity at the galactocentric radius R. κ_e is between ω for a disk with keplerian rotation and 2ω for a disk with solid rotation.

Actually, the galactic disk is made of two "fluids": the interstellar matter and the stars, that are gravitationally coupled. This leads to a somewhat different stability criterion (Jog & Solomon [266]; Elmegreen [160]). The growth time for the instability is then of the order of $v_{rms}/(\pi G \sigma)$, i.e. some 5×10^7 years locally. Hydrodynamic simulations like those of Wada & Norman [538] show that if the Toomre criterion is useful in defining the *global* stability of a disk, this disk can be locally unstable even if the Toomre criterion is met, so that condensations can arise and possibly lead to star formation. This throws some doubt on the usefulness of this criterion as a star formation criterion; it is however often used for this purpose.

The previous discussion neglected any magnetic field effects, despite their very important role. Parker [397] showed the existence of a large-scale instability which can result in the condensation of large gas complexes. This *Parker instability* is probably the dominating mechanism in the galactic disk. It is a variant of the Rayleigh–Taylor instability, which arises in a disk supported against gravity by the pressures of the gas, of the cosmic rays and of the magnetic fields. The equation of hydrostatic equilibrium for the system is (2.2), the scale height is given by (2.5). The field lines of the ordered component of the galactic magnetic fields are parallel to the galactic plane, and the instability results from the fact that the gas is forced, by gravity, to slip along any irregularity in the magnetic field for which the field lines are closer to the symmetry plane (Fig. 15.1). The rising parts of the field lines are then unloaded, allowing them to blow further out. The gas condensations resulting from this instability have sizes similar to the scale height of the disk and are separated from each order by a distance of the same order, as in the case of a gravitational instability. The Parker instability corresponds to an effective increase in gravity in the Jeans instability criterion, which reinforces the tendency of the medium towards inhomogeneity.

Whatever the exact mechanism, the formation of these large structures takes place in a time comparable to the free-fall time, which is of the order of 4×10^7 yr for the mean density of the disk, 1 atom cm^{-3} (14.81).

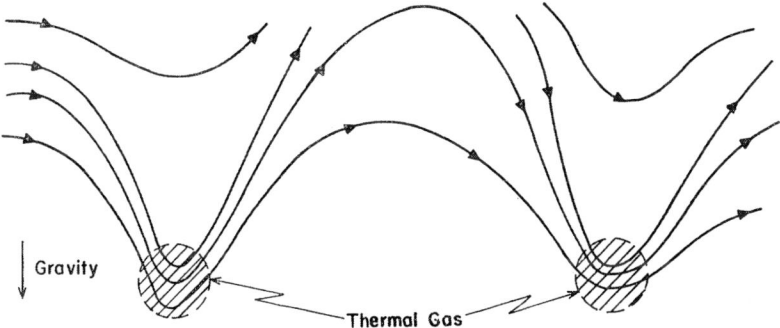

Fig. 15.1. Scheme illustrating the Parker instability. The system of conducting gas + magnetic field experiences an instability which deforms the lines of force perpendicularly to the galactic plane. The gas slips along these lines of force through the effect of gravity and accumulates near the plane of symmetry, while the pressure of cosmic rays inflates the other regions. Reproduced from Parker [397], with the permission of the AAS.

There are different reasons for the gravitational and Parker instabilities to be more efficient inside spiral arms than between them: the density is larger in the arms ($n_H \geq 10$ cm^{-3}) and the characteristic time is shorter ($< 10^7$ years), less than the crossing time for the gas to traverse a spiral arm. The magnetic fields might be larger (a somewhat controversial point), which would enhance the loss of the angular momentum of the complexes and thus favor their contraction. This was studied by Elmegreen, who showed that the instability produces complexes of the order of 10^7 M$_\odot$ separated by roughly three times the width of the arm, as is observed.

The formation of smaller structures from these large complexes is a poorly understood phenomenon. A possible, non-gravitational mechanism for dynamically condensing a tenuous medium into denser clouds in the presence of magnetic fields is described by Hennebelle & Pérault [230]. On the other hand, we saw in Chap. 14 that the gravitational fragmentation of a cloud is natural as long as it stays cold, because at constant temperature and macroscopic velocities the Jeans mass (14.14) decreases with increasing density. This property remains valid in the presence of magnetic fields but is affected by turbulence (Sect. 14.1). The density being larger in the fragments, their free-fall time is smaller than that of the initial cloud if they are self-gravitating (14.81). However, while fragmentation can play a role, it is likely that the mechanical action of supernovae, and to a lesser extent that of stellar winds, is more important. We expect that the gas layers accumulated behind the shocks, in particular around the large bubbles, cool by radiation and form "clouds", or rather layers or filaments, that have colder temperatures if they are denser. Some of these structures will be self-gravitating, others not, and the whole ensemble is likely to be turbulent. We also expect that the condensations tend to merge into more massive structures, which will either collapse or be fragmented and destroyed. The direct compression of a low-density gas or the effects of turbulence and magnetic fields could also temporarily form stable clouds (for a review see Elmegreen [158]). A recent numerical study of these processes, applied to the conditions in the Large Magellanic Cloud, was undertaken by Wada et al. [539]. This work showed that thermal and gravitational instabilities in the gas, triggered by the mechanical perturbations, lead to the formation of cavities, filaments and condensations, resembling the structures observed in H I, including large bubbles.

Formation and Destruction of Giant Molecular Clouds

Giant molecular clouds form in a natural way during contraction of H I clouds when the density is sufficiently high for the efficient formation of H_2 on grains and the column density large enough to shield it from UV radiation. Other processes than the collapse of isolated clouds can also contribute and are probably more efficient. Inelastic collisions between H I clouds lead to a decrease in their bulk random velocities and ease the gravitational collapse over wide size scales. These collisions have been studied by many authors, for example Chièze & Lazareff [94] and more recently Chièze et al. [98]. If there is no magnetic field, the trend is towards fragmentation. However, magnetic fields tend to reinforce coalescence because the field lines of the two clouds mix and interconnect (Falgarone & Puget [170], Clifford & Elmegreen [102]). Also, clouds can grow by accretion, in particular in shocks. Finally, Elmegreen [157] concludes that cloud complexes with masses between 10^2 and 10^7 $M_\odot$ can form and collapse to the typical densities of molecular clouds within a characteristic time equivalent to a few times the characteristic collision time between the initial substructures.

These phenomena are faster in the spiral arms of galaxies because the density is higher. It is possible to explain, in terms of the coalescence of smaller structures, why giant molecular clouds are more numerous in the arms (Casoli & Combes [79]).

15.1 Atomic, Molecular and Warm Ionized Gas

We saw that magnetic fields are necessary for this process to be efficient. For the corresponding calculation, see Elmegreen [156] Sect. 3.2.1.

The destruction of molecular clouds occurs with the formation of massive stars, in particular during the champagne phase (Sect. 12.3). They become partially ionized and fragmented. The lifetime of molecular clouds that have formed massive stars can be empirically determined by investigating the time after which the associations and star clusters no longer emit CO lines. Leisawitz et al. [314] in this way find a destruction time of about 3×10^7 years. However, not all molecular clouds form massive stars. In this case their lifetime can certainly be much longer.

Quantitatively, the cycling of the gas through the molecular phase is a very important phenomenon. Franco et al. [188] show that stars formed with the standard IMF in a molecular cloud ionize, on the average, a mass of gas equal to 25 times their own mass (the rest of the cloud is fragmented and dispersed). With a star formation rate of 3 $M_\odot$/year in the Galaxy, the ionization rate is thus 75 $M_\odot$/year, implying that the interstellar medium with a total mass of about 4×10^9 $M_\odot$ is, on average, ionized once every 5×10^7 years. If there is as much dispersed mass as ionized mass, the cycling time is reduced by a factor 2. The ionized gas from classical H II regions is probably a major source for the diffuse ionized medium (Larson [301]). The recombination of the latter can produce the warm neutral medium, but this medium can also be formed by the heating and evaporation of denser neutral clouds (Wolfire et al. [555]). Finally, the non-ionized but photodissociated fragments of the molecular clouds are expected to be a major source of H I clouds at various temperatures. According to Larson [301] the diffuse ionized gas coming from the ionization of molecular clouds by newly-formed stars, whose thickness perpendicular to the galactic plane is larger than that of the neutral gas due to the champagne effect, should fall back towards the plane with a velocity of a few tens of km s^{-1}, while at the same time undergoing recombination and cooling. This appears to be observed in the 21-cm line (see Plate 2) and the observations are not inconsistent with the expected flux, which is of the order of several tens of $M_\odot$/year for the galactic disk as a whole (see above).

To conclude this section, let us emphasize that the characteristic time for the formation and collapse of the kpc-size interstellar complexes, which is longer than the time scales for all the phenomena that follow this collapse (fragmentation, formation of molecular clouds, star formation and evolution of supernova remnants) is not very different from the time for the destruction of molecular clouds. This implies that the masses of the two phases cannot be very different. In fact, observations show that the total mass of CO-line emitting molecular clouds in the Galaxy is about one quarter of the total mass of the interstellar medium (Table 1.1) indicating that, averaged over the Galaxy, the first time scale might be somewhat longer than the second one. However, the uncertainties are so large that we do not consider it useful to elaborate on this point further.

15.2 Hot Gas and the Galactic Fountain

A large fraction of the interstellar medium is filled with hot ($\sim 10^{5-6}$ K, very low-density gas (see Sect. 5.3). We recall that the existence of this gas was predicted by Spitzer [486] in order to produce a pressure P such that $P/k \approx 2\,000$ cm^{-3} K, which seemed to be necessary to confine those H I clouds which are not self-gravitating. The warm ($\sim 10^4$ K), diffuse ionized gas, with $P/k \approx 0.03 \times 8\,000 = 240$ cm^{-3} K, even if uniform, would be insufficient for this purpose[3]. We will now discuss the formation and evolution of the hot gas. Our knowledge of this gas is still incomplete in spite of a large amount of work. However, the situation is changing thanks to observations with the FUSE satellite. An excellent review on this subject is given by Spitzer [491]; see also Spitzer [492].

The hot gas originates in supernova remnants and large bubbles. We explained Sect. 12.1 how it is heated by the shock during the adiabatic phase of the supernova remnants. This gas encounters and engulfs dense regions which it compresses, heats by compression but mostly by thermal conduction, and possibly evaporates or destroys (Sect. 12.1). It eventually spreads out over large distances from the galactic plane, forming a halo from which it can either escape or fall back to the plane after cooling and condensation. In fact, all these phenomena overlap so much that they are not easy to separate, making the study particularly difficult.

The fact that magnetic fields reduce thermal conduction introduces major uncertainties into the nature of the final isothermal phase of the supernova remnant evolution. This effect is assumed to be very small in the three-component model of the interstellar medium of McKee & Ostriker [359]. In this model evaporation increases the mass of the hot gas, which occupies a considerable fraction of the interstellar medium. If, conversely, conduction is not efficient, then this fraction is reduced. The reduction is even larger if magnetic fields play a major role in the isothermal phase of supernova remnants: it decreases the compression by the shock, which results in a smaller compression of the engulfed neutral interstellar medium. This neutral medium then fills a larger part of the volume of the remnant, accordingly reducing the fraction of volume occupied by the hot gas. It would seem possible to observe these effects by observationally determining the fraction of the volume of the galactic disk filled by the hot gas. Unfortunately this determination is as uncertain as the predictions of the theory. Furthermore, the respective contributions of the individual supernova remnants and of the large bubbles are poorly known. Ferrière [178] considers that bubbles dominate and estimates a filling factor for the hot gas of the order of 30% in the galactic plane.

In any case, conductive evaporating envelopes should contain multi-charged ions such as O VI, N V, C IV and Si IV, which produce absorption lines in the far-UV. Observations of some of these lines with the FUSE satellite (Sembach et al. [465]) show that such envelopes exist around the high-velocity H I clouds, which are at large distances from the galactic plane. It is possible to calculate the column density

---

[3] Recent observations seem to suggest a smaller filling factor for this medium and a density closer to 0.1 cm^{-3}, but this would still be insufficient.

of these ions in an envelope, where they are not in ionization equilibrium (Ballet et al. [17]; cf. Figs. 5.11 and 5.12 for the differences between the results of models with and without ionization equilibrium). Calculations with zero magnetic fields give column densities far larger than the observed ones, indicating once more that magnetic fields are probably present and lead to a reduction of the conduction, and hence evaporation. The agreement is much better for models with magnetic fields (Borskowski et al. [56]). Another suggestion that can explain these moderately hot envelopes is turbulent mixing between the surrounding hot gas and the cold or warm gas observed in the 21-cm line (Slavin et al. [476]). They could even be the cooling fossils of hot structures from the time of formation of the Galaxy (Cen & Ostriker [81]).

The hot gas coming from supernova remnants, and above all from large bubbles, can rise to large heights above the galactic plane, forming chimneys (Sect. 12.2, Plate 32 and Fig. 12.1). This gas can either escape from the Galaxy, forming a wind which has been observed in some external galaxies seen edge-on (see e.g. Dahlem et al. [110]), or cool down and condense into clouds which will later fall back onto the galactic plane, producing a *galactic fountain* (Fig. 15.2, Bregman [67]). The corresponding physics contain two characteristic time scales. The first one is the characteristic time τ_s for the hot gas coming from the disk to reach pressure equilibrium with the ambient medium, which is roughly the time taken for sound waves to cross the scale height h of the halo[4] with the adiabatic velocity c_s

$$\tau_s \approx h/c_s \approx 2.8 \times 10^7 (T/10^6 \text{K})^{1/2} \text{ year}, \qquad (15.7)$$

assuming that the temperatures of the gas of the bubble and the halo are approximately the same. The other characteristic time scale of interest is the radiative cooling time for the hot gas τ_c, which is given by

$$\tau_c = \frac{3nkT}{2\Lambda(T)} \approx 2.4 \; 10^8 \left(\frac{n}{10^{-3} \text{ cm}^{-3}}\right)^{-1} \left(\frac{T}{10^6 \text{ K}}\right) \text{ year}, \qquad (15.8)$$

for temperatures close to 10^6 K, where $\Lambda(T)$ is the non-equilibrium cooling function ($\Lambda(T)$ is considerably larger at temperatures lower than 5×10^5 K: see Fig. 1 of Houck & Bregman [249], which actually gives Λ/n^2 with the definition of Λ of Chap. 8). If $\tau_c \ll \tau_s$ the gas cools before reaching the distance from the plane where it would have been in equilibrium, forms cold clouds and falls back onto the galactic plane. If $\tau_c \gg \tau_s$ the gas sets in hydrostatic equilibrium before cooling, and contributes to the halo. If chimneys have been opened, due to the expansion of large bubbles, the gas spreads out directly into the halo with a velocity of the order of the sound velocity in the hot gas, and can possibly escape from the Galaxy (cf. Norman & Ikeuchi [386]).

---

[4] The scale height for the hot halo gas is approximately given by (2.5) in which we take $\alpha = \beta = 0$, and $\langle v_z^2 \rangle$ as the r.m.s. velocity of protons at the temperature T of the hot gas. For $T = 10^6$ K, $h \approx 6$ kpc. This is considerably more than the scale heights measured with FUSE for different multicharged ions, see Sect. 5.4.2.

HIGH-VELOCITY CLOUDS

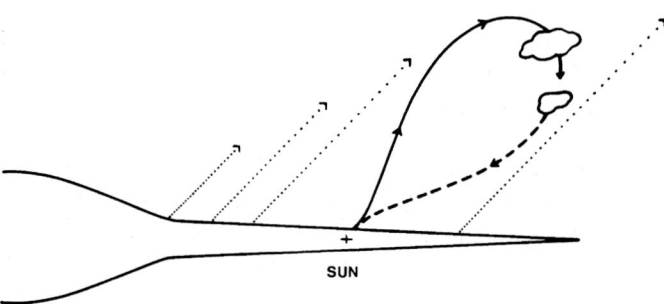

Fig. 15.2. The galactic fountain. The hot gas from the galactic disk rises into the halo (dotted lines with arrow). The solid line and the dashed line illustrate the cycle during which the gas ascends towards the outer part of the Galaxy before condensating through thermal instability. It forms ionized, and then neutral clouds which fall back onto the plane, not far from their original radius. From Bregman [67], with the permission of the AAS.

The dynamical calculations of Houck & Bregman [249] predict that for a model in relatively good agreement with observations, with a hot gas at $T_0 = 3 \; 10^5$ K and $n_0 = 10^{-3}$ cm^{-3} in the galactic plane, clouds condense at about 1.5 kpc above the plane. Their ascending velocity is still about 30 km s^{-1}. Then they have an approximatively ballistic trajectory (i.e., viscous braking by the ambient gas is negligible). They keep rising, stop and then fall back reaching a velocity of some -100 km s^{-1} near the galactic plane. Then, they are decelerated by the viscosity of the interstellar medium, however their dynamical effect is negligible with respect to that of supernova remnants. Recombination occurs during these motions. Such a velocity field for the atomic gas is suggested by 21-cm observations at high galactic latitudes (Plate 3) and by observations of the interstellar absorption lines in front of halo stars, thus giving some support for the galactic fountain model.

The galactic fountain is an important source for the recycling of the interstellar gas. In the model of Houck & Bregman, the flux of gas on either side of the galactic plane is of the order of 0.2 M$_\odot$/year, 0.4 M$_\odot$/year in total, to be compared to the formation rate of stars of about 3 M$_\odot$/year. If we accept the idea of a hot halo extending to several kpc above the galactic plane, whose radiative cooling would locally be compensated for by the energy of Type Ia supernovae, some parts of the halo must experience thermal instabilities and fall back onto the galactic plane. In the halo, the mixing of gas coming from different parts of the disk, and its subsequent redistribution, might smooth out the radial abundance gradient of elements in the galactic disk. However, we do not understand these phenomena well enough to be able to give credible numerical values and to draw firm conclusions about the evolution of the Galaxy. A further complication comes from the discovery of neutral gas extending to a scale height as high as 4.4 kpc and which might be in hydrostatic equilibrium (Kalberla et al. [276]). Moreover, gas may be infalling from intergalactic space, leading to the idea that the Galaxy is an open system.

15.3 Gas–Dust Exchange

Interstellar dust grains are subject to constructive or destructive processes in which matter is exchanged with the gas. Constructive processes are the nucleation, the condensation of volatile molecules, the accretion of atoms and grain coagulation. They generally occur in dense neutral regions, mainly in circumstellar envelopes, which are not discussed in this book, and in molecular clouds. Of course, the coagulation of grains leads to a change in their size distribution, but without exchange with the gas. It was studied in detail by Dominik & Tielens [128]. Destructive processes are the sputtering by the impact of ions or atoms at relatively high energies, and the shattering and vaporization in grain–grain collisions. They mainly occur in the diffuse medium where the effects of supernova shocks and stellar winds dominate. We saw in Sect. 13.3 that turbulence can favor, in important but poorly known ways, the collisions between small grains. Below some relative velocity, collisions between grains tend to produce coagulation, and above this velocity they tend to produce shattering or even vaporization. Sputtering and vaporisation return to the gas a part of the mass of the grains, while the other processes only change their size distribution.

There are many studies of the physics of these processes. Observations show that they are efficient. The important variations in the heavy element abundances in the gas from one line of sight to another (Sect. 4.1 and Table 4.2) show that grain sputtering or vaporization are important in the diffuse interstellar medium and return a large part of the dust mass to the gas. We saw in Sect. 7.4 that ice mantles form on grains in the depths of molecular clouds. These mantles are destroyed when molecular clouds are disrupted, either through thermal evaporation, or through photodesorption (the ejection of a molecule that is electronically excited by the absorption of a UV photon), or through the sudden dissipation of accumulated chemical energy (cf. Sect. 9.2). We also described, in Sect. 11.3, some observations which show that shocks can partly or entirely destroy the grains, returning matter to the gas. However, it is not easy to make a quantitative estimate, and there are controversial points.

Theoretical studies of dust destruction in shocks lead to lifetimes of about 10^6 years for volatile materials like ice, whose binding energy is $E_b \simeq 0.5$ eV, and of the order of 5×10^8 years for refractory materials such as silicates or graphite, for which $E_b \simeq 5$ eV (Jones et al. [269], [270]). The short lifetime of volatile materials does not raise great difficulties because these materials recondense rapidly inside molecular clouds whose lifetimes are of the order of several 10^7 years. However, the refractory materials raise a problem that we will now examine.

Refractory dust grains form in the envelopes of cold, evolved stars, and also in novae and supernovae. For quantitative estimates of the production rates of these different sources, see Table 4 in Boulanger et al. [63]. The dust formation rate in the Galaxy can be estimated if the density and mass loss rate of the producers is known. From this table, it is between 8×10^{-6} and 3×10^{-5} $M_\odot$ kpc^{-2} yr^{-1}, or roughly 6×10^{-3} $M_\odot$ yr^{-1} for the whole Galaxy. The large uncertainty is due to our poor knowledge of the dust production by supernovae[5]. The mass of the dust being approximately 5×10^7 $M_\odot$, the characteristic injection time for dust is of the order of 3×10^9 years. Even if this number is uncertain, it is definitely larger than the lifetime of grain refractory materials in the interstellar medium. This implies that dust has to be re-formed *in situ* in the interstellar medium. It does not seem possible to escape this problem despite the large uncertainties on the formation time and on the lifetime of the dust. The way this re-formation occurs is not understood in detail. Greenberg [209] suggested that the ices in volatile grain mantles can be photolysed by UV photons, producing a refractory organic mantle on the grains. This might perhaps explain the formation of carbonaceous grains from CH_4 and other interstellar molecules containing carbon, but not the formation of silicate grains. It is of interest to tackle the problem in a purely empirical way, based on the observed abundance deficiencies in the interstellar medium. This was done by Tielens [514] and by Jones [271]. Let us give an outline of the study by Tielens.

Assuming that dust is destroyed in the diffuse medium and re-formed in the dense medium, it is sufficient to consider two ISM phases, forgetting the ice mantles: a dense "cloud" phase (atomic or molecular) and an "intercloud" phase, with respective masses M_c and M_i. These phases transform into each other with respective rates k_1 and k_2 such that

$$\frac{M_i}{M_c} \approx \frac{k_2}{k_1} \simeq 0.1 \text{ in the galactic plane,} \qquad (15.9)$$

neglecting the matter which goes into star formation and that injected by the stars at the end of their life. To a first approximation, it is legitimate to neglect the formation of dust in stars, due to the very long time scale for this process. Assuming steady-state, we can also neglect dust destruction during star formation. Grains are destroyed by shocks in the intercloud medium with a rate k_3, and accrete atoms or molecules in the cloud medium with a rate k_4. Calling δ_i and δ_c the respective depletions in the intercloud medium and in the clouds, i.e., the fraction of heavy elements in grains, we have

$$\frac{d\delta_c}{dt} \approx -k_2 \delta_c + k_4(1 - \delta_c) + k_1 \delta_i \frac{M_i}{M_c}, \text{ and} \qquad (15.10)$$

$$\frac{d\delta_i}{dt} \approx -(k_1 + k_3)\delta_i + k_2 \delta_c \frac{M_c}{M_i}. \qquad (15.11)$$

In steady-state, and taking (15.9) into account, this yields

$$\frac{\delta_c}{1 - \delta_c} = \frac{k_4}{k_2}\left(1 + \frac{k_1}{k_3}\right), \qquad (15.12)$$

$$\frac{\delta_c}{\delta_i} = 1 + \frac{k_3}{k_1}. \qquad (15.13)$$

---

[5] Dust formation is proven only in the case of Cassiopeia A with about 10^{-3} $M_\odot$ of dust formed (Douvion et al. [130], [131]).

Cloud destruction occurs on a time scale of about 3×10^7 years, hence $k_2 = 3 \times 10^{-8}$ yr^{-1}, and $k_1 = 3 \times 10^{-7}$ yr^{-1}. The destruction rate of grains in the intercloud medium is estimated from theory to be $k_3 = 2 \times 10^{-8}$ yr^{-1}: this is 10 times the rate of $1/(5 \times 10^8$ yr) that we gave earlier, which is relative to the entire interstellar medium. The accretion rate k_4 in the clouds can now be estimated from measurements of the depletions in the intercloud medium. However, the ratio between the depletions in the clouds and in the intercloud medium differs from element to element. Equation (15.13) gives, from these depletions, values for k_3/k_1 equal to 0.5 ± 0.1, 0.14 ± 0.04 and 0.06 ± 0.003, respectively, for Si, Mg and Fe. The corresponding values of k_4/k_1 are 0.6 ± 0.1, 0.3 ± 0.2 and 0.8 ± 0.1, respectively, not significatively different from one another. We are thus lead to believe that silicon is less resilient to shocks than magnesium, itself less resilient than iron. The fact that silicon is not very resilient is confirmed by the large abundances of gaseous Si and SiO in shocks (see Sect. 11.3). Unfortunately there are currently no similar data for Mg and Fe. It is interesting to note that the value of k_3 for Fe is close to that predicted by theory for refractory materials. This shows that Si and to a lesser extent Mg are not as refractory as Fe, and are therefore expected to exist as mantles deposited on pre-existing grains in the clouds. The depletion of carbon in the diffuse medium is not accurately known, it is thus not possible to say to what extent carbonaceous grains are affected by the destruction/re-formation processes.

It is of interest to compare the accretion rate $k_4 \approx 0.6 k_1 \approx 2 \times 10^{-7}$ yr^{-1} that was empirically determined above with a theoretical estimate. The average time for atoms with density n_i in the interstellar medium to stick on a grain with radius a is (cf. (9.17))

$$t_{acc} = (S\pi a^2 n_i \langle v \rangle)^{-1}, \tag{15.14}$$

where S is a sticking probability and $\langle v \rangle$ is the mean thermal velocity of the atoms. Taking $n_i = 10^{-5} - 10^{-4} n_H$ for abundant atoms, $\langle v \rangle = 1$ km s^{-1}, $S = 1$ and $\langle \pi a^2 \rangle = 1.1 \times 10^{-21}$ cm^2 per H atom for an MRN size distribution (7.7) we find

$$t_{acc} \approx 3 \times 10^9 n_H^{-1} \text{ year}, \tag{15.15}$$

hence a rate of $3 \times 10^{-10} n_H$ yr^{-1}. Accretion occurs mostly in molecular clouds. Letting f be the fraction of the interstellar mass in molecular clouds, the comparison of this rate with the empirical rate implies $3 \times 10^{-10} f n_H \approx 2 \times 10^{-7}$. With $f \approx 0.25$ the required density of molecular hydrogen in molecular clouds would be 1 400 cm^{-3}, a reasonable number. The empirical accretion rate thus causes no real problem.

15.4 Evolution of Interstellar Dust

We have just approached this topic in the preceding section. Here, we bring some details on the nature of dust as encountered in circumstellar envelopes, in the interstellar medium, in protostellar environments and in the Solar system. This is presently a very active research field thanks to results obtained with the ISO satellite and to progress on the laboratory studies of meteorites. Most of this section is inspired by the reviews of Boulanger et al. [63], Jones [272] and Lequeux [319] where the most useful references will be found.

15.4.1 Dust in Circumstellar Envelopes and Planetary Nebulae

The spectra of the envelopes of intermediate-mass stars (approximately $1.5 - 8 \, M_\odot$) at the end of their evolution on the *asymptotic giant branch* (AGB) show that some are oxygen-rich (O/C> 1) and others, the *carbon stars*, are carbon-rich (C/O> 1). The chemistry is very different in the two types of envelopes. C and O combine preferentially with each other to form CO, so that the O-rich sources contain free oxygen but no free carbon, and in the C-rich sources the opposite situation holds. The spectra of oxygen-rich stars exhibit the characteristic bands of amorphous silicates at 9.7 and 18 μm, and also numerous narrower bands of crystalline silicates. Grains of amorphous and crystalline silicates are thus formed in their envelopes. The spectra of carbon stars do not exhibit the aromatic bands discussed in Sect. 7.2, but only a moderately broad band at 11.3 μm due to SiC grains, with perhaps some contribution of amorphous hydrogenated carbon grains. The band at 3.4 μm characteristic of amorphous hydrogenated carbon is indeed seen in a few envelopes. A broad band at 30 μm is probably due to MgS.[6] We also expect that carbon stars produce iron grains but there is no spectroscopic signature of iron in the infrared, thus no possibility for a check, as indeed the case for all metals.

The next step in the evolution of some of these stars is the *proto-planetary nebula* stage, in which the material of the circumstellar envelope detaches from the star which is still cold. It is followed by the planetary nebula stage, where the central star has become very hot and ionizes the surrounding ejected matter, except for very dense condensations which remain neutral. The bands of amorphous and crystalline silicates remain visible in the spectra of oxygen-rich objects[7]. Spectacular changes occur in the infrared spectra of carbon-rich objects. We first observe a spectrum very similar to that of amorphous, but partly aromatic hydrogenated carbon, or to that of natural coals. Thereafter the aromatic infrared bands (the AIBs) appear, which are quite intense in the spectra of young carbon-rich planetary nebulae such as NGC 7027. However, more evolved planetary nebulae such as the Helix nebula do not show these bands anymore, probably because their carriers have been destroyed by the strong UV radiation of the central star. Planetary nebulae are therefore probably not the places where the carriers of the aromatic infrared bands are formed.

Novae and supernovae, as well as some Wolf–Rayet stars, also produce silicates and/or carbonaceous grains, but their contributions to interstellar dust are quite uncertain.

---

[6] Another strong band near 21 μm cannot be due to TiC as often assumed : see Li [324].

[7] Some proto-planetary nebulae such as the famous Red rectangle HD 44179 *simultaneously* exhibit silicate bands and the bands of carbonaceous materials, showing that the surface composition of the star changed during its evolution and that we are observing ejecta produced at different epochs.

15.4.2 Dust in the Interstellar Medium

The nature of interstellar grains was discussed in Chap. 7 and in the preceding section. We expect that grains formed around stars at the end of their lives, as just mentioned, comprise only a small fraction of interstellar grains. They can perhaps form condensation nuclei. It is interesting to mention that there is evidence for the transformation of grains in different interstellar environments. We have already mentioned in Sect. 11.3 and 15.3 the release of silicon from grains in shocks. As far as silicates are concerned, we observe the disappearance of the spectral features of crystalline silicates in mid-IR absorption spectra such as that of Fig. 7.18. It is possible that shocks and the impact of cosmic rays play a role in the amorphisation of silicates, but clearly those which re-form on the grains must be amorphous. This is natural because the crystallisation of silicates requires temperatures higher than 1 000 K for many hours. In some H II regions, however, we observe the reappearance of crystalline silicates, which can be explained by the heating of grains by the intense radiation of the exciting stars.

Carbon grains also experience transformations in the interstellar medium. There is evidence for production of the AIB carriers (probably PAHs) from grains of hydrogenated amorphous carbon which mostly emit a continuum in the mid-infrared. This could occur in photodissociation regions (Cesarsky et al. [87]). Conversely, these carriers are definitely destroyed in an intense UV radiation field. They can also condense on other grains or agglomerate among themselves, the latter processes probably being reversible. The important spatial variations in the infrared colours of the cirrus clouds, which are illustrated in Plates 19 and 21, are produced by these phenomena. Condensation onto the grains and coalescence between small grains are probably the reasons for the spectacular decrease of the intensity in the aromatic bands in regions where the (big) grains are colder than average, due to an increase in their size through condensation or coalescence. This increase in the size of the grains, together with the disappearance of the smaller grains, is necessary in order to explain the observed variations in the UV extinction law (Fig. 7.3) and in the $R = A_V/E(B - V)$ ratio in molecular clouds. The formation of ice mantles is insufficient to account for these observations. A particularly surprising observation is that of the considerable differences between the mid-IR interstellar spectra of the Andromeda galaxy M 31 and galactic spectra (Cesarsky et al. [85], Pagani et al. [393]). In M 31, the 11.3 µm band is extremely strong with respect to the other AIBs. It is difficult to interpret this observation.

Close to compact or ultracompact H II regions, which surround very young, massive stars, the very small carbonaceous grains undergo various transformations that are obviously related to the very strong UV radiation field. These transformations are evidenced by considerable variations in the infrared spectra that are still not well understood. The disappearance of the classical carriers of the AIBs due to photodestruction allows us to see other bands which are probably due to more resilient grains of amorphous carbon, which are also responsible for a strong continuum. These grains might be widespread in the interstellar medium. To complicate the situation even more, let us recall the presence of the 3.4 µm band of hydrogenated aliphatic carbon that is seen in absorption towards the Galactic centre. The carrier of this band is a grain component that appears to exist only in the diffuse interstellar medium and not in molecular clouds.

15.4.3 Dust Around Protostars and in the Solar System

Mid-infrared observations of protostellar disks with ISO yielded surprising results. The spectra are dominated by the emission of silicates, and the crystalline silicates have reappeared, implying that the interstellar grains have been annealed at temperatures higher than 1 000 K.

The infrared spectra of Solar-system comets are very similar. Comets are believed to be formed from pristine material which has not experienced important modification during the 4.6×10^9 years of existence of the Solar system. Their spectra also show the aliphatic hydrogenated carbon band at 3.4 µm, as well as the 3.3 µm aromatic band, both of which are encountered in the interstellar medium. Grains probably coming from comets have been collected in the Earth's atmosphere by U2 aircrafts flying at altitudes of some 20 km. The collected mass is very small, but it has already allowed a detailed mineralogical study that shows the presence of amorphous and crystalline silicates.

Meteorites are other remnants of the primitive Solar system, which have however experienced more violent phenomena than comets. The smallest meteorites (micrometeorites) can be collected from on or inside arctic and antarctic ices. Large masses of meteoritic material are available for investigation, and rare components can be extracted in sufficient quantity for detailed analyses. The methods of mineralogical, elemental and isotopic analyses applied to meteorites have reached a considerable degree of sophistication. However, fragile silicate materials that were in any case modified before the meteorites fell to the Earth are lost for the analysis in the separation of the various meteoritic components. Carbon grains which are undoubtly of interstellar origin, given their elemental and isotopic composition, have been isolated in some meteorites called the *carbonaceous chondrites* (for reviews see Anders & Zinner [4] and Hoppe & Zinner [248]). The most abundant by far of these grains are diamond nanoparticles. There are also grains of SiC, some of which are very big (10 µm or more). The analysis of these big grains shows that they come from AGB stars, a small part also coming from supernovae. Graphite grains are also found, with similar origins. Some have a TiC core. These big grains are the rare ones that have never been exposed to the destructive effect of interstellar shocks or turbulence intermittency, and which come directly from their sources without modification. Unfortunately, it is not possible to determine what proportion of the grains have never been modified by interstellar processes.

Interstellar grains directly penetrate the Solar system. The direction of arrival (apex) for the very big grains is similar to that of the neutral H and He atoms, which also enter the solar system and are not affected by the magnetic field. Their direction and mass distribution have been measured by the ULYSSES and GALILEO space probes. Similar measurements have also be performed using radar reflection from the ionized gas produced by these grains along their trajectories when they penetrate the

Earth's atmosphere. The mass (or size) spectrum of these grains was measured from 0.1 to 1 μm and connects well with the extrapolation of the MRN size distribution (7.7) (Frisch et al. [190]), a confirmation of their interstellar origin. It should be emphasized that grains that big are very difficult to detect in the interstellar medium through extinction or emission measurements. This will perhaps be possible through X-ray scattering observations (Sect. 7.1). These are the same rare grains found imbedded in meteorites.

A Designation of the Most Used Symbols

Where necessary, the number of the chapter where this designation can be found is indicated in parentheses.

a	radius	recombination coefficient (5, 12)
a_0	atomic unit (0,529 Å)	
A	extinction (1, 7, 9, 10, 13, 14)	spontaneous emission probability (3, 4, 5, 8, 9)
$A_{u,l}$ etc.	spontaneous emission probability	
$\mathcal{A}$	atomic mass number	
b	galactic latitude	line broadening coefficient (4)
b_n	coefficient of departure from LTE	
B	magnetic field or induction	
$B_{u,l}$ etc.	coefficients of absorption or of stimulated emission	
B_ν	blackbody brightness	
c	velocity of light	
c_s	adiabatic sound velocity	
c_S	isothermal sound velocity	
C	heat capacity (7)	specific heat
C_{ul} etc.	coefficients of collisional excitation or de-excitation	
d	distance	element depletion (8)
D	binding energy (9)	diffusion coefficient (12)
	fractal dimension (13)	
e	elementary charge	
E	energy	
E_a	activation energy	
E_n	energy per nucleon	
$E(B-V)$, etc.	color excess	
f	fraction	oscillator strength (4)
F	hyperfine quantum number (3)	
g	phase function (7)	
G	constant of gravitation	
h	Planck constant	scale height
H	hamiltonian (4)	photoelectric heating rate (8)
i	index	
I	radiation intensity (2, 3, 15)	moment of inertia
I_ν	radiation intensity	

A Designation of the Most Used Symbols

$\mathscr{J}$	spectral power	
j	index	
$j(\nu)$	emissivity (per steradian)	
J	rotational quantum number	ionizing flux (12)
$\mathscr{J}$	emission or accretion rate	matter flux (11, 12)
	electron flux	
k	index	Boltzmann constant
	wavenumber ($2\pi/\lambda$)	reaction rate (9, 15)
K	force (1)	quantum number (4)
l	index (lower energy level)	galactic longitude (1)
	turbulence scale (13)	length
L	length	grammage (6,12)
	turbulence scale (13)	
m	mass	apparent magnitude (1,6)
m_r	reduced mass	
M	mass	absolute magnitude (1)
M_J	Jeans mass	
n	index (energy level)	number density
N	column density	
$\mathscr{N}$	radiative transition rate (5)	number of particles (6, 7)
p	impact parameter	momentum (12)
P	pressure	
$\mathscr{P}$	momentum flux	
q	electric charge	
Q	partition function	
Q_a, Q_s, Q_e	absorption, scattering, extinction efficiencies	
r	distance, radius	index (ionization degree)
R	distance, radius	Rydberg constant (5)
Re	Reynolds number	
R_S	Strömgren radius	
s	length	
S	ionizing flux of a star	entropy (8)
S_ν	source function	
t	time	
T	temperature	
$\mathscr{T}$	kinetic energy	
u	index (upper energy level)	energy density
	internal energy (11, 12, 14)	
U	ionizing power of a star (5)	velocity (11, 12)
v	velocity	vibration quantum number (4)
v_A	Alfvén velocity	
V	volume	electric potential (4)
w	enthalpy	
W	equivalent width (4)	line intensity (6)
$\mathscr{W}$	energy flux	
x	coordinate	fractional ionization
X	coordinate (4)	conversion factor CO$\to$H$_2$

A Designation of the Most Used Symbols 373

y	coordinate	
Y	coordinate (4)	
z	coordinate	electric charge number (6, 12)
Z	coordinate (4)	electric charge number
α	right ascension (1)	polarisability (6, 9)
	index (11)	accomodation coefficient (8)
	recombination coefficient (5, 12, 15)	
β	escape probability (3, 8)	v/c (6, 8, 11)
	index (11)	
γ	damping coefficient (4)	continuous emission coefficient (5)
	Lorentz factor $1/\sqrt{1-v^2/c^2}$ (6, 8)	C_P/C_V (11, 12, 14)
	braking coefficient (14)	
Γ	rate of energy gain per cm^3	
δ	declination (1)	Kronecker symbol
ϵ	emissivity	heating efficiency (8)
	energy (12)	rate of energy transfer (13)
ζ	cosmic ray ionization rate	
η	efficiency (3, 12)	
θ	angle	
κ	rate of thermal conductivity (12)	
κ_ν	absorption coefficient	
λ	wavelength	mean free path (11, 12, 13)
Λ	rate of energy loss per cm^3	
μ	cosine of the angle with the normal	
ν	frequency	
ξ	coordinate (12)	
ρ	mass density	
σ	cross section	velocity dispersion (1, 4, 13)
	Stefan–Boltzmann constant	
σ_{jk}, etc.	shear tensor	
τ	optical thickness or depth	time (10 to 15)
ϕ	angle	line profile (3, 4, 7)
ϕ_B	magnetic flux	
ϕ_i, etc.	particle flux	
Φ	gravitational potential	
φ	angle	
ψ	angle	
Ψ	electrostatic potential	
ω	angular frequency ($2\pi\nu$)	angular velocity (1, 2, 4, 12)
	vorticity (13)	
ω_p	plasma angular frequency	
Ω	solid angle (2)	potential energy
Ω_{ul}, etc.	collision strength	

B Principal Physical Constants

Velocity of light (by definition)	$c = 2.99792458 \times 10^{10}$ cm s^{-1}
Constant of gravitation	$G = 6.67259(85) \times 10^{-8}$ dyne cm^2 g^{-2}
Planck constant	$h = 6.6260755(40) \times 10^{-27}$ erg s^{-1}
Boltzmann constant	$k = 1.380658(12) \times 10^{-16}$ erg K^{-1}
Stefan–Boltzmann constant	$\sigma = 5.67051(19) \times 10^{-5}$ erg cm^{-2} K^{-4} s^{-1}
Elementary charge	$e = 4.8032068(15) \times 10^{-10}$ E.S.U.
Mass of electron	$m_e = 9.1093897(54) \times 10^{-28}$ g
Mass of proton	$m_p = 1.6726231(10) \times 10^{-24}$ g
Masse of ^{1}H atom	$m_H = 1.6735344 \times 10^{-24}$ g
Rydberg constant for ^{1}H	$R_H = 1.0967758306(13) \times 10^5$ cm^{-1}
Rydberg energy	ryd $= 2.1798741(13) \times 10^{-11}$ erg
	$= 13.606$ eV
Wavelength associated to 1 ryd	911.763 Å
Radius of first Bohr orbit	$a_0 = 0.529 \times 10^{-8}$ cm
Mass energy of electron	0.51099906(15) MeV $= 8.187 \times 10^{-7}$ erg
Mass energy of proton	938.27330(30) MeV $= 1.5032 \times 10^{-3}$ erg
Bohr magneton	$\mu_B = 9.2740154(31) \times 10^{-21}$ erg gauss^{-1}
Magnetic moment of electron	$\mu_e = 1.001 \, \mu_B$
Magnetic moment of proton	$\mu_p = 1.521 \times 10^{-3} \, \mu_B$

The Electron–Volt

Wavelength associated with 1 eV	12 398 Å $= 1.2398$ μm
Frequency associated with 1 eV	2.418×10^{14} Hz
Energy of 1 eV	1.602×10^{-12} erg
Temperature associated with 1 eV	11 604 K

Principal Astronomical Constants

Astronomical unit	AU $= 1.496 \times 10^{13}$ cm
Parsec	pc $= 3.086 \times 10^{18}$ cm $= 3.262$ light years
Solar mass	$M_\odot = 1.989 \times 10^{33}$ g
Solar radius	$R_\odot = 6.955 \times 10^{10}$ cm
Solar luminosity	$L_\odot = 3.845 \times 10^{33}$ erg s^{-1}
Solar absolute bolometric magnitude	$M_{bol,\odot} = 4.72$
Solar absolute V magnitude	$M_{V,\odot} = 4.83$
Solar absolute B magnitude	$M_{B,\odot} = 5.48$
Received power for $m_{bol} = 0$	2.488×10^{-5} erg cm^{-2} s^{-1}
Received flux density for zero magnitude	U: 1 780 Jy; B: 4 000 Jy; V: 3 600 Jy; R: 3 060 Jy; I: 2 420 Jy; J: 1 570 Jy; H: 1 020 Jy; K: 636 Jy

Received flux density for UV magnitude = 0	3.36×10^{-9} erg cm^{-2} s^{-1} Å^{-1}
Tropical year (from equinox to equinox)	365.242 days = 3.156×10^7 s

Units and Unit Conversions

Length

metre (S.I. unit)	1 m = 100 cm
angström	1 Å ≡ 1 A = 10^{-8} cm = 10^{-10} m

Mass

kilogram (S.I. unit)	1 kg = 10^3 g

Energy

joule (S.I. unit)	1 J = 10^7 erg
calorie	1 cal = 4.1855×10^7 erg = 4.1855 J (by definition)

Power

watt (S.I. unit)	1 W = 10^7 erg s^{-1} = J s^{-1}

Flux density

jansky (S.I. sub–unit)	1 Jy = 10^{-26} W m^{-2} Hz^{-1} = 10^{-23} erg s^{-1} cm^{-2} Hz^{-1}

Force

newton (S.I. unit)	1 N = 10^5 dyne

Pressure

pascal (S.I. unit)	1 pascal = N m^{-2} = 10 dyne cm^{-2} = 10^{-5} bar

Dynamical viscosity

poise	1 P = 1 g cm^{-1} s^{-1}

Kinematical viscosity

stoke	1 stoke = 1 cm^2 s^{-1}

Electric charge

coulomb (S.I. unit)	1 C = 2.9979×10^9 E.S.U. = 0.10 E.M.U.

Electric potential

volt (S.I. unit)	1 V = 3.3357×10^{-3} E.S.U. = 10^8 E.M.U.

Electric field

volt per metre (S.I. unit)	1 V/m = 3.3357×10^{-5} E.S.U. = 10^6 E.M.U.

Electric current

ampere (S.I. unit)	1 A = 2.9979×10^9 E.S.U.

Magnetic field or induction

tesla (S.I. unit)	1 T = 10^4 G (gauss: E.M.U.)

Conversion of electromagnetic quantities from the cgs (Gauss) system to the international system (MKSA ≡ S.I.)

In order to perform this conversion, replace all quantities of the second column (cgs) by those of the third column (MKSA). For the electric charge q, current density J, intensity I and polarisation P the conversion is the same as for the charge density ρ. For the impedance Z the conversion is the same as for the resistance R. Of course, all non–electromagnetic quantities (length, mass, etc.) must then be expressed in S.I. units.

Quantity	cgs (Gauss)	MKSA (international)
Velocity of light	c	$\sqrt{\mu_0 \epsilon_0}$
Electric field	E	$\sqrt{4\pi \epsilon_0}$
Electric displacement	D	$\sqrt{\frac{4\pi}{\epsilon_0}}\, D$
Charge density	ρ	$\frac{1}{\sqrt{4\pi\epsilon_0}}\, \rho$
Magnetic induction	B	$\sqrt{\frac{4\pi}{\mu_0}}\, B$
Magnetic field	H	$\sqrt{4\pi\mu_0}\, H$
Magnetic moment	M	$\sqrt{\frac{\mu_0}{4\pi}}\, M$
Conductivity	σ	$\frac{\sigma}{4\pi\epsilon_0}$
Dielectric constant	ϵ	$\frac{\epsilon}{\epsilon_0}$
Magnetic permeability	μ	$\frac{\mu}{\mu_0}$
Resistance	R	$4\pi\epsilon_0\, R$
Inductance	L	$\pi\epsilon_0\, L$
Capacitance	C	$\frac{1}{4\pi\epsilon_0}\, C$

$\mu_0 = 4\pi\, 10^{-7} = 1.2566 \times 10^{-6}\, \frac{\text{A s}}{\text{V m}}$

$\epsilon_0 = \frac{1}{c^2 \mu_0} = 8.8543 \times 10^{-12}\, \frac{\text{V s}}{\text{A m}}$

C Journal Titles Abbreviations

A&A	Astronomy and Astrophysics
A&AS	Astronomy and Astrophysics Supplement Series
AJ	The Astronomical Journal
ApJ	The Astrophysical Journal
ApJS	The Astrophysical Journal Supplements
ARAA	Annual Review of Astronomy and Astrophysics
At. Data Nc. Tables	Atomic Data and Nuclear Data Tables
Bull. Astron. Inst. Netherlands	Bulletin of the Astronomical Institutes of the Netherlands
JETP	Journal of Experimental and Theoretical Physics (Russia)
J. Fluid Mech.	Journal of Fluid Mechanics
J. Geoph. Research	Journal of Geophysical Research
MNRAS	Monthly Notices of the Royal Astronomical Society
Nucl. Phys.	Nuclear Physics
PASP	Publications of the Astronomical Society of the Pacific
Phys. Fluids	Physics of Fluids
Phys. Rev.	Physical Review
Phys. Rev. Letters	Physical Review Letters
Physik. Zeitschr.	Physikalische Zeitschrift
Proc. Roy. Soc. London	Proceedings of the Royal Society of London
Rev. Mod. Phys.	Reviews of Modern Physics

References

1. Aannestad P.A., Purcell E.M. 1973, ARAA 11, 309
2. Abbott D.C. 1982, ApJ 263, 723
3. Abgrall H., Le Bourlot J., Pineau des Forêts G., Roueff E., Flower D.R., Heck L. 1992, A&A 253, 525
4. Anders E., Zinner E. 1993, Meteoritics 28, 490
5. Allamandola L.J., Tielens A.G.G.M., Barker J.R. 1985, ApJ 290, L25
6. Allamandola L.J., Tielens A.G.G.M., Barker J.R. 1989, ApJS 71, 733
7. Allamandola L.J., Hudgins D.M., Sandford S.A. 1999, ApJ 511, L115
8. Allen R.J., Le Bourlot J., Lequeux J., Pineau des Forêts G., Roueff E. 1995, ApJ 444, 157
9. Alves J., Lada C.L., Lada E.A. 1999, ApJ 515, 265
10. Andriesse C.D. 1978, A&A 66, 169
11. Arendt R.G., Dwek E., Moseley S.H. 1999, ApJ 521, 234
12. Axford W.I. 1981, *Proc. Int. School and Workshop on Plasma Astrophysics, Varenna, European Space Agency*, SP-161, p. 425
13. Bakes E.L.O., Tielens A.G.G.M. 1994, ApJ 427, 822
14. Bakes E.L.O. 1997, *The Astrochemical Evolution of the Interstellar Medium*, Twin Press, Veddler
15. Bakes E.L.O., Tielens A.G.G.M. 1998, ApJ 499, 258
16. Baldwin J.A., Ferland G.J., Martin P.G., Corbin M.R., Cota S.A., Peterson B.M., Slettebak A. 1991, ApJ 374, 580
17. Ballet J., Arnaud M., Rothenflug R. 1986, A&A 161, 12
18. Bally J., Stark A.A., Wilson R.W., Langer W.D. 1987, ApJ 312, L45
19. Baluteau J.-P., Zavagno A., Morisset C., Péquignot D. 1995, A&A 303, 173
20. Baring, M.G., Ogilvie K.W., Ellison D.C., Forsyth R.J. 1997, ApJ 476, 889
21. Baring M.G., Ellison D.C., Reynolds S.P., Grenier I.A., Goret P. 1999, ApJ 513, 311
22. Barstow M.A., Wesemael F., Holberg T.B., *et al.* 1994, MNRAS 267, 647
23. Bastiaansen P.A. 1992, A&A 93, 449
24. Bates D.R., Herbst E. 1988, in *Rate Coefficients in Astrochemistry*, ed. T.A. Millar & D.A. Williams, Kluwer, Dordrecht, p. 17
25. Bazell D., Désert F.-X. 1988, ApJ 333, 353
26. Beckert T., Duschl W.J., Mezger P.G. 2000, A&A 356, 1149
27. Bel N., Leroy B. 1989, A&A 224, 206
28. Bell A.R., 1978, MNRAS 182, 147
29. Belloche A., André P., Despois D., Blinder S. 2002, A&A 393, 927
30. Benjamin R.A., Skillman E.D., Smits D.P. 1999, ApJ 514, 307
31. Bennett C.L., Fixsen D.J., Hinshaw G., *et al.* 1994, ApJ 434, 587
32. Berezhko E.G., Völk H.J. 2000, A&A 357, 283

33. Berezhko E.G., Völk H.J. 2004, A&A 419, L27
34. Berezinskii V.S., Bulanov S.V., Dogiel V.A., Ginzburg, V.L., Ptuskin V.S., 1990, *Astrophysics of Cosmic Rays*, North-Holland
35. Bergin E.A., Langer W.D., Goldsmith P.F. 1995, ApJ 441, 222
36. Bergin E.A., Langer W.D. 1997, ApJ 486, 316
37. Bertoldi F., Draine B.T. 1996, ApJ 458, 222
38. Bettens R.P.A., Lee H.-H., Herbst E. 1995, ApJ 443, 664
39. Bieber J.W., Matthaeus W.H. 1997, ApJ 485, 655
40. Binney J., Tremaine S. 1988, *Galactic Dynamics*, Princeton University Press, Princeton, NJ
41. Binney J., Merrifield M. 1998, *Galactic Astronomy*, Princeton University Press
42. Blandford R.D., Ostriker J.P. 1980, ApJ 237, 793
43. Bloemen H. 1989, ARAA 27, 469
44. Blottiau P., Bouquet S., Chièze J.-P. 1988, A&A 207, 24
45. Blumenthal G.R, Gould R.J. 1970, Rev. Mod. Phys. 42, 23
46. Bodenheimer P. 1995, ARAA 33, 199
47. Bohlin R. 1975, ApJ 200, 402
48. Bohlin R.C., Savage B.D., Drake J.F. 1978, ApJ 224, 132
49. Bohren C.F., Huffman D.R. 1983, *Absorption and Scattering of Light by Small Particles*, Wiley, New York
50. Böhringer H. 1998, in *The Local Bubble and Beyond*, ed. D. Breitschwerdt, M.J. Freyberg, J. Trümper, Springer-Verlag, Berlin, p. 341
51. Boissé P. 1990, A&A 228, 483
52. Bolatto A.D., Jackson J.M., Ingalls J.G. 1999, ApJ 513, 275
53. Bonazzola S., Falgarone E., Heyvaerts J., Pérault M., Puget J.-L. 1987, A&A 172, 293
54. Bonazzola S., Pérault M., Puget J.L., Heyvaerts J., Falgarone E., Panis J.F. 1992, J. Fluid Mech. 245, 1
55. Bonnell I.A., Bate M.R., Clark C.J., Pringle J.E. 1997, MNRAS 285, 201
56. Borskowski K.J., Balbus S.A., Fristom C.C. 1990, ApJ 355, 501
57. Born M. 1962, *Atomic Physics*, 7th. ed., Blackie & Son, London
58. Boss A.P. 1980, ApJ 237, 866
59. Boss A.P. 1999, ApJ 520, 744
60. Boulanger F., Prévot M.L., Gry C. 1994, A&A 284, 956
61. Boulanger F., Abergel A., Bernard J.-P., Burton W.B., Désert F.-X., Hartmann D., Lagache G., Puget J.-L. 1996, A&A 312, 256
62. Boulanger F., Boissel P., Cesarsky D., Ryter C. 1998, A&A 339, 194
63. Boulanger F., Cox P., Jones A.P. 2000, in *Space Infrared Astronomy, today and tomorrow*, ed. Casoli F., Lequeux J., David F., EDP Sciences & Springer-Verlag, p. 253
64. Bowyer C.S., Field G.B., Mack J.E. 1968, Nature 217, 3
65. Bowyer S. 1998, in *The Local Bubble and beyond*, ed. D. Breitschwerdt, M.J. Freyberg, J. Trümper, Springer-Verlag, Berlin, p. 45
66. Brandner W., Grebel E., You-Hua Chu, *et al.* 2000, AJ 119, 292
67. Bregman J.N. 1980, ApJ 236, 577
68. Breitschwerdt D., Schmutzler T. 1994, Nature 371, 774
69. Brocklehurst M. 1971, MNRAS 153, 471
70. Brown R.L., Gould R.J. 1970, Phys. Rev. D 1, 2252
71. Brown R.L., Mathews W.G. 1970, ApJ 160, 939
72. Bruhweiler F.C, Gull T.R., Kafatos M., Sofia S. 1980, ApJ 238, L27
73. Burke J.R., Hollenbach D.J. 1983, ApJ 265, 223

74. Bykov A.M., Chevalier R.A., Ellison D.C., Uvarov Yu. A. 2000, ApJ 538, 203
75. Calzetti D., Bohlin R.C., Gordon K.D., Witt A.N., Bianchi L. 1995, ApJ 446, L97
76. Caplan J., Deharveng L. 1986, A&A 155, 297
77. Cardelli J.A., Clayton G.C., Mathis J.S. 1988, ApJ 329, L33
78. Cardelli J.A., Savage B.D., Bruhweiler F.C., Smith A.M., Ebbets D.C., Sembach K.R., Sofia U.J. 1991, ApJ 377, L57
79. Casoli F., Combes F., 1982 A&A 110, 287
80. Castets A., Duvert G., Dutrey A., Bally J., Langer W.D., Wilson R.W. 1990, A&A 234, 469
81. Cen R., Ostriker J.P. 2000, ApJ 514, 1
82. Cesarsky C.J. 1980, ARAA 18, 289
83. Cesarsky C.J. 1992, Nucl. Phys. B (Proc. Suppl.) 25A, 1
84. Cesarsky D., Lequeux J., Abergel A., Pérault M., Palazzi E., Madden S., Tran D. 1996, A&A 315, L309
85. Cesarsky D., Lequeux J., Pagani L., Ryter C. 1998, A&A 337, L35
86. Cesarsky D., Jones A., Lequeux J., Verstraete L. 2000a, A&A 358, 708
87. Cesarsky D., Lequeux J., Ryter C., Gérin M. 2000b, A&A 354, L87
88. Chandrasekhar S. 1951, Proceedings of the Royal Society A210, 18 and 26
89. Chandrasekhar S. 1961, *Hydrodynamic and Hydromagnetic Stability*, Oxford University Press, Oxford
90. Chevalier R.A. 1974, ApJ 188, 501
91. Chevalier R.A. 1975, ApJ 198, 355
92. Chevalier R.A., Oegerle W.R. 1979, ApJ 227, 398
93. Chevalier R.A. 1994, in *Supernovae*, ed. S.A. Bludman *et al.*, Ecole de Physique des Houches Session 54, North Holland, Amsterdam.
94. Chièze J.-P., Lazareff B. 1980, A&A 91, 290
95. Chièze J.-P. 1987, A&A 171, 225
96. Chièze J.P., Pineau des Forêts G. 1987a, A&A 183, 98
97. Chièze J.P., Pineau des Forêts G. 1989, A&A 221, 89
98. Chièze J.P., Pineau des Forêts G., Flower D.R. 1998, MNRAS 295, 672
99. Chini R., Reipurth Bo, Ward-Thompson D., Bally J., Nyman L.-A., Sievers A., Billawala Y. 1997, ApJ 474, L135
100. Chupp E.L. 1976, *Gamma-ray astronomy*, Reidel, Dordrecht
101. Clegg A.W., Cordes J.M., Simonetti J.H., Kulkarni S.R. 1992, ApJ 386, 143
102. Clifford P., Elmegreen B.G. 1983, MNRAS 202, 629
103. Contursi A., Lequeux J., Hanus M., *et al.* 1998, A&A 336, 662
104. Cook D.J., Saykally R.J. 1998, ApJ 493, 793
105. Cowie L.L., McKee C.F. 1977, ApJ 211, 135
106. Crawford M.K., Genzel R., Townes C.H., Watson D.M. 1985, ApJ 291, 755
107. Crézé M., Chereul E., Bienaymé O., Pichon C. 1998, A&A 329, 920
108. Crutcher R.M. 1999, ApJ 520, 706
109. Culhane J.L. 1969, MNRAS 144, 375
110. Dahlem M., Petr M.G., Lehnert M.D., Heckman T.M., Ehle M. 1997, A&A 320, 731
111. Dalgarno A., McCray R.A. 1972, ARAA 10, 375
112. Dame T.M., Thaddeus P. 1994, ApJ 436, L173
113. Dame T.M., Hartmann D., Thaddeus P. 2001, ApJ 547, 792
114. de Boer K.S., Kerp J. 1998, in *The Local Bubble and Beyond*, ed. D. Breitschwerdt, M.J. Freyberg, J. Trümper, Springer-Verlag, Berlin, p. 65
115. Decourchelle A., Sauvageot J.L., Audard M., Aschenbach B., Sembay S., Rothenflug R., Ballet J., Stadlbauer T., West R.G. 2001, A&A 365, L218.

116. De Jong T., Chu Shih-I., Dalgarno A. 1975, ApJ 199, 69
117. De Jong T., Dalgarno A., Boland W. 1980, A&A 91, 68
118. Dennerl K., Haberl F., Aschenbach B., et al. 2001, A&A 365, L 102
119. Dermer C.D. 1986, A&A 157, 223
120. Désert F.-X., Boulanger F., Puget J.-L. 1990, A&A 237, 215
121. Désert F.-X., Jenniskens P., Dennefeld M. 1995, A&A 303, 223
122. d'Hendecourt L.B., Allamandola L.J., Greenberg J.M. 1985, A&A 152, 130
123. d'Hendecourt L., et al. 1989, in *Interstellar Dust*, ed. L. Allamandola & A.G.G.M. Tielens, Kluwer, Dordrecht; p. 207
124. Dickey J.M., Lockman F.J. 1990, ARAA 28, 215
125. Digel S.W., Grenier I.A., Heithausen A., Hunter S.D., Thaddeus P. 1996, ApJ 463, 609
126. Digel S.W., Aprile E., Hunter S.D., Mukherjee R., Xu C. 1999, ApJ 520, 196
127. Dixon W.V.D., Hurwitz M., Bowyer S. 1998, ApJ 492, 569
128. Dominik C., Tielens A.G.G.M. 1997, ApJ 480, 647
129. Domgörgen H., Mathis J.S. 1994, ApJ 428, 647
130. Douvion T., Lagage P.O., Pantin E. 2001, A&A 369, 589
131. Douvion T., Lagage P.O., Cesarsky C.J., Dwek E. 2001, A&A 373, 281
132. Draine B.T. 1978, ApJS 36, 595
133. Draine B.T. 1980, ApJ 241, 1021 (erratum 246, 1045)
134. Draine B.T., Lee H.M. 1984, ApJ 285, 89
135. Draine B.T., Anderson N. 1985, ApJ 292, 494
136. Draine B.T., Sutin B. 1987, ApJ 320, 803
137. Draine B.T., McKee C.F. 1993, ARAA 31, 373
138. Draine B.T., Bertoldi F. 1996, ApJ 468, 269
139. Draine B.T., Lazarian A. 1998, ApJ 58, 157
140. Draine B.T., Bertoldi F. 1999, in *The Universe as seen by ISO*, ed. P. Cox, M. Kessler, ESA SP-427, p. 553
141. Draine B.T., Li A. 2001, ApJ 554, 778
142. Draine B.T., Tan J.C. 2003, ApJ 594, 347
143. Drury L.O'C. 1983, Report on Progress in Physics 46, 973
144. Duffy P., Kirk J.R., Gallant Y.A., Dendy R.O. 1995, A&A 302, L21
145. Duley W.W., Williams D.A. 1986, MNRAS 223, 177
146. Duric N., Gordon S.M., Goss W.M., Viallefond F., Lacey C. 1995, ApJ 445, 173
147. Duvernois M.A., Simpson J.A., Thayer M.R., 1996, A&A 316, 555
148. Dwek E. 1986, ApJ 302, 363
149. Dwek E., Arendt R.G., Fixsen D.J., et al. 1997, ApJ 475, 565
150. Dyer P., Bodansky D., Seamster A.G., Norman E.B., Maxson D.R., 1981, Phys. Rev. C23, 1865
151. Dyer P., Bodansky D., Leach D.D., Norman E.B., Seamster A.G., 1985, Phys. Rev. C32, 1873
152. Egger R.J., Aschenbach B. 1995, A&A 294, L25
153. Eichler D., 1980, ApJ 237, 809
154. Ellison D.C., Drury L.O'C., Meyer J.-P. 1997, ApJ 487, 197
155. Ellison D.C., Berezhko E.G., Baring M.G. 2000, ApJ 540, 292
156. Elmegreen B.G. 1987, in *Interstellar Processes*, ed. D.J. Hollenbach & H. A. Thronson, Reidel, Dordrecht, p. 259
157. Elmegreen B.G. 1989, ApJ 344, 306
158. Elmegreen B.G. 1991 in *Protostars and Planets III*, ed. E.H. Levy & M.S. Matthews, University of Arizona Press, Tucson, p. XXX

159. Elmegreen B.G. 1994, ApJ 433, 39
160. Elmegreen B.G. 1995, MNRAS 275, 944
161. Elmegreen B.G., Falgarone E. 1996, ApJ 471, 816
162. Elmegreen B.G. 1997, ApJ 486, 944
163. Elmegreen, B.G. 1998, in *Origins of Galaxies, Stars, Planets and Life*, ed. C.E. Woodward, H.A. Thronson & M. Shull, Astronomical Society of Pacific Conference Series, San Francisco, 148, 149
164. Elmegreen B.G. 2000, ApJ 539, 342
165. Elmegreen B.G. 2000a, MNRAS 311, L5
166. Encrenaz T., Bibring J.-P., Blanc M., Barucci M.A., Roques F., Zarka P. 2003, *Le Système Solaire*, EDP Sciences/CNRS Editions
167. English J., Taylor A.R., Mashchenko S.Y., Irwin J.A., Basu S., Johnstone D. 2000, ApJ 533, L25
168. Esteban C., Peimbert M., Torres-Peimbert S., Escalante V. 1998, MNRAS 295, 401
169. Falgarone E., Lequeux J. 1973, A&A 25, 253
170. Falgarone E., Puget J.-L. 1985, A&A 142, 157
171. Falgarone E., Phillips T.G., Walker C.K. 1991, ApJ 378, 186
172. Falgarone E., Puget J.-L. 1995, A&A 293 840
173. Falgarone E., Pineau des Forêts G., Roueff E. 1995, A&A 300, 870
174. Falgarone E., Panis J.-F., Heithausen A., Pérault M., Stutzki J., Puget J.-L., Bensch F. 1998, A&A 331, 669
175. Federman S.R., Glassgold A.E., Kwan J. 1979, ApJ 227, 466
176. Ferland G., Koriska K.T., Verner D.A., Ferguson J.W., Ingdon J.B.K., Verner E.M. 1998, PASP 110, 761
177. Fermi E. 1949, Phys. Rev. 75, 1169
178. Ferrière K.M. 1995, ApJ 441, 281
179. Ferrière K.M., Zweibel E.G., Shull M. 1988, ApJ 332, 984
180. Field G.B. 1965, ApJ 142, 531
181. Field G.B., Goldsmith D.W., Habing H.J. 1969, ApJ 155, L149
182. Fisk L.A., Jokipii J.R., Simnett G.M., von Steiger R., Wenzel K.-P., eds. 1998, *Cosmic Rays in the Heliosphere,* Space Science Series of ISSI, Kluwer, Dordrecht
183. Fitzpatrick E.L., Massa D. 1985, ApJS 72, 163
184. Fitzpatrick E.L., Massa D. 1988, ApJ 328, 734
185. Flower D.R., Pineau des Forêts G., Hartquist T.W. 1985, MNRAS 216, 775
186. Flower D.R., Pineau des Forêts G., Walmsley C.M. 1995, A&A 294, 815
187. Flower D.R., Pineau des Forêts G. 1999, MNRAS 308, 271
188. Franco J., Shore S., Tenorio-Tagle G. 1994, ApJ 436, 795
189. Frisch U., Sulem P.L., Nelkin M. 1978, J. Fluid Mech. 87, 719
190. Frisch P., Dorschner J.M., Geiss J. *et al.* 1999, ApJ 525, 492
191. Galli D., Shu F.H. 1993, ApJ 417, 220
192. Garnett D.R., Skillman E.D., Dufour R.J., Peimbert M., Torres-Peimbert S., Terlevich R., Terlevich E., Shields G.A. 1995, ApJ 443, 64
193. Gérin M., Falgarone E., Joulain K., Kopp M., Le Bourlot J., Pineau des Forêts G., Roueff E., Schilke P. 1997, A&A 318, 579
194. Gérin M., Phillips T.G. 2000, ApJ 537, 644
195. Giacalone J., Jokipii J.R. 1999, ApJ 520, 204
196. Gibson S.J., Taylor A.R., Higgs L.A, Dewdney P.E. 2000, ApJ 540, 851
197. Gies D.R. 1987, ApJS 64, 545
198. Ginzburg L.V., Syrovatzkii S.I. 1964, *The Origin of Cosmic Rays*, Pergamon Press, London

199. Gispert R., Lagache G., Puget J.L. 2000, A&A 360, 1
200. Goldberg L., Dupree A.K. 1967, Nature 215, 41
201. Goldreich P., Kwan J. 1974, ApJ 189, 441
202. Goldsmith D.W., Habing H.J., Field G.B. 1969, ApJ 158, 173
203. Goldsmith P.F. 1972, ApJ 176, 597
204. Goldsmith P.F., Langer W.D. 1978, ApJ 222, 881
205. Goldsmith P.F., Langer W.D. 1999, ApJ 517, 209
206. Gondhalekar P.M., Philips A.P., Wilson R. 1980, A&A 88, 272
207. Gordon K.D., Witt A.N., Friedmann B.C. 1998, ApJ 498, 522
208. Gredel R., Lepp S., Dalgarno A., Herbst E. 1989, ApJ 347, 289
209. Greenberg J.M. 1979, in *Stars and Stellar Systems*, ed. B. Westerlund, Reidel, Dordrecht, p. 173
210. Gry C., Lequeux J., Boulanger F. 1992, A&A 266, 457
211. Gry C., Boulanger F., Falgarone E., Pineau des Forêts G., Lequeux J. 1998, A&A 331, 1070
212. Guélin M., Langer W.D., Snell R.L., Wootten H.A. 1977, ApJ 217, L165
213. Guhathakurta P., Draine B.T. 1989, ApJ 345, 230
214. Guibert J., Lequeux J., Viallefond F. 1978, A&A 68, 1
215. Gull S.F. 1973, MNRAS 161, 47
216. Guyon E., Hulin J.-P., Petit L., Mitescu C. 2001, *Physical Hydrodynamics*, Oxford University Press
217. Habing H.J. 1968, Bull. Astron. Inst. Netherlands 19, 421
218. Haffner L.M., Reynolds R.J., Tufte S.L. 1999, ApJ 523, 223
219. Hartmann D. 1994, *The Leiden/Dwingeloo Survey of Galactic Neutral Hydrogen*, Ph.D. Thesis, University of Leiden
220. Hartmann D., Burton W.B. 1997, *Atlas of Galactic Neutral Hydrogen*, Cambridge University Press
221. Hartquist T.W. 1977, ApJ 217, L45
222. Hasegawa T.I, Herbst E. 1993, MNRAS 261, 83
223. Hébrard G., Péquignot D., Vidal-Madjar A., Walsh J.R., Ferlet R. 2000, A&A 354, L79
224. Heiles C. 1987, ApJ 229, 533
225. Heiles C. 1994, ApJ 436, 720
226. Heiles C., Reach W.T., Koo B.-C. 1996, ApJ 466, 191
227. Heithausen A., Bensch F., Stutzki J., Falgarone E., Panis J.F. 1998, A&A 331, L65
228. Heitler W. 1954, *The Quantum Theory of Radiation*, Clarendon Press, Oxford
229. Henke B.L., Gullison E.M., Davis J.C. 1993, At. Data Nc. Tables 54, 181 (updated tables available at: http://www-cxro.lbl.gov)
230. Hennebelle P., Pérault M. 2000, A&A 359, 1124
231. Henyey L.G., Greenstein J.L. 1941, ApJ 93, 70
232. Herbig G.H. 1994, ARAA 33, 19
233. Herbst E. 1980, ApJ 237, 462
234. Herbst E., Lee H.-H. 1997, ApJ 485, 689
235. Herzberg G. 1945, *Infrared and Raman Spectra of Polyatomic Molecules*, Van Nostrand, New York
236. Herzberg G. 1960, *Spectra of Diatomic Molecules*, Reinhold, Princeton
237. Herzberg G. 1966, *Electronic Spectra and Electronic Structure of Polyatomic Molecules*, Reinhold, Princeton
238. Hess V. 1912, Physik. Zeitschr. 13, 1084
239. Heydari-Malayeri M. 1993, in *New Aspects of Magellanic Cloud Research*, ed. B. Baschek, G. Klare, J. Lequeux, Springer-Verlag, Berlin, p. 245

240. Hillenbrand L.A., Carpenter J.M. 2000, ApJ 540, 236
241. Hollenbach D.H., Salpeter E.E. 1971, ApJ 163, 155
242. Hollenbach D., McKee C.F. 1979, ApJS 41, 555
243. Hollenbach D., McKee C.F. 1989, ApJ 342, 306
244. Hollenbach D.H., Tielens A.G.G.M. 1997, ARAA 35, 179
245. Hollenbach D.H., Tielens A.G.G.M. 1999, Rev. Mod. Phys., 71, 173
246. Holzer T.E. 1989, ARAA 27, 199
247. Hoopes C.G., Walterbos R.A.M. 2000, ApJ 541, 597
248. Hoppe P., Zinner E. 2000, J. Geoph. Research 105, A5, 10371
249. Houck J.C., Bregman J.N. 1990, ApJ 352, 506
250. Howk J.C., Savage B.D., Fabian D. 1999, ApJ 525, 253
251. Hummer D.G., Seaton M.J. 1963, MNRAS 125, 437
252. Hummer D.G., Storey P.J. 1987, MNRAS 224, 801
253. Hunter S.D., Bertsch D.L., Catelli J.R., *et al.* 1997, ApJ 481, 205
254. Hurwitz M., Bowyer S. 1996, ApJ 465, 296
255. Hutchings J.B., Giasson J. 2001, PASP 113, 1205
256. Inoue H., Koyama K., Matsuoka M., Ohashi T., Tanaka Y., Tsunemi H. 1979, ApJ 227, L85
257. Ip W.-H., Axford W.I. 1985, A&A 149, 7
258. Jackson J.D. 1975, *Classical Electrodynamics*, 2nd edition, John Wiley, New York
259. Jenkins E.B., Meloy D.A. 1974, ApJ 193, L12
260. Jenkins E.B., Shaya E.J. 1979, ApJ 231, 55
261. Jenkins E.B., Jura M., Loewenstein M. 1983, ApJ 270, 88
262. Jenkins E.B., Peimbert A. 1997, ApJ 477, 265
263. Jenniskens P., Désert F.-X. 1994, A&AS 106, 39
264. Jenniskens P., Porceddu I., Benvenuti P., Désert F.-X. 1996, A&A 313, 649
265. Joblin C., Léger A., Martin P. 1992, ApJ 393, L79
266. Jog C.J., Solomon P.M. 1984, ApJ 276, 144
267. Johnstone D., Wilson C.D., Moriarty-Schieven G., Joncas J., Smith G., Gregersen E., Fich M. 2000, ApJ 545, 327
268. Jokipii, J.R. 1973, ARAA 11, 1
269. Jones A.P., Tielens A.G.G.M., Hollenbach D.J., McKee C.F. 1994, ApJ 433, 797
270. Jones A.P., Tielens A.G.G.M., Hollenbach D.J.1996, ApJ 469, 740
271. Jones A.P., d'Hendecourt L. 2000, A&A, 355, 1191
272. Jones A.P. 2000, J. Geoph. Research 105, A5, 10257
273. Joulain K., Falgarone, E., Pineau des Forêts G., Flower D. 1998, A&A 340, 241
274. Jura M. 1974, ApJ 191, 375
275. Jura M. 1975, ApJ 197, 575 (erratum 202, 561)
276. Kalberla P.M.W., Westphalen G., Mebold U., Hartmann D., Burton W.B. 1998, A&A 332, L61
277. Kamper K., van den Bergh S. 1976, ApJS 32, 351
278. Kaplan S.A. 1966, *Interstellar Gas Dynamics*, edit. F.D. Kahn, Pergamon Press, Oxford
279. Karzas W.J., Latter R. 1960, ApJS 6, 167
280. Katz N., Furman I., Biham O., Pirronello V., Vidali G. 1999, ApJ 522, 305
281. Kaufman M.J., Wolfire M.G., Hollenbach D.J., Luhman M.L. 1999, ApJ 427, 795
282. Kim S.H, Martin P.G 1996, ApJ 462, 296
283. Klessen R.S., Heitsch F., Mac Low M.-M. 2000, ApJ 535, 887
284. Knödlseder J., Dixon D., Bennett K. *et al.* 1999, A&A 345, 813

285. Kolmogorov A.N. 1941, reproduced in 1991 in Proceedings of the Royal Society of London 434, 9; see also other articles in this issue
286. Kolmogorov A.N. 1962, J. Fluid Mech. 13, 82
287. Kozlovsky B., Ramaty R., Lingenfelter R.E 1997, ApJ 484, 286
288. Kramer C., Alves J., Lada C.J., Lada E.A., Sievers A., Ungerechts H., Walmsley C.M. 1999, A&A 342, 257
289. Krügel E. 2002, *The Physics of Interstellar Dust*, IoP Publishing, Bristol
290. Kuijken K., Gilmore G. 1989, MNRAS 239, 605
291. Kulessa A.S., Lynden-Bell D. 1992, MNRAS 255, 105
292. Kulkarni S.R., Heiles C. 1987, in *Interstellar Processes*, ed. D.J. Hollenbach & H.A. Thronson, Reidel, Dordrecht, p. 87
293. Kulsrud R., Pierce W.P. 1969, ApJ 156, 445
294. Kutner M.L., Ulich B.L. 1981, ApJ 250, 341
295. Lagache G., Abergel A., Boulanger F., Désert F.-X., Puget J.-L. 1999, A&A 344, 322
296. Lagage P.O., Claret A., Ballet J., Boulanger F., Cesarsky C.J., Cesarsky D., Fransson C., Pollock A. 1996, A&A 315, L273
297. Landau L.D., Lifshitz E.M. 1987, *Fluid Mechanics*, 2nd. edition, Pergamon Press, London
298. Lang C.C., Anantharamaiah K.R., Kassim N.E., Lazio T.J.W. 1999, ApJ 521, L41
299. Lang K.R. 1999, *Astrophysical Formulae*, 3rd edition, Vol. 1, Springer-Verlag, Berlin
300. Larson R.B. 1981, MNRAS 194, 809
301. Larson R.B. 1996, in *The Interplay between Massive Star Formation, the ISM and Galaxy Evolution*, eds. D. Kunth, B. Guiderdoni, M. Heydari-Malayeri, T.X. Thuan, Editions Frontières, Gif sur Yvette, p. 3
302. Le Bourlot J., Pineau des Forêts G., Roueff E., Flower D.R. 1993, A&A 267, 233
303. Le Bourlot J., Pineau des Forêts G., Roueff E., Schilke P. 1993, ApJ 416, L87
304. Le Bourlot J., Pineau des Forêts G., Roueff E., Dalgarno A., Gredel R. 1995, ApJ 449, 178
305. Le Bourlot J., Pineau des Forêts G., Roueff E., Flower D. 1995, A&A 302, 870
306. Le Bourlot J., Pineau des Forêts G., Flower D.R. 1999, MNRAS 305, 802
307. Le Bourlot J. 2000, A&A 360, 656
308. Ledoux G., Ehbrecht M., Guillois O., *et al.* 1998, A&A 333, L39
309. Lee H.-H., Bettens R.P.A., Herbst E. 1996, A&AS 119, 111
310. Lee H.-H., Herbst E., Pineau des Forêts G., Roueff E., le Bourlot J. 1996, A&A 311, 690
311. Lee H.-H., Roueff E., Pineau des Forêts G., Shalabiea O.M., Terzieva R., Herbst E. 1998, A&A 334, 1047
312. Léger A., Puget J.-L. 1984, A&A 137, L5
313. Leinert Ch., Bowyer S., Haikala L.K., *et al.* 1998, A&AS 127, 1
314. Leisawitz D., Bash F.N., Thaddeus P. 1988, ApJS, 70, 731
315. Lequeux J. 1983, A&A 125, 394
316. Lequeux J., Roueff E. 1991, Physics Reports 200, 241
317. Lequeux J., Le Bourlot J., Pineau des Forêts G., Roueff E., Boulanger F., Rubio M. 1994, A&A 292, 371
318. Lequeux J., Kunth D., Mas-Hesse J.M., Sargent W.L.W. 1995, A&A 301, 18
319. Lequeux J. 2000, J. Geoph. Research 105, A5, 10249
320. Leroy J.L. 1999, A&A 346, 955
321. Lesko K.T., Norman E.B., Larimer R.-M., Kuhn S., Meekhopf D.M., Crane S.G., Bussel H.G. 1988, Phys. Rev. C37, 1808

322. Li A., Greenberg J.M. 1997, A&A 323, 566
323. Li A., Draine B.T. 2001, ApJ 550, L213
324. Li A. 2003, ApJ 599, L45
325. Linsky J.L., Diplas A., Wood B.E., *et al.* 1995, ApJ 451, 335
326. Lis D.C., Pety J., Phillips T.G., Falgarone E. 1996, ApJ 463, 623
327. Lis D.C., Keene J, Dowell C.D., Benford D.J., Phillips T.G., Hunter T.R., Wang N. 1998, ApJ 509, 299
328. Liszt H., Lucas R. 1998, A&A 339, 561
329. Liszt H., Lucas R. 2000, A&A 355, 333
330. Liszt H., Lucas R. 2002, A&A 391, 693
331. Litvak M.M. 1969, ApJ 156, 471
332. Lizano S., Shu F.H. 1989, ApJ 342, 834
333. Lockman F.J. 1976, ApJ 209, 429
334. Loinard L., Castets A., Ceccarelli C., Tielens A.G.G.M., Faure A., Caux E., Duvert G. 2000, A&A 359, 1169
335. Loinard L., Castets A., Ceccarelli C., Caux E., Tielens A.G.G.M. 2001, ApJ 552, L163
336. Longair M. 1992, *High Energy Astrophysics*, Cambridge University Press, Cambridge
337. Lucas R. 1980, A&A 84, 36
338. Lucas R., Liszt H. 1996, A&A 307, 237
339. Lucas R., Liszt H. 1998, A&A 337, 246
340. Lucas R., Liszt H. 1998, A&A 358, 1069
341. Luhman K.L., Rieke G.H., Young E.T., Cotera A.S., Chen H., Rieke M.J., Schneider G., Thompson R.I. 2000, ApJ 540, 1016
342. Lund N. 1986, in *Cosmic Radiation in Contemporary Astrophysics*, ed. M.M. Shapiro, Reidel, Dordrecht, p. 1.
343. Lutz D. 1999, in *The Universe as seen by ISO*, ed. P. Cox, M. Kessler, ESA SP-427, p. 623
344. Madden S.C., Geis N., Genzel R., Herrmann F., Jackson J., Poglitsch A., Stacey G.J., Townes C.H. 1993, ApJ 407, 579
345. MacLaren I., Richardson K.M., Wolfendale A.W. 1988, ApJ 333, 821
346. Malhotra S. 1995, ApJ 448, 138
347. Martin P.G., Whittet D.C.B. 1990, ApJ 357, 113
348. Martin P.G., Schwartz D.H., Mandy M.E. 1996, ApJ 461, 265
349. Martin P.G., Clayton G.C., Wolff M.J. 1999, ApJ 510, 905
350. Massey P., Hunter D.A. 1998, ApJ 493, 180
351. Mathis J.S., Rumpl W., Nordsieck K.H. 1977, ApJ 217, 425
352. Mathis J.S., Mezger P.G., Panagia N. 1983, A&A 128, 212
353. Mathis J.S. 1990, ARAA 28, 37
354. May P.W., Pineau des Forêts G., Flower D. R., Field D., Allan N.L., Purton J.A. 2000, MNRAS 318, 808
355. McClure-Griffiths N.M., Dickey J.M., Gaensler B.M., Green A.J. 2003, ApJ 594, 833
356. McCray R., Kafatos M. 1987, ApJ 317, 190
357. McCrea W.H. 1957, MNRAS 117, 562
358. McKee C.F., Cowie L.L. 1977, ApJ 215, 213
359. McKee, C.F. Ostriker J.P. 1977, ApJ 218, 148
360. McKee C.F., Hollenbach D.J., Seab C.G., Tielens A.G.G.M. 1987, ApJ 318, 674
361. McKee C.F., Holliman J.H. II 1999, ApJ 522, 313
362. Mebold U., Dürstenberg C., Dickey J.M., Staveley-Smith L., Kalberla P. 1997, ApJ 490, L65

363. Mebold U., Hills D.L. 1975, A&A 42, 187
364. Meixner M., Tielens A.G.G.M. 1993, ApJ 405, 216
365. Meneguzzi M., Audouze J., Reeves H. 1971, A&A 15, 337
366. Meyer D.M., Roth K.C., Hawkins I. 1989, ApJ 343, L1
367. Meyer J.-P., Drury L.O'C., Ellison D.C. 1997, ApJ 487, 182
368. Meynet G., Arnould M., Prantzos N., Paulus G. 1997, A&A 320, 460
369. Mezger P.G., Henderson A.P. 1967, ApJ 147, 471
370. Miesch M.S., Bally J. 1994, ApJ 429, 645
371. Millar T.J., Farquhar P.R.A., Villacy K. 1997, A&AS 121, 139 *(The UMIST data base)*
372. Miville-Deschênes M.-A., Joncas G., Durand D. 1995, ApJ 454, 316
373. Miville-Deschênes M.-A., Joncas G., Falgarone E., Boulanger F. 2003, A&A 411, 109
374. Morrison R., McCammon D. 1983, ApJ 270, 119
375. Morton, D.C. 1975, ApJ 197, 85
376. Motte F., André P., Neri R. 1998, A&A 336, 150
377. Mouschovias T.Ch., Spitzer L. 1976, ApJ 210, 326
378. Muench A.A., Lada E.A., Lada C.J. 2000, ApJ 533, 358
379. Myers P.C., Evans N.J. II, Ohashi N. 2000, in *Protostars and Planets IV*, ed. V. Mannings, A. Boss, S. Russell, University of Arizona Press, Tucson, p. 217
380. Nagano M., Watson A.A. 2000, Reviews of Modern Physics, 72, 689
381. Nakano T. 1998, ApJ 494, 587
382. Narayan R. 1992, Proc. Roy. Soc. London A, 341, 151
383. Neininger N., Guélin M., García-Burillo S., Zylka R., Wielebinski R. 1996, A&A 310, 725
384. Neufeld D.A. 1991, ApJ 370, L85
385. Nguyen-Q-Rieu, Winnberg A., Guibert J., Lépine J.R.D., Johansson L.E.B., Goss W.M. 1976, A&A 46, 413
386. Norman C.A., Ikeuchi S. 1989, ApJ 345, 372
387. Novak G., Dotson J.L., Dowell C.D., Goldsmith P.F., Hildebrand R.H., Platt S.R., Schleuning D.A. 1997, ApJ 487, 320
388. Olling R.P., Merrifield M.R. 1998, MNRAS 297, 943
389. Osterbrock D.E. 1989, *Astrophysics of Gaseous Nebulae and Active Galactic Nuclei*, Freeman, San Francisco
390. Ostriker E.C., Stone J.M., Gammie C.F. 2001, ApJ 546, 980
391. Pak S., Jaffe D.T., van Dishoeck E.F., Johansson L.E.B., Booth R.S. 1998, ApJ 498, 735
392. Pagani L. 1998, A&A 333, 269
393. Pagani L., Lequeux J., Cesarsky D., Donas J., Milliard B., Loinard L., Sauvage M. 1999, A&A 351, 447
394. Pagel B.E.J. 1997, *Nucleosynthesis and Chemical Evolution of Galaxies*, Cambridge University Press
395. Panagia N. 1973, AJ 73, 929
396. Parker E.N. 1965, Planetary and Space Sciences 13, 9
397. Parker E.N. 1966, ApJ 145, 811
398. Parizot E.M.G. 1998, A&A 331, 726
399. Pauzat F., Talbi D., Ellinger Y. 1995, A&A 293, 263
400. Payne H.E., Anantharamaiah K.R., Erickson W.C. 1994, ApJ 430, 690
401. Peimbert M. 1967, ApJ 150, 825
402. Pendleton Y. 1999, in *Solid Interstellar Matter: the ISO Revolution*, ed. L. d'Hendecourt, C. Joblin, A. Jones, EDP Sciences, Paris & Springer-Verlag, Berlin, p. 119
403. Péquignot D. 1990 A&A 231, 499

404. Petrie S., Herbst E. 1997, ApJ 491, 210
405. Pety J., Falgarone E. 2000, A&A 356, 279
406. Pety J., Falgarone E. 2003, A&A 412, 417
407. Pierini D., Lequeux J., Boselli A., Leech K.J., Völk H.J. 2001, A&A 373, 827
408. Pfenniger D., Combes F., Martinet L. 1994, A&A 285, 79
409. Pfenniger D., Combes F. 1994, A&A 285, 94
410. Pineau des Forêts G., Flower D.R., Chièze J.-P. 1992, MNRAS 256, 247
411. Porter D., Pouquet A., Woodward P. 1994, Phys. Fluids 6, 2133
412. Pottasch S.R., Wesselius P., van Duinen R.J. 1979, A&A 74, L15
413. Prantzos N., Cassé M., Vangioni-Flam E. 1993, ApJ 403, 630
414. Prantzos N. 1996, A&AS 120, 303
415. Prasad S.S., Tarafdar S.P., Villere K.R., Huntress W.T., Jr. 1987, in *Interstellar Processes*, ed. D.J. Hollenbach & H.A. Thronson, Reidel, Dordrecht, p. 631
416. Preibisch T., Balega Y., Hofmann K.-H., Weigelt G., Zinnecker H. 1999, New Astronomy 4, 531
417. Priest E.R. 1982, *Solar Magnetohydrodynamics*, Reidel, Dordrecht
418. Puget J.-L., Guiderdoni B. 2000, in *Space Infrared Astronomy, today and to-morrow*, ed. Casoli F., Lequeux J., David F., EDP Sciences & Springer-Verlag, p. 417
419. Purcell E.M. 1969, ApJ 158, 433
420. Purcell W.R., Bouchet L., Johnson W.N., et al. 1996, A&AS 120, 389
421. Rachford B.L., Snow T.P., Tumlinson J., et al. 2002, ApJ 577, 221
422. Ramaty R., Kozlovsky B., Lingenfelter R.E. 1979, ApJS 40, 487
423. Rand R.J., Kulkarni S.R. 1989, ApJ 343, 760
424. Rand R.J., Lyne A.G. 1994, MNRAS 268, 497
425. Raymond J.C., Cox D.P., Smith B. 1976, ApJ 204, 290
426. Read S.M., Viola V.E. Jr. 1984, At. Data Nc. Tables 31, 359
427. Reeves H. 1974, ARAA 12, 437
428. Reif F. 1965, *Fundamentals of Statistical and Thermal Physics*, McGraw Hill, New York
429. Reynolds R.J. 1992, ApJ 392, L35
430. Reynolds R.J., Tufte S.L., Kung D.T., McCullough P.R., Heiles C. 1995, ApJ 448, 715
431. Reynolds R.J., Hausen N.R., Tufte S.L., Haffner L.M. 1998, ApJ 494, L99
432. Reynolds R.J., Haffner L.M., Tufte S.L. 1999, ApJ 525, L21
433. Reynolds R.J., Sterling N.C., Haffner L.M., Tufte S.L. 2001, ApJ 548, L221
434. Reynolds S.P., Chevalier R.A. 1984, ApJ 278, 630
435. Rieke G.H., Lebovsky M.J. 1985, ApJ 288, 618
436. Ristorcelli I., Serra G., Lamarre J.-M., et al. 1999, in *Solid Interstellar Matter: the ISO Revolution*, ed. L. d'Hendecourt, C. Joblin, A. Jones, EDP Sciences, Paris & Springer-Verlag, Berlin, p. 49
437. Roberge W.G., Jones D., Lepp S., Dalgarno A. 1991, ApJS 77, 287
438. Roberge W.G., Lazarian A. 1999, MNRAS 305, 615
439. Rohlfs K., Wilson T.L. 2004, *Tools of Radio Astronomy*, 4th. edition, Springer-Verlag, Berlin
440. Rubin R.H. 1989, ApJS 69, 897
441. Rubio M., Garay G., Probst R. 1998, The Messenger 93, 38
442. Rumph T., Bowyer S., Vennes S. 1994, AJ 107, 2108
443. Russeil D. 2003, A&A 397, 133
444. Ryter C. 1996, ApSS 236, 285
445. Ryter C., Cesarsky C.J, Audouze J. 1975, ApJ 198, 103

446. Salama F., Bakes E.L.O., Allamandola L.J., Tielens A.G.G.M. 1996, ApJ 458, 621
447. Salpeter E.E. 1955, ApJ 121, 161
448. Sanders W.T. 1998, in *The Local Bubble and Beyond*, ed. D. Breitschwerdt, M.J. Freyberg, J. Trümper, Springer-Verlag, Berlin, p. 83
449. Savage B.D., Bohlin R.C., Drake J.F., Budich W. 1977, ApJ 216, 291
450. Savage B.D., Mathis J.S. 1979, ARAA 17, 73
451. Savage B.D., Sembach K.R. 1996, ARAA 34, 279
452. Savage B.D., Sembach K.R., Jenkins E.B., *et al.* 2000, ApJ 538, L27
453. Scalo J.M. 1977, ApJ 213, 705
454. Scalo J.M. 1984, ApJ 277, 556
455. Scalo J.M. 1998, in *The Stellar Initial Mass Function*, ed. G. Gilmore, I. Parry, S. Ryan, Cambridge University Press, Cambridge, p. 201
456. Schaerer D., de Koter A. 1997, A&A 322, 598
457. Schilke P., Groesbeck T.D., Blake G.A., Phillips T.G. 1997, ApJS 108, 301
458. Schmutzler T., Tscharnuter W.M. 1993, A&A 273, 318
459. Schnopper H.W., Delvaille J.P., Rocchia R., *et al.* 1982, ApJ 253, 131
460. Schutte W.A., van der Hucht K.A., Whittet D.C.B. *et al.* 1998, A&A 337, 261
461. Schutte W.A. 1999, in *Solid Interstellar Matter: the ISO Revolution*, ed. L. d'Hendecourt, C. Joblin, A. Jones, EDP Sciences, Paris & Springer-Verlag, Berlin, p. 188
462. Scoville N.Z., Solomon P.M. 1973, ApJ 180, 31
463. Sedov L.I. 1959, *Similarity and Dimensional Methods in Mechanics*, Academic Press, New York
464. Sellgren K. 1984, ApJ 277, 652
465. Sembach K.R., Savage B.D., Shull J.M., *et al.* 2000, ApJ 538, L31
466. Sfeir D.M., Lallement R., Crifo F., Welsh B.Y. 1999, A&A 346, 785
467. Shalabiea O.M., Caselli P., Herbst E. 1998, ApJ 502, 652
468. Shklovsky J.S. 1960, Soviet Astronomy 4, 243
469. Shu F.H. 1977, ApJ 214, 488
470. Shu F.H., Adams F.C., Lizano S. 1987, ARAA 25, 23
471. Shull J.M., van Steenberg M.E. 1985, ApJ 298, 268
472. Signore M., Dupraz C. 1993, A&AS 97, 141
473. Silberberg R., Tsao C.H. 1988, in *Genesis and Propagation of Cosmic Rays*, ed. M.M. Shapiro & J.P. Wefel, Reidel, Dordrecht, p. 41
474. Simpson, J.A., Garcia-Munoz, M., 1988, Space Science Reviews 46, 205
475. Sivan J.-P. 1974, A&AS 16, 163
476. Slavin J.D., Shull J.M., Begelman M.C. 1993, ApJ 407, 83
477. Smith L.F., Mezger P.G., Biermann P. 1978, A&A 66, 65
478. Smith R.K., Dwek E. 1998, ApJ 503, 831 (erratum 541, 512)
479. Snow T.P., Witt A.N. 1996, ApJ 468, L65
480. Snowden S.L., Egger R., Finkbeiner D.P., Freyberg M.J., Plucinsky P.P. 1998, ApJ 493, 715
481. Sofia U.J., Cardelli J.A., Guerein K.P., Meyer D.M. 1997, ApJ 482, L105
482. Sokolov A.A., Ternov J.M. 1969, *Synchrotron radiation*, Pergamon Press, London
483. Sokoloff D.D., Bykov A.A., Shukurov A., Berkhuijsen E.M., Beck R., Poezd A.D. 1998, MNRAS 299, 189
484. Spaans M. 1996, A&A 307, 271
485. Spaans M., van Dishoeck E.F. 1997, A&A 323, 953
486. Spitzer L., Jr. 1956, ApJ 124, 20
487. Spitzer L., Jr. 1962, *Physics of Fully Ionized Gases*, 2nd edition, Wiley & Sons, New York

488. Spitzer L., Jr., Scott E.H. 1969, ApJ 158, 161
489. Spitzer L., Jr., Cochran W.D. 1973, ApJ 186, L23
490. Spitzer L., Jr. 1978, *Physical processes in the interstellar medium*, Wiley & Sons, New York
491. Spitzer L., Jr. 1990, ARAA 28, 71
492. Spitzer L., Jr. 1996, ApJ 458, L29
493. Sreekumar P. A. 1993, Phys. Rev. Letters 70, 127
494. Stasinska G., Schaerer D. 1997, A&A 322, 615
495. Sternberg A., Dalgarno A. 1995, ApJS 99, 565
496. Störzer H., Hollenbach D. 1999, ApJ 515, 669
497. Storey P.J., Hummer D.G. 1995, MNRAS 272, 41
498. Stutzki J., Stacey G.J., Genzel R., Harris A.I., Jaffe D.T., Lugten J.B. 1988, ApJ 332, 379
499. Stutzki J., Güsten R. 1990, A&A 356, 513
500. Stutzki J., Bensch F., Heithausen A., Ossenkopf V., Zielinski M. 1998, A&A 336, 697
501. Surdej J. 1977, A&A 60, 303
502. Syrovatskii S.I. 1961, JETP 13, 1257
503. Takeda, M., Hayashida, N., Honda K., *et al.* 1998, Phys. Rev. Letters 81, 1163
504. Takahashi T., Hollenbach D.J., Silk J. 1983, ApJ 275, 145
505. Tatischeff V., 1996, Thèse, Université de Caen, France
506. Taylor J.H., Cordes J.M. 1993, ApJ 411, 674
507. Tenorio-Tagle G., Bodenheimer P. 1988, ARAA 26, 145
508. Tenorio-Tagle G., Różycza M., Bodenheimer P. 1990, A&A 237, 207
509. Tenorio-Tagle G., Palous J. 1987, A&A 186, 287
510. Thoraval S., Boissé P., Duvert G. 1997, A&A 319, 948
511. Thoraval S., Boissé P., Duvert G. 1999, A&A 351, 1051
512. Tielens A.G.G.M., Hagen W. 1982, A&A 114, 245
513. Tielens A.G.G.M., Hollenbach D. 1985, ApJ 291, 722
514. Tielens A.G.G.M. 1998, ApJ 499, 267
515. Tiné S., Roueff E., Falgarone E., Gérin M., Pineau des Forêts G. 2000, A&A 356, 1039
516. Toomre A. 1964, ApJ 139, 1217
517. Torres-Peimbert S., Lazcano-Araujo A., Peimbert M. 1974, ApJ 191, 401
518. Tsamis Y.G., Barlow M.J., Liu X.-W., Danziger J., Storey P.J. 2003, MNRAS 338, 687
519. Tsamis Y.G., Barlow M.J., Liu X.-W., Danziger J., Storey P.J. 2003, MNRAS 345, 186
520. Turner B. 1990, ApJ 362, L29
521. Turner B.E., Terzieva R., Herbst E. 1999, ApJ 518, 699
522. Uchida K.I., Sellgren K., Werner M. 1998, ApJ 493, L109
523. Ungerechts H., Umbanhowar P., Thaddeus P. 2000, ApJ 537, 221
524. van de Hulst H.C. 1957, *Light Scattering by Small Particles*, Wiley, New York
525. Vandenbussche B., Ehrenfreund P., Boogert A.C.A., *et al.* 1999, A&A 346, L57
526. van der Laan H. 1962, MNRAS 124, 127
527. van Dishoeck E.F., Black J.H. 1986, ApJS 62, 109
528. van Dishoeck E.F., Black J.H. 1988, ApJ 334, 771
529. van Dishoeck E.F., Black J.H. 1989, ApJ 340, 273
530. van Dishoeck E.F., in 1998 *The Physics and Chemistry of the Interstellar Medium*, ed. V. Ossenkopf, Shaker-Verlag, Aachen, p. 53
531. van Dishoeck E.F., Black J.H., Boogert A.C.A., *et al.* 1999, in *The Universe as seen by ISO*, ed. P. Cox, M. Kessler, ESA SP-427, p. 437
532. Verstraete L., Léger A., d'Hendecourt L., Dutuit O., Defourneau D. 1990, A&A 237, 436

533. Verstraete L., Puget J.-L., Falgarone E., Drapatz S., Wright C.M., Timmermann R. 1996, A&A 315, L337
534. Verstraete L., Pech C., Moutou C., Sellgren K., Wright C.M., Giard M., Léger A., Timmermann R., Drapatz S. 2001, A&A 372, 981
535. Viala Y.P., Roueff E., Abgrall H. 1988, A&A 190, 215
536. Völk H.J., Jones F.C., Morfill G.E., Röser S. 1980, A&A 85, 316
537. Vuong M.H., Montmerle T., Grosso N., Feigelson E.D., Verstraete L., Ozawa H. 2003, A&A 408, 581
538. Wada K., Norman C.A. 1999, ApJ 516, L13
539. Wada K., Spaans M., Kim S. 2000, ApJ 540, 797
540. Walmsley M. 1990, A&AS 82, 201
541. Warin S., Benayoun J.-J., Viala Y.P. 1996, A&A 308, 535
542. Watson W.D. 1972, ApJ 176, 103
543. Weiler K.W., Sramek R.A. 1988, ARAA 26, 295
544. Weingartner J.C., Draine B.T. 2001, ApJ 548, 296
545. Webber W.R., Simpson G.A., Cane H.V. 1980, ApJ 236, 448
546. Whittet D.C.B. 1992, *Dust in the Galactic Environment*, IoP, Bristol
547. Whittet D.C.B., Schutte W.A., Tielens A.G.G.M. *et al.*1996, A&A 315, L357
548. Wilgenbus D., Cabrit S., Pineau des Forêts G., Flower D. 2000, A&A 356, 1010
549. Williams J.P., Myers P.C., Wilner D.J., di Francesco J. 1999, ApJ 513, L61
550. Wilson O.C., Münch G., Flather E.M., Coffeen M.F. 1959, ApJS 4, 199
551. Wilson T.L., Rood R.T. 1994, ARAA 32, 191
552. Witt A.N., Gordon K.D. 1996, ApJ 463, 681
553. Witt A.N., Gordon K.D., Furton D.G. 1998, ApJ 501, L111
554. Witt A.N., Smith R.K., Dwek E. 2001, ApJ 550, L201
555. Wolfire M.G., Hollenbach D., McKee C.F., Tielens A.G.G.M., Bakes E.L.O. 1995, ApJ 443, 152
556. Wolfire M.G., McKee C.F., Hollenbach D., Tielens A.G.G.M. 1995, ApJ 453, 673
557. Woltjer L. 1972, ARAA 10, 129
558. Wright E.L., Mather J.C., Bennett C.L. *et al.* 1991, ApJ 381, 200
559. Xie T., Allen M., Langer W.D. 1995, ApJ 440, 674
560. Xu C., Helou G. 1996, ApJ 456, 163
561. Yan M., Sadeghpour H.R., Dalgarno A. 1998, ApJ 496, 1044
562. York D.G. 1974, ApJ 193, L127
563. Yorke H.W., Tenorio-Tagle G., Bodenheimer P. 1983, A&A 127, 313
564. Yorke H.W. 1986, ARAA 24, 49
565. Yorke H.W., Tenorio-Tagle G., Bodenheimer P., Różycza M. 1989, A&A 216, 207
566. Yoshioka S., Ikeuchi S. 1990, ApJ 360, 352
567. Zsargó J., Sembach K.R., Howk J.C., Savage B.D. 2003, ApJ 586, 1019

Index

Ar II   53, 109, 169
Ar III   53, 109, 169
C I   52, 53, 60, 200, 204, 206, 208, 223, 226, 234, 237, 240
C II   52–54, 59, 197, 198, 203, 204, 227, 228, 235, 237–239, 258
C III   109
C IV   63, 113, 117, 207, 360
Fe II   53, 197, 232, 257
H II region   4, 9, 11, 61, 62, 87, 110, 162, 163, 191, 195, 227, 257, 354, 367, 405, 413, 414, 418, 421, 422, 428, 434, 435
   cooling   197, 198, 200, 202
   dynamics   244, 274, 282, 301
   evolution   283–285, 310, 355
   heating   191, 200
   thermal equilibrium   207, 208
N II   53, 108–112, 197, 413
N III   53, 109, 111, 197
N V   63, 113, 117, 207, 360
Ne II   53, 109, 169, 197
Ne III   53, 109, 169, 197, 426
O I   16, 52, 53, 55, 112, 197, 198, 204, 227, 228, 232–234, 236, 238, 257, 258
O II   53, 90, 110, 197, 208, 257
O III   53, 90, 105–107, 109, 197, 208, 257, 413
O VI   63, 112, 113, 117, 360
S II   53, 111, 197, 232, 257, 420, 429
S III   53, 109, 257
S IV   53, 109, 169
Si II   53, 197, 263
Si IV   117, 207, 360
21-cm line   11, 45, 51, 118, 144, 204, 362, 406–408, 410, 427, 436
30 Doradus   349, 350, 415, 420, 421
30 Doradus C   415

absorbance   175, 176
absorption
   21-cm line   47, 49, 50, 410
   atmospheric   16
   band   67, 72, 73, 84, 151, 155, 156, 173–175, 177, 229, 230, 367, 368
   coefficient   28, 29, 32, 89, 92, 102, 104
   continuum   18, 89, 110, 117, 145
   cross-section   143, 154, 170, 172, 187
   efficiency   154, 161
   line   35, 45, 55, 57, 63, 70, 72, 110, 112, 117, 207, 221, 222, 229, 230, 254, 257, 263, 360, 362
abundance
   H II regions   101, 105, 107, 108, 110, 208
   cosmic rays   122, 124, 126, 127, 132, 133, 184, 291, 300
   dust   62, 144, 157, 170, 176, 240, 363, 365
   gas   58, 60–63, 70, 75, 77, 85, 118, 144, 157, 215, 221–223, 226, 234, 240, 254, 256, 313–315, 317, 362, 363, 365
   Solar system   18
   solar system   61, 62, 118
   stars   5, 61, 62
acceleration
   charged particles   123, 127, 265, 272, 284, 300
   grains   126, 300
accomodation coefficient   195, 318
accretion   188, 329, 348, 350, 351, 358, 363, 365
activation barrier   209, 213, 218, 222, 254
adiabatic invariant   288
advection   305

AIB   167, 170–172, 227, 366, 367, 425–427
albedo   155
Alfvén
  velocity   250, 294, 297, 311, 345
  wave   131, 196, 249, 291, 298, 300, 337
antenna
  beam   31
  efficiency   31
  gain   31, 32
  temperature   30, 74
asymptotic giant branch   63, 366, 368
ATCA   436
AUGER project   129
auroral line   108, 112

B 68   416
B 72   417
Baker & Menzel cases A and B   96, 99
Balmer
  continuum   96, 97
  decrement   100
  discontinuity   95, 96
  line   96–98, 100, 233, 413, 414, 420
band
  absorption   176
  diffuse   83, 85
beam dilution   31
Bethe–Bloch formula   185, 291
Bohm diffusion coefficient   290
Bohr magneton   21
Boltzmann law   29, 32
Brackett
  line   100
Brackett line   100, 421
braking coefficient   336, 340
branching ratio   213
Bremsstrahlung   91, 139
brightness temperature   30, 32, 38, 47, 75, 76, 94, 102
bubble   51, 87, 113, 272, 274, 284, 355, 358, 360, 361, 407, 409, 414, 420, 426, 433, 436
  Local   8, 19, 117, 193
  local   274
bulge (galactic)   2
Burgers vortex   317

C^+ 52, 183, 188, 192, 221, 227, 232, 234, 235, 237, 313, 339
C_2 55, 66, 221
Canadian Galactic Plane Survey   408–410, 436
carbon
  grain   155, 157, 163, 167, 173, 174, 187, 188, 191, 364, 365
  star   366
Cassiopeia A   270, 271, 364, 430, 431
Ced 201   170
CFHT   405, 417, 422, 424
CH   22, 35, 55, 63, 66, 73, 221
CH^+ 55, 63, 66, 221, 222, 230, 254, 317
CH_4 66, 149, 176, 364
Chamaeleon   9, 423
champagne effect   91, 110, 284, 285, 355, 359
CHANDRA   115, 144, 269, 429
charge exchange   112, 222, 252
chemisorption   217
chemistry   209, 226
  in photodissociation regions   230, 280, 281
  in shocks   254, 256, 257
  in turbulent regions   312–315, 317
  kinetics   219
  on grains   215, 219–221
chimney   26, 273, 361, 436
chondrite   368
cirrus cloud   15, 17, 170, 307, 367, 423
cloud
  dark   9, 416–418
  high-velocity   51, 117
  molecular   4, 26, 51, 63, 66, 72, 83, 91, 140–144, 162, 174–176, 194–196, 200, 208, 214, 215, 223, 226, 255, 256, 302, 303, 310, 312–316, 332, 333, 344, 358, 359, 363, 365, 410–412, 428
  translucent   214, 221
CN   35, 55, 63, 66, 73, 221
CO   12, 38–40, 63, 66, 67, 72, 74, 76, 77, 140, 141, 143, 144, 149, 175, 176, 200–202, 214, 221–223, 228–230, 232–235, 237, 239, 301, 316, 359, 366, 406, 409–412
CO_2 72, 149, 175, 176, 215

coalescence 348, 350, 358, 367
COBE 19, 55, 59, 110, 157, 162, 166, 204
collapse 21, 196, 302, 308, 325, 328, 332, 335, 340, 348, 358, 359
collision
   gas–grain 150, 194, 196, 208, 318, 319
   grain–grain 318, 319, 363
   strength 53, 54
   with electrons 77, 180, 183, 197
   with neutrals 40, 77, 81, 180, 183, 197
colour excess 100, 118, 144, 151, 153
comet 368
compression wave 276, 277, 279, 283, 342
Compton effect 137, 139
contact discontinuity 244, 260, 261, 271
cooling
   of gas 59, 197, 199, 202–204, 207, 208, 232–234
   of grain 159
   of grains 166
COPERNICUS 70, 112, 118, 145, 153
core
   dense 303, 310–312, 316, 342, 343, 346, 347
   hot 215, 256
COS-B 135, 138, 141
cosmic rays 5, 119, 133, 140, 145, 146, 150, 183, 186, 196, 208, 222, 326
   confinement 130, 133
   electrons 129, 130, 136, 140, 271
   galactic 122, 123, 127
   light elements 124, 125
   origin 120
   solar 120, 121, 123
   very high energy 128, 129
Crab nebula 269–272, 429
CS 75–77, 221
curve of growth 57, 58

dark matter 3, 4, 6, 8, 140
de-excitation
   collisional 33, 34, 179, 194, 231
   of nuclei 145
   radiative 28, 33, 54, 95, 211, 212, 214
Debye screening factor 181
density
   column 29
   critical 34, 46, 53, 54, 76, 77, 81, 109, 236, 237

flux 31
depletion 62, 157, 173, 198, 340, 364
diamond 172, 368
diffuse gas
   ionized 4, 11, 110, 112, 205, 354, 355, 359, 360, 414
   neutral 4, 12, 54, 55, 59, 61, 62, 70, 105, 145, 167, 174, 179, 183, 186, 187, 189, 191, 192, 205, 214, 221, 223, 229, 253, 310, 312, 317, 339, 368
diffusion
   ambipolar 197, 250, 309, 334, 335, 340, 346
   coefficient 294, 312
   of cosmic rays 131, 288, 291, 294, 298
   turbulent 312, 314, 315
disk
   accretion 350
   galactic 2, 6, 8, 9, 20, 174, 356
   protostellar 281, 346, 368, 434
dissociation front 280, 281
distance
   kinematic 9, 11
   to stars 9
Draine's radiation field 17
drift velocity 287, 336, 339
Drude function 153, 170
dust
   electric charge 187, 188
   emission 159, 173, 406, 409, 410, 418, 422, 423, 425, 431
   model 173, 174
Dwingeloo radiotelescope 407
dynamics
   equations of 244, 245

Eddington approximation 38
Eddington stability limit 350
Einstein's coefficient, probability, relation 28
elephant trunk 281
emission
   expected 48
   spontaneous 27
   stimulated 27, 42
emission measure 93
emissivity 91
epicyclic frequency 356

equation of state   203, 205, 325, 327, 328, 330, 331, 333, 342
equilibrium
   gravitational   327, 335
   hydrostatic   7, 20, 261, 331
escape probability   36, 38, 201, 233, 295
ESO   411, 415, 416, 418, 421, 427–429, 432, 433, 435
EUVE   118
evaporation   216, 218, 270, 280
excitation
   collisional   33, 185, 198, 199, 202
   emperature   33
   of nuclei   145
   radiative   27, 33, 179
   temperature   32
extensive air shower   128
extinction   9, 18, 100, 143, 149, 150, 155–158, 170, 195, 227, 235, 416, 417, 422
   cross-section   154
   curve   84, 150–153, 172, 173, 175
   efficiency   154, 155
   law   150, 367

Faraday rotation   22, 23, 25, 110
Fermi mechanism   292, 293, 297, 299
fine-structure line   51–53, 55, 59, 105, 108–110, 115, 163, 169, 197, 198, 203, 207, 208, 232, 257, 259
fluorescence   68, 72, 109, 232, 233
   infrared   177
fountain (galactic)   207, 360, 362
fractal   51, 302, 351, 411
   dimension   304, 309
fragmentation   301, 330, 335, 340, 348, 351, 358
free–bound emission   94, 95, 117
free–free emission   91, 93, 95, 110, 198, 207, 406
free-fall time   340, 341, 357
FUSE   55, 70, 117, 145, 153, 360, 361

GALILEO   368
gamma radiation   406
gamma-rays   120, 133, 135, 137, 138, 141, 143, 270
   line   145, 148
Gaunt factor   91, 93, 116, 118

globule
   Bok   416
   cometary   281
   neutral   270, 279, 281, 282
grain
   big   173
   destruction   152, 255, 363
   small, very small   172, 173
grammage   130, 132, 136, 139, 291
graphite   156, 157, 160, 162, 164–167, 174, 188, 363, 368
GRO   138, 141, 147
Grotrian diagram   52, 96
Gum nebula   281
gyration radius   121, 128–130, 264, 286, 295, 299, 300

H_2 55, 63, 66–70, 72, 75, 76, 78, 144, 145, 163, 194, 210, 214, 215, 227, 228, 230–232, 234, 235, 244, 252, 254, 258, 280, 421
   formation   211, 215, 216, 218, 230, 240
H_2CO 63, 81, 82, 221
H_2O 40, 63, 79, 82, 83, 149, 175–177, 194, 200, 223, 225, 226, 228, 254, 314, 315
H_3^+ 213, 225, 231, 281
Habing's radiation field   16–18
halo (galactic)   2, 3, 6, 8, 15, 19, 25, 51, 112, 117, 131, 133, 207, 360–362
HCN   73, 76, 77, 221
HCO^+ 77, 221, 222, 231, 254, 317, 339
heating
   dust   174
   of gas   51, 112, 117, 150, 180, 197, 203, 204, 207, 208, 230, 232, 236, 242, 249, 251, 312, 317
   of grains   367
helium   59, 90, 95, 96, 101, 109, 117, 118, 136
Helix nebula   366
Helmholtz
   equation   309, 317
   instability   261
HIPPARCOS   8, 9
HNC   77, 221
Horsehead   424, 425
hot gas   3, 19, 112, 117, 202, 207, 266, 270, 273, 360, 362, 415

Hubble Space Telescope   55, 61, 413, 419, 420, 426, 434
hydrogenated amorphous carbon   366–368

IC 349   419
ice   83, 144, 174, 176, 177, 200, 215, 256, 363
initial mass function   348, 351
instability
　gravitational   21, 312, 321, 340, 347
　thermal   205, 259
INTEGRAL   145
intermittency   183, 213, 308, 309, 313, 319
ion–molecule reaction   210, 211
ionization
　by cosmic rays   87, 183, 186, 214, 225, 339
　collisional   87, 113, 115, 247
　energy losses by   292
　equilibrium   59
　front   228, 240, 275, 277–281, 283, 284
IRAM   302
IRAS   19, 423
IRTS   168
ISO   19, 41, 55, 76, 83, 109, 145, 152, 168, 169, 171, 174, 175, 204, 233, 252, 365, 368, 422, 425–427, 431
isobaric   204, 205
isochoric   205

jansky   31
Jeans
　length   323, 325
　mass   323, 325, 330, 335, 340, 351, 358

Kolmogorov
　law   289, 309, 318
　turbulence   305, 307, 312
Kramers–Kronig relations   158
Kuiper airborne observatory   55, 109

L 1512   316
L 977   143
Landé factor   21
Langevin rate   210, 337
large velocity gradient approximation   35, 39, 75, 76
Larmor frequency   285, 286, 337
local standard of rest   9, 47

local thermal equilibrium   27, 29
Lorentz function   170
luminosity function   142, 349
Lyman
　continuum   18, 45, 89, 90, 117, 275, 354
　contiunuum   89
　discontinuity   19, 45, 89
　line   16, 55, 60, 67, 70, 96, 199

M 16   426
M 17   229, 240, 428
M 31   5, 47, 168, 170, 171, 272, 367, 427
M 33   131, 355, 414
M 51   405
Mach number   248
Magellanic clouds   48, 49, 115, 120, 174, 239, 240, 272, 349, 358, 415, 420, 421, 433
magnetic
　bottle   288
　field   5, 20, 26, 110, 196, 249, 252, 261, 285, 289, 291, 417
　measurement of magnetic field   21, 25
　mirror   288, 289, 293
　precursor   253, 298
magnetohydrodynamics   197, 241, 242, 335
magnetosonic wave   249
magnitude   9
main-beam temperature   31
maser   39, 44, 82, 83, 102, 104, 202
　saturated   43
　unsaturated   43
mass
　critical   323, 325, 329, 333, 334, 345, 346
　of the Galaxy   2, 3
　of the ISM   4, 50, 110, 137, 145, 149, 205, 359
　umbral   346
Mathis, Mezger & Panagia radiation field   16, 17
Maxwell equations   241, 242
mean free path   248, 264, 266, 289, 295, 299, 311
meteorite   147, 368
Mie theory   154, 155, 159, 160
molecule
　Λ doubling   73

deuterated 215, 220, 226, 256
diatomic 39, 71, 73, 78
electronic transition 66, 71
inversion transition 73
list 64, 65
polyatomic 71, 78, 83
rotational transition 14, 73, 83
vibrational transition 71, 73
Moseley law 115
MRN 156, 157, 159, 160, 166, 173, 187, 217, 365, 369
multi-fluid medium 241, 243, 250–252, 258
mushroom 436

N_2 210
N 157b 48–50
N 70 433
Navier–Stokes equation 304, 305, 309
nebula
  planetary 91, 108, 109, 354, 366
  proto-planetary 366
  reflection 149, 155, 164, 170, 419, 424, 425
nebular line 107, 112
neutral–neutral reaction 182, 213, 214
NEWTON 115, 144, 269, 415, 432
NGC 2023 424, 425
NGC 206 272
NGC 3603 272
NGC 7027 108, 366
NGC 7538 175, 176
NH_3 73, 79–81, 149, 176, 256
nucleation 363
Nyquist theorem 30

O_2 15, 16, 66, 200, 223, 225, 228, 313–315
ODIN 83, 200
OH 15, 16, 22, 40, 41, 55, 66, 73, 221, 222, 225, 254, 317
Oort parameter 10
Ophiuchi (ρ) 9, 162, 351
optical depth 28
optical thickness 28
Orion
  bar 233, 234, 413, 435
  molecular cloud 63, 66, 146, 215, 256, 412

nebula 102, 104, 106, 146, 162, 163, 233, 240, 281, 302, 349, 350, 413, 434, 435
ORPHEUS 70
ortho 63, 70, 76, 80–83
oscillator strength 56, 61, 85, 98, 101

PAH 84, 171, 172, 174, 188, 189, 230, 235, 367, 427
PAH^+ 84, 171, 187, 189
PAH^- 171, 189, 230
para 63, 70, 76, 80–83
Parker instability 357
partition function 39, 75, 77, 187, 211, 212
Paschen
  discontinuity 95
  line 100, 106
Perseus arm 408–411
Pfund line 100, 163
phase function 155
photodissociation 18, 145, 214, 215
  region 52, 55, 72, 75, 101, 105, 109, 163, 168, 189, 191, 194, 199, 209, 213, 227, 240, 258, 259, 279–281, 367, 413, 420, 421, 426, 428, 435
photoelectric effect 150, 187, 194, 203, 204, 228, 230
photoionization 87, 91, 187, 188, 191, 193, 194, 214, 215
photon trapping 38
physisorption 216
Planck function 29, 30
plasma
  dispersion 23
  frequency 185
  wave 183, 242, 249, 250
Pleiades 419
plerion 263, 268, 269, 272, 429
Poisson equation 6, 324
polarization
  by dust 24, 25, 158
  by Zeeman effect 22
polytropic index 327, 328, 330, 333, 342
potential barrier 187, 209
potential curve 67, 68, 71
precursor
  magnetic 250
  radiative 247, 248, 257

pressure
  cosmic rays   20, 123, 130, 357
  kinetic   20, 204
  magnetic   20
  radiation   237, 240, 275
  turbulent   141, 307, 308
pulsar   22, 23, 25, 110, 263, 268, 272

radiation
  cosmological   15, 47, 75, 201
  extragalactic   13–15
  galactic   15–17, 19
radiation length   136
radiative association   211, 212, 223
radiative transfer   27, 44, 229, 230
radical   213, 218, 219
Rankine–Hugoniot relations   246
rarefaction wave   276, 277, 279, 342
rayleigh   111, 355
Rayleigh–Gans approximation   159, 160
Rayleigh–Jeans approximation   30, 32–34, 38, 43, 74, 93, 102, 143
Rayleigh–Taylor instability   259, 261, 265, 270, 271, 274, 357
RCW 108   418
recombination
  dielectronic   105
  dissociative   212, 213
  line (optical)   96, 97, 100, 101, 108, 111, 151, 152, 233, 257
  line (radio)   98, 101, 105, 110
Reynolds number   305, 306, 311
rigidity   286
rotation
  galactic   3, 6, 8–11, 273, 310, 356
  of clouds   322, 343, 345
  of grains   25
Rydberg
  constant   98
  state   101

Saha law   98, 102, 113
SAS-2   138
scale height
  gas disk   11, 51, 111, 117, 129, 174, 273, 356, 357, 361, 362
  stellar disk   7, 8
scaling law   301, 302, 309
scattering

cross-section   154
  efficiency   154, 155
  resonant   96
scintillation   110
screening constant   115
self-shielding   214, 215, 230
self-similar solution   265, 266, 278, 342, 343
shattering   319, 363
shock   55, 63, 69, 72, 87, 112, 121, 127, 129, 150, 183, 192, 194, 197, 199, 207, 209, 213, 218, 241, 261, 263, 300, 317, 342, 351, 358, 360, 363, 365, 367
  C   251, 253, 254, 258
  C*   251, 254
  isothermal   248, 269
  J   245, 251, 253–259, 271, 293, 297
  modified   298
  non-stationary   252, 254
  radiative   248, 250, 258, 267
  reverse   267, 278
Si   177, 256, 365
SiC   177, 366, 368
silicate   18, 151, 152, 155–158, 160, 162–164, 166, 173, 175, 187, 218, 256, 363, 366, 368
SiO   221, 256, 365
snowplough   267, 268
SOFIA airborne observatory   109
solar modulation   120–123, 129, 135, 141, 184, 186
Solar system   291, 368, 369
sound velocity   206, 244, 250, 276, 278, 280, 284, 325, 361
source function   28, 30, 36, 104
spallation   124, 125, 132, 133, 146
spin temperature   46, 50
Spitzer (satellite)   109, 145
stability
  gravitational   308, 321, 340, 356
  thermal   203, 207
star
  association   272, 359
  binary   55, 263, 344, 348, 354
  cluster   272, 335, 349, 351, 359, 435
  formation   302, 310, 335, 347, 348, 351, 354, 356, 359, 405

population   5, 8
runaway   354
statistical weight   29, 42, 46, 77, 79, 98
Strömgren
  radius   275, 283
  sphere   87, 91
structure function   307
supernova   5, 26, 51, 127, 131, 146–148, 206, 207, 273, 274, 310, 355, 358, 363, 366
  remnant   19, 87, 113, 115, 117, 129, 147, 150, 195, 197, 249, 253, 257, 259–261, 263, 272, 278, 284, 298, 355, 360, 362, 368, 408, 415, 429–432
SWAS   55, 83, 200
synchrotron radiation   22–26, 110, 129, 131, 133, 263, 265, 269, 270, 272, 406, 408, 429, 430, 432

Taurus   9, 347
Taylor scale   311
TD1   16
thermal conduction   117, 207, 266, 267, 270, 360, 361
thermalization   179, 180, 183, 249
three-component model   270, 360
Toomre criterion   356
transfer
  equation   27, 33, 42, 191, 229
Trifid nebula   422
turbulence   51, 112, 213, 261, 272, 289, 298–301, 319, 322, 325, 326, 330, 332, 340, 347, 351, 358, 363
  atmospheric   16, 112
  compressible   307, 309, 314, 325
  dissipative scale   301, 305
  incompressible   304, 308, 309, 313
  inertial scale   305
  integral scale   305, 310
two-component model   206
two-photon radiation   95, 96
Tycho supernova   432

ULYSSES   298, 368

velocity dispersion
  gas   4, 12, 51, 59, 233, 302, 303
  stars   2, 5, 6, 8
virial   140–142, 308, 321, 330
viscosity   242, 251, 305
  dynamic   305
  kinematic   305
  turbulent   307, 308
VLA   430
Voigt function   58
vorticity   306, 309, 314, 316, 317

X-rays
  absorption   18, 117, 118, 144, 145
  emission   19, 112, 115–117, 138, 266, 406, 415, 429, 431, 432
  line   113, 115, 116, 415, 432
  lines   266
  scattering   158–160

Zanstra method   96
Zeeman effect   21, 22

Color Plates

Color Plates

Plate 1. The galaxy M 51 (NGC 5195)

This image results from the combination of images in the B band (blue) and in the V band (green) with a monochromatic image in the Hα line (red). It was obtained with the 12K CCD camera at the Canada-France-Hawaii (CFHT) telescope. The spiral arms of this galaxy are particularly evident and result from the gravitational action of the companion NGC 5194, located to the North. They correspond to the compression zones of the interstellar matter. Absorption due to interstellar dust is clearly visible along the arms, but also elsewhere, in particular in the companion. The many H II regions visible in the Hα line are preferentially located along the arms, on the convex side of the dust band. The arms sweep up interstellar matter which enters through the concave side. This matter is then compressed and forms massive stars which appear on the convex side and ionize the gas.

In this plate and all others, with a few exceptions which are indicated in the corresponding caption, North is to the top and East to the left. (Courtesy Jean-Charles Cuillandre, CFHT)

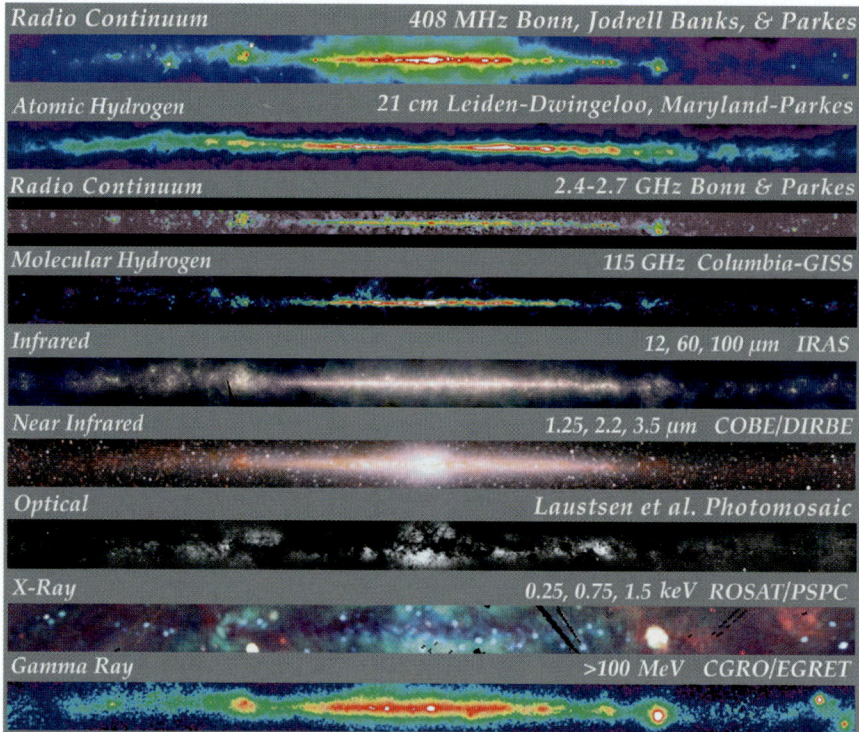

Plate 2. The Milky way at a number of wavelengths

These images show a band of ±5° on each side ot the galactic plane, at frequencies increasing from top to bottom. The galactic centre is at the centre. Galactic longitude l increases from $-180°$ to $180°$ from left to right. The Cygnus region is at $l \approx 90°$ and that of Vela at $l \approx 270°$. Several images are in false colors, color indicating intensity. The colours in the infrared and X-ray images, which are combinations of three wavelengths, are coded from blue to green and to red with increasing wavelength. The principal emission mechanisms are as follows:

- for the radio continuum at 408 MHz, the synchrotron radiation from relativistic cosmic electrons in the galactic magnetic field dominates;
- the 21-cm line is due to atomic hydrogen (see details in Plates 3, 4, 5, 6 et 31);
- for the high-frequency radio continuum (2.4–2.7 GHz), the free–free emission of the electrons in the ionized gas dominates: note that the disk thickness is smaller than for the synchrotron continuum emission at lower frequencies;
- at 115 GHz, the line of the CO molecule traces the molecular gas (see details in Plates 5, 6, 7 et 8): the disk thickness is approximately the same as for the high-frequency radio continuum;
- the far-infrared emission at 100 μm (red) is due to relatively large dust grains, that at 12 μm (blue) to PAHs and very small grains and that at 60 μm (green) to a combination of these (see details in Plate 19);
- the near-infrared emission is dominated by that of the stars, a diffuse emission is due to the free–free and free–bound emission of the electrons in the ionized gas and to recombination lines; emission by very small dust grains contributes at 3.3 μm (red) and 2.2 μm (green);

- in the visible, the emission is essentially due to stars, but we can also see the absorption which is dominated by relatively large dust grains;
- the X-rays are emitted by stellar objects and by the diffuse radiation from the hot, diluted gas in supernova remnants, bubbles and the general interstellar medium (details on Plate 11). The interpretation of this image is difficult due to the strong interstellar absorption which is important at lower X-ray energies;
- the gamma-ray emission above 100 MeV is dominated by the decay of pions created by the interaction between cosmic-ray protons and interstellar matter.

(Document NASA/Goddard Space Flight Center)

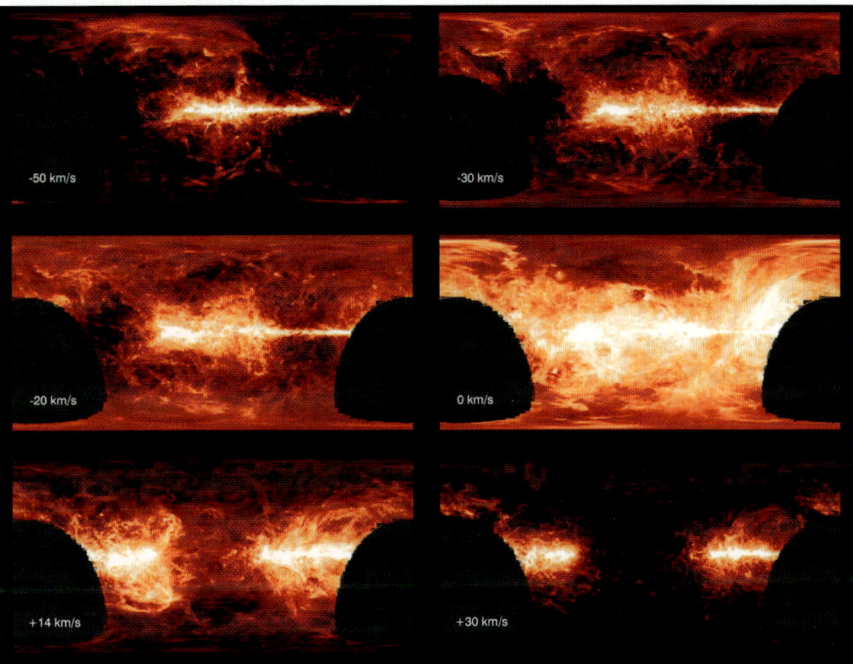

Plate 3. The emission of the Milky way in the 21-cm line of atomic hydrogen

These maps of the whole sky result from a systematic survey with the 25-m radiotelescope at Dwingeloo (Netherlands), made with an angular resolution of 0.5°. They are presented in galactic coordinates and correspond to different radial velocities with respect to the Local standard of rest (LSR). Galactic longitude increases from 300° ($\equiv -60°$) to the right, to 300° to the left, and galactic latitude increases from $-90°$ (bottom) to $+90°$ (top). The missing part cannot be observed from the Netherlands. At the low velocities, the emission is dominated by nearby gas. Near the galactic plane, at high velocities (in absolute value), distant regions contribute. At high galactic latitudes (mainly positive, to the top), we also observe gas falling onto the galactic plane. Note the filamentary structure of the gas, and the empty bubbles. (From Hartmann [219] and Hartmann & Burton [220], courtesy Dap Hartmann)

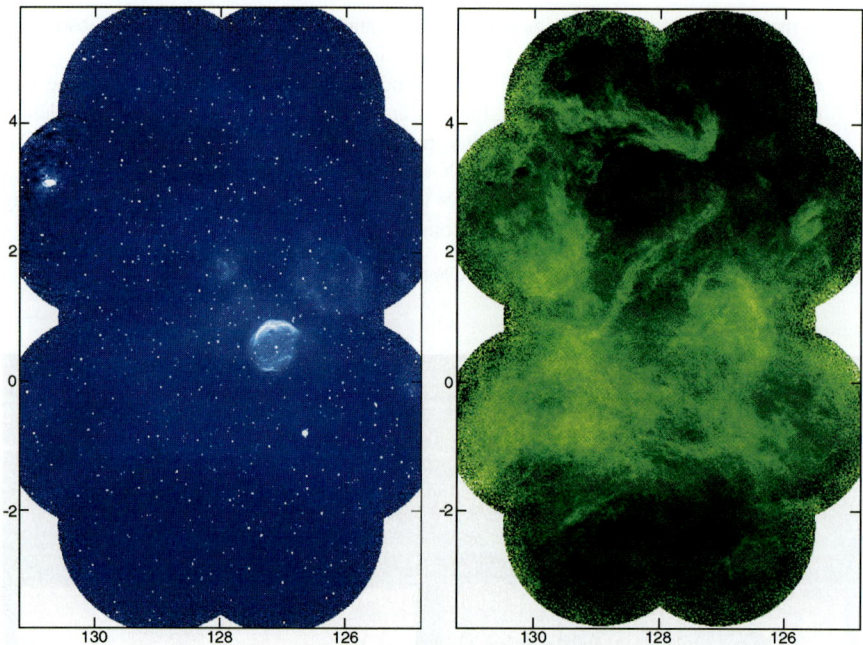

Plate 4.
A region of the Perseus arm in the radio continuum at 21 cm and in the 21-cm line

These maps were obtained as a part of the Canadian Galactic Plane Survey made with the interferometer and the 26-m radiotelescope of the Dominion Radio Astrophysical Observatory at Penticton (B.C, Canada). The angular resolution is $1'$. Abcissae are galactic longitudes and ordinates galactic latitudes. The map to the left, in the 21-cm continuum, is dominated by the synchrotron radiation of three supernova remnants: SNR G130.7+3.1 = 3C 58 (top left), whose strong emission produces circular artefacts, SNR G127.1+0.5 = R 5 (to the right of the centre), and SNR G126.2+1.6, weaker and more extended. The distances to these objects are very uncertain. The objects with smaller angular diameters throughout the map are extragalactic. The map to the right is made in the 21 cm line, at a radial velocity of $-35,3$ km s^{-1} relative to the LSR which corresponds to the Perseus arm. Note the filamentary structure of the gas. A cloud can be seen in the direction of SNR G127.1+0.5 and could be associated if this supernova remnant is in the Perseus arm. Absorption in front of 3C 58 is conspicuous (small dark dot), showing that this supernova remnant is located inside or behind the Perseus arm. (CGPS document, courtesy J. English & R. Taylor)

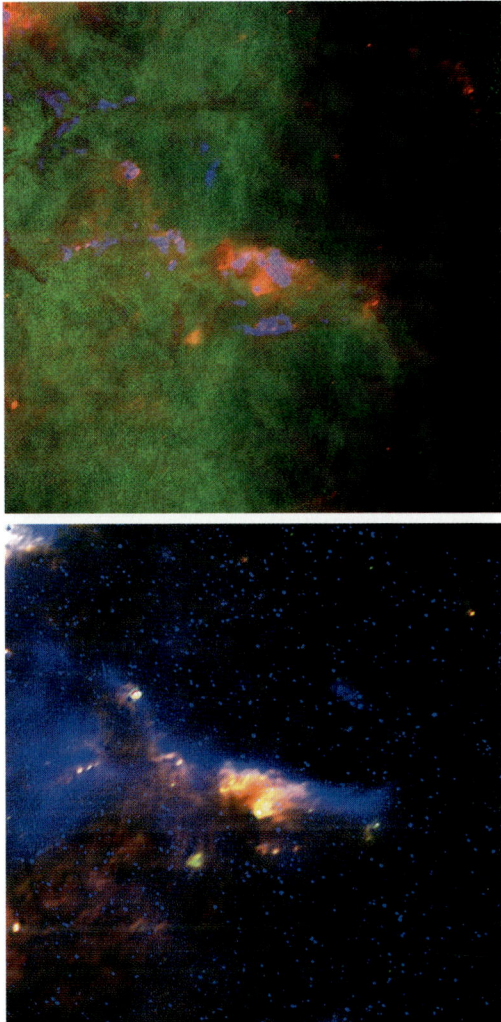

Plate 5. An optically obscured region seen in infrared and radio

These two composite images show the same region of the sky between galactic longitudes 138° (left) and 143° (right) and between galactic latitudes −3,5° (bottom) and +1,5° (top). In the top image, the 21-cm line emission at a velocity of −40 km s^{-1} with respect to the LSR, corresponding to the Perseus arm, is displayed in green. The CO(1-0) line emission at the same velocity is in blue, and the dust emission at 60 μm is in red. In the bottom image, the emission in the radio continuum at 21 cm is in blue, that of the dust at 60 μm in green and the dust emission at 100 μm in red. The radio observations are from the Canadian Galactic Plane Survey and the infrared data are from the IRAS satellite. The interpretation of this region is made difficult by the absence of any visible emission, probably due to strong extinction by dust. The general impression is that we are dealing with the edge of a large bubble which compresses an active zone, where molecular clouds and young stars are responsible for the heating of the dust. The origin of the radio continuum emission is uncertain. (CGPS document, courtesy J. English & R. Taylor)

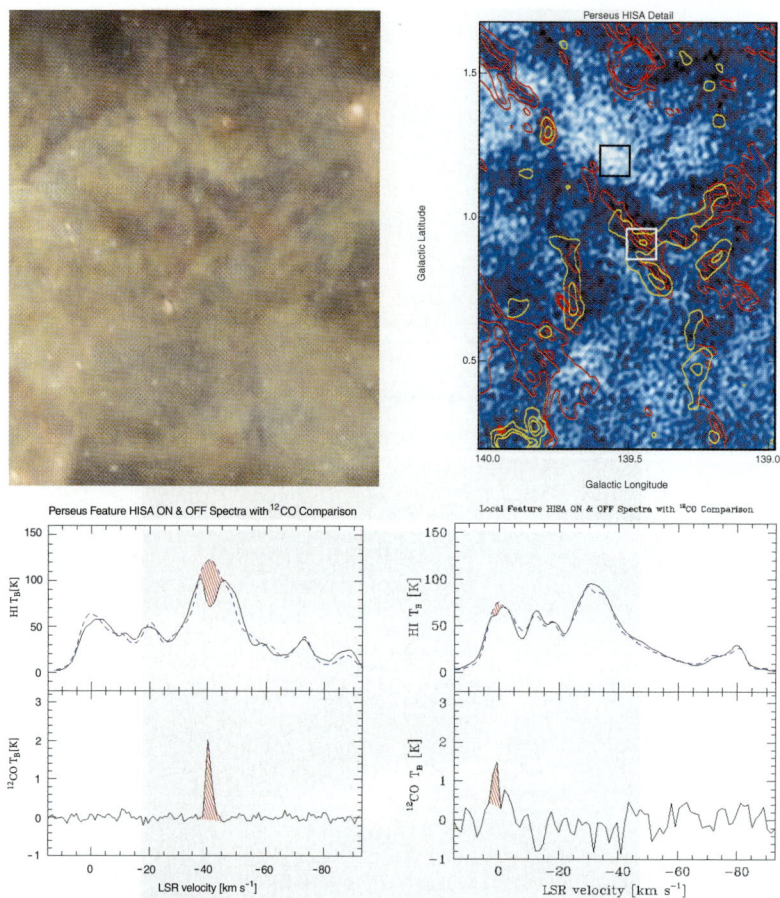

Plate 6. Self-absorption in the 21-cm line and molecular clouds

The map at the top left shows a region of the Perseus arm near the galactic plane in the 21-cm line, at a radial velocity of -41 km s^{-1} with respect to the LSR. This map was obtained as a part of the Canadian Galactic Plane Survey (angular resolution $1'$). Galactic longitudes are from $138.3°$ (left) to $141.1°$ (right) and galactic latitudes from $-1°$ (bottom) to $+2.2°$ (top). The presence of self-absorptions can be seen as dark filaments, due to cold atomic hydrogen in front of warmer hydrogen. The map at the top right shows a part of this region: the 21-cm emission is in blue, molecular clouds are indicated as yellow contours and the dust emission at 60 μm as red contours. A number of atomic clouds seen in absorption correspond to molecular clouds seen through their emission in the CO(1-0) line. Some clouds emit in the far-infrared but not in CO. Below these maps to the left is the 21-cm spectrum of the region delineated by a white square (full line) with the spectra of nearby regions for comparison (dashed line), and the lower plot shows the CO spectrum. To the right, the same for the region delineated by a black square. The 21-cm self-absorption is indicated in red and correspond to CO emission at the same velocity. Even within the black square, we see a small self-absorption at 21 cm in the local gas (radial velocity close to zero), with a weak CO emission. (CGPS document from Gibson et al. [196], courtesy A.R. Taylor)

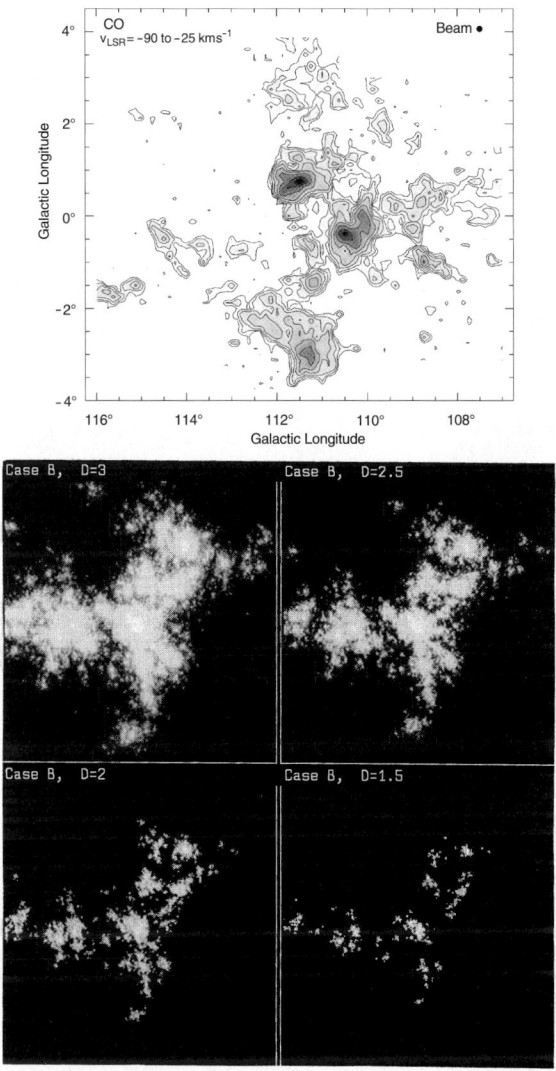

Plate 7. Fractal structure of the molecular gas

To the top, map of a molecular complex in the Perseus arm, from Ungerechts et al. [523] (with the permission of the AAS), obtained in the CO(1-0) line with the 1-m radiotelescope of the Center for Astrophysics at Harvard. The angular resolution is 9' ("beam" represented at the top left) corresponding to 6 pc at the distance of 2.2 kpc of the gas. Below, a simulation of a fractal cloud with 3-dimension fractal dimension D, reproduced from Pfenniger & Combes [409] with the permission of ESO. The elementary structures in this simulation are spheres with a radial distribution of density $\rho(r, r_L) \propto \frac{1}{(r/r_L)^2 \times [1+(r/r_L)^5]}$, (isothermal spheres truncated in a relatively smooth fashion). The morphological similarity with the CO map is striking for $D = 2.5$. The fractal structure of the molecular gas has been observed to much smaller scales in other regions.

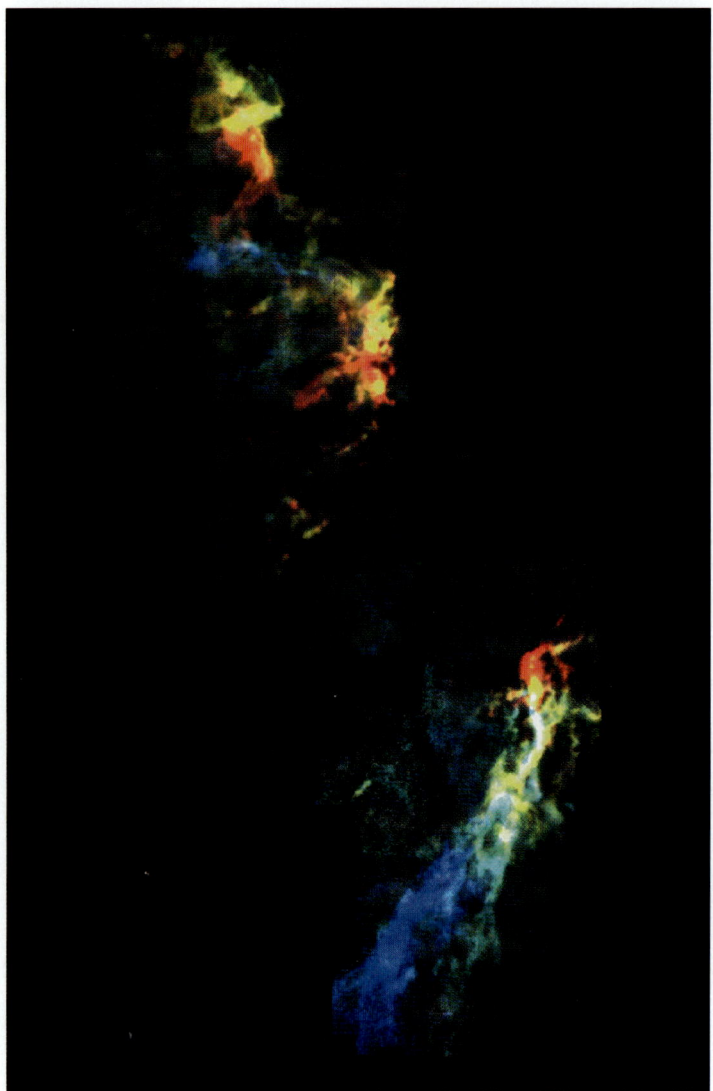

Plate 8. Map of the ^{13}CO(1-0) line in Orion

This map was obtained with the 7-m radiotelescope of the AT&T Bell Laboratories with an angular resolution of 1′. It covers 10° in declination. The contours represent the radial velocity with respect to the LSR from 4 km s^{-1} (blue) to 13 km s^{-1} (red). All the gas is at a distance of about 500 pc. The Orion Nebula is in the cloud at the bottom of the red part of the southern cloud. Notice the filamentary structure and the many condensations, some of which are more or less regularly aligned along the filaments. For a discussion, see Bally et al. [18]. Detailed maps of a part of the southern cloud in other millimeter lines and in the dust emission at 1.3 mm can be found in Chini et al. [99]. (Courtesy John Bally)

Plate 9. The Orion Nebula, observed with the Hubble Space Telescope

The best studied of all H II regions, the Orion Nebula, is at a distance of 550 pc. This false-colour image, with dimensions about 0.8×0.8 pc, results from the combination of 45 monochromatic images taken with the Wide Field and Planetary Camera 2 in the lines of [O III] (blue), Hα (green) et [N II] (red). A zoom on the central part showing the four exciting stars forming the Trapezium is presented in Plate 30, and an infrared view is in Plate 32. The Orion bar located to the bottom left of the image is a photodissociation interface between the H II region and a molecular cloud. The ionized gas resulting from the surface ionization of this cloud exhibits a conspicuous ionization structure: in the region closest to the cloud, the [N II] lines dominate with respect to the [O III] lines, while the ratio is inverted closer to the exciting stars. The region contains many young stars at various stages of evolution, some of which show jets (one of which is visible close to the photodissociation region), and many protoplanetary disks can be seen in emission or in absorption, although not outstanding in this image because of their very small sizes. (Document C.R. O'Dell & S.K. Wong, Rice University, NASA)

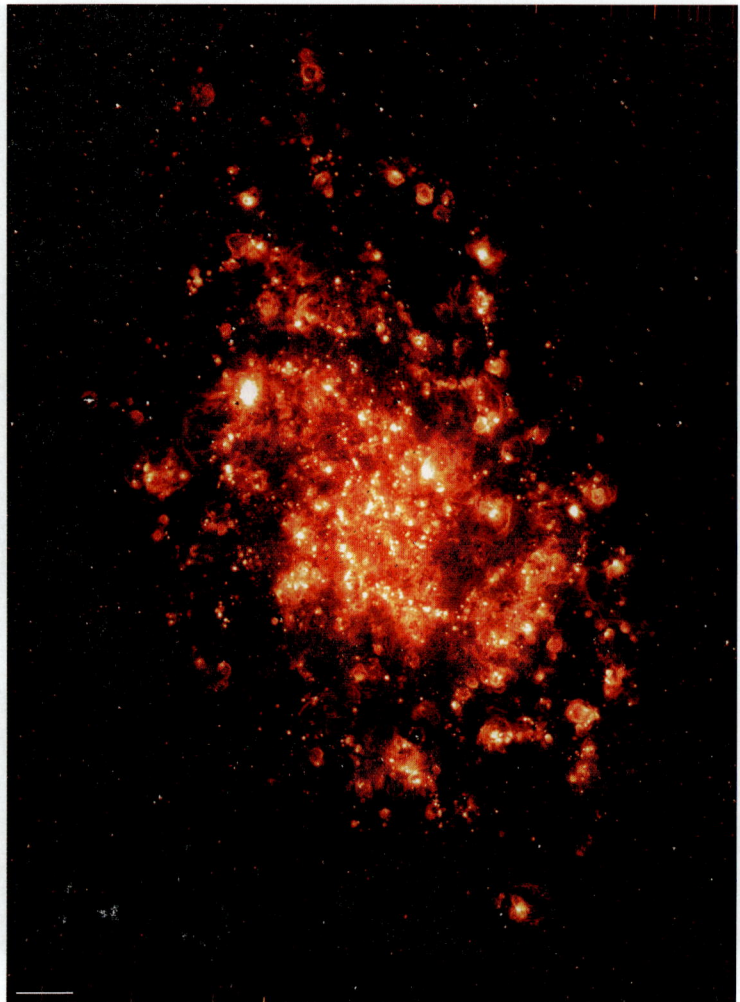

Plate 10. Ionized gas in the galaxy M 33

This image was obtained at the Burrell Schmidt telescope through a filter centred on the Hα line. The most intense H II region, to the north-east, is NGC 604. Note the alignment of the H II regions along the spiral arms, in particular to the south of the centre, the many bubbles and the diffuse ionized gas present almost everywhere in the central region. The ionization of this gas is entirely due to the hot stars (Hoopes & Walterbos [247]). (Courtesy Charles Hoopes)

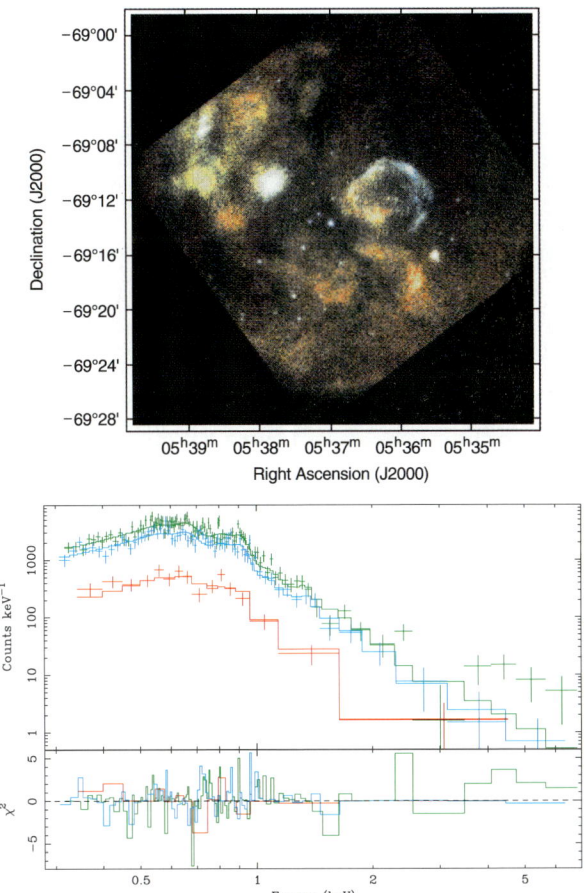

Plate 11. The X-ray emission of the region to the south-east of 30 Doradus, in the Large Magellanic Cloud

The image at the top was obtained with the EPIC instrument on board the XMM-NEWTON european satellite. The false colours correspond to energies from 0.3 keV (red) to 5 keV (blue). The shell clearly visible to the middle right of the image is 30 Dor C, a well-developed supernova remnant. $6'$ to the south-west, the compact object is the very young remnant of Supernova SN 1987A. To the east of 30 Dor C, the bright extended object is N 157 B, a supernova remnant with intermediate properties between a classical remnant and a plerion. The extended emission visible over most of the image is due to the hot gas created by previous generations of supernova remnants. The spectrum of this diffuse gas in the northern region is displayed on the lower figure. We can see the lines of several multi-ionized elements. This spectrum is fitted by a model with a temperature of about 0.3 keV (3.5×10^6 K; the quality of the fit is indicated below), with an overabundance of O, Ne, Mg and Si with respect to the solar abundances (the Large Magellanic Cloud is in fact *underabundant* by a factor ≈ 4): this suggests that the hot gas is dominated by the matter produced in, and injected by supernovae. (From Dennerl et al. [118], with the permission of ESO)

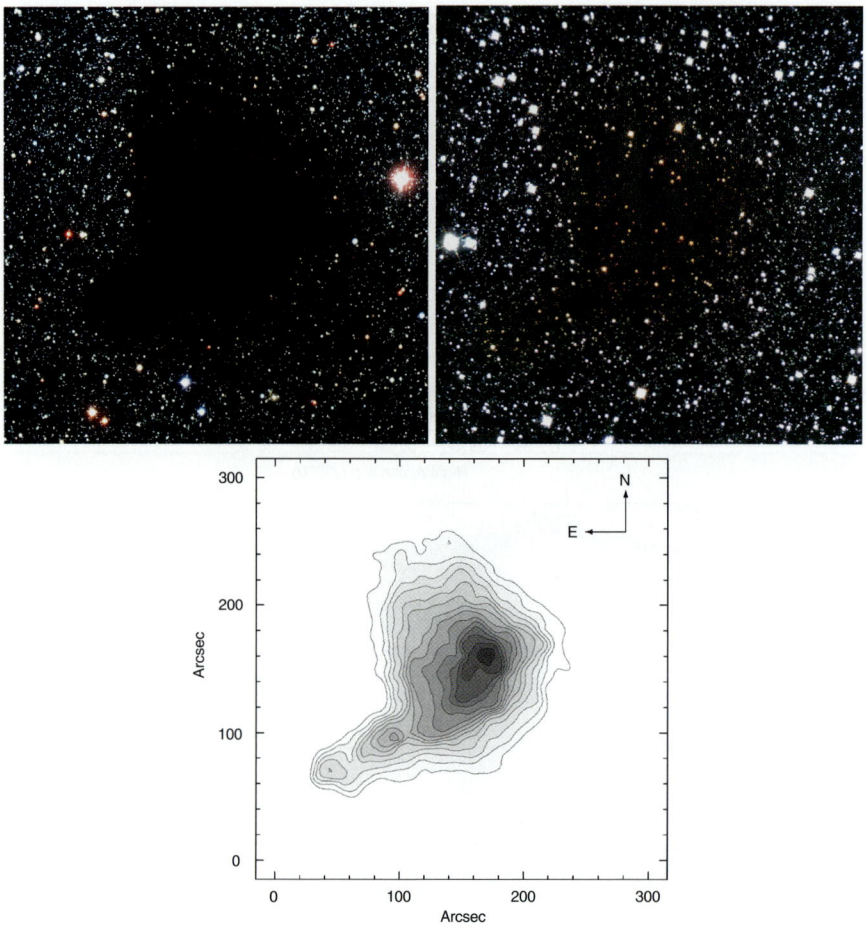

Plate 12. The dark cloud B 68 in the visible and the infrared

At the top are two composite images of this nearby cloud (a *Bok globule*). At left is a composite image in the visible and the near infrared in the bands B (blue), V (green) and I (red), taken with one of the 8-m telescopes of the Very Large Telescope of the European Southern Observatory. The cloud is totally opaque and no star can be seen through it, except a few very reddened stars near the edge. At right is a composite image in the near-infrared bands J (1.25 μm, blue), H (1.65 μm, green) and K_S (2.17 μm, red) taken with the 3.5 m New Technology Telescope of the European Southern Observatory. Extinction being much smaller in the infrared than in the visible, we can see stars through the cloud, especially in the K_S band. The photometry of these stars allowed to derive the extinction map to the bottom, which has an angular resolution of 10″. The outer contour of this map corresponds to a visible extinction $A_V = 4$ magnitudes; extinction then increases by 2 mag. from contour to contour. The extinction at the centre is as high as 35 mag., corresponding to an attenuation of visible light by a factor 10^{14}. (Document ESO Education and Public Relations Department)

Plate 13. The dark cloud B 72 *(the Snake)*

Most interstellar clouds are filaments or sheets. This is a nice example observed with the 12 K CCD camera at the prime focus of the Canada–France–Hawaii telescope. The image dimensions are $30' \times 20'$. The blue, green and red colours correspond to the B, V and R bands, respectively, and are the true colours. The dark cloud absorbs almost all the light of the background stars (compare this with Plate 12). These filamentary structures are probably due to the magnetic field. Like the cloud B 68, displayed Plate 12, the present cloud does not seem to form stars. (Courtesy Jean-Charles Cuillandre, CFHT)

Plate 14. The RCW 108 complex

This infrared image results from 33 images taken with the New Technology Telescope of the European Southern Observatory. The false colours correspond to the J (1.25 µm, blue), H (1.65 µm, green) and K_S (2,17 µm, red) bands. The size of the field is 12.8×12.8 arc minutes. The dark cloud is rather complex. It is very opaque in its central regions. Many background stars are very reddened, even in the infrared, even though extinction is much smaller than in the visible. An H II region is partly embedded in the molecular cloud. Dust heated by the young stars radiates strongly in the mid- to far-infrared. This produces the infrared source IRAS 16362-4845. (Document ESO Education and Public Relations Department)

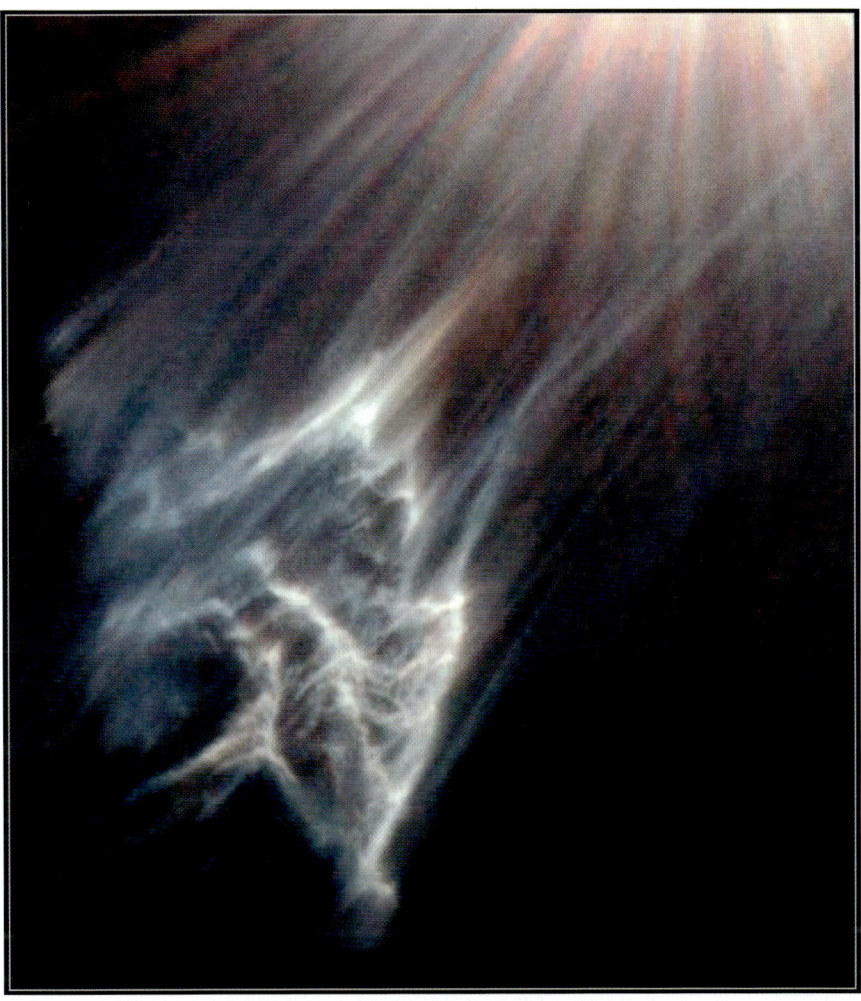

Plate 15. The reflection nebula IC 349 in the Pleiades

The bright star Merope (23 Tauri), which belongs to the Pleiades open cluster at a distance of 120 pc, is outside the field. But it produces, by scattering in the camera, radial rays some 0.02 pc from the star in the upper half of this Hubble Space Telescope image taken with the Wide Field and Planetary Camera 2. The rest of the structures is real. The star and an interstellar cloud containing dust are approaching each other with a relative velocity of about 11 km s^{-1}. The radiation pressure of the starlight pushes the dust, forming the observed filamentary structure. (Document G. Herbig & Th. Simon, NASA)

Plate 16. Hot stars and interstellar matter in the region of 30 Doradus, in the Large Magellanic Cloud

This image is a composite of many frames obtained using 5 different filters of the Wide Field and Planetary Camera 2 of the Hubble Space Telecope. It covers a field of $4.53' \times 2.98'$, corresponding to 68×45 pc at the distance of the Large Magellanic Cloud (about 52 kpc). The blue colour corresponds to images in the broad U filter centred at 3 360 Å. Green corresponds to images in the broad V filter centred at 5,550 Å and through a narrow filter centred on the Hα line. Red corresponds to images in the broad I filter centred at 8 140 Å and through a narrow filter centred on the [S II] lines near 6 730 Å. This set up shows the stars in roughly their true colours. We see that the main stellar cluster (30 Dor) is made only of blue stars and is thus very young. Conversely, the Hodge 301 cluster at the bottom right contains relatively fewer blue stars, but does contain several red supergiants, and is thus older. The diffuse emission in this image is essentially that of lines emitted by the gas ionized by the stars of 30 Dor. The Hα line (green) traces all this gas, while the S II doublet is mainly emitted in zones of weak excitation, deeper inside the photodissociation regions. Supernova explosions and stellar winds have produced bubbles almost empty of gas, except for the hot gas whose X-ray emission can be seen in the top left of Plate 11.

North is towards an angle of 61,86° from the top of the image, clockwise. (Courtesy R. Barbá, J. Maîz-Apellániz and N. Walborn, Space Telescope Science Institute)

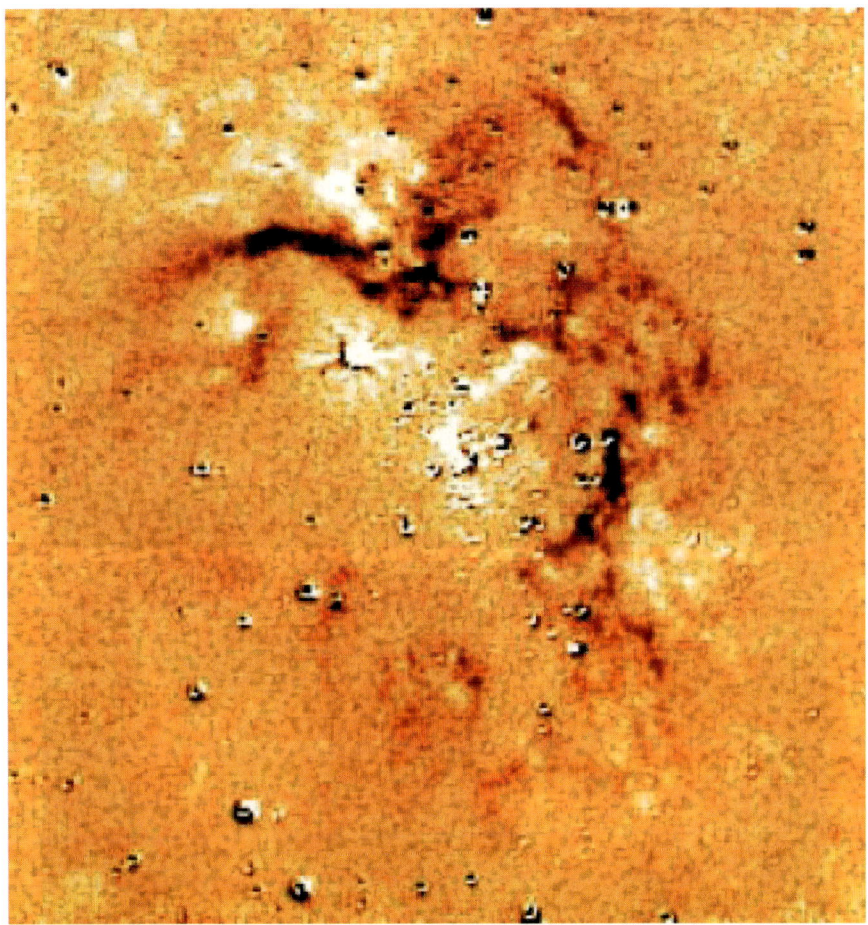

Plate 17. Emissions in the Brackett γ line and in the v(1-0) S(1) line of H_2 in the region of 30 Doradus

The dimensions of this composite image are 4, 4′ × 4, 5′. It was made by the superimposition of a negative image obtained through a narrow filter centred at 2.16 μm (Brγ) and a positive image obtained through another narrow filter centred at 2.12 μm (H_2), taken with the 1.5-m telescope of the Interamerican Observatory in Cerro Tololo, Chile. Stars are visible due to the imperfect subtraction of the continuum. The ionized gas appears as dark brown and the H_2 emission as light yellow. The ionized arc is visible on Plate 16. The emission in the H_2 line is due to fluorescence in the photodissociation regions: this line is an excellent tracer of such regions.

In this image, contrary to Plate 16, north is to the top. (From Rubio et al. [441], with permission of ESO)

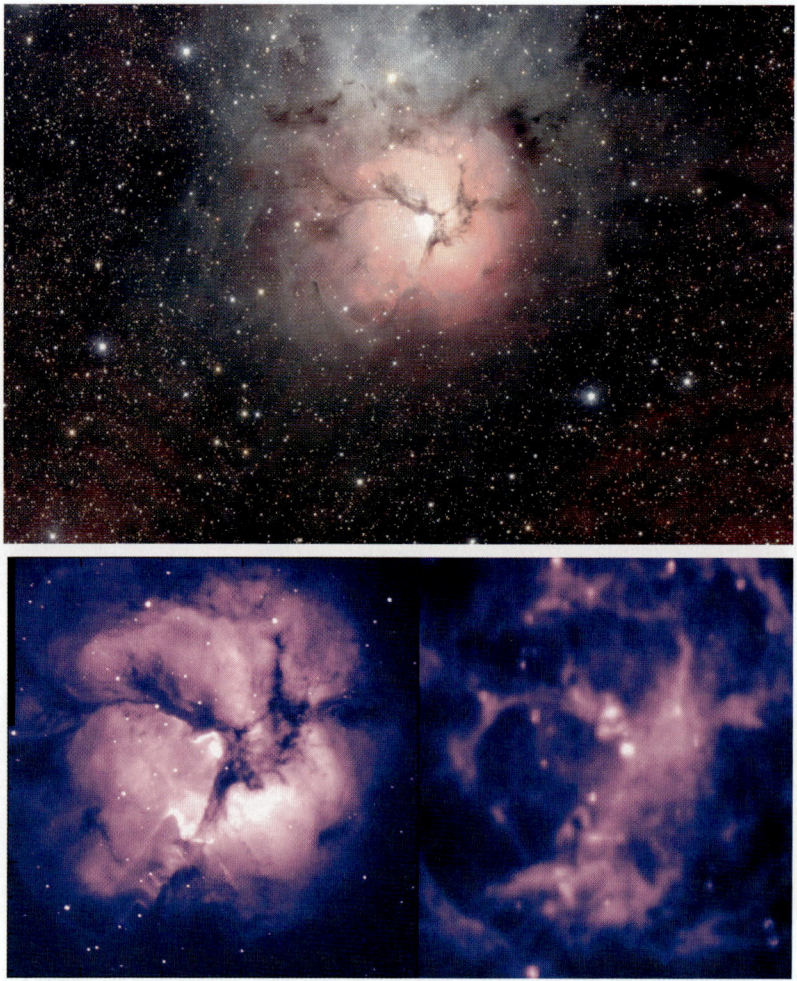

Plate 18. The Trifid nebula in the optical and in the mid-infrared

The Trifid nebula is an H II region associated with a molecular cloud. The figure to the top is a composite image obtained with the 12 K CCD camera at the prime focus of the Canada–France–Hawaii telescope. The image size is $21' \times 7'$. The colours are blue for the V band and red for the I band. The absorption of background stars by the dark cloud surrounding the H II region is clearly visible, and also the scattered light from the exciting stars (these stars are not visible in this saturated image), which forms a diffuse blue extension to the north. The figure at the bottom left is an Hα image obtained through a narrow filter using the 80-cm telescope of the Astrophysical Institute of the Canary Islands. It shows only the ionized gas, with dust absorption bands dividing the image into three parts. The image at the bottom right is the same field seen at 12 µm with the ISOCAM camera on board the ISO satellite (angular resolution about $6''$). The image is dominated by dust emission and looks like a negative of the previous image. (Courtesy Jean-Charles Cuillandre and European Space Agency)

Plate 19. The Chamaeleon constellation seen in the far infrared

This image covers a field of $12.5° \times 12.5°$ (33×33 pc at the 150 pc distance where essentially all the matter along the line of sight can be found). It is a composite image obtained with the IRAS satellite in three bands: 12 μm (blue), 60 μm (green) and 100 μm (red). The stars visible on this image are brighter at 12 μm and appear in blue colour. Outside some regions of intense emission, the image is dominated by the "cirrus" clouds which are heated by the general interstellar radiation field. The local colour variations are conspicuous and correspond to real changes in the composition or size distribution of the very small grains. These changes are probably due in part to the passage of shock waves (see Gry et al. [211]); see Chapter 15. (Courtesy IPAC, California Institute of Technology and NASA)

Plate 20. The Horsehead nebula region in visible light

Bottom: a photograph obtained at the Anglo–Australian Observatory, in approximately real colours. The Horsehead is the prominence of a dark cloud which extends over all of the lower half of the image. It stands out against a red background, which is due to the Hα line emitted by a diffuse H II region ionized by the hot star σ Orionis (O9.5V) located outside the field. The bright blue object is the reflection nebula NGC 2023, illuminated by the B1.5V star HD 37903, not visible due to saturation. The dimensions of the image are $30' \times 33'$, corresponding to 4.8×5.3 pc at the distance of 550 pc.

Top: an image of the Horsehead obtained with the 12 K CCD camera at the prime focus of the Canada–France–Hawaii telescope. The colours are blue for the V band and red for the I band. This image emphasises the light scattered by the surface of the dark cloud and by some isolated filaments to the right of the Horsehead illuminated by σ Orionis.

Exceptionally, north is to the left and east to the bottom in these images. (Documents Anglo-Australian Telescope and CFHT, courtesy D. Malin and J.-C. Cuillandre)

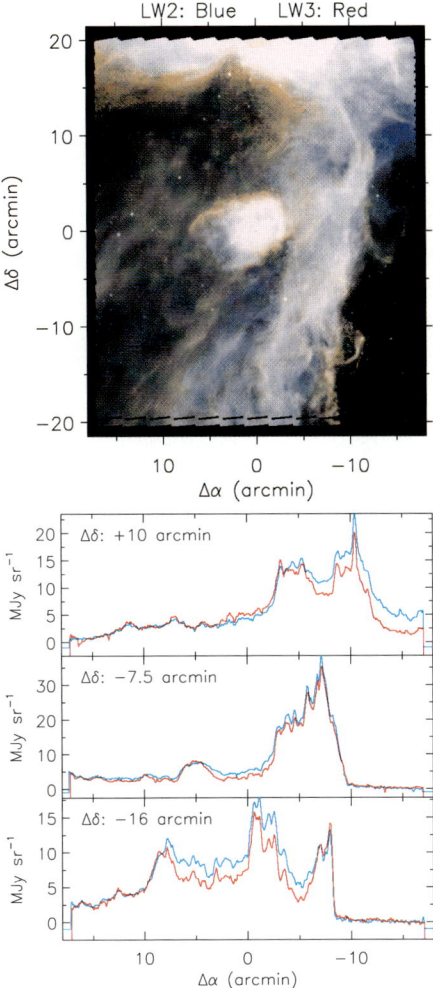

Plate 21. Variations in the composition of very small dust grains in the region of the Horsehead nebula

This image covers a slightly larger field than the image at the bottom of Plate 20. It was obtained at 7.7 μm (blue) and 15.5 μm (red) with the ISOCAM camera on board the ISO satellite. The dust, which absorbs the visible and UV light, is now seen in emission, as for Plate 18 (actually we are not really dealing with the same dust grains, the visible extinction being dominated by rather large grains and the emission in the mid-infrared by much smaller ones). In the infrared, we only see the superficial layer of the Horsehead, the regions of the dark cloud exposed to the radiation of σ Orionis (outside the field), and the reflection nebula NGC 2023, also seen in emission. To the bottom of this image, three cuts through the field show the differences between the emission at 7.7 μm (dominated by the aromatic infrared bands probably emitted by PAHs), and the emission at 15.5 μm due to very small, 3-D carbonaceous grains. The origin of the observed variations is not well understood. (Courtesy A. Abergel, Institut d'Astrophysique Spatiale)

Plate 22. Ionized gas and aromatic infrared bands in the region of M 16

This composite image was obtained at 7.7 μm (blue) and 15.5 μm (red) with the ISOCAM camera on board the ISO satellite. The angular dimensions are $30' \times 30'$, corresponding to 17.5×17.5 pc at a distance of 2 kpc. The image is centred on the young star cluster M 16 (NGC 6611) whose hot stars are barely visible here. They are responsible for ionizing a part of the gas. In red, we see the emission of the ionized gas in the [Ne III] line at 15.56 μm. Stellar winds have dug clearly visible bubbles in the interstellar medium. The sharp edges of these bubbles are dense photodissociation regions where the very small grains responsible for the aromatic infrared band at 7,7 μm are concentrated and excited. Some of these photodissociation regions look like pillars, fingers or other structures. Remarkable visible images of these structures have been obtained with the Hubble Space Telescope (see http://oposite.stsci.edu/pubinfo/PR/95/44.html). Their appearance is similar to that of the Horsehead in Plate 20, top.

In this image, north is at 58,4° from the top, anticlockwise. (ISOGAL document, courtesy M.J. McCaughrean, A. Moneti and A. Omont). The angular resolution is that of the orginal data

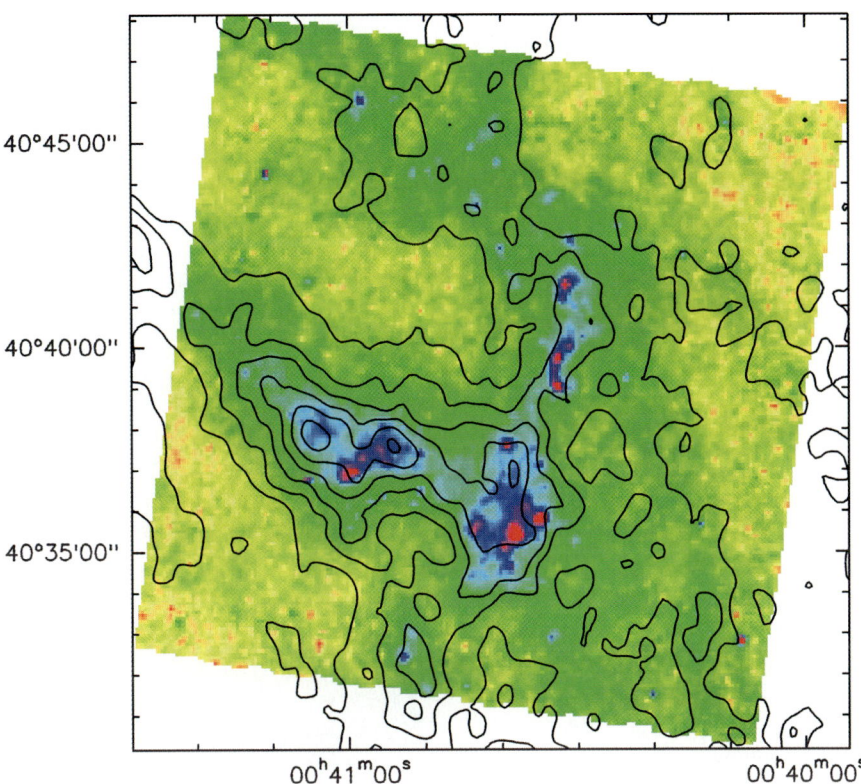

Plate 23. Comparison between the emission in the mid-infrared and the distribution of atomic hydrogen in the Andromeda galaxy M 31

This image covers the south-west part of the Andromeda galaxy M 31 approximately 50′ (10 kpc) from its centre. The coordinates are J2000. The false colours represent the intensity observed near 7.7 μm with the ISO satellite, at an angular resolution of approximately 6″. The contours display the intensity of the 21-cm line observed with the Westerbork interferometer at a resolution of 25″. We see that the emission at 7.7 μm, which is due to PAHs heated transiently by the absorption of individual photons, follows precisely the distribution of the gas. However, while the visible radiation field, which is comparable to that near the Sun in our Galaxy, is approximately uniform over the field, this is not the case for the UV radiation field, which is non-uniform but weaker than near the Sun. This demonstrates that the heating of the grains is due to visible photons in this case, rather than to UV photons. (From Pagani et al. [393], with the permission of ESO)

Plate 24. The interface between the H II region M 17 and the adjacent molecular cloud

This composite infrared image was taken with the New Technology Telescope of the European Southern Observatory. The colours correspond to the J (1.25 μm, blue), H (1.65 μm, green) and K_S (2.17 μm, red) bands. The image size is $5' \times 5'$, corresponding to 3.5×3.5 pc at the distance of the object (2.2 kpc). A molecular cloud covers the right part of the image and absorbs most of the starlight. The left part contains an H II region ionized by stars located to the left, out of the field. The diffuse emission is dominated by the free–free and free–bound continua of the ionized gas. The molecular cloud is photodissociated and ionized by the UV radiation and is evaporating to the left. The photodissociation region formed in this way has a very fragmented structure which reflects that of the molecular cloud. This has been much studied at a variety of wavelengths: see e.g. Stutzki et al. [498], Cesarsky et al. [84] and Verstraete et al. [533]. An ultra-compact H II region, not visible here, is embedded inside the photodissociation region. It testifies to a secondary star formation process (see Cesarsky et al. [84]). (Document ESO Education and Public Relations Department)

Plate 25. The Crab Nebula in the visible, radio and X-rays

The Crab nebula M 1 is the remnant of the explosion of a supernova which occured in 1054. It is a typical example of a plerion, continuously supplied by relativistic particles from a central pulsar. Top left, a visible image obtained with one of the 8-m units of the Very Large Telescope of the European Southern Observatory (dimensions 6.8′ × 6.8′, or 4 × 4 pc at the distance of the nebula, 2 kpc). It results from the combination of images in the B (blue) and R (green) bands and in the [SII] λ6716,6731 lines (red). The filaments are remnants of the matter ejected during the explosion; they form a rapidly expanding shell. The diffuse blue light is due to the synchrotron emission from relativistic electrons accelerated by the pulsar, interacting with the local magnetic field. Top right, a radio image obtained with the Very Large Array (size 4.7′ × 5.5′). It shows the pulsar and the synchrotron emission, which has a similar morphology to the diffuse visible emission. There is a rather close relation between this morphology and that of the filaments, probably because the magnetic field is more intense in the filaments. Bottom, an X-ray image obtained with the CHANDRA satellite (size 2.5′ × 2.5′). It shows the central pulsar and the synchrotron emission from very high-energy electrons, the morphology is different from that of the visible and radio emissions which are due to lower-energy electrons. Note the jets emitted more or less symmetrically by the pulsar. (Documents ESO Education and Public Relations Department, National Radio Astronomy Observatory and NASA/Harvard-Smithsonian Center for Astrophysics)

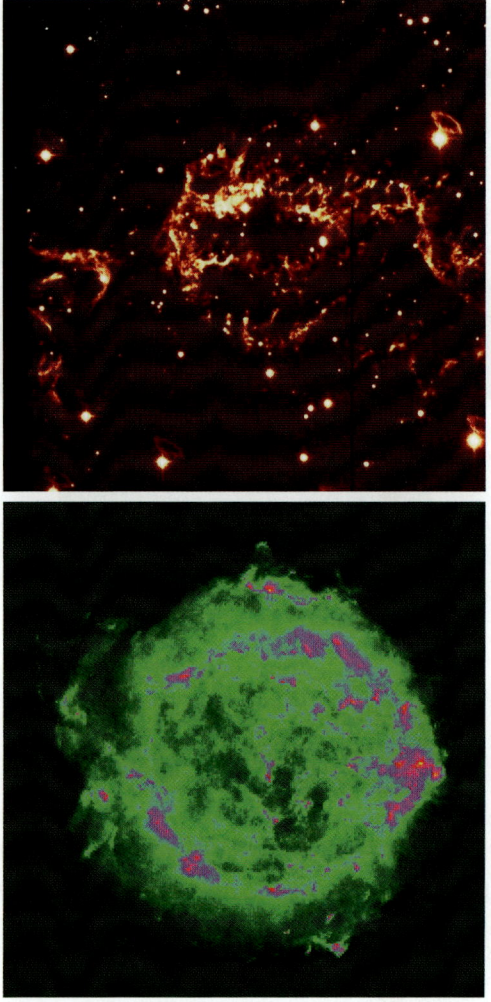

Plate 26. The supernova remnant Cassiopeia A in the radio and visible

Bottom, a radio image in false colours obtained with the Very Large Array (dimensions $6' \times 6'$, or 4.9×4.9 pc at the distance of the object, 2.8 kpc). The emission is synchrotron. Its structure is globally that of a spherical shell with many irregularities, nodules and diffuse regions. Top, a visible image obtained at the 3.6 m Canada–France–Hawaii telescope (dimensions $2.8' \times 2.8'$). We observe complex filaments. A spectroscopic study reveals that they belong to two families. Some filaments have relatively slow velocities and a normal chemical composition: they are composed of interstellar matter and of matter ejected by the supernova progenitor. Others have very high velocities and a very anomalous chemical composition, showing that they are composed of matter resulting from the thermonuclear processes that occurred during the explosion and ejected following the supernova explosion. The elliptical rings above and to the right of the bright stars are artefacts produced by reflections in the instrument. (Documents National Radio Astronomy Observatory and CFHT)

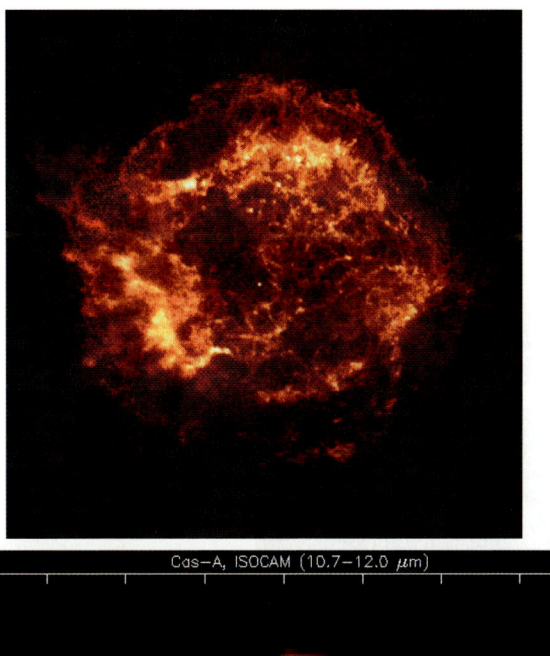

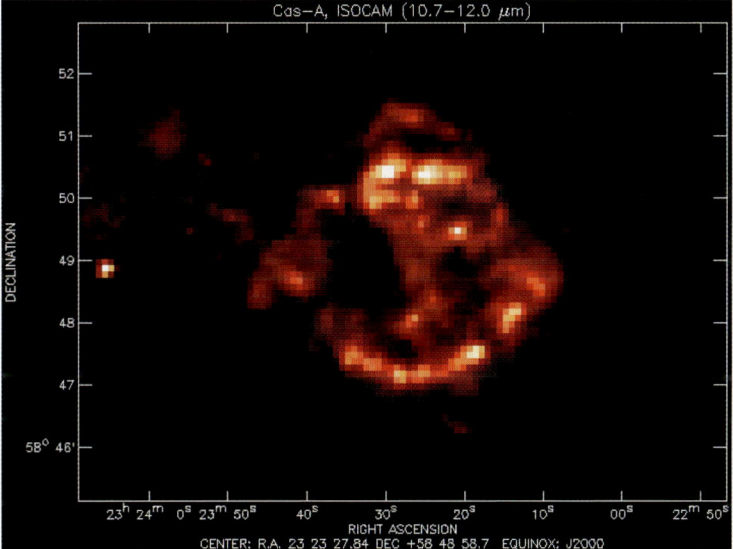

Plate 27. The supernova remnant Cassiopeia A in X-rays and in the mid-infrared

Top, an X-ray image obtained with the CHANDRA satellite (same size as the radio image of Plate 26). Although the emission mechanisms are different, this image is remarkably similar to the radio image. In both images we see the morphology of the optical filaments. The point source at the centre might be a neutron star, a remnant of the explosion, but this hypothesis requires confirmation. Bottom, a mid-infrared image (10.7–12.0 µm) obtained with the ISOCAM camera on board the ISO satellite. It shows the emission of dust condensed from the gas ejected by the explosion. This dust contains silicates (see Lagage et al. [296], Arendt et al. [11] and Douvion et al. [130]). (Documents NASA, and courtesy P.-O. Lagage)

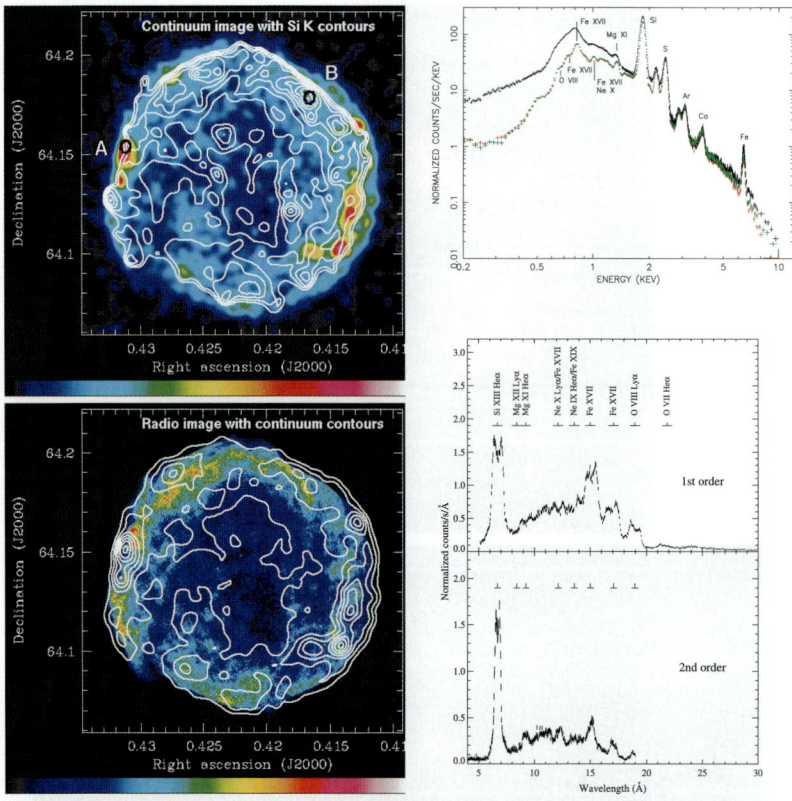

Plate 28. The remnant of the Tycho Supernova in X-rays

The explosion of this supernova was reported by Tycho Brahé in 1572. Its X-ray emission is dominated by the thermal radiation from the ejecta heated by shocks. To the right, an X-ray spectrum obtained with three different spectrographs of the EPIC instrument on board XMM–NEWTON: MOS 1 (green), MOS 2 (red) et PN (black) (more detailed spectra at low energy are shown below). The most important lines are identified. The line marked "Si" is a blend of the triplet of Si XIII between 1.67 and 2.00 keV. In the image at the top left, the contours in this blend are superimposed on an image of the X continuum between 4.5 and 5.8 keV. The spherical shell structure is conspicuous, but there are differences between the distributions of the continuum and the lines. The continuum image is more regular and more extended than that of the lines. This continuum is interpreted as the emission from the shocked interstellar gas. The structure seen in the lines probably results from Rayleigh–Taylor instabilities at the contact discontinuity between the ejected material and the ambient medium. In the image at the bottom left, the contours correspond to the emission in the X-ray continuum between 4.5 and 5.8 keV, superimposed on the radio image observed at 22 cm with the Very Large Array. These two emissions are limited to the same radius (taking into account the lower angular resolution in X-rays), but the radio emission is correlated with the line emission (compared to the upper image), confirming the importance of Rayleigh–Taylor instabilities in the acceleration of relativistic electrons and/or in the amplification of the magnetic field. (From Decourchelle et al. [115], with the permission of ESO)

Plate 29. The bubble N 70 in the Large Magellanic Cloud

This composite image was obtained with one of the 8-m unit telescopes of the Very Large Telescope of the European Southern Observatory. It results from the combination of images in B (blue), V (green) and in the Hα line (red). The image dimensions are $6.8' \times 6.8'$, or 103×103 pc at the distance of 52 kpc. The filamentary structure of this bubble is spectacular. It is ionized by the radiation of many young, hot stars present in the field, but these stars cannot be those which have (perhaps) produced the bubble. The small red object to the top left of the image is probably a compact H II region. (Document ESO Education and Public Relations Department)

Plate 30. The Trapezium region in the Orion Nebula

In this image obtained with the Wide Field & Planetary Camera of the Hubble Space Telescope, we see the four hot stars which ionize the H II region (see also Plate 32). North is at about 45° anticlockwise from the top of the image. The star at the bottom right, θ^1C Orionis (O6pe), is more luminous and hotter than the others. The diffuse filamentary structures are due to the emission from the ionized gas. Notice the many small-diameter objects with tails pointing away from the star. They are protostellar disks which are evaporating under the effects of the intense UV radiation of the star. Such objects are often called *cometary nebulae*. North is approximately at 45° from the top, anticlockwise. (Document J. Bally, D. Devine and R. Sutherland, NASA–HST)

Plate 31. The Orion Nebula and its cluster of young stars in the infrared

The appearance of the Orion Nebula in the infrared is very different from its appearance in the visible (Plates 9 and 30). This composite image has been obtained with one of the 8-m unit telescopes of the Very Large Telescope of the European Southern Observatory in the J_S (1.24 µm, blue), H (1.65 µm, green) and K_S (2.17 µm, red) bands. It reveals a rich cluster of about 1 000 young stars, aged about 1 million years. Some are very reddened, even in the infrared. Only the brightest stars, in particular the four Trapezium stars (see Plate 30), are easily visible optically. The others are either buried in the emission of the H II region or in the light scattered by dust, or extinct due to the residual dust of the regions where they formed. In the infrared the emission of the ionized gas is essentially free–free and free–bound, with a weak contribution of some lines. Note the Orion bar, a photodissociation region which crosses the field at the bottom left and separates the H II region from a molecular cloud. In this region as elsewhere, the diffuse emission is due to the ionized gas. Also note the many regions of absorption due to condensations of matter which could form new stars. (Document ESO Education and Public Relations Department)

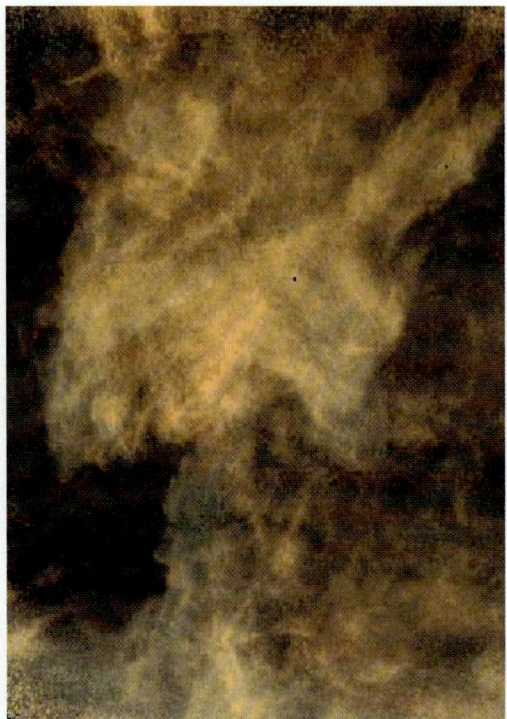

Plate 32. Chimneys and mushrooms emerging from the galactic disks

To the right, six 21-cm line velocity channel maps of the galactic bubble and chimneys GSH 277+00+36, located at a distance of about 6.5 kpc. The observations are from the Australia Telescope Compact Array (ATCA). The gray scale is linear and runs from 0 to 80 K as shown on the wedge at the left. The LSR velocity of each channel in km s^{-1} is given for each image. The coordinates are galactic. Note the empty bubble at the centre and several chimneys on either side of the galactic disk which appear to open into the halo.

To the left, an astonishing structure in our Galaxy which reminds us of the atomic mushrooms formed by aerial explosions of A or H bombs. It was discovered during the 21-cm line Canadian Galactic Plane Survey. The image shows the distribution of atomic hydrogen and extends from galactic latitudes $-1°$ (bottom) to $-6.5°$ (top). The galactic longitude of the middle of the image is $124°$. At the assumed distance of 3.8 kpc, that of the Perseus arm in this direction, the cloud extends at least to 350 pc from the galactic plane. While it is clear that this ascending structure is due to supernova explosions which have pushed the gas out of the galactic plane, the details are uncertain. A possibility is that a supernova explosion at some distance from the plane creates a bubble of hot gas which rapidly rises. This bubble leaves behind a rarefied zone into which the surrounding gas falls, forming the foot of the mushroom. The bubble rises supersonically with respect to the gas at rest, it sweeps up this gas in the snowplough regime behind a radiative shock, while slowing down (see Sect. 12.1.3) and forming the head of the mushroom. The other structures arise from the interactions between the different components (cold, warm and hot) of the ambient gas and the magnetic field. This is a somewhat more complex equivalent of what occurs in an atomic explosion mushroom cloud. (From McClure-Griffiths et al. [355] and English et al. [167])

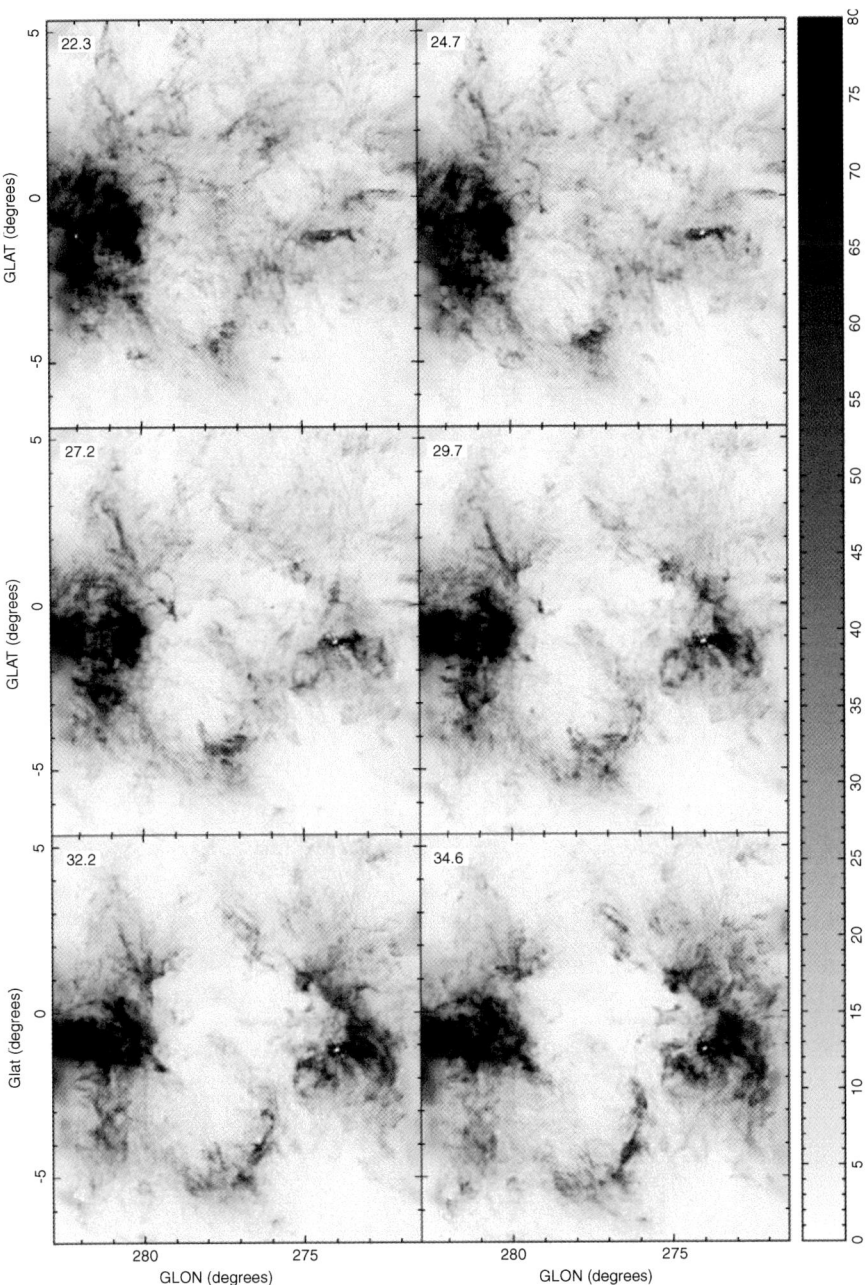

Printing: Mercedes-Druck, Berlin
Binding: Stein+Lehmann, Berlin

ASTRONOMY AND ASTROPHYSICS LIBRARY

Series Editors: I. Appenzeller · G. Börner · A. Burkert · M. A. Dopita
T. Encrenaz · M. Harwit · R. Kippenhahn · J. Lequeux
A. Maeder · V. Trimble

The Stars By E. L. Schatzman and F. Praderie

Modern Astrometry 2nd Edition
By J. Kovalevsky

The Physics and Dynamics of Planetary Nebulae By G. A. Gurzadyan

Galaxies and Cosmology By F. Combes, P. Boissé, A. Mazure and A. Blanchard

Observational Astrophysics 2nd Edition
By P. Léna, F. Lebrun and F. Mignard

Physics of Planetary Rings Celestial Mechanics of Continuous Media
By A. M. Fridman and N. N. Gorkavyi

Tools of Radio Astronomy 4th Edition
By K. Rohlfs and T. L. Wilson

Astrophysical Formulae 3rd Edition (2 volumes)
Volume I: Radiation, Gas Processes and High Energy Astrophysics
Volume II: Space, Time, Matter and Cosmology
By K. R. Lang

Tools of Radio Astronomy Problems and Solutions By T. L. Wilson and S. Hüttemeister

Galaxy Formation By M. S. Longair

Astrophysical Concepts 2nd Edition
By M. Harwit

Astrometry of Fundamental Catalogues
The Evolution from Optical to Radio Reference Frames
By H. G. Walter and O. J. Sovers

Compact Stars. Nuclear Physics, Particle Physics and General Relativity 2nd Edition
By N. K. Glendenning

The Sun from Space By K. R. Lang

Stellar Physics (2 volumes)
Volume 1: Fundamental Concepts and Stellar Equilibrium
By G. S. Bisnovatyi-Kogan

Stellar Physics (2 volumes)
Volume 2: Stellar Evolution and Stability
By G. S. Bisnovatyi-Kogan

Theory of Orbits (2 volumes)
Volume 1: Integrable Systems and Non-perturbative Methods
Volume 2: Perturbative and Geometrical Methods
By D. Boccaletti and G. Pucacco

Black Hole Gravitohydromagnetics
By B. Punsly

Stellar Structure and Evolution
By R. Kippenhahn and A. Weigert

Gravitational Lenses By P. Schneider, J. Ehlers and E. E. Falco

Reflecting Telescope Optics (2 volumes)
Volume I: Basic Design Theory and its Historical Development. 2nd Edition
Volume II: Manufacture, Testing, Alignment, Modern Techniques
By R. N. Wilson

Interplanetary Dust
By E. Grün, B. Å. S. Gustafson, S. Dermott and H. Fechtig (Eds.)

The Universe in Gamma Rays
By V. Schönfelder

Astrophysics. A New Approach 2nd Edition
By W. Kundt

Cosmic Ray Astrophysics
By R. Schlickeiser

Astrophysics of the Diffuse Universe
By M. A. Dopita and R. S. Sutherland

The Sun An Introduction. 2nd Edition
By M. Stix

Order and Chaos in Dynamical Astronomy
By G. J. Contopoulos

Astronomical Image and Data Analysis
By J.-L. Starck and F. Murtagh

ASTRONOMY AND ASTROPHYSICS LIBRARY

Series Editors: I. Appenzeller · G. Börner · A. Burkert · M. A. Dopita
T. Encrenaz · M. Harwit · R. Kippenhahn · J. Lequeux
A. Maeder · V. Trimble

The Early Universe Facts and Fiction
4th Edition By G. Börner

The Design and Construction of Large Optical Telescopes By P. Y. Bely

The Solar System 4th Edition
By T. Encrenaz, J.-P. Bibring, M. Blanc, M. A. Barucci, F. Roques, Ph. Zarka

General Relativity, Astrophysics, and Cosmology By A. K. Raychaudhuri, S. Banerji, and A. Banerjee

Stellar Interiors Physical Principles, Structure, and Evolution 2nd Edition
By C. J. Hansen, S. D. Kawaler, and V. Trimble

Asymptotic Giant Branch Stars
By H. J. Habing and H. Olofsson

The Interstellar Medium
By J. Lequeux